Astrophysics and Space Science Library

Volume 468

The Astrophysics and Space Science Library is a series of high-level monographs and edited volumes covering a broad range of subjects in Astrophysics, Astronomy, Cosmology, and Space Science. The authors are distinguished specialists with international reputations in their fields of expertise. Each title is carefully supervised and aims to provide an in-depth understanding by offering detailed background and the results of state-of-the-art research. The subjects are placed in the broader context of related disciplines such as Engineering, Computer Science, Environmental Science, and Nuclear and Particle Physics. The ASSL series offers a reliable resource for scientific professional researchers and advanced graduate students.

Rosaly M. C. Lopes • Katherine de Kleer •
James Tuttle Keane
Editors

Io: A New View of Jupiter's Moon

Second Edition

Springer

Editors
Rosaly M. C. Lopes
Jet Propulsion Laboratory
California Institute of Technology
Pasadena, CA, USA

Katherine de Kleer
California Institute of Technology
Pasadena, CA, USA

James Tuttle Keane
Jet Propulsion Laboratory
California Institute of Technology
Pasadena, CA, USA

ISSN 0067-0057 ISSN 2214-7985 (electronic)
Astrophysics and Space Science Library
ISBN 978-3-031-25669-1 ISBN 978-3-031-25670-7 (eBook)
https://doi.org/10.1007/978-3-031-25670-7

Previous edition originally published in the series: "Springer Praxis Books in Astronomy and Planetary Sciences" with the title: "Io After Galileo - A New View of Jupiter's Volcanic Moon" under: Lopes, R. M. C., Spencer, J. R.

Cover illustration: A montage of New Horizons images of Jupiter and its volcanic moon Io, taken during the spacecraft's Jupiter flyby in early 2007. *Credit: NASA/JHU/APL.* https://www.nasa.gov/mission_pages/newhorizons/news/nh_jupiter_oct09.html

This Springer imprint is published by the registered company Springer Nature Switzerland AG
The registered company address is: Gewerbestrasse 11, 6330 Cham, Switzerland

Contents

Contributors

Fran Bagenal Laboratory for Atmospheric & Space Physics, University of Colorado, Boulder, CO, USA

A. C. Barr Planetary Science Institute, Tucson, AZ, USA

C. J. Bierson Arizona State University, Phoenix, AZ, USA

R. Brasser Origins Research Institute, Research Centre for Astronomy and Earth Sciences, Budapest, Hungary

Katherine de Kleer California Institute of Technology, Pasadena, CA, USA

I. de Pater Department of Astronomy, University of California, Berkeley, CA, USA

V. Dobos Kapteyn Astronomical Institute, University of Groningen, Groningen, The Netherlands
MTA-ELTE Exoplanet Research Group, Szombathely, Hungary

Vincent Dols Laboratory for Atmospheric & Space Physics, University of Colorado, Boulder, CO, USA

D. Goldstein Department of Aerospace Engineering, University of Texas at Austin, Austin, TX, USA

Amanda F. Haapala Johns Hopkins University Applied Physics Laboratory, Laurel, MD, USA

James Tuttle Keane Jet Propulsion Laboratory, California Institute of Technology, Pasadena, CA, USA

Laszlo P. Keszthelyi U.S. Geological Survey, Astrogeology Science Center, Flagstaff, AZ, USA

E. Lellouch Laboratoire d'Études Spatiales et d'Instrumentation en Astrophysique (LESIA), Observatoire de Paris, Meudon, France

Rosaly M. C. Lopes Jet Propulsion Laboratory, California Institute of Technology, Pasadena, CA, USA

Kathleen E. Mandt Johns Hopkins University Applied Physics Laboratory, Laurel, MD, USA

I. Matsuyama University of Arizona, Tucson, AZ, USA

Alfred S. McEwen Lunar and Planetary Laboratory, University of Arizona, Tucson, AZ, USA

William B. McKinnon Department of Earth and Planetary Sciences and McDonnell Center for the Space Sciences, Washington University, Saint Louis, MO, USA

L. C. Quick NASA Goddard Space Flight Center and The Nexus for Exoplanet System Science (NExSS), Greenbelt, MD, USA

Jani Radebaugh Brigham Young University, Provo, UT, USA

J. A. Rathbun Planetary Science Institute, Tucson, AZ, USA

Paul M. Schenk Lunar and Planetary Institute, Houston, TX, USA

Nicholas M. Schneider Laboratory for Atmospheric and Space Physics, University of Colorado, Boulder, CO, USA

John R. Spencer Department of Space Studies, Southwest Research Institute, Boulder, CO, USA

Terry-Ann Suer Department of Earth and Planetary Sciences, Harvard University, Cambridge, MA, USA

A. Trinh University of Arizona, Tucson, AZ, USA

David A. Williams Arizona State University, Tempe, AZ, USA

Chapter 1
Introduction

Rosaly M. C. Lopes, Katherine de Kleer, and James Tuttle Keane

Abstract The Galilean satellite Io is a dynamic body in the solar system and a prime location to study volcanism. We summarize the content of the chapters in this book and provide a table of basic orbital and physical properties.

Io was discovered by Galileo Galilei on 8 January 1610 and has fascinated scientists ever since. The innermost of Jupiter's four Galilean satellites (Fig. 1.1, Table 1.1), Io is the only body in our solar system besides Earth known to have large-scale active volcanism (Fig. 1.2). Io has an important role in our understanding of the Solar System, as it is one world where we can observe extreme processes *in action*, including tidally-powered volcanism, tectonism, and atmospheric and magnetospheric interactions. Io's heat flow is much higher than the Earth's, its interior may contain a magma ocean (Fig. 1.3), and its lavas may be hotter than any erupted on the Earth today. Io's intense volcanism makes it the best present-day analogue for the early Earth and other rocky worlds, and likely for some present-day exoplanets and exomoons.

Since the discovery of active volcanism on Io from *Voyager 1* images in 1979, our knowledge of Io has evolved considerably (Fig. 1.4). While the *Voyager 1* and *Voyager 2* flybys gave us a glimpse of a world that few had imagined, the *Galileo* mission in the 1990s and early 2000s provided a much deeper knowledge of Io and the Jupiter system. Despite complications with the mission, *Galileo* revolutionized our understanding of the Jupiter system, including revealing the tantalizing possibility that tidal dissipation not only fuels the volcanoes on Io, but also supports subsurface water oceans beneath the icy shells of Europa and Ganymede. The amazing findings of *Galileo* at Io inspired a previous book, *Io After*

R. M. Lopes (✉) · J. T. Keane
Jet Propulsion Laboratory, California Institute of Technology, Pasadena, CA, USA
e-mail: Rosaly.m.lopes@jpl.nasa.gov; James.t.keane@jpl.nasa.gov

K. de Kleer
California Institute of Technology, Pasadena, CA, USA
e-mail: dekleer@caltech.edu

R. M. C. Lopes et al. (eds.), *Io: A New View of Jupiter's Moon*, Astrophysics and Space Science Library 468, https://doi.org/10.1007/978-3-031-25670-7_1

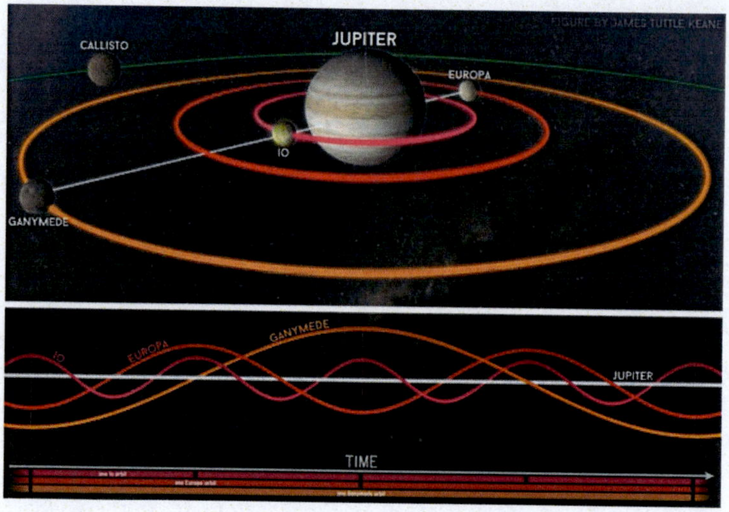

Fig. 1.1 A schematic of the Jupiter system. The top panel shows a perspective view of the Jupiter system. The relative sizes of the orbits are to scale, as are the relative sizes of the moons, although the orbits and moons are not to scale with each other, nor is Jupiter to scale with anything. The bottom panel shows a simplified view of the configuration of the three satellites in the Laplace resonance (Io, Europa, and Ganymede) as a function of time. Thin, vertical dashed lines indicate different conjunctions and oppositions of the satellites. Figure credit: James Tuttle Keane, NASA's Eyes on the Solar System (https://eyes.nasa.gov/)

Table 1.1 Io's basic orbital and physical properties

Mean radius:	1821.6 ± 0.5 km
Bulk density:	3528 ± 3 kgm^{-3}
Orbital period:	1.769 days (42.459 h)
Orbital eccentricity:	0.0041 (forced)
Orbital inclination	0.037°
Orbital semimajor axis:	421,800 km
Rotational period:	Synchronous (identical to orbital period)
Mass:	$(8.9320 \pm 0.0013) \times 10^{22}$ kg
Surface gravity:	1.796 ms^{-2} (18.3% Earth gravity)
Global average heat flow:	>2.5 Wm^{-2}
Core dynamo magnetic field strength:	<50 nT
Geometric albedo:	0.62
Local topographic relief:	<17 km
Number of active volcanic centers	>166
Typical surface temperatures:	85 K (night)140 K (day)1000~2000 K (erupted lavas)
Atmospheric pressure:	$<10^{-9}$ bar

Source: Lopes, R. and D. Williams: Io after Galileo. *Reports on Progress in Physics*, Institute of Physics Publishing, 68, 303–340

Fig. 1.2 Images of Io and its dramatic activity. (**a**) A New Horizons view of Io and, with the Tvashtar plume prominently extending above the limb. The left-hand side of Io is illuminated by sunlight, while the right-hand-side of Io is partially illuminated by Jupiter-shine. (**b**) A Galileo false color image mosaic of Io. (**c**) A Voyager 1 image mosaic of Io's south polar terrain. The terminator (line between daylight and night) runs diagonally across the frame. On the right is Haemus Mons, a 10-km tall mountain. Volcanic plains, plateus, and crater-like depressions (patera) cover the rest of the scene. Figure credit: James Tuttle Keane, NASA, JHUAPL, SwRI, JPL, University of Arizona, USGS

Galileo (Editors: R.M.C. Lopes and J.R. Spencer), published in 2007. That same year saw new observations of Io and the Jupiter system from the *New Horizons* spacecraft on its way to Pluto. The *Juno* spacecraft has been in orbit of Jupiter since 2016, and while the primary goal of the *Juno* mission is to understand Jupiter, *Juno* has provided exciting serendipitous views of Io. *Juno* is expected to have several close flybys of Io late in its extended mission. Simultaneously, ground- and space-based astronomy has advanced rapidly: adaptive optics observations have provided sharper views of Io, and synoptic monitoring is revealing intriguing patterns in the cadence of Io's volcanic activity. New observatories have enabled detailed views of Io at wholly new wavelengths, including the Atacama Large Millimeter Array (ALMA). The combination of observations have allowed great strides to be made on understanding Io, and more recent models have gained us knowledge of Io's formation and interior.

This past decade of advances motivated the need for an updated review book of Io, to build on and complement *Io After Galileo*. The scope, and chapters, are

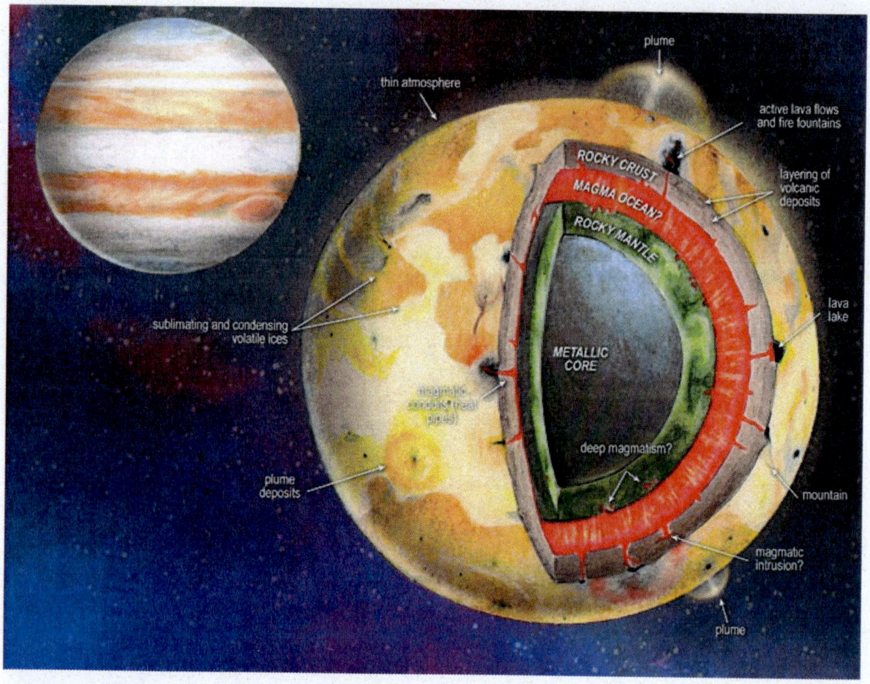

Fig. 1.3 Schematic illustration of the possible interior structure of Io, and the various processes shaping Io. Figure credit: James Tuttle Keane and Aaron Rodriquez

different from those of the 2007 book, as we have chosen to highlight the areas where significant progress has been made since then:

In *Chap.* 2, Nick Schneider and John Spencer take a thematic approach to the extraordinary discoveries made about Io, from early telescopic observations to present-day telescopic and space-based observations. It complements the comprehensive review of pre-*Galileo* results in *Io After Galileo*.

In *Chap.* 3, William McKinnon discusses the formation and earliest evolution of Io, reviewing the new models that have caused significant changes in our prior understanding. Since Galileo data were acquired in the 1990s and early 2000s, astronomical observations of protoplanetary disks, including those by ALMA, plus advances in theoretical and numerical models have enabled significant progress in our understanding satellite formation and evolution scenarios.

In *Chap.* 4, James Tuttle Keane, Isamu Matsuyama, Carver J. Bierson, and Antony Trinh review the major advances in our understanding of Io's interior structure and the fundamental process of tidal heating evolution since the *Galileo* era. Advances in geophysics provide the context for interpreting Earth-based observations, which are often designed to interrogate Io's interior structure and evolution, and have seen tremendous advances in the last couple of decades.

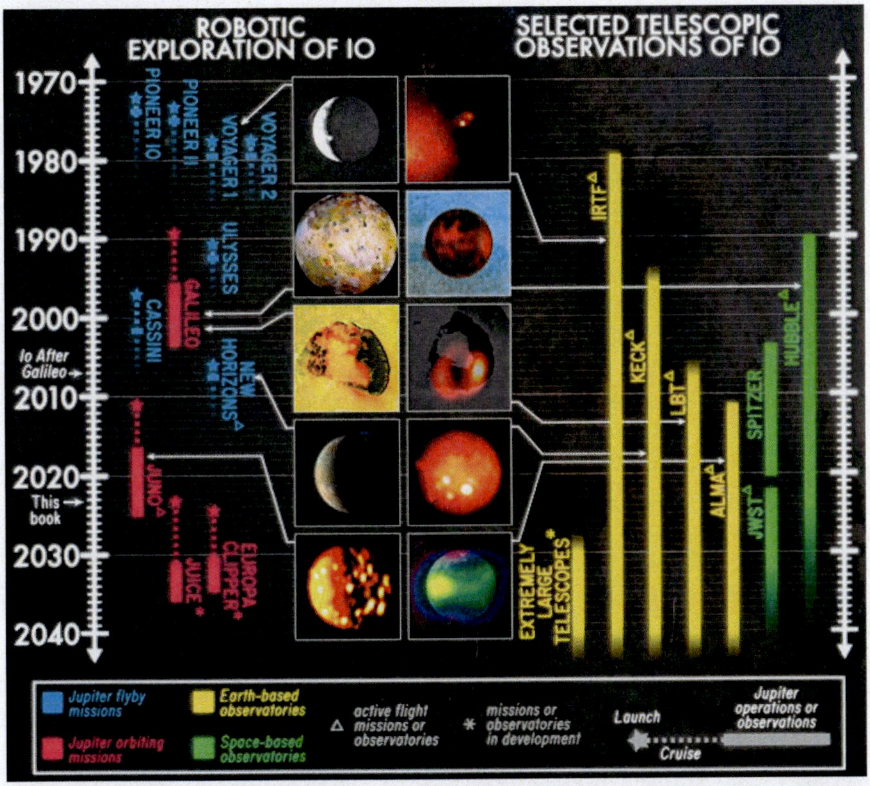

Fig. 1.4 Timeline of robotic and telescopic observation of Io. The left-hand side shows robotic exploration of Io, and all spacecraft that have flown through the Jupiter system and observed Io. The right-hand side shows a subset of ground- and space-based telescopic observations of Io. There is an extensive history of ground-based observations prior to the 2000s, which is more thoroughly detailed in *Io After Galileo*. All future dates should be taken as notional. "Extremely Large Telescopes" include the European Extremely Large Telescope (EELT), Giant Magellan Telescope (GMT), and Thirty Meter Telescope (TMT). Figure credit: James Tuttle Keane, and references therein

In *Chap.* 5, David Williams, Paul Schenk, and Jani Radebaugh review Io's surface geology, which is unique in the solar system as it is completely dominated by volcanic and tectonic features. Io's extreme volcanism causes a resurfacing rate that has effectively erased all impact craters from the surface, making Io the only object in the Solar System on which no impact craters have been identified.

In *Chap.* 6, Katherine de Kleer and Julie Rathbun discuss Io's thermal emission and heat flow. Io's surface shows many sources of thermal emission (hot spots) and, although many of these were detected from *Galileo* data, our understanding of the hot spots and heat flow, both volcanic and passive, has progressed substantially since the end of the *Galileo* mission due to new telescopic datasets, continuing analyses of spacecraft data, and improvements in theoretical models.

In *Chap.* 7, Laszlo P. Keszthelyi and Terry-Ann Suer review the many different, but indirect, constraints on the bulk composition Io. The chapter focuses on bulk composition rather than surface composition, since there have been few new observations of Io's surface composition since Galileo. In this chapter, the authors use a detailed consideration of Io's lavas to illustrate how decades of research have bounded, but not pinned down, the chemistry of Io.

In *Chap.* 8, Imke de Pater, David Goldstein, and Emmanuel Lellouch review our latest knowledge of the plumes and atmosphere of Io with an emphasis on research conducted since the *Galileo* era. While the primary source of Io's atmosphere is sublimation of SO_2 frost, volcanoes can have a substantial effect on the atmosphere as shown both via observations and model simulations. Although considerable progress has been made towards both a characterization and understanding of Io's atmosphere, there are some fundamental questions that are still unanswered.

In *Chap.* 9, Fran Bagenal and Vincent Dols review the major role that Io plays in Jupiter's giant magnetosphere and how, in turn, magnetospheric particles and fields affect Io. They discuss the physical processes that shape the space environment around Io and the impact from Jupiter out into interplanetary space. Since *Galileo* observations, data from *New Horizons*, *Hubble Space Telescope*, the Japanese *Hisaki* satellite, and the *Juno* spacecraft have made significant contributions to our understanding of the space environment of Io.

In *Chap.* 10, Amy Barr, Ramon Brasser, Vera Dobos, and Lynnae C. Quick discuss how Io can be an analogue for tidally heated exoplanets. The conditions we see at Io—a rocky body orbiting close to its parent planet, in resonant orbit with its sibling satellites and experiencing intense tidal heating, also occur in at least one system of a star and its planets, TRAPPIST-1. The chapter discusses the use of simple geophysical models, which reproduce observed behaviors of Io, to show that the TRAPPIST-1 bodies may be in a similar geophysical regime as Io.

In *Chap.* 11, Alfred McEwen, Amanda F. Haapala Chalk, Laszlo P. Keszthelyi, and Kathleen E. Mandt review the key outstanding questions and future observations of Io, including future telescopic and spacecraft observations. The chapter reviews what instruments and observations might be made by a future mission and why Io is so important as a target for future exploration.

NASA's long-term goals are defined by Decadal Surveys conducted by the National Academies of Science, Engineering, and Medicine. The most recent Decadal Survey, *Origins, Worlds, and Life: A Decadal Strategy for Planetary Science and Astrobiology 2023–2032* prioritized Io science. Origins, Worlds, and Life identified 12 priority science questions, and Io features prominently in the majority—with particular strong connections to priority science questions related to the evolution of solid body interiors, circumplanetary systems, and dynamic habitability.

At the time of writing, we look forward to future spacecraft observations of Io by NASA's *Juno* spacecraft, currently in orbit around Jupiter, and the European Space Agency's *JUpiter ICy moons Explorer* (*JUICE*) spacecraft, scheduled to arrive at the Jupiter system in 2031. Dedicated Io missions have been, and we expect will continue to be, proposed under NASA's competitive programs, including NASA's New Frontiers 5 opportunity and Discovery.

It is our hope that this book will serve as inspiration for researchers and students to familiarize themselves with the state of our understanding of the most extreme and unique worlds in our Solar System.

Acknowledgements J.T.K. carried out a portion of this research at the Jet Propulsion Laboratory, California Institute of Technology, under a contract with the National Aeronautics and Space Administration (80NM0018D0004). R.M.C.L.'s work was done as a private venture and not in the author's capacity as an employee of the Jet Propulsion Laboratory, California Institute of Technology.

Chapter 2
Understanding Io: Four Centuries of Study and Surprise

Nicholas M. Schneider and John R. Spencer

Abstract We now know Io to be a world of superlatives among solar system bodies. It experiences the strongest orbital resonances, exhibits the greatest volcanic activity, sustains the most rapidly escaping atmosphere, and lies deep within the most powerful magnetosphere. This chapter synopsizes the centuries of studies that revealed Io's remarkable properties, but highlights how the fundamental interconnectedness between these distinct properties were revealed only in recent decades. In fact, the revelation of links between seemingly unrelated planetary phenomena placed Io in the position to revolutionize planetary science. Before Io, who might have hypothesized that orbital peculiarities could drive volcanoes, shrink moons and power aurora? Io's example forces planets and moons to be studied as coupled as systems, from celestial mechanics through interiors, surfaces, and atmospheres to magnetospheres.

2.1 Introduction

The history of Io studies shows that many of its exceptional properties were quick to reveal themselves through observation, but the connections between them took time to appreciate. In the pages that follow, we'll see repeated cases where the "superlatives" of Io's properties lead to the earliest discoveries of new planetary processes, starting with its very discovery along with the other Galilean moons (Fig. 2.1a). The trend continues to this day, establishing Io as the prototype for tidal heating, atmospheric escape and many other processes. In this role it has inspired explorations of countless other objects where similar processes may play out in less dramatic fashion. Beyond our solar system, Io informs studies of planets around

N. M. Schneider (✉)
Laboratory for Atmospheric and Space Physics, University of Colorado, Boulder, CO, USA
e-mail: Nick.Schneider@lasp.colorado.edu

J. R. Spencer
Department of Space Studies, Southwest Research Institute, Boulder, CO, USA
e-mail: spencer@boulder.swri.edu

R. M. C. Lopes et al. (eds.), *Io: A New View of Jupiter's Moon*, Astrophysics and Space Science Library 468, https://doi.org/10.1007/978-3-031-25670-7_2

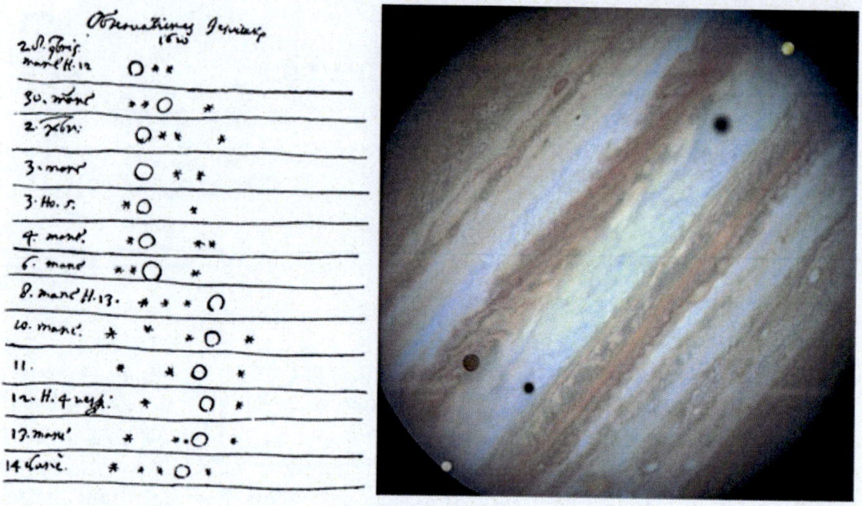

Fig. 2.1 Galileo's discovery drawings of Jupiter's moons from 1610 (left) contrasted with a Hubble Space Telescope image for (right) showing three moons against Jupiter

other stars, where Io's superlatives must certainly be outdone within the astonishing diversity of exoplanets.

The interconnectedness of Io phenomena present a challenge to any attempt to summarize the history of its exploration. A strictly chronological exposition can be as confusing to the reader as it was to those who were making the discoveries. And a topical structure inevitably loses focus on the links between the topics. We therefore have taken an approach highlighting our understanding of the connections themselves, which often spanned a decade or more to come in focus.

In the sections that follow we'll review the early days of discovery, and then explore six fundamental breakthroughs in Io science and planetary science:

- Celestial mechanics drives tidal heating
- Tidal heating controls internal structure and heat flow
- Heat flow generates diverse volcanic styles
- Volcanism creates a unique surface
- Volcanism supplies an atmosphere out of balance
- Rampant atmospheric escape fuels the magnetosphere

Each section follows the initial decades or centuries of study from groundbased observations, and is then abruptly punctuated by the flood of results from the Voyager spacecraft. Figures have been selected to pair the initial discoveries with the current state-of-the-art, demonstrating how far we have come. For instance, each of the six stories below is transformed by the Io plume discovery photo (Fig. 2.2a, described in more detail in Sect. 2.5), arguably among the most revolutionary images in all of planetary science. The accompanying results from other instruments were as profound, if less visually staggering. The ensuing years of spacecraft visits,

Fig. 2.2 Voyager's plume discovery image from 1979 (left; Morabito et al. 1979), compared to a Galileo plume image from 1997 (right; NASA PIA01081)

extraordinary observations from Earth, critical theoretical and modeling efforts, bring us to our current understanding.

This chapter benefits tremendously from prior histories of Io exploration which take more traditional chronological and topical approaches, specifically chapters in *Satellites of Jupiter* (Morrison 1982) *Satellites* (Burns and Matthews 1986), *Jupiter: The Planet, Satellites and Magnetosphere* (Bagenal et al. 2004), and reviews by Spencer and Schneider (1996), Cruikshank and Nelson (2007) and de Pater et al. (2021). In addition, each chapter within this volume lays out the critical history of their subjects.

Countless Io presentations over the decades have included the cartoon of six blind men examining different parts of an elephant, each with a contradictory interpretation and none getting a sense of the whole being. The analogy would be better if the six blind people were examining an alien, something we still don't know quite what it looks like.

2.2 Discovery and Bulk Properties

It is not surprising that Galileo Galilei and Simon Marius, the first two people to point telescopes to the heavens, should quickly discover moons around Jupiter. Their size and proximity render them technically bright enough to be naked-eye objects, though Jupiter's glare prevents clear detections. They travel farther in the sky from Jupiter than moons of other planets, and their rapid orbital motion quickly differentiates them from the fixed stars.

Galileo in Italy and Marius in Germany were contemporaries and competitors in this endeavor, and in the late days of 1609 were both observing the moons of Jupiter and documenting their positions. It was Galileo who published first, honoring his sponsor by naming them the Medicean Moons. Marius published later, offering the individual names (as suggested to him by Kepler) that we still use today. The confusing fact that Galileo's observing logs were dated in the "new" Gregorian calendar while Marius' were recorded in the old Julian calendar is a commentary on the religious and societal schisms of the time. It is perhaps fitting that the moons today are collectively named after Galileo while they carry the individual names proposed by Marius.

The importance of Jupiter's moons in subverting geocentrism cannot be overstated, but it is not central to the discoveries to come. What is relevant is the recognition that these newly discovered objects were potentially useful in both practical and exploratory ways. Galileo noticed that the repeatability of the moons' appearances and disappearances against Jupiter's disk offered the possibility of measuring "absolute time", by which observers anywhere on earth could witness the events and agree on the time. This could potentially solve the longstanding "longitude" problem in navigation at sea. Solar time, relative to the ship's location, could readily be observed; the addition of an absolute time reference would permit calculation of longitude. The concept was sound when applied through measurements on land, but in the end it proved impractical to make the necessary astronomical observations of these occasional phenomena from shipboard.

Increasingly complete and accurate tabulations of orbital phenomena established a long baseline which proves valuable even today. The orbit sizes and periods validated Kepler's Third Law when it was published a decade after the moons' discovery. In 1675, Roemer also noticed that the time between repeating phenomena was stable when Jupiter was closest or farthest, but lost or gained time in between. He recognized that could only happen if light took a finite time to cross the diameter of Earth's orbit, about 16 min. Modern-day readers will recognize this as twice the 8 min for the Sun's light to reach Earth, though an actual measurement of the speed of light in physical units needed to wait a century for a measurement of the Astronomical Unit. Roemer he never published his results, but his time estimate leads to an error of less than 2% in the speed of light.

The determination of the Astronomical Unit in the 1700s would have permitted an early insight into Jupiter's nature, thanks again to Io's orbital motion. The absolute size of Jupiter would have been known to be about ten times Earth from its telescopic angular diameter and distance from Earth. Similarly, Io's orbit would be recognized as nearly the same as that of Earth's moon. If Jupiter and Earth were composed of the same materials, Newton's law of gravitation would predict Io's orbital period to be about 22 h. Its actual value of 42 h is incontrovertible evidence of Jupiter's low density compared to Earth, even without knowing the value of the gravitational constant. It's not clear who, if anyone made this leap in understanding.

By the late 1800s, telescopes were capable of resolving Io's 1.2 arc-second disk frequently enough to investigate the nature of the body itself. The appearance was initially puzzling: against empty space, Io appeared elongated parallel to Jupiter's belts, while against Jupiter's disk, Io appeared as two distinct dark spots displaced perpendicularly to the belts. Barnard (1891a) originally favored the idea that Io was a double object, but eventually concluded (Barnard 1891b, 1894) that both phenomena could be explained if Io's equator were bright and its poles darker. Against the sky, the bright equatorial band would dominate the image, while against Jupiter's bright disk the band would blend in, leaving the dark poles as distinct objects. Barnard (1897) soon measured Io's angular diameter which translated to 3950 km, less than 10% above the modern value. Better accuracy only came much later, when Io occulted the star Beta Scorpii, and yielded a diameter of 3656 ± 5 km (Taylor 1972).

With Io's physical size first constrained by Barnard (albeit a slight overestimate), Laplace's dynamically-deduced mass (see Sect. 2.3) was used to derive Io's density. Initial calculations (Russell et al. 1945) found values around 2.7–2.9 g/cm^3, lower than the modern value of 3.5 g/cm^3 but high enough to correctly conjecture that Io must be made mostly of rock and metal. The same method gave lower densities for Ganymede and Callisto, hinting at the trend which drives theories on the formation of these moons. The Pioneer spacecraft flyby's gave definitive measurements of Io's density at 3.53 g/cm^3 and the declining trend with distance from Jupiter (Andersson et al. 1974).

Small, repeatable brightness variations with orbital phase demonstrated that Io and the other Galilean moons orbited synchronously (Stebbins 1927). As higher precision became possible, Binder and Cruikshank (1964) reported the phenomenon of post-eclipse brightening, in which Io's brightness exceeded its pre-eclipse value by 10% for about 10–20 min before lowering again. They proposed that an atmosphere was partially condensing on the surface as it cooled in Jupiter's shadow. Many subsequent efforts could not reproduce effects of this amplitude, though Nelson et al. (1993) did find occasional brightenings of a few percent in some cases. Disk-resolved imaging from Voyager, Galileo and Hubble (Veverka et al. 1981; Burrati et al. 1995; Secosky and Potter 1994, respectively) showed negligible effects at a global level though regional variations could not be ruled out. Fanale et al. (1981) and Nelson et al. (1993) concluded that the timescales for condensation was marginally plausible, but that the quantity necessary to brighten the surface globally (several mm thick) was not available in the atmosphere nor could it all sublime quickly enough. Nonetheless, the supposition that some atmosphere must condense during eclipse continued to drive observational searches and remains a candidate mechanism for many variable phenomena at Io.

By-eye assessments of Io's color and early photometric measurements confirmed Io's reddish color, especially remarkable in contrast to its nearest neighbors. Ever-improving photoelectric measurements allowed multi-bandpass filter measurements (Harris 1961; Johnson and McCord 1971; Morrison et al. 1974), which quantified the red slope of the spectrum and suggested a broad, shallow absorption centered around 600 nm. Io's overall high albedo suggested water ice or frost on the surface.

But the extension of observations to the near-infrared (Kuiper 1957; Moroz 1966; Johnson and McCord 1971) showed Io's reflected solar spectrum to be brighter at 1.6 and 2.2 μm than other moons. This led these observers to conclude that water ice was not a major constituent of Io's surface, contrary to general expectations for moons of the outer solar system.

Io's low reflectance from the blue into the ultraviolet suggested a different constituent: sulfur. Elemental sulfur (S_8) was a good match to the spectrum below 500 nm (Wamsteker 1972; Kuiper 1973). Additional ultraviolet absorptions confirmed by spacebased- (Caldwell 1975) and groundbased-observations (Nelson and Hapke 1978) supported the case for sulfur, including the possibility of other allotropes affected by outgassing or irradiation (Nelson and Hapke 1978).

The identification of sodium and potassium escaping Io (see Sect. 2.5), and Io's overall high red reflectivity led to salt (NaCl) as a potentially major constituent (Fanale et al. 1974). With sulfur and oxygen soon discovered beyond Io, sulfate salts became plausible (Nash and Fanale 1977), and other evaporite deposits not exhibiting water absorptions. Cruikshank et al. (1978) and Pollack et al. (1978) detected a distinct absorption feature at 4.07 μm but were unable to identify candidates for matching surface materials. There was relatively little speculation at the time on how surface processes might have concentrated these constituents in surface layers.

Even from a distance, Voyager imagery quickly confirmed the blurry groundbased view: a bright equator, darkened poles, and an abundance of yellow, orange and red. Laboratory work soon showed that the 4.07 μ absorption feature unidentified in groundbased spectra was wholly consistent with SO_2 frost (Fanale et al. 1979). Soderblom et al. (1980) concluded from Voyager imagery than the principal surface constituents were sulfur dioxide frost and allotropes of sulfur. The explanations for this unique state of affairs are described in Sects. 2.5 and 2.6, and corroborating evidence from atmospheric measurements are covered in Sect. 2.7.

2.3 Celestial Mechanics Drives Tidal Heating

The orbit tabulations by Roemer and later observers caused consternation for Giovanni Cassini and others working towards predictive tables in the late 1600s. Even when corrected for changing Earth-Jupiter distances, the orbits were not sufficiently fixed to be predictable by Kepler's laws (due to effects we now attribute to precession and libration). Tabulations by Pehr Wargentin in 1743 showed conclusively that Io, Europa and Ganymede were locked into a 4:2:1 orbital resonance to astonishingly high accuracy. In 1788, Pierre-Simon Laplace published a mathematical explanation for this fundamental relationship, and was even able to derive approximate masses for the moons based on their mutual gravitational perturbations. Laplace's theory required the moons' orbits to maintain "forced eccentricities" which were so well explained that astronomers became more interested in the minor "free eccentricity" deviations away from those values.

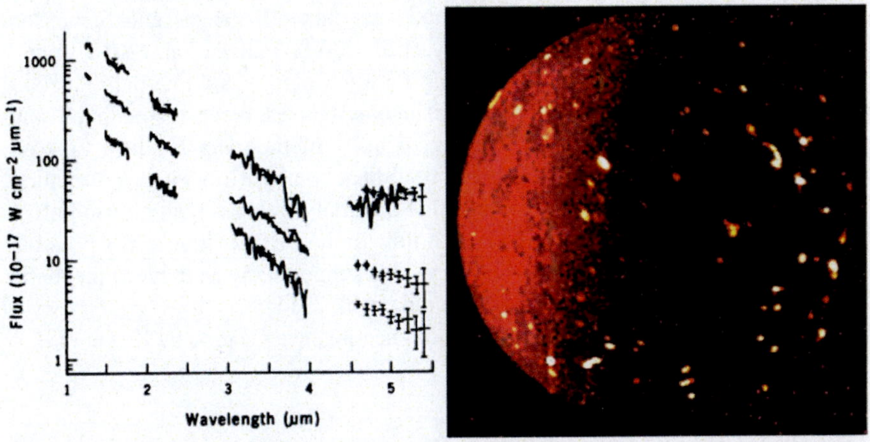

Fig. 2.3 The first strong indication of Io's excess thermal emission from Witteborn's 5 μm brightening (1979, left), compared to hotspots observed by Juno in 2019 (right, NASA PIA25698). In the plot at left, spectra taken over two nights are offset vertically to show the difference at long wavelengths. In the image at right, hotspots appear as points, some with diffraction spikes, with thermal emission from the dayside at left

The orbital resonances were little more than a curiosity for near two centuries. In the late 1970s, the attention of astronomers and planetary scientists was turning to the Jupiter system in anticipation of the arrival of the Voyager 1 spacecraft in March 1979. In February of that year, Witteborn et al. (1979) published observations of an unexpected brightening of 5 μm emission over a period of hours (Fig. 2.3a). They considered thermal emission as an explanation, with 0.01% of Io's surface at 600 K, but rejected it based on experience with objects in the inner solar system. They favored an explanation involving Jupiter's magnetosphere, known by this time to be unusual as we will discuss in Sect. 2.8.

On March 2, 1979, mere days before the Voyager 1 encounter with the Jupiter system, Peale et al. (1979) published their work showing that tidal heating might melt Io's interior and cause active volcanism. Their paper prophetically concludes that "Voyager images of Io may reveal a structure and history different from any previously observed." Their fundamental realization was that tidal heating is driven by the large forced eccentricity, not the nearly-negligible free eccentricity.

Voyager and Galileo stereo imaging bore out the tidal distortion: the bulges were measured at 13 km (Gaskell et al. 1988; Thomas et al. 1998), the largest tidal deformation of any solar system object of that size. Models indicated that Io's orbital motion closer and farther from Jupiter would lead to tens of meters daily variation, greater even that the fluid tides of Earth's oceans.

Io also raises tides on Jupiter, and dissipation inside Jupiter drives orbital evolution of the three resonant moons. Conservation of energy and angular momentum link orbital evolution with tidal heating. Thanks to nearly four centuries of orbital observations, the orbital evolution can be measured and thereby permit an estimate of dissipation in Io and the resulting total heat flux (Ojakangas and Stevenson 1986; Hussman and Spohn 2004; Lainey et al. 2009). Fuller et al. (2016) found that the phenomenon of resonance locking between moons and internal planetary oscillations had the potential to resolve inconsistencies between predicted and measured heat flow, and allows for the possibility of time-variable heat flow on geological timescales. The deposition of tidal heat controls Io's interior structure. Models by Seagatz et al. (1988) revealed that the tidal heating is not uniform in Io's interior, which can significantly affect heat flow to the surface. In fact, it's possible that tidal heating creates a magma ocean, an idea supported by an induced magnetic field potentially observed by Galileo (Khurana et al. 2011).

For greater detail on Io's interior and tidal heating, see Chap. 4 by Keane et al.

2.4 Tidal Heating Controls the Interior and Heat Flow

Following the Peale et al. (1979) theoretical proof of tidal heating, the full picture rapidly emerged: orbital resonances cause tidal heating in the three innermost Galilean moons, with Io most affected. Witteborn's infrared excess was a large volcanic eruption caused by this heating. Prior observations of unexpectedly high heat flux in eclipse could be properly attributed to volcanic activity and not unusual thermal inertia properties (Hansen 1973; Morrison and Cruikshank 1973). The "heat pipe" model (O'Reilly and Davis 1981) provided a key insight into Io's heat transport, in which heat is primarily advected at hotspots and not conducted through the lithosphere. This allows for a thicker and more rigid lithosphere capable of supporting the rugged topography observed in some locations. Lithospheric thickness may also be controlled by magmatic intrusions (Spencer et al. 2020).

Voyager's IRIS instrument confirmed the hotspot concept (Hanel et al. 1979), though initial estimates of the hotspot temperature were complicated by the instrument's limited wavelength coverage and the large fraction of field of view containing cooler surrounding areas. Modeling of a two-temperature surface suggested hotspot temperatures of 290 K. Hanel et al. recognized that using three temperature regions might allow a better fit with higher temperatures, potentially as high as sulfur's melting temperature of 385 K.

Thus began decades of productive and insightful infrared observations from Earth and interplanetary spacecraft. The Galileo mission orbited Jupiter from 1995 to 2003, carrying a powerful near infrared mapping spectrometer (NIMS). NIMS could spatially resolve hotspots during Io encounters and measure their thermal emission with enough spectral coverage to identify multiple components. The SSI CCD imager identified volcanic centers system and provided visible wavelength context. The Cassini mission performed a gravity assist at Jupiter and obtained low spatial resolution visible images of Io; like the Galileo SSI, its longest visible wavelength bandpass was sensitive to the hottest lava flows. New Horizons also took advantage of a Jupiter gravity assist, and obtained visible and IR imaging of Io. The

Juno spacecraft now orbits Jupiter, and its infrared instrument has obtained stunning images of Io hotspot thermal emission (e.g., Fig. 2.3b).

The long gaps between spacecraft observations were used to great advantage with groundbased telescopes. Initially, low spatial resolution prevented identification of more than a few of the very brightest features, but ingenious observing plans extracted a wealth of information. Jupiter's disk could be used as an occulting edge while Io passed behind, and the infrared brightness could be measured at sufficient time resolution to tie the drops in emission with their location in one dimension across the disk (Spencer et al. 1991). The geometry was limited to the Jupiter-facing hemisphere, but opportunities presented themselves many times per week, and hundreds of observations have been taken over the years (Fig. 2.4a). Similarly, Io's occasional occultation by other jovian moons allowed the occulting disk to cover and uncover Io's hotspots (e.g., Goguen et al. 1988). This gave more precise location information, thanks to the sharper edge of the satellite disk and the combined ingress and egress. But opportunities were rare and came in seasons separated by 6 years.

Adaptive optics changed the game for Io hotspot studies, allowing a dozen or more hotspots to be mapped and measured in a single observation. Since the geometry didn't require a particular orbital geometry, all of Io could be mapped and much greater temporal coverage obtained (Fig. 2.4b).

The Earth-based and space-based observations strove to address the same key issues: What is Io's total heat flow, and how much can be attributed directly to the volcanic hotspots? Where are Io's volcanic hotspots located, and how do they vary? Can their spectra constrain temperature and magma composition?

The first question has the most straightforward answers: Io emits about 100 TW of power (Johnson et al. 1984; Veeder et al. 1994; Rathbun et al. 2004; McEwen et al. 2004), with just over half coming from the 242 identified volcanic hotspots (Veeder 2015, and prior work described therein). A single volcanic center, Loki, may be responsible for 20% of the hotspot emission. Curiously, the other ~50%

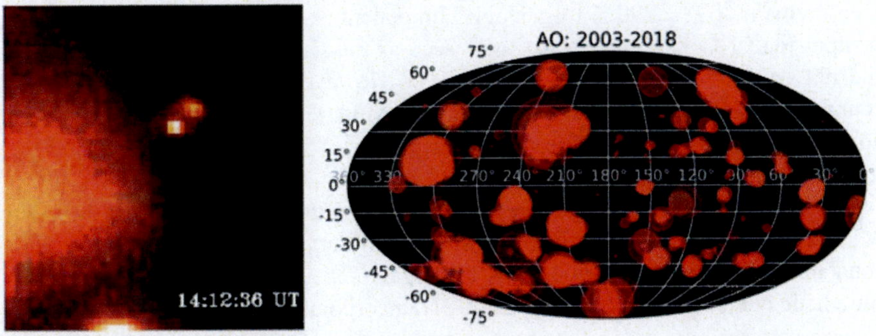

Fig. 2.4 A 3.8 μm NASA/IRTF image from Spencer et al. (1991) of Io with two visible hotspots passing behind Jupiter's limb (left), compared to de Kleer et al.'s (2019a) cumulative hotspot map (right). Power is indicated by spot size

of the total heat flow cannot be traced to regions on the surface or other heat transport processes according to the analysis of Veeder et al. (2012). The average heat flux on Io is ~2.5 W/m^2, compared to Earth's value of ~0.1 W/m^2. Earth's total power output is only ~50 TW, less than Io's despite 12 times great surface area. Heat sources other than tidal heating are negligible, contributing only about 1% in comparison (McEwen et al. 2004).

The holy grail for volcanologists was an accurate measurement of the highest temperature possible of the erupting lava. This would constrain what material was being erupted: elemental sulfur, as originally proposed, could not exceed 600 K, a silicate of basaltic composition was limited to 1475 K, and ultramafic silicates could reach 1800 K. The nature of the lava composition plays into every aspect of Io's solid body: the internal composition and temperatures, the "plumbing" and pressure that supplies magma to the surface, the viscosity and material properties of erupting lava, and the strength and rheological properties of lava long after it has solidified. Central to interpreting measurements were models of the thermal emission predicted from erupting and cooling lava flows (Davies 1996; Howell 1997; Keszthelyi and McEwen 1997).

Even on the Earth it's not easy to measure the hottest temperature a lava can reach, so remote sensing methods face an even greater challenge. On Earth, a volcanologist would have to make a measurement of hot lava and none of the adjacent materials: the narrow orange-hot crack in a lava flow or lava lake, without the adjacent solidified material still at very high temperature. Remote sensing methods at best combine hot and warm materials in the field of view; since lava can cool 400 K in 2 min on Io, it's likely that the warm areas dominate the signal. Furthermore, the hottest lava's emissions extend into the visible wavelength range and can be challenging to quantify in the presence of sunlight. In principle, multiple components and backgrounds can be modeled and fit to the observations, but the uncertainty of the highest temperatures can be too large to draw a definitive conclusion on the magma composition. There's little substitute for a direct measurement of the hottest component.

The quest for Io's hottest temperatures began with the Pearl et al. (1979) measurement of T = 290 K for a single component or possibly 385 K for multiple components at Loki. Both values were seen as consistent with molten sulfur but did not constitute evidence for silicate lava. However, Pearl and Sinton (1982) found much higher temperatures, up to 650 K, in IRIS spectra of the volcano Pele. Johnson et al. (1988) inferred temperatures of at least 900 K, requiring silicates, in ground-based observations of a large eruption in 1986, and Stansberry et al. (1997), using shorter-wavelength ground-based data, found temperatures of at least 1400 K. McEwen et al. (1998) examined Galileo SSI observations of an apparent fire-fountain eruption at the Pillan hotspot, finding a temperature of >1600 K, requiring ultramafic composition. Davies et al. (2001), incorporating NIMS data and cooling models increased the estimated eruption temperature to >1870 K. Keszthelyi et al. (2007) reanalyzed the observations and found a peak observed temperature closer to 1340 K, more consistent with basaltic composition, though eruption temperatures could be higher.

The hotspot locations from Earth and space together give a sense of fairly uniform coverage in longitude, and a slight preference for equatorial and mid-latitudes (perhaps due to observational bias). There are slightly more volcanoes on the Jupiter-facing hemisphere, and a handful of volcanoes at high latitudes. These may be clues to where and how tidal heating is generated and transported in the interior, but the evidence does not yet favor or rule out any such models.

The Loki hotspot, which is likely a lava lake 200 km in diameter, is large enough that spatial temperature variations across its surface can be mapped and observed to vary with time. Occultation measurements allowed determination of where within the lava lake the highest temperatures were observed. As on terrestrial lava lakes, the cool crust is likely denser than the underlying molten lava, so the lake is unstable against foundering. The changing location of the highest temperatures has been interpreted as a wave of foundering circulating around the perimeter of the lake, taking more than a year to do so (Rathbun et al. 2004; Davies 2003).

Succinctly summarizing the temporal behavior is considerably harder. Hotspots and plumes were originally categorized as persistent or transient (Lopes-Gautier et al. 2000), only to later find that some volcanic centers changed categories in both directions on timescale of a decade. Some hotspots have been observed more than a hundred times, and others only once.

Loki again offers the best temporal study as its substantial brightness allows frequent measurements. Periodicity studies initially identified a period of 540 days (Rathbun et al. 2002), but a longer baseline and new analysis shows that a shorter period of 460–480 days is also consistent with the observed variability (de Kleer and de Pater 2017; de Pater et al. 2017; de Kleer et al. 2019a). These values coincide with periodic changes in Io's eccentricity and semimajor axis, suggesting that celestial mechanics affects tidal flexing in a manner that affects an active volcanic eruption (de Kleer et al. 2019a). More observations will be required to verify this result.

Juno's JIRAM instrument is capturing the most recent close-up imaging of Io's hotspots. Mura et al. (2020) have taken advantage of Juno's polar orbit to identify the first of south polar hotspots on Io, and five more in previously unimaged areas. Juno's extended mission offers even more opportunities, and the last views until the arrival of Europa Clipper anticipated in 2030.

For a closer look at Io's individual hotspots and their diverse behaviors, see Chap. 6 by de Kleer & Rathbun.

2.5 Heat Flow Generates Diverse Volcanic Styles

The story behind Fig. 2.2a's plume discovery image from Morabito et al. (1979) reveals the stochastic process of science as well as anything else in this chapter, and should be required knowledge for anyone studying Io. While the predictions of volcanism by Peale et al. (1979) had appeared days before the Jupiter encounter, no observation changes were possible. So while the surface did look truly bizarre, and the total lack of impact craters was highly suggestive, the science team could not immediately identify features that were unequivocally due to active volcanism.

The plumes themselves were not discovered until after the encounter. Standard spacecraft operational procedures in fact revealed the surprising behavior. Linda Morabito, a JPL navigation engineer, was using a set of deliberately overexposed images of Io and other moons to identify background stars. Such images could be used to triangulate Voyager's position and thereby fine-tune the spacecraft's onward trajectory. What she saw puzzled her: was another object hiding behind Io, what today we would call "photobombing"? No other moon should be there. The bright spot on the terminator also had no clear explanation. Only after locating these unexpected features on the brand-new maps of Io did the volcanic context makes sense.

Special processing of Voyager images eventually revealed nine faint plumes against the darkness of space (Strom and Schneider 1982). McEwen and Soderblom (1983) found these fell into two distinct classes: the smaller, stable plumes like Prometheus and larger, transient type like Pele. The Galileo spacecraft 20 years later observed four of these, plus eight others previously unseen. Cassini and New Horizons also observed multiple plumes. All told, active plumes have been observed in association with 17 different volcanic centers, all associated with identifiable surface vents and thermal hotspots. Additional plumes sites can be inferred from ring-shaped deposits, presumably not active or not observable by any of the five spacecraft capable of imaging them. Plume heights range from 50 to nearly 500 km, linked to their source processed discussed below. Examples are shown in Figs. 2.2b, 2.3a, b and 2.5a, b. SO_2 and S_2 are thought to be the primary gases driving the volcanic eruptions and creating the plumes (Kieffer 1982). This conclusion derives both from direct detection of these gaseous species in the plumes, their presence on the surface in condensed form, and the complete absence of terrestrial volcanic driving gases H_2O or CO_2, neither in the plumes nor the surrounding magnetosphere (see Sect. 2.8).

The driving gases of plumes are not visible by themselves, so plumes are visible due to condensates forming within the flow and entrained pyroclastic particles originating at the vent. Since the particles do not achieve escape velocity, they fall back to the surface forming rings near the limit of ballistic trajectories or radial streaks closer to the vents. Condensates formed of SO_2 or S_2 tend to be white, yellow or orange, while pyroclastic particles were assumed to be dark (Geissler et al. 2004).

Imaging of volcanic centers reveals volcanic constructs with recognizable terrestrial counterparts. Lava flows and lava lakes closely resemble those formed from low viscosity mafic magmas on Earth, and in fact the resemblance to basaltic flows in Hawaii is uncanny. There are no steep-sided stratovolcanoes and few shallow-sloped shield volcanoes, indicating the range of volcanic styles on Io is more restricted than that on Earth. Many volcanic centers have dark circular features resembling terrestrial calderas. Since their appearance could alternatively result from a variety of non-volcanic processes, the more generic term "patera" is used.

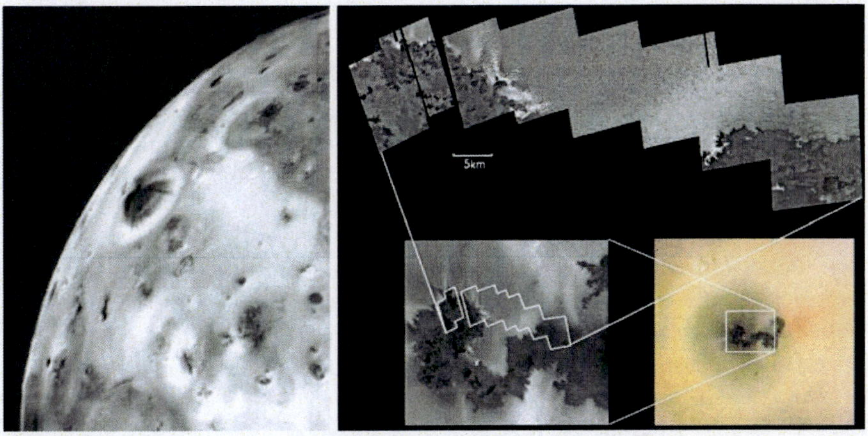

Fig. 2.5 Voyager's best view of the Prometheus plume near the limb (left), compared to Galileo views showing how the plume results from lava vaporizing SO_2 frost (right)

Building on the plume classifications by McEwen and Soderblom (1983), integrated studies of plumes, plume deposits, volcanic features and thermal emissions allowed identification of three classes of volcanic eruption (reviewed in McEwen et al. 2004, Williams and Howell 2007):

- Flow-dominated eruptions, with Prometheus as the archetype (Fig. 2.5b): Imaging shows lava flows 100–300 km long, with advancing flows vaporizing subsurface volatiles or volatile frosts to drive plumes (Kieffer et al. 2000; Milazzo et al. 2001). These eruptions are steady and long-lasting, from months to years.
- Explosion-dominated eruptions, with Pillan as the archetype: These are typified by vigorous, short-lived eruptions lasting days to months. Lava effusion rates rival terrestrial flood basalts on Earth, occurring more frequently but not lasting as long. Hotspot temperatures are the highest observed, and the eruptive volume suggests fire-fountaining as seen in terrestrial volcanoes.
- Intra-patera eruptions, with Loki as the archetype: These eruptions are interpreted as active lava lakes, in which the overturning crust releases gas for plumes and infrared emission. Typical lava lakes on Io have about ten times the volume of their largest terrestrial counterparts, and Loki, which is likely to be a lava lake, is larger still. Lava is confined within the patera, and flows are rare or non-existent.

While the behavior of most volcanic centers fits consistently into one of these categories, others change categorization over time. Loki and Tvashtar, for example, have had episodes of both flow-dominated and explosion-dominated eruptions. Exception from the classes also do exist, e.g., if silicate lava flows over certain sulfur-bearing species it causes melting instead of vaporization, giving rise to rare but distinct sulfur-rich flows.

Plumes are associated with many active eruptions. Plume characteristics generally depend on the style of eruption. Flow-dominated eruptions form the most predictable plumes, as flowing lava vaporizes surface frosts. Loki, a long-lasting lava lake, only forms a plume on rare occasions when lava reaches the surface outside the lava lake. These plumes are effectively composed of snow created during the expansion of vaporized gases into the vacuum of space. Relatively small umbrella-shaped plumes form that are largely consistent with ballistic trajectories over a range of launch angles. Ring-shaped plume deposits appear white, yellow or orange depending on the composition of the vaporized materials.

Explosion-dominated eruptions (e.g., Pillan) and some intra-patera eruptions (e.g., Pele) form plumes which can carry microscopic entrained particles either formed in the fire fountain or spalled from the vent. These eruptions create the largest observed plumes, and the bright outer plume edge suggests a shock forms in a relatively dense gas flow. Plume deposits include both radially dark pyroclastic streak and a lighter distal ring, usually red or orange. The red deposits are attributed to condensed S_2 from the plume (Spencer et al. 2000). Red deposits have been observed to revert to yellow on a timescale of months, attributed to the instability of red sulfur (Geissler et al. 2004).

It's possible that some classes of eruption have been missed, given limited instrumentation resolution or sensitivity. Johnson et al. (1995) proposed the existence of "stealth plumes" whose hotspots cannot be detected and whose plumes are too small or lacking in particulates to be seen. Such eruption would have negligible effects on Io's geology, but significant effects on the atmosphere (to be discussed in Sect. 2.7).

Readers interested in a deeper examination of Io's volcanoes and plumes are encouraged to see McEwen et al. (2004), Williams and Howell (2007), Geissler and Goldstein (2007), and Chap. 8 by de Pater et al.

2.6 Volcanism Creates a Unique Landscape

The dramatic plumes were correctly identified as a manifestation of Io's tidal heating, though the connection proved less direct than initially supposed. With strong evidence for sulfur on the surface and "warm" hotspots, attention focused on sulfur volcanism instead of silicates as found in the inner solar system. Sagan (1979) championed the idea that most of Io's surface colors and features could be explained by liquid sulfur quenched at different temperatures on Io's surface. Initial interior scenarios for the plumes (Fig. 2.6a) proposed that molten silicates at depth heated a sulfur ocean beneath a sulfur-SO_2 crust (Smith et al. 1979a).

Problems soon arose with sulfur as the dominant surface and crustal ingredient. Its rheological properties are much weaker than silicates, and the observed topography near hotspots could not be maintained at the assumed temperatures (Carr et al. 1979; Clow and Carr 1980). Young (1984) undertook a detailed examination of laboratory work on sulfur compounds, going so far as to use commercial paint chips as spectral analogs to Io's surface. His title "No sulfur flows on Io" foreshadowed the end of this early paradigm; new models came to the fore in which a patina of sulfur compounds might overlay a strong silicate crust with active silicate

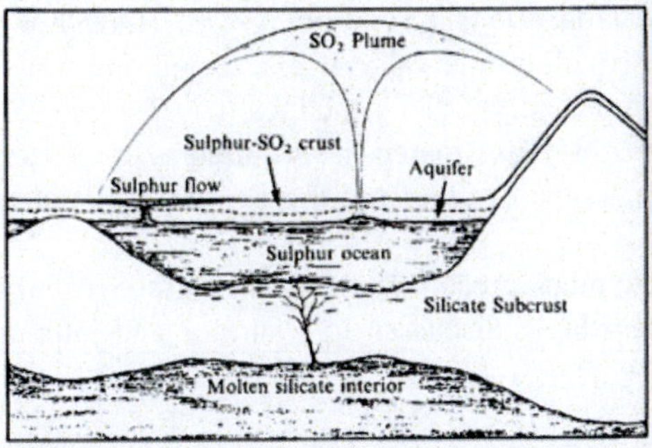

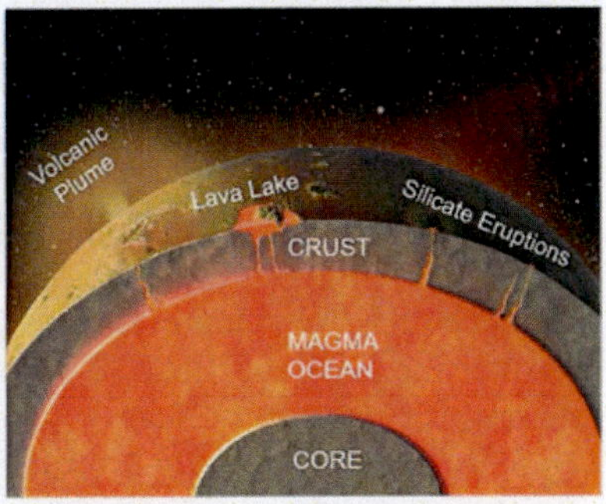

Fig. 2.6 Smith's 1979b model for Io volcanism involving sulfur (left), compared to a more recent model emphasizing silicate volcanism (McEwen et al. 2019)

volcanism (Carr 1986). Evidence for lava temperatures too hot for sulfur continued to accumulate from both Earth-based observations and Galileo, as discussed in Sect. 2.4. Figure 2.6b shows a current cutaway view of volcanic processes.

Geologic mapping ultimately provided the context by which we understand the profound means by which volcanism at a few discrete locations cause a literally global shaping and reshaping of Io's surface. Schaber (1980) identified five primary types of geologic features: pateras, lava flows, diffuse deposits (from explosive volcano deposits and condensed gases), plains and mountains. The complete absence of cratered terrains set Io apart from all other solid solar system bodies studied so far. These same categories were confirmed in more recent and complete maps (Williams et al. 2011). Ultimately, all terrains derive from Io's volcanic activity, directly or indirectly. Since three of the categories (paterae, lava flows and diffuse deposits) have been discussed in preceding sections, we'll discuss only the last two below.

Plains are the most common terrain type, and constitute a catch-all category of varied buried volcanic deposits of lava, pyroclastics and condensates from multiple volcanic centers, many presumably long extinct. Surface colors correspond to the more recent and potentially closest volcanic activity, with the color indicating the likely composition as above. Douté et al. (2001) used Galileo NIMS observations to show that the extensive bright plains are covered in abundant SO_2 frost. Note that this frost can condense either from a plume or the atmosphere, a distinction we'll discuss in the next section.

Io's mountains are predominantly tectonic in origin (reviewed by Schenk et al. 2001), but the tectonic stresses are directly linked to volcanism. Io's tallest mountain reaches 17 km altitude above the surrounding plains, and there are more than 100 mountains greater than 1 km. Schenk and Bulmer (1998) proposed that ongoing resurfacing buries underlying layers to greater depths where the circumference of the planet is smaller. The resulting compression causes the faulting which creates mountains.

Volcanism not only creates surface features at a prodigious rate, it also buries all others. Most importantly, the complete burial of even recent small craters requires a resurfacing rate of at least 0.1 cm/year (Johnson and Soderblom 1982). Lava flows (Phillips 2000) appear sufficient for providing this value if distributed uniformly, but much volcanic resurfacing occurs repeatedly in the same locations (Lopes et al. 2004). Plume deposits (Geissler et al. 2004) also match the required rates and are more broadly distributed, making them a better candidate for the global coverup of impact craters.

Chapter 5 by Williams et al. goes into much greater depth on Io's surface features and their origins, and Chap. 7 by Keszthelyi discusses Io's surface and interior composition.

2.7 Volcanism Sustains an Atmosphere Out of Balance

Initial expectations for atmospheres of outer solar system moons were low, extrapolating from Earth's moon and considering the low vapor pressure of ices in the frigid outer solar system. The strict upper limits placed by the Beta Scorpii stellar occultation therefore came as no real surprise (Smith and Smith 1972; Bartholdi and Owen 1972). So the Pioneer radio occultation soon after that detected an ionosphere was surprising, and allowed the inference of a 1 nbar atmosphere (Kliore et al. 1974).

The discovery of atoms and ions along Io's orbit in the mid-70s (discussed in detail in the next section) forced the conclusion that Io must possess at least an exosphere or corona to supply atmospheric escape, though even its atomic vs. molecular nature was uncertain. The debate over whether there must be a collisionally thick bound component continued for two decades, and a parallel debate centered on whether the atmosphere was supplied and maintained by surface frosts, volcanic vents, or some combination.

The post-Voyager view of the atmosphere was "nonuniform, turbulent, young, restless, and molecular" (Spencer and Schneider 1996; Lellouch 1996). Characterizing such an atmosphere presented unique observational challenges. Voyager readily detected gaseous SO_2 at 7 μm against thermal emission at the Loki volcanic center, inferring 100 nbar pressure in that region (Pearl et al. 1979, Fig. 2.7a), but could not measure it against Io's cold surface. After Voyager, astronomical observations at UV, IR and millimeter wavelengths broadened our understanding, though the methods were sensitive to different aspects of the atmosphere. Millimeter observations later detected ~10 nbar of SO_2 in emission at temperatures of hundreds of K (Lellouch et al. 1990). The emission must have been limited to ~10% of the surface, presumably associated with the plumes. Hubble UV absorption spectra also showed SO_2, consistent either with a uniform atmosphere of 0.2 nbar, or a patchy atmosphere at 4 nbar covering 10% of the dayside.

Techniques developed later were more sensitive to the more global—if not necessarily uniform—atmosphere away from the plumes. Hubble imaging of Io's disk in reflected solar Lyman alpha light (Roesler et al. 1999; Strobel and Wolven 2001) allowed measurements of the spatial distribution of the absorbing SO_2 gas.

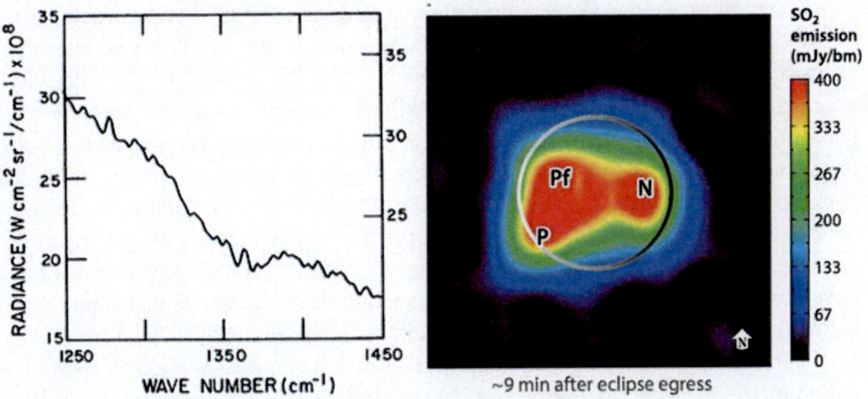

Fig. 2.7 Spectral detection of SO_2 against the Loki (left; Pearl et al. 1979) compared to SO_2 emission across Io's disk after it emerged from eclipse (right; de Pater et al. 2020b, 2021)

Spencer et al. (2005) used infrared observations of SO_2 at 19 μm against the thermal emission of Io's frigid disk to map the longitudinal distribution of SO_2, establishing a practical means of studying variability on many timescales from the ground. They found that SO_2 diminishes significantly on the sub-jovian hemisphere. Feaga et al. (2009), building on prior Lyman alpha studies, confirmed the longitudinal asymmetry and showed that SO_2 peaks over the bright equatorial band of SO_2 frost, and effectively vanishes over the dark polar regions. Spatially resolved HST observations also showed the non-uniformity directly using spectroscopy (McGrath et al. 2000; Jessup et al. 2004). The longitudinal variation at this point could be equally well explained by the broader distribution of frost there, or by the greater volcanic outgassing responsible for the frost in the first place.

The discovery of minor species offered further insights into chemical processes at Io. The discovery of SO (Lellouch et al. 1996) at millimeter wavelengths was anticipated both from photochemical and thermochemical (volcanic) processes. The detection of SO emission IR at 1.7 μm in Jupiter eclipse (de Pater et al. 2002) offered a different insight: the favored explanation was direct outgassing of SO in an excited state which promptly decayed and radiated. Images of the SO emission did not correlate with volcanic hotspots, supporting the stealth plume hypothesis (de Pater et al. 2020a). The discovery of S_2 in the Pele plume (Spencer et al. 2000) and others (Jessup and Spencer 2012) placed chemical constraints on magmatic properties. NaCl and KCl detected in Io's atmosphere (Lellouch et al. 2003; Moullet et al. 2013) were also interpreted as volcanic in origin, since the vapor pressure of these salts under ambient surface conditions is negligible. This offered a good answer to the 30-year-old question about the source of sodium in Io's neutral clouds, and solved additional mysteries in Io's neutral clouds and torus discussed the next section. It's noteworthy, though, that sodium, potassium and chlorine are apparently the only elements outgassed in significant quantities from silicate magmas (Na et al. 1998).

Isolated observations proved inconclusive in distinguishing between outgassing or sublimation as the dominant atmospheric source, but variability studies had promise in identifying the roles played by each driver. Tsang et al.'s (2012) decades-long timeline of 19 μm measurements finally demonstrated that the pressure over the anti-Jupiter hemisphere does respond to Jupiter's slightly varying heliospheric distance, establishing sublimation as primarily responsible for maintaining the background atmosphere on that hemisphere. Furthermore, the lack of correlation between volcanic activity and atmospheric pressure argues against outgassing as the primary control of the atmosphere (Roth et al. 2020). Curiously, correlations have been observed between volcanic outbursts and atmospheric escape that supplies the neutral clouds and plasma torus, discussed in the next section.

If sublimation controlled the background atmosphere, does it collapse under the extreme cold of night? No nightside observations have yet been possible, lacking an appropriately instrumented spacecraft at Jupiter, but observations of Io eclipsed by Jupiter offer the next best chance. Both 19 μm and millimeter observations confirmed the rapid decrease in atmospheric pressure in the first few minutes of eclipse (Tsang et al. 2015; de Pater et al. 2020b). Curiously, the millimeter measurements witnessed the decline leveling off, indicating either a noncondensing gas inhibiting the last stages of SO_2 condensation, or the underlying contribution of volcanic outgassing from plumes, perhaps the stealth plumes proposed by Johnson in 1995 but still eluding observational confirmation. Figure 2.8 shows visible wavelength images from Galileo and New Horizons of Io in eclipse, demonstrating that Io's atmosphere persists at some level in Jupiter's shadow, and exhibits auroral emission (Geissler et al. 1999; Spencer et al. 2007).

Atmospheric models developed for other solar system bodies offered little guidance on the dynamics of an atmosphere driven by such diverse processes such as volcanic and sublimation sources leading to condensation and escape sinks. The

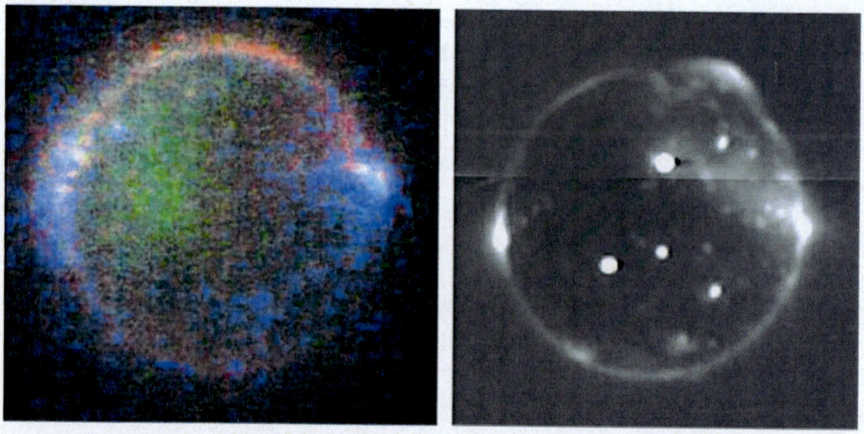

Fig. 2.8 Galileo visible-light image of Io's atmosphere glowing in eclipse by Jupiter (left; Geissler et al. 1999), compared to New Horizons (right; PIA09354)

extreme spatial contrasts, with virtually no atmosphere in polar regions or on the nightside drives supersonic winds not seen elsewhere in the solar system. Thermal structure normally driven by solar heating and radiative transfer is complicated by the addition of hot volcanic gases and plasma heating from above. Atmospheric chemistry is similarly affected by the localized injection of hot volcanic gases far from equilibrium and the plasma interactions at higher levels and potentially with the surface itself. Short timescales, both the diurnal freeze/sublime cycle and ~10-day residence time dictated by the ton-per-second loss rate, prevent the use of steady-state assumptions and force intense computational approaches.

Modelers tackled the challenge incrementally, initially limiting the scope to SO_2 chemistry (e.g., Kumar 1982), subsequently adding radiative transfer (Summers and Strobel 1996), and volcanic gases (Moses et al. 2002). Fanale et al. (1981) first modeled a sublimation-supported atmosphere, leading Ingersoll et al. (1985) to model the supersonic winds that must blow towards the night side. Walker et al. (2012) developed a more rigorous coupling between surface microphysics and the sublimation atmosphere. Zhang et al. (2004) developed plume dynamical models, which McDoniel et al. (2017) combined with Walker's sublimation model. A key finding was that the plume atmosphere could rapidly expand and flow over the top of the sublimation atmosphere, resulting in a volcanic atmosphere, perhaps warmer and chemically different, overlying and exerting pressure on a sublimation atmosphere which might partially condense back onto the surface.

Readers wanting a deeper review of the history of Io atmosphere studies are directed to excellent reviews by McGrath et al. (2004) and Lellouch et al. (2007). Chapter 8 by de Pater, Goldstein and Lellouch expands on advances since Galileo in much greater detail.

2.8 Rampant Atmospheric Escape Fuels the Magnetosphere

Jupiter's magnetosphere was discovered through its radio emissions in the 1950s (Burke and Franklin 1955), and Bigg (1964) identified Io's peculiar role in triggering radio bursts at particular orbital phases (Fig. 2.9a). It would be another decade before new observations would begin to reveal how Io could exert this influence, and even longer before the unusual asymmetry evident in the figure could be explained.

In 1972, a new interferometric spectrograph was undergoing checkout on Io, a bright target. Expecting just a reflected solar spectrum, Brown (1974) instead saw an anomalous bump at the wavelengths of atomic sodium (Fig. 2.9a). Higher resolution spectra the following year confirmed the emission appeared within the sodium D Fraunhofer lines of reflected sunlight. Though the emission was originally attributed to sodium on Io's surface, Trafton et al. (1974) soon showed the emission came from an extended region surrounding Io. In fact, Brown's interferometer had employed an entrance aperture about 500 times larger than Io's disk for ease of tracking (in contrast to narrow spectrograph slits), and therefore captured light from the entire

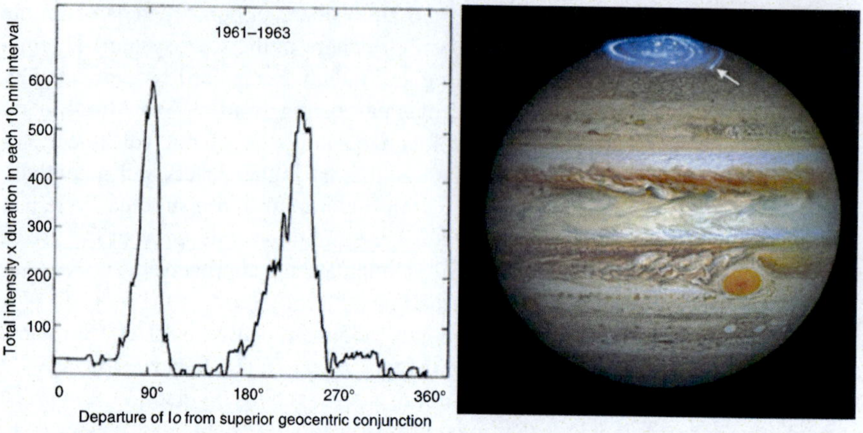

Fig. 2.9 Discovery of Io-controlled radio emission from Jupiter (left: Bigg 1964) compared to a Hubble image of Jupiter's aurora (right) including Io's footprint indicated with an arrow

region around Io. Had the observations used a standard narrow spectrograph slit matched to Io's diameter, the discovery might have had to wait for years. Bergstralh et al. (1975) demonstrated that resonant scattering of sunlight was the process responsible for the bright emission.

Pioneer 10 offered a puzzling result in the same year: its UV photometer had apparently detected a vast cloud of hydrogen accompanying Io (Carlson and Judge 1974). But no such cloud exists, and to this date it is not clear whether the instrument was somehow seeing emissions from other species, or whether noise from the harsh radiation environment (that Pioneer's plasma instruments also detected near Io's orbit) had masqueraded as a hydrogen cloud.

Brown's (1974) detection of sodium inspired an "Io week" for further observations from radio to IR to optical wavelengths. Planetary astronomers searching the circumjovian space for other species soon made spectroscopic identifications of atomic potassium (Trafton 1975), distributed about Io similarly to sodium, and atomic oxygen (Brown 1981; Durrance et al. 1983) as far away as the opposite side of Io's orbit. Sulfur was detected in ionized form (Kupo et al. 1976): the first optical detection of a planet's magnetosphere. Brown (1976) showed the plasma behaved like astronomical nebulas, and used spectroscopic diagnostics to make initial determination of the density. The discovery of oxygen ions (Pilcher and Morgan 1979) rounded out the inventory of major constituents. Imaging observations (Goldberg et al. 1984) soon showed that the neutral species for a "banana-shaped" cloud which co-orbited with Io, and that ionized species picked up by the corotating magnetic field formed a ring or torus (Pilcher 1980) slightly tilted relative to Jupiter's equator by its asymmetric magnetic field. Cummings et al. (1980) showed that the magnetosphere's rapid rotation and great size give rise to substantial centrifugal forces which constrain plasma to oscillate about the centrifugal equator.

While the known ingredients of the neutral clouds and plasma torus were used to constrain Io's surface composition (Sect. 2.6), their mere existence remote from Io drove investigations into how material could be liberated from the surface and escape from Io. Matson et al. (1974) proposed that energetic charged particle sputtering could eject atoms from Io's surface, especially if the atmosphere were collisionally thin as assumed at the time. Ejected atoms would at least form a bound exosphere, and models showed that more energetic escaping atoms could form the observed spatial and spectral properties of the neutral clouds (Smyth and McElroy 1977, and many studies to follow). Comparable structures were anticipated for atomic oxygen and sulfur, the dominant species, but their brightest emissions lie in the ultraviolet and the first imaging of the oxygen torus had to wait for the Hisaki Earth-orbiting telescope (Koga et al. 2018).

The Voyager flyby's in 1979 validated much of the prior groundbased work described above, and further showed that most of the torus energy was radiated in the extreme ultraviolet (Broadfoot et al. 1979). In situ measurements constrained the density and composition of the plasma (Bridge et al. 1979; Bagenal 1994).

The post-Voyager era was rich with Earth-based observations and modeling efforts. Together these jointly came to reveal additional ways by which material could escape Io. Brown and Schneider (1981) identified a high-speed population of sodium atoms escaping Io, which Mendillo et al. (1990) imaged to the astonishing distance of 400 Rj. They hypothesized that the fast sodium was created by atomic sodium charge exchange in Io's corona. Schneider et al. (1991) detected the signature of a fast sodium source uniquely tied to ions of sodium-bearing molecules, showing that Io's upper atmosphere was not decomposed into atoms as elsewhere in the solar system. Figure 2.10b shows Io's sodium features over a range of spatial scales. Progressively more sophisticated models of Io's sodium cloud started with Smyth and McElroy (1977) and continued though Smyth and Marconi (2003), with extensions to other neutral species and the plasma torus.

Many efforts to model the plasma torus focused on understanding the flow of mass and energy through the system. Shemansky (1988) established the Neutral Cloud Theory framework explaining first-order effects in the torus. Specifically, it assumes that ionization from the extended neutral clouds was the sole source of mass, and energy from ion pickup was the sole source of power. Shemansky's steady-state "cubic centimeter" model showed that these assumptions could not simultaneously explain the brightness of the plasma and its heavily ionized state, under any plausible assumptions of supply and loss, residence time, etc. Atomic and molecular cross-sections and rate coefficients from laboratory and theory were critical (e.g., Johnson and Strobel 1982). The conclusion that fresh ions did not bring enough energy to maintain the torus was dubbed an energy crisis. Subsequent work (e.g., Delamere et al. 2004; Tsuchiya et al. 2019) have validated the primary result, while expanding the model to multiple dimensions and allowing time variability. Proposed solutions to the energy crisis include hot electrons, perhaps originating from the outer magnetosphere (Barbosa 1994).

The sodium cloud and plasma torus were inherently time-variable due to geometrical changes associated with Io's orbit and Jupiter's rotation, but significant

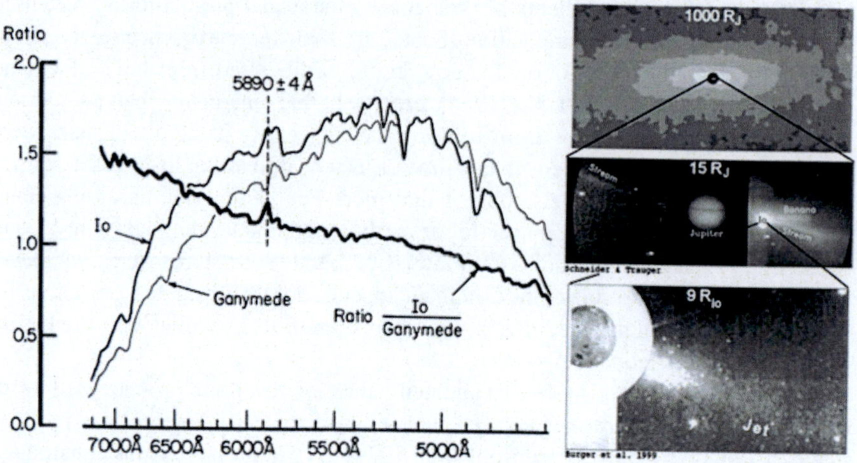

Fig. 2.10 Spectroscopic discovery of sodium emission (left: Brown 1974) compared to sodium cloud images at multiple spatial scales (right: Thomas et al. 2004)

intrinsic variations superposed on geometrical effects were well documented (see review by Thomas et al. 2004). These were presumably caused by variations in volcanic activity leading to changes in the escape rate from Io, but establishing correlations (let alone cause-and-effect relationships) proved challenging.

Multiple groundbased attempts sought to find correlations between two or more of the primary variable observable phenomena, volcanoes, the neutral clouds, and the plasma torus. Brown and Bouchez (1997) used systematic long-term observations of neutral sodium and ionized sulfur to demonstrate that a significant brightening of the sodium cloud was followed by a significant brightening in the plasma torus, consistent with the hypothesis that atmospheric escape first increased densities in the neutral clouds, which then were ionized to build up densities in the plasma torus. Mendillo et al. (2004) sought to link volcanic activity to atmospheric escape, finding a correlation between the distant sodium brightness and the infrared flux from Io's brightest thermal emission from the Loki lava lake. Subsequent studies could not confirm this correlation (Roth et al. 2020), and greater understanding of volcanic styles and atmospheric escape raised doubts over whether Loki would be expected to drive the production of fast sodium.

Two space missions offered the best insights into long-term correlation studies. The Cassini flyby of Jupiter offered an opportunity for months of observation of the plasma torus through its UVIS ultraviolet spectrograph, during which the imaging systems captured a dramatic increase in volcanic activity. Steffl et al. (2004, 2005) documented not only a brightening in the torus, but changes in the ionization states and even drifting longitudinal structures. JAXA's Hisaki mission recorded a similar response in the torus (Yoshikawa et al. 2017) following a volcanic outburst identified by groundbased telescopes (de Kleer and de Pater 2016).

Io's influence extends beyond the plasma torus and neutral clouds. Iogenic plasma inflates Jupiter's magnetosphere, exceeding the planet's ability to maintain corotation. The result is intense aurora ovals encircling Jupiter's poles (Fig. 2.9b). Furthermore, five Io flyby's by the Galileo spacecraft studied the intense electrodynamic perturbations as the corotating plasma flows around Io (Kivelson et al. 2004). This interaction drives Alfvén waves, carrying electrical currents up magnetic field lines which accelerate electrons and cause an additional aurora spot at the footprint of the field line threading through Io (Clarke et al. 2004; Bonfond et al. 2008). The spot can also be seen in Fig. 2.9b at upper right, and the radio emissions associated with the precipitation are shown in Fig. 2.9a.

Readers seeking a broader retrospective are directed to Thomas et al. (2004) and Schneider and Bagenal (2007). The modern perspective is reviewed in Chap. 9 by Bagenal, and in greater detail in Bagenal and Dols (2020).

2.9 Summary and Outlook

Five decades of increasing intense study have shown that Io's remarkable nature is richer and more varied than could have been imagined from the first puzzling results. Even the six surprising cause-and-effect relationships Fig. 2.11 captures our current understanding, though even this ambitious cartoon can show neither the full extent of our knowledge nor the scope of unanswered questions. Those are well laid out in the chapters ahead.

Answers to the questions may be a long time in coming. The Juno mission will focus on Io and other moons in coming years, and Europa Clipper and ESA's JUICE mission will observe Io occasionally while they focus on Europa and Ganymede. But no mission directed at Io's fundamental questions has been approved. The compelling nature of those questions will certainly drive innovative Earth-based observations with technology not foreseen, and ever-more sophisticated and realistic models will yield new insights. Io remains a target of priority in the Decadal Survey of the National Research Council, encouraging mission proposals in the Discovery and New Frontiers mission class (e.g., McEwen et al. 2021). Chapter 11 by McEwen et al. gives a more complete view of future plans for Io studies.

Figure 2.11 evokes a broader perspective on Io, a vertical construct actively sorting its constituents by volatility. It acts almost as a refinery in which volcanic energy from tidal heating creates layers of silicates, sulfur-bearing compounds, a bound atmosphere—and ultimately drives escape and loss. Io is effectively in a process of taking itself apart. Diverse observational and modeling approaches have quantified the rate of disassembly at a ton per second (Spencer and Schneider 1996; Schneider and Bagenal 2007). In this chapter we have focused on how this rate constitutes a prodigious supply of mass and energy to the neutral torus and plasma torus. But the escape rate also points to Io's declining mass: over the age of the solar system, Io has lost the equivalent of a 2 km thick layer of solid SO_2. While this is not sufficient to significantly deplete Io or SO_2 at the current rate, it begs the question of

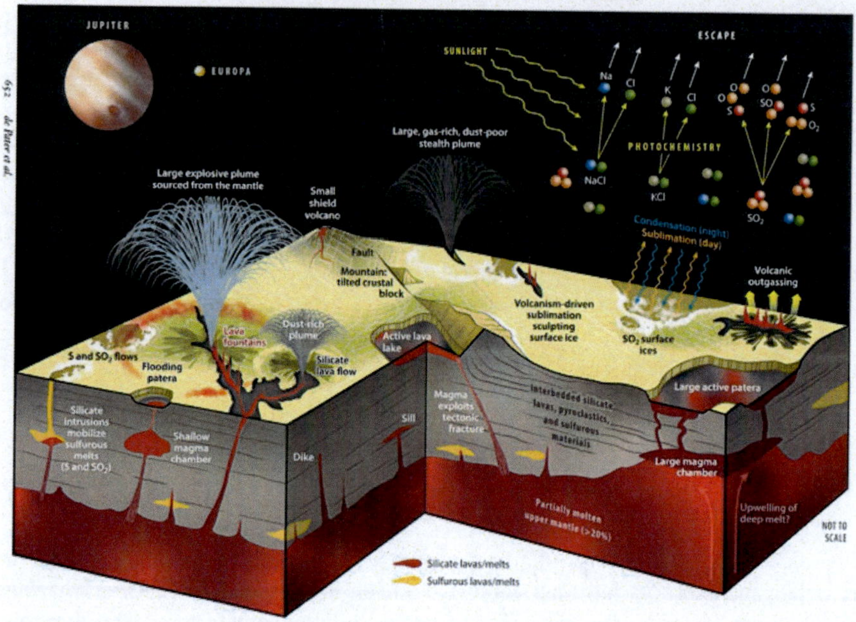

Fig. 2.11 This cartoon summarizes our current thinking of how Io works, as of 2021 (de Pater et al. 2021)

whether other volatile species (such as H_2O) might have been completely depleted early in Io's history (Spencer and Schneider 1996).

While Io is clearly unique in our solar system, it may be the archetype for "Exo-Io's" orbiting exoplanets. The existence of moons affected in one way or another by tidal heating around all our jovian planets suggests such moons could be common in the galaxy. (See Chap. 10 by Barr) Oza et al. (2019) make the case that some may already have been found through their sodium signatures. Similarly, exoplanet observers are considering the possibility that exoplanets may have auroral radio emissions, possibly powered by an orbiting moon as in Jupiter's case. It is no coincidence that these two types of searches follow in the footsteps of Bigg (1964) and Brown (1974) whose startling discoveries alerted us to Io's unusual nature in the first place.

References

Anderson, J.D., Null, G.W., Wong, S.K.: Gravitational parameters of the Jupiter system from the Doppler tracking of Pioneer 10. Science. **183**, 322–323 (1974)

Bagenal, F.: Empirical model of the Io plasma torus: voyager measurements. J. Geophys. Res. **99**, 11043–11062 (1994)

Bagenal, F., Dols, V.: The space environment of Io and Europa. J. Geophys. Res. **125** (2020)
Bagenal, F., Dowling, T., McKinnon, W. (eds.): Jupiter: The Planet, Satellites, Magnetosphere. Cambridge University Press (2004)
Barbosa, D.D.: Neutral cloud theory of the Jovian nebula: anomalous ionization effect of suprathermal electrons. Astrophys. J. **430**, 376–376 (1994)
Barnard, E.E.: Jupiter and his satellites. Mon. Not. Royal Astron. Soc. **51**(Supplementary Number), 543–549 (1891a)
Barnard, E.E.: On the phenomena of the transits of the first satellite of Jupiter. Mon. Not. Royal Astron. Soc. **9**(Supplementary Number), 557–558 (1891b)
Barnard, E.E.: On the dark poles and bright equatorial belt of the first satellite of Jupiter. Mon. Not. Royal Astron. Soc. **54**, 134–136 (1894)
Barnard, E.E.: A micrometrical determination of the dimensions of the planets and satellites of the solar system made with the 36-inch refractor of the Lick Observatory. Popular Astronomy. **5**, 285–302 (1897)
Bartholdi, P. Owen, F.: The occultation of beta Scorpii by Jupiter and Io, II: Io. Astron. J. **77**, 60–65 (1972)
Bergstralh, J.T., Matson, D.L., Johnson, T.V.: Sodium D-line emission from Io: synoptic observations from Table Mountain Observatory. Astrophys. J. **195**, L131–L135 (1975)
Bigg, E.K.: Influence of the satellite Io on Jupiter's decametric emission. Nature. **203**, 1008–1010 (1964)
Binder, A.B., Cruikshank, D.P.: Evidence for an atmosphere on Io. Icarus. **3**, 299–305 (1964)
Bonfond, B., Grodent, D., Gerard, J.-C., Radioti, A., Saur, J., Jacobsen, S.: UV Io footprint leading spot: a key feature for understanding the UV Io footprint multiplicity? Geophys. Res. Lett. **35**, L05107 (2008)
Bridge, H.S., Belcher, J.W., Lazarus, A.J., Sullivan, J.D., McNutt, R.L., Bagenal, F., et al.: Plasma observations near Jupiter: initial results from Voyager 1. Science. **204**, 987–991 (1979)
Broadfoot, A.L., Belton, M.J.S., Takacs, P.Z., Sandel, B.R., Shemansky, D.E., Holberg, J.B., Ajello, J.M., Atreya, S.K., Donahue, T.M., Moos, H.W., Bertaux, J.L., Blamont, J.E., Strobel, D.F., McConnell, J.C., Dalgarno, A., Goody, R., McElroy, M.B.: Extreme ultraviolet observations from Voyager 1 encounter with Jupiter. Science. **204**, 979–982 (1979)
Brown, R.A.: Optical line emission from Io. In: Brown, R.A. (ed.) Exploration of the Planetary System, pp. 527–531. Reidel (1974)
Brown, R.A.: A model of Jupiter's sulfur nebula. Ap. J. **206**, L179–L183 (1976)
Brown, R.A.: The Jupiter hot plasma torus: observed electron temperatures and energy flows. Astrophys. J. **244**, 1072–1080 (1981)
Brown, M.E., Bouchez, A.H.: The response of Jupiter's magnetosphere to an outburst on Io. Science. **278**, 268–271 (1997)
Brown, R.A., Schneider, N.M.: Sodium remote from Io. Icarus. **48**, 519–535 (1981)
Buratti, B.J., Mosher, J.A., Terrile, R.J.: First observational evidence for condensation of Io's SO_2 atmosphere on the nightside. Icarus. **118**, 418–422 (1995)
Burke and Franklin: Observations of a variable radio source associated with the planet Jupiter. JGR. **60**, 213 (1955)
Burns, J., Matthews, M. (eds.): Satellites. University of Arizona Press, Tucson, AZ (1986)
Caldwell, J.: Ultraviolet observations of small bodies in the Solar System by OAO-2. Icarus. **25**, 384–396 (1975)
Carlson, R.W., Judge, D.L.: Pioneer 10 ultraviolet photometer observations at Jupiter encounter. J. Geophys. Res. **79**, 3623–3633 (1974)
Carr, M.H.: Silicate volcanism on Io. J. Geophys. Res. **91**, 3521–3532 (1986)
Carr, M.H., Masursky, H., Strom, R.G., Terrille, R.J.: Volcanic features on Io. Nature. **280**, 729–723 (1979)
Clarke, J.T., Grodent, D., Cowley, S.W.H., Bunce, E.J., Zarka, P., Connerney, J.E.P., Satoh, T.: Jupiter's Aurora. In: Bagenal, F., Dowling, T.E., McKinnon, W.B. (eds.) Jupiter. The Planet, Satellites and Magnetosphere, pp. 537–560. Cambridge University Press (2004)
Clow, G.D., Carr, M.H.: Stability of sulfur slopes on Io. Icarus. **44**, 268–279 (1980)

Cruikshank, D.P., Nelson, R.M.: A history of the exploration of Io. See Lopes & Spencer 2007, pp. 5–33. (2007)
Cruikshank, D.P., Jones, T., Pilcher, C.B.: Absorption bands in the spectrum of Io. Astrophys. J. Lett. **225**, L89–L92 (1978)
Cummings, W.D., Dessler, A.J., Hill, T.W.: Latitudinal oscillations of plasma within the Io torus. J. Geophys. Res. **85**, 2108–2114 (1980)
Davies, A.G.: Io's volcanism: thermo-physical models of silicate lava compared with observations of thermal emission. Icarus. **124**, 45–61 (1996)
Davies, A.G.: Temperature, age and crust thickness distributions of Loki Patera on Io from Galileo NIMS data: implications for resurfacing mechanism. Geophys. Res. Lett. **30** (2003)
Davies, A.G., et al.: Thermal signature, eruption style, and eruption evolution at Pele and Pillan on Io. J. Geophys. Res. **106**, 33079–33104 (2001)
de Kleer, K., de Pater, I.: Spatial distribution of Io's volcanic activity from near-IR adaptive optics observations on 100 nights in 2013–2015. Icarus. **280**, 405–414 (2016)
de Kleer, K., de Pater, I.: Io's Loki Patera: modeling of three brightening events in 2013–2016. Icarus. **289**, 181–198 (2017)
de Kleer, K., de Pater, I., Molter, E.M., et al.: Io's volcanic activity from time domain adaptive optics observations: 2013–2018. Astron. J. **158**, 29 (2019a)
de Kleer, K., Nimmo, F., Kite, E.: Variability in Io's volcanism on timescales of periodic orbital changes. Geophys. Res. Lett. **46**, 6327–6332 (2019b)
de Pater, I., Roe, H.G., Graham, J.R., Strobel, D.F., Bernath, P.: Detection of the forbidden SO $a^1\Delta \rightarrow X^3\Sigma^-$ rovibronic transition on Io at 1.7μm. Icarus. **156**, 196–301 (2002)
de Pater, I., de Kleer, K., Davies, A.G., Ádámkovics, M.: Three decades of Loki Patera observations. Icarus. **297**, 265–281 (2017)
de Pater, I., de Kleer, K., Adamkovics, M.: High spatial and spectral resolution observations of the forbidden 1.707 μm rovibronic SO emissions on Io: evidence for widespread stealth volcanism. Planet. Sci. J. **1** (2020a)
de Pater, I., Luszcz-Cook, S., Rojo, P., Redwing, E., de Kleer, K., Moullet, A.: ALMA observations of Io going into and coming out of Eclipse. Planet. Sci. J. **1**, 60 (2020b)
de Pater, I., Keane, J.T., de Kleer, K., Davies, A.G.: A 2020 observational perspective of Io. Annu. Rev. Earth Planet. Sci. **49**, 633–668 (2021)
Delamere, P.A., Steffl, A., Bagenal, F.: Modeling temporal variability of plasma conditions in the Io torus during the Cassini era. J. Geophys. Res. **109**, A10216 (2004)
Douté, Â.S., Lopes-Gautier, R., Carlson, R., Schmitt, B., Soderblom, L., the Galileo NIMS Team: Mapping the SO_2 frost on Io by the modeling of NIMS hyperspectral images. Icarus. **149**, 107–132 (2001)
Durrance, S.T., Feldman, P.D., Weaver, H.A.: Rocket detection of UV emission from neutral oxygen & sulfur in the Io torus. Ap. J. **267**, L125–L129 (1983)
Fanale, F.P., Johnson, T.V., Matson, D.L.: Io: a surface evaporate deposit. Science. **186**, 922–924 (1974)
Fanale, F.P., Brown, R.H., Cruikshank, D.P., Clark, R.N.: Significance of absorption features in Io's IR reflectance spectrum. Nature. **280**, 761–763 (1979)
Fanale, F.P., Banerdt, W.B., Cruikshank, D.P.: Io: Could SO_2 condensation/ sublimation cause the sometimes reported post-eclipse brightening? Geophys. Res. Lett., **8**, 625–628 (1981)
Feaga, L.M., McGrath, M., Feldman, P.D.: Io's dayside SO_2 atmosphere. Icarus. **201**, 570–584 (2009)
Fuller, J., Luan, J., Quataert, E.: Resonance locking as the source of rapid tidal migration in the Jupiter and Saturn moon systems. MNRAS. **458**(4), 3867–3879 (2016)
Gaskell, R.W., Synnott, S.P., McEwen, A.S., Schaber, G.G.: Geophys. Res. Lett. **15**, 581 (1988)
Geissler, P.E., McEwen, A.S., Ip, W., Belton, J.S., Johnson, T.V., Smythe, W.H., Ingersoll, A.P.: Galileo imaging of atmospheric emissions from Io. Science. **285**, 870–874 (1999)
Geissler, P., McEwen, A.S., Phillips, C.B., Keszthelyi, L.P., Spencer, J.: Surface changes on Io during the Galileo mission. Icarus. **169**, 29–64 (2004)

Geissler, P.E., Goldstein, D.B.: Plumes and their deposits. In: Lopes, R.M.C., Spencer, J.R. (eds.) Io After Galileo, A New View of Jupiter's Volcanic Moon. Springer/Praxis (2007)

Goguen, J.D., Matson, D.L., Sinton, W.M., Howell, R.R., Dyck, H.M.: Io hot spots: infrared photometry of satellite occultations. Icarus. **76**, 465–484 (1988)

Goldberg, B.A., Garneau, G.W., LaVoie, S.K.: Io's sodium cloud. Science. **226**, 512–516 (1984)

Hanel, R., Conrath, B., Flasar, M., Kunde, V., Lowman, P., Maguire, W., Pearl, J., Pirraglia, J., Samuelson, R., Gautier, D., et al.: Infrared observations of the Jovian system from Voyager 1. Science. **204**, 972–976 (1979)

Hansen, O.L.: Ten-micron eclipse observations of Io, Europa, and Ganymede. Icarus. **18**, 237–246 (1973)

Harris, D.L.: Photometry and colorimetry of planets and satellites. In: Kuiper, G.P., Middlehurst, B.M. (eds.) Planets and Satellites, pp. 272–342. University of Chicago Press, Chicago (1961)

Howell, R.R.: Thermal emission from lava flows on Io. Icarus. **127**, 394–407 (1997)

Hussmann, H., Spohn, T.: Thermal-orbital evolution of Io and Europa. Icarus. **171**(2), 391–410 (2004)

Ingersoll, A.P., Summers, M.E., Schlipf, S.G.: Supersonic meteorology of Io: sublimation-driven flow of SO_2. Icarus. **64**, 375–390 (1985)

Jessup, K.L., Spencer, J.R.: Characterizing Io's Pele, Tvashtar and Pillan plumes: lessons learned from Hubble. Icarus. **218**, 378–405 (2012)

Jessup, K.L., Spencer, J.R., Ballester, G.E., Howell, R.R., Roessler, F., Vigel, M., Yelle, R.: The atmospheric signature of Io's Prometheus plume and anti-Jovian hemisphere: evidence for a sublimation atmosphere. Icarus. **169**, 197–215 (2004)

Johnson, T.V., McCord, T.B.: Spectral geometric albedos of the Galilean satellites. Astrophys. J. **169**, 589–593 (1971)

Johnson, T.V., Soderblom, L.A.: Volcanic eruptions on Io: implications for surface evolution and mass loss. In: Morrison, D. (ed.) Satellites of Jupiter, pp. 634–646. University of Arizona Press, Tuscon, AZ (1982)

Johnson, R.E., Strobel, D.F.: Charge exchange in the Io torus and exosphere. J. Geophys. Res. **87**, 10385–10393 (1982)

Johnson, T.V., Morrison, D., Matson, D.L., Veeder, G.J., Brown, R.H., Nelson, R.M.: Volcanic hotspots on Io: stability and longitudinal distribution. Science. **226**, 134–137 (1984)

Johnson, T.V., Veeder, G.J., Matson, D.L., Brown, R.H., Nelson, R.M., Morrison, D.: Io: evidence for silicate volcanism in 1986. Science. **242**, 1280–1283 (1988)

Johnson, T.V., Matson, D.L., Blaney, D.L., Veeder, G.J., Davies, A.: Stealth plumes on Io. Geophys. Res. Lett. **22**, 3293–3296 (1995)

Keszthelyi, L., McEwen, A.: Thermal models for basaltic volcanism on Io. Geophys. Res. Lett. **24**, 2463–2466 (1997)

Keszthelyi, L., Jaeger, W., Milazzo, M., et al.: Icarus. **192**, 491 (2007)

Khurana, K.K., Jia, X., Kivelson, M.G., Nimmo, F., Schubert, G., Russell, C.T.: Evidence of a global magma ocean in Io's interior. Science. **332**, 1186 (2011)

Kieffer, S.W.: Ionian volcanism. In: Morrison, D. (ed.) Satellites of Jupiter, pp. 647–723. University of Arizona Press (1982)

Kieffer, S.W., Lopes-Gautier, R., McEwen, A.S., Keszthelyi, L., Carlson, R.: Prometheus, the wanderer. Science. **288**, 1204–1208 (2000)

Kivelson, M.G., Bagenal, F., Kurth, W.S., Neubauer, F.M., Paranicas, C., Saur, J.: Magnetospheric interactions with satellites. In: Bagenal, F., Dowling, T.E., McKinnon, W.B. (eds.) Jupiter: Planet, Satellites, Magnetosphere, pp. 513–536. Cambridge University Press, Cambridge (2004)

Kliore, A., Cain, D.L., Fjeldbo, G., Seidel, B.L., Rasool, S.I.: Preliminary results on the atmospheres of Io and Jupiter from the Pioneer 10 S-Band occultation experiment. Science. **183**, 323–324 (1974)

Koga, R., Tsuchiya, F., Kagitani, M., Sakanoi, T., Yoneda, M., Yoshioka, K., Yoshikawa, I., Kimura, T., Murakami, G., Yamazaki, A., Smith, H.T., Bagenal, F.: Spatial distribution of

Jovian Moon Io's oxygen neutral cloud observed by Hisaki. J. Geophys. Res. **123**, 3764–3776 (2018)
Kuiper, G.P.: Infrared observations of planets and satellites. Astron. J. **62**, 295 (1957) (abstract)
Kuiper, G.P.: Comments on the Galilean satellites. Comm. Lunar. Planet. Lab. **10**, 28–34 (1973)
Kumar, S.: Photochemistry of SO_2 in the atmosphere of Io and implications on atmospheric escape. J. Geophys. Res. **87**, 1677–1684 (1982)
Kupo, I., Mekler, Y., Eviatar, A.: Detection of ionized sulphur in the jovian magnetosphere. Astrophys. J. **205**, L51–L54 (1976)
Lainey, V., Arlot, J., Karatekin, Ö., van Hoolst, T.: Strong tidal dissipation in Io and Jupiter from astrometric observations. Nature. **459**, 957–959 (2009)
Lellouch, E.: Urey Prize Lecture. Io's atmosphere: not yet understood. Icarus. **124**, 1–21 (1996)
Lellouch, E., Belton, M.J.S., de Pater, I., Gulkis, S., Encrenaz, T.: Io's atmosphere from microwave detection of SO_2. Nature. **346**, 639–641 (1990)
Lellouch, E., Strobel, D., Belton, M.J.S., Summers, M.E., Paubert, G., Moreno, R.: Detection of sulfur monoxide in Io's atmosphere. Astrophys. J. Lett. **459**, L107–L110 (1996)
Lellouch, E., Paubert, G., Moses, J.I., Schneider, N.M., Strobel, D.F.: Volcanically-emitted sodium chloride as a source for Io's neutral clouds and plasma torus. Nature. **421**, 45–47 (2003)
Lellouch, E., McGrath, M.A., Jessup, K.L.: Io's atmosphere. See Lopes & Spencer 2007, pp. 231–264. (2007)
Lopes, R.M.C., Kamp, L.W., Smythe, W.D., Mouginis-Mark, P., Kargel, J., Radebaugh, J., Turtle, E.P., Perry, J., Williams, D.A., Carlson, R.W., et al.: Lava lakes on Io: observations of Io's volcanic activity from Galileo NIMS during the 2001 ȳ-bys. Icarus. **169**, 140–174 (2004)
Lopes-Gautier, R., Doute, Â.S., Smythe, W.D., Kamp, L.W., Carlson, R.W., Davies, A.G., Leader, F.E., McEwen, A.S., Geissler, P.E., Kieffer, S.W., et al.: A close-uplook at Io in the infrared: results from Galileo's near-infrared mapping spectrometer. Science. **288**, 1201–1204 (2000)
Matson, D.L., Johnson, T.V., Fanale, F.P.: Sodium D-line emission from Io: sputtering and resonant scattering hypothesis. Ap. J. **192**, L43–L46 (1974)
McDoniel, W.J., Goldstein, D.B., Varghese, P.L., Trafton, L.M.: The interaction of Io's plumes and sublimation atmosphere. Icarus. **294**, 81–97 (2017)
McEwen, A.S., Soderblom, L.A.: Two classes of volcanic plumes on Io. Icarus. **55**, 191–198 (1983)
McEwen, A.S., Keszthelyi, L., Spencer, J.R., Schubert, G., Matson, D.L., et al.: High-temperature silicate volcanism on Jupiter's Moon Io. Science. **281**, 87–90 (1998)
McEwen, A.S., Keszthelyi, L.P., Lopes, R., Schenk, P.M., Spencer, J.R.: The lithosphere and surface of Io. In: Bagenal, F., Dowling, T.E., McKinnon, W.B. (eds.) , vol. 1, pp. 307–328 (2004)
McEwen, A., de Kleer, K., Park, R.: Does Io have a magma ocean? EOS. **100** (2019)
McEwen, A.S., Keszthelyi, L.P., Mandt, K.E., the IVO Team: The Io Volcano Observer (IVO). 52nd Lunar Planet. Sci. Conf., Abstract #1352. (2021)
McGrath, M.A., Belton, M.J.S., Spencer, J.R., Sartoretti, P.: Spatially resolved spectroscopy of Io's Pele plume and SO_2 atmosphere. Icarus. **146**, 476–493 (2000)
McGrath, M.A., Lellouch, E., Strobel, D.F., Feldman, P.D., Johnson, R.E.: Satellite atmospheres. In: Bagenal, F., Dowling, T.E., McKinnon, W. (eds.) Jupiter: Planet, Satellites & Magnetosphere, pp. 457–483. Cambridge University of Press, Cambridge (2004)
Mendillo, M., Baumgardner, J., Flynn, B., Hughes, W.J.: The extended sodium nebula of Jupiter. Nature. **348**, 312–314 (1990)
Mendillo, M., Wilson, J., Spencer, J., Stansberry, J.: Io's volcanic control of Jupiter's extended neutral clouds. Icarus. **170**, 430–442 (2004)
Milazzo, M.P., Keszthelyi, L.P., McEwen, A.S.: Observations and initial modeling of lava-SO_2 interactions at Prometheus, Io. J. Geophys. Res. **106**, 33121–33128 (2001)
Morabito, L.A., Synnott, S.P., Kupferman, P.N., Collins, S.A.: Discovery of currently active extraterrestrial volcanism. Science. **204**, 972 (1979)
Moroz, V.I.: Infrared spectroscopy of the Moon and Galilean Satellites. Soviet Astron. A. J. **9**, 999–1006 (1966) (original article appeared in 1965)
Morrison, D. (ed.): Satellites of Jupiter. University of Arizona Press, Tuscon, AZ (1982)

Morrison, D., Cruikshank, D.P.: Thermal properties of the Galilean satellites. Icarus. **18**, 224–236 (1973)

Morrison, D., Morrison, N.D., Lazarewicz, A.: Four-color photometry of the Galilean satellites. Icarus. **23**, 399–416 (1974)

Moses, J.I., Zolotov, M.Y., Fegley, B.: Alkali and chlorine photochemistry in a volcanically driven atmosphere on Io. Icarus. **156**, 107–135 (2002)

Moullet, A., Lellouch, E., Moreno, R., Gurwell, M., Black, J.H., Butler, B.: Exploring Io's atmospheric composition with APEX: first measurement of $34SO_2$ and tentative detection of KCl. Astrophys. J. **776**, 32 (2013)

Mura, A., Adriani, A., Tosi, F., et al.: Icarus. **341**, 113607 (2020)

Na, C.Y., Trafton, L.M., Barker, E.S., Stern, S., Alan, A.: Search for new species in Io's extended atmosphere. Icarus. **131**, 449–453 (1998)

Nash, D.B., Fanale, F.P.: Io's surface composition based on re ēctance spectra of sulfur/salt mixtures and proton irradiation experiments. Icarus. **31**, 40–80 (1977)

Nelson, R.M., Hapke, B.W.: Spectral reflectivities of the Galilean satellites and Titan 0.32-0.86 micrometers. Icarus. **36**, 304–329 (1978)

Nelson, R.M., Lane, A.L., Morrill, M.E., Wallis, B.D., Gibson, J., Smythe, W.D., Horn, L.J., Buratti, B.: The brightness of Jupiter's satellite Io following emergence from eclipse: selected observations, 1981-1989. Icarus. **101**, 223–233 (1993)

Ojakangas, G.W., Stevenson, D.J.: Episodic volcanism of tidally heated satellites with application to Io. Icarus. **66**, 341–358 (1986)

O'Reilly, T.C., Davies, G.F.: Magma transport of heat on Io: a mechanism allowing a thick lithosphere. Geophys. Res. Lett. **8**, 313–316 (1981)

Oza, A., et al.: Sodium and potassium as remnants of volcanic satellites orbiting close-in gas giant exoplanets. Astrophys. J. **885**, 168 (2019)

Peale, S.J., Cassen, P., Reynolds, R.T.: Melting of Io by tidal dissipation. Science. **203**, 892–894 (1979)

Pearl, J.C., Sinton, W.M.: Hot spots of Io. In: Morrison, D. (ed.) Satellites of Jupiter, pp. 724–755. University of Arizona Press (1982)

Pearl, J., Hanel, R., Kunde, V., Maguire, W., Fox, K., Gupta, S., Ponnamperuma, C., Raulin, F.: Identification of gaseous SO_2 and new upper limits for other gases on Io. Nature. **280**, 755–758 (1979)

Phillips, C.B.: Voyager and Galileo SSI views of volcanic resurfacing on Io and the search for geologic activity on Europa. Ph.D. thesis (2000)

Pilcher, C.B.: Images of Jupiter's sulfur ring. Science. **207**, 181–183 (1980)

Pilcher, C.B., Morgan, J.S.: Detection of singly ionized oxygen around Jupiter. Science. **205**, 297 (1979)

Pollack, J.B., Witteborn, F.C., Erickson, E.F., Strecker, D.W., Baldwin, B.J., Bunch, T.E.: Near-infrared spectra of the Galilean satellites: observations and compositional implications. Icarus. **36**, 271–303 (1978)

Rathbun, J.A., Spencer, J.R., Davies, A.G., Howell, R.R., Wilson, L.: Loki, Io: A periodic volcano. Geophys. Res. Lett. **29**(10), 1443 (2002)

Rathbun, J.A., Spencer, J.R., Tamppari, L.K., Martin, T.Z., Barnard, L., Travis, L.D.: Mapping of Io's thermal radiation by the Galileo photopolarimeter-radiometer (PPR) instrument. Icarus. **169**, 127–139 (2004)

Roesler, F.L., Moos, H.W., Oliversen, R.J., Woodward Jr., R.C., Retherford, D.K., Scherb, F., et al.: Far UV imaging spectroscopy of Io's atmosphere, with HST/STIS. Science. **283**, 353–357 (1999)

Roth, L., Boissier, B., Moullet, A., Sánchez-Monge, Á., de Kleer, K., et al.: An attempt to detect transient changes in Io's SO_2 and NaCl atmosphere. Icarus. **350**, 113925 (2020)

Russell, H.N., Dugan, R.S., Stewart, J.Q.: Astronomy, vol. 1, p. 371. Ginn & Co., New York (1945)

Sagan, C.: Sulphur flows on Io. Nature. **280**, 750–753 (1979)

Schaber, G.G.: The surface of Io: geologic units, morphology and tectonics. Icarus. **43**, 302–333 (1980)

Schenk, P.M., Bulmer, M.H.: Origin of mountains on Io by thrust faulting and large-scale mass movements. Science. **279**, 1514–1517 (1998)

Schenk, P., Hargitai, H., Wilson, R., McEwen, A., Thomas, P.: The mountains of Io: global and geological perspectives from Voyager and Galileo. J. Geophys. Res. **106**(33), 201–33222 (2001)

Schneider, N.M., Bagenal, F.: Io's neutral clouds, plasma torus, and magnetospheric interaction. In: Lopes, R.C., Spencer, J.R. (eds.) Io After Galileo, pp. 265–286. Springer, New York (2007)

Schneider, N.M., Trauger, J.T., Wilson, J.K., Brown, D.I., Evans, R.W., Shemansky, D.E.: Molecular origin of Io's fast sodium. Science. **253**, 1394–1397 (1991)

Secosky, J.J., Potter, M.: A Hubble Space Telescope study of post-eclipse brightening and albedo changes on Io. Icarus. **111**, 73–78 (1994)

Segatz, M., Spohn, T., Ross, M.N., Schubert, G.: Tidal dissipation, surface heat ōw, and figure of viscoelastic models of Io. Icarus. **75**, 187–206 (1988)

Shemansky, D.E.: Energy branching in the Io plasma torus: the failure of neutral cloud theory. J. Geophys. Res. **93**, 1773–1784 (1988)

Smith, B.A., Smith, S.A.: Upper limits for an atmosphere on Io. Icarus. **17**, 218–222 (1972)

Smith, B.A., Soderblom, L.A., Johnson, T.V., et al.: The Jupiter System through the eyes of Voyager 1. Science. **204**, 951 (1979a)

Smith, B.A., Shoemakert, E.M., Kieffert, S.W., Cook, A.F.: The role of S02 in volcanism on Io. Nature. **280**, 738 (1979b)

Smyth, W.H., Marconi, M.L.: Nature of the iogenic plasma source in Jupiter's magnetosphere. I. Circumplanetary distribution. Icarus. **166**, 85–106 (2003)

Smyth, W.H., McElroy, M.B.: The sodium and hydrogen gas clouds of Io. Planet. Space Sci. **25**, 415–431 (1977)

Soderblom, L.A., Johnson, T.V., Morrison, D., Danielson, G.E., Smith, B., Veverka, J., Cook, A., Sagan, C., Kupferman, P., Pieri, D., et al.: Spectrophotometry of Io: preliminary Voyager 1 results. Geophys. Res. Lett. **7**, 963–966 (1980)

Spencer, J.R., Schneider, N.M.: Io on the eve of the Galileo Mission. Annu. Rev. Earth Planet. Sci. **24**, 125–190 (1996)

Spencer, J.R., Toomey, D., Kaminski, C.: Jupiter's occultation of Io's hot spots. EOS. **72**, 33 (1991)

Spencer, J.R., Jessup, K.L., McGrath, M.A., Ballester, G.E., Yelle, R.: Discovery of gaseous S_2 in Io's Pele plume. Science. **288**, 1208–1210 (2000)

Spencer, J.R., Lellouch, E., Richter, M.J., Lopez-Valverde, M.A., Lea Jessup, K., et al.: Mid-infrared detection of large longitudinal asymmetries in Io's SO_2 atmosphere. Icarus. **176**, 283–304 (2005)

Spencer, J.R., et al.: Io volcanism seen by new horizons: a major eruption of the Tvashtar volcano. Science. **318**, 240–224 (2007)

Spencer, D.C., Katz, R.F., Hewitt, I.J.: Magmatic intrusions control Io's crustal thickness. J. Geophys. Res. Planets. **125** (2020)

Stansberry, J.A., Spencer, J.R., Howell, R.R., Dumas, C., Vakil, D.: Violent silicate volcanism on Io in 1996. Geophys. Res. Lett. **24**, 2455–2458 (1997)

Stebbins, J.: The light variations of the satellites of Jupiter and their applications to measures of the solar constant. Lick Obs. Bull. **13**, 1–11 (1927)

Steffl, A.J., Stewart, A.I.F., Bagenal, F.: Cassini UVIS observations of the Io plasma torus. I. Initial results. Icarus. **172**, 78–90 (2004)

Steffl, A.J., Delamere, P.A., Bagenal, F.: Cassini UVIS observations of the Io plasma torus: III. Observations of temporal and azimuthal variability. Icarus. (2005)

Strobel, D.F., Wolven, B.C.: The atmosphere of Io: abundances and sources of sulfur dioxide and atomic hydrogen. Astrophys. Space Sci. **277**, 271–287 (2001)

Strom, R.G., Schneider, N.M.: Volcanic eruptions on Io. In: Morrison, D. (ed.) Satellites of Jupiter, pp. 598–633. University of Arizona Press, Tucson, AZ (1982)

Summers, M.E., Strobel, D.F.: Photochemistry and vertical transport in Io's atmosphere and ionosphere. Icarus. **120**, 290–316 (1996)

Taylor, G.E.: The determination of the diameter of Io from its occultation of β-Scorpii on May 14, 1971. Icarus. **17**, 202–208 (1972)

Thomas, P.C., Davies, M.E., Colvin, T.R., Oberst, J., Schuster, P., Neukum, G., Carr, M.H., McEwen, A., Schubert, G., Belton, M.J.S.: The shape of Io from Galileo limb measurements. Icarus. **135**, 175–180 (1998)

Thomas, N., Bagenal, F., Hill, T.W., Wilson, J.K.: The Io neutral clouds and plasma torus. In: Bagenal, F., Dowling, T.E., McKinnon, W.B. (eds.) Jupiter. The Planet, Satellites and Magnetosphere, pp. 561–591. Cambridge University Press (2004)

Trafton, L.M.: Detection of a potassium cloud near Io. Nature. **258**, 690–692 (1975)

Trafton, L., Parkinson, T., Macy, W.: The spatial extent of sodium emission around Io. Ap. J. **190**, L85 (1974)

Tsang, C.C.C., Spencer, J.R., Lellouch, E., Lopez-Valverde, M.A., Richter, M.J., Greathouse, T.K.: Io's atmosphere: constraints on sublimation support from density variations on seasonal timescales using NASA IRTF/TEXES observations from 2001 to 2010. Icarus. **217**, 277–296 (2012)

Tsang, C.C.C., Spencer, J.R., Jessup, K.L.: Non-detection of post-eclipse changes in Io's Jupiter-facing atmosphere: Evidence for volcanic support? Icarus. **248**, 243–253 (2015)

Tsuchiya, F., Arakawa, R., Misawa, H., Kagitani, M., Koga, R., Suzuki, F., et al.: Azimuthal variation in the Io plasma torus observed by the Hisaki satellite from 2013 to 2016. J. Geophys. Res. Space Phys. **124**, 3236–3254 (2019)

Veeder, G.J., Matson, D.L., Johnson, T.V., Blaney, D.L., Goguen, J.D.: Io's heat flow from infrared photometry: 1983-1993. J. Geophys. Res. **99**, 17095–17162 (1994)

Veeder, G.J., Davies, A.G., Matson, D.L., et al.: Icarus. **219**, 701 (2012)

Veeder, G.J.: Io: Heat flow from small volcanic features. Icarus. **252**, 121–128 (2015)

Veverka, J., Simonelli, D., Thomas, P., Morrison, D., Johnson, T.V.: Voyager search for post-eclipse brightening on Io. Icarus. **47**, 60–74 (1981)

Walker, A.C., Moore, C.H., Goldstein, D.B., Varghese, P.L., Trafton, L.M.: A parametric study of Io's thermophysical surface properties and subsequent numerical atmospheric simulations based on the best fit parameters. Icarus. **220**, 225–253 (2012)

Wamsteker, W.: Narrow-band photometry of the Galilean satellites. Comm. Lunar Planet. Lab. **9**(167), 171–177 (1972)

Williams, D.A., Howell, R.H.: Active volcanism: effusive eruptions. In: Lopes, R.C., Spencer, J.R. (eds.) Io After Galileo, pp. 265–286. Springer, New York (2007)

Williams, D.A., Keszthelyi, L.P., Crown, D.A., Yff, J.A., Jaeger, W.L., Schenk, P.M., Geissler, P.E., Becker, T.L.: Geologic map of Io, U.S. Geological Survey Scientific Investigations Map 3168, scale 1:15,000,000, 25 p. http://pubs.usgs.gov/sim/3168/ (2011)

Witteborn, F.C., Bregman, J.D., Pollack, J.B.: Io: an intense brightening near 5 microns. Science. **203**, 643–646 (1979)

Yoshikawa, I., Suzuki, F., Hikida, R., Yoshioka, K., Murakami, G., Tsuchiya, F., Tao, C., Yamazaki, A., Kimura, T., Kita, H., Nozawa, H., Fujimoto, M.: Volcanic activity on Io and its influence on the dynamics of the Jovian magnetosphere observed by EXCEED/Hisaki in 2015. Earth Planets Space. **69**, 110 (2017)

Young, A.T.: No sulfur flows on Io. Icarus. **58**, 197–226 (1984)

Zhang, J., Goldstein, D., Varghese, P., Trafton, L., Moore, C., Miki, K.: Numerical modeling of Ionian volcanic plumes with entrained particulates. Icarus. **172**, 479–502 (2004)

Chapter 3
Setting the Stage: Formation and Earliest Evolution of Io

William B. McKinnon

Abstract The Galilean satellites—Io, Europa, Ganymede, and Callisto—form a relatively closely spaced, coplanar, prograde set of bodies of similar mass orbiting in Jupiter's equatorial plane. As such they must have formed from a dissipative disk of gas and/or solids in orbit around Jupiter. The ice-rich compositions of Europa and, especially, Ganymede and Callisto, indicate that this circumplanetary disk must have been cool enough for water ice stability. Beyond these fundamental facts there is limited agreement. Nevertheless, spurred by *Galileo* results, ALMA and other astronomical observations of protostellar disks, theoretical advances, and increasingly sophisticated numerical models, significant progress has been made in detailing possible satellite formation and evolution scenarios. The leading hypothesis for the formation of the Galilean satellites posits inflow of gas and solids across a tidal gap in the protosolar nebula, once Jupiter has itself formed in the "core accretion" model of giant planet formation. This leads to a relatively low-mass, circumjovian accretion and/or decretion disk, in which Io and the other satellites accrete while the gap exists, and are not lost due to gas drag or tidal torques, either because they are the last formed or, possibly, because their inward evolution is halted by a magnetospheric cavity in the disk close to Jupiter. Satellite-forming solids may also have been delivered, or even dominated, by heliocentric planetesimals, either through capture or ablation. Accordingly, Io may or may not have been water-rich originally, depending on the range of radial distance over which it accreted, and whether planetesimals or pebbles were its major building blocks. The latter also determined whether Io accreted hot or warm, but early core formation is predicted regardless. Io likely migrated inwards as it accreted in the protojovian nebula, and the Laplace resonance is predicted by most models to be primordial.

W. B. McKinnon (✉)
Department of Earth and Planetary Sciences and McDonnell Center for the Space Sciences, Washington University in St. Louis, Saint Louis, MO, USA
e-mail: mckinnon@wustl.edu

R. M. C. Lopes et al. (eds.), *Io: A New View of Jupiter's Moon*, Astrophysics and Space Science Library 468, https://doi.org/10.1007/978-3-031-25670-7_3

3.1 Formation of Jupiter and the Galilean Satellites

Any model for the formation of Io and the other Galilean satellites must start with the formation of Jupiter. The leading model for giant planet formation in our Solar System is the *core accretion* model: formation of a massive ice-rock-gas core by coagulation of planetesimals and pebbles (small millimeter-to-decimeter scale solids) in the protosolar nebula followed by an accelerating gravitational capture of a massive gas and dust envelope from the protosolar nebula (e.g., Mizuno 1980; Stevenson 1982; Bodenheimer and Pollack 1986; Pollack et al. 1996; Inaba et al. 2003; Alibert et al. 2005a; Hubickyj et al. 2005; Klahr and Bodenheimer 2006; Lissauer and Stevenson 2007; Lissauer et al. 2009; D'Angelo et al. 2010; Helled et al. 2014). This model has received strong empirical support from results of the *Juno* and *Cassini* missions, which indicate that Jupiter and Saturn, respectively, indeed contain central cores of high-Z (non-H-and-He) elements, albeit distributed in a dilute or "fuzzy" fashion (e.g., Wahl et al. 2017; Mankovich and Fuller 2021). Such cores are a natural prediction of the core accretion model (see Helled et al. 2022).

The alternative *disk instability* model proposes that the solar nebula was sufficiently gravitationally unstable that a massive clump or subcondensation collapses directly, forming a "giant gaseous protoplanet" (e.g., Cameron 1978; Boss 2002; Mayer et al. 2004; Durisen et al. 2007; Boley 2009). Heavy element cores are not a natural first-order prediction of this model, however, though later (and arguably ad hoc) processes may create such a core (see discussions in D'Angelo et al. 2010; Helled et al. 2014; Helled and Morbidelli 2021). The now classic correlation of giant exoplanet frequency with stellar metallicity for solar-type stars (e.g., Fischer and Valenti 2005) lends further credence to the plausibility of the core accretion model over the disk instability model for Jupiter's formation. Accordingly, while disk instability is not ruled out as a formation mechanism for some extrasolar giant planets (especially ones distant from their host stars; e.g., Inderbitzi et al. 2020), essentially all modern work on the origin of the Galilean satellites has taken core accretion of Jupiter as a starting point.

In this chapter I first briefly review pre-*Galileo* work on giant planet satellite formation (Sect. 3.1.1), noting constraints on Io's time of formation (Sect. 3.1.2). The major focus of ongoing research is on accretion and/or decretion[1] disk models, and the roles of pebbles vs. "satellitesimals." These are discussed in more detail in Sect. 3.2, along with major issues or challenges to such models. The implications for Io's composition and initial thermal state, especially in the case of slow-inflow, "gas-starved" (or at least gas-limited) accretion/decretion disks (and their many variations), are taken up in Sect. 3.3. The issue of whether Io accreted dry or wet, and if the latter, how it might have lost its water, is discussed in Sect. 3.3.2.

[1] I will use the term "decretion" in the sense proposed by Batygin and Morbidelli (2020), for disk outflows away from Jupiter. Classic "accretion" disk models in astrophysics spread viscously and thus decrete in this sense at sufficiently large radial distances from the primary body in question. But by using decretion I am not, however, implying that Jupiter itself is losing mass.

Finally, I summarize how these new results may change our view of Io's long-term evolution (Sect. 3.4), and offer some concluding remarks and prospects for improved constraints (Sect. 3.5).

3.1.1 Classes of Satellite-Forming Disks

As reviewed by Pollack et al. (1991) and Coradini et al. (1995), early giant planet satellite formation models traditionally broke down into four classes (cf. Lunine et al. 2004). In the *accretion disk* model, a circumjovian disk forms as solar nebula gas and entrained solid particles (dust to boulder-size) flow through the growing planet's Roche lobes to feed Jupiter during the rapid (runaway) gas-capture phase of core accretion (Coradini et al. 1989; Pollack et al. 1996; cf. Lubow et al. 1999; Bate et al. 2003; D'Angelo et al. 2003a; Papaloizou and Nelson 2005). In the *spinout disk* model, a circumjovian disk forms after gas capture terminates and the distended, hot Jupiter cools and contracts, stranding material in orbit in order to conserve angular momentum (Korycansky et al. 1991; Magni and Coradini 2004). A *blowout* or impact-generated circumjovian disk is also conceivable, and super-giant impacts into Jupiter have recently been modeled (Liu et al. 2019), but because Jupiter's obliquity is so small (3°) compared with say, that of Uranus, it is highly unlikely that such an impact—even if it occurred—would have provided the angular momentum necessary for a blowout disk. Finally, collisions of solid bodies within Jupiter's Hill sphere or tidal disruption of bodies approaching Jupiter could have formed a gas-free *co-accretion* or *debris disk* (Safronov et al. 1986).

The traditional accretion disk model was conceived in terms of the runaway growth of Jupiter. Jupiter's protoplanetary envelope is greatly distended as this phase initiates, but as the nebular gas and solid supply increases, the protoplanet reaches a critical mass where it begins to contract, and ultimately gas accretes hydrodynamically into the collapsing proto-Jupiter in a runaway (e.g., Lissauer and Stevenson 2007). This runaway persists as long as there is gas in the vicinity of proto-Jupiter's orbit, but may last no longer than ~10^4–10^5 year, depending on protosolar disk viscosity (e.g., models of Hubickyj et al. 2005; D'Angelo et al. 2010). Any circumjovian disk left after the end of this hydrodynamic phase might be better thought of as a spinout disk. Such a spinout disk is seen explicitly in the final thermal contraction phase of Magni and Coradini's (2004) 3-D numerical hydrodynamical model of Jupiter's formation by core accretion-gas capture. An actual accretion (or decretion, see Sect. 3.2.4 below) disk, and one that can form satellites, requires late-inflowing gas and solids containing sufficient angular momentum for centrifugal force balance at orbital distances compatible with satellite formation, and after proto-Jupiter has contracted to a scale of less than a few R_J (1 R_J = the current jovian radius).

In this regard, it is now established that Jupiter's accretion almost certainly did not terminate cleanly, but rather lasted as long as the protosolar disk persisted. Even if Jupiter opens a gap in the protosolar nebula around its orbital position, either

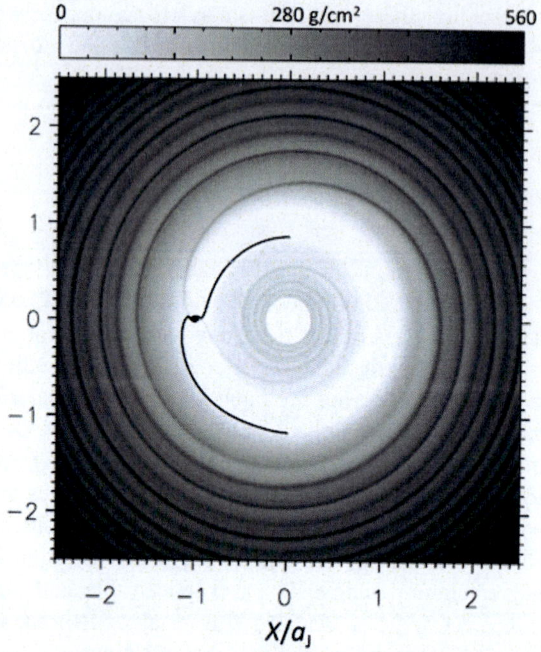

Fig. 3.1 Global surface density around a 1 M_J (Jupiter mass) planet orbiting a 1 $M_\odot$ (solar mass) star at $a_J = 5.2$ AU (technically, the planet/star mass ratio is 10^{-3}), modified from D'Angelo et al. (2010). In this numerical 3D hydrodynamic model, the relative disk thickness (scale height/radial distance) is $H/a = 0.05$ and the Shakura-Sunyaev turbulence parameter $\alpha = 4 \times 10^{-3}$. The disk is rotating counterclockwise, and the plot shows the density gap along the planet's orbit and the wave pattern generated by disk-planet interactions (features typical of similar simulations in the literature; see Baruteau and Masset 2013). The two black lines represent trajectories of gas in the co-moving frame of the planet. These gas parcels move along the inner and outer gap edges, become gravitationally bound to the planet, and are eventually accreted

through tidal torques or by simply drawing down the available nebular gas, inflow of protosolar nebula gas and entrained (small) solids across the gap continues, and potentially at a much reduced (though still non-negligible) rate (Lubow et al. 1999; Bryden et al. 1999; Bate et al. 2003; D'Angelo et al. 2003a, b) (Fig. 3.1). A circumplanetary disk (hereafter CPD) or envelope forms around Jupiter, and should last as long as the protosolar nebula exists to feed it (e.g., Stevenson 2001; Canup and Ward 2002; D'Angelo et al. 2003a; Alibert et al. 2005b; Ayliffe and Bate 2009, 2012; Tanigawa et al. 2012; Szulágyi et al. 2016), and see Coradini et al. (2010) and Peale and Canup (2015). This is the genesis of the *gas-starved* disk model for the formation of the Galilean satellites (Canup and Ward 2002, 2006, 2009; Ward and Canup 2010)—a modern version of the traditional accretion disk model. This model self-consistently solves or resolves a number of long-standing satellite formation issues (see Stevenson et al. 1986), though leaves some unaddressed, and is discussed in detail in the next section.

An alternative satellite formation model was proposed by Mosqueira and Estrada (2003a, b), and can be viewed as either a spinout or accretion disk model, depending on the timing of disk creation, but the important point is that they argue for a more massive circumjovian disk than do Canup and Ward (2002), one closer to the classic minimum-mass (protojovian) subnebula (or MMsN[2]), wherein the rock+ice in the Galilean satellites is augmented by enough H and He gas to match solar composition (for a total of ~0.02 M_J, where M_J is the mass of Jupiter; Lunine and Stevenson 1982).[3] Such a relatively massive accretion disk could in principle be created *after* proto-Jupiter contracted for sufficient continuing inflow of solar nebula gas and dust and low enough CPD turbulent viscosity, and is also discussed below.

For completeness I note that the gas-free, coaccretion model for the Galilean satellites was reconsidered by Estrada and Mosqueira (2006). The principal difficulties with the coaccretion model are (1) an adequate supply of solid bodies to feed the growing satellites after the solar nebula has dispersed is questionable (given that most local solids have presumably already been accreted or scattered by the giant planets), and (2) the mean angular momentum of collisionally captured material is low (~zero). Estrada and Mosqueira (2006) acknowledge these difficulties, and given that there is no obvious or natural explanation for the compositional gradient among the Galilean satellites in the coaccretion model (cf. Sect. 3.3.2 below), I do not consider it further.

3.1.2 *Pebbles vs. Planetesimals*

Over the past 20 years there has been a decisive shift in the planetary accretion paradigm. Whereas previously planets were thought to form from the bottom up, by binary or hierarchical coagulation of planetesimals of ever increasing size and mass, now it is thought that ensembles of much smaller, mm-to-dm sized bodies, termed pebbles, can through collective aerodynamics in the presence of protosolar gas leap the problematic "meter size accretion barrier" and form large (~100-km scale) planetesimals via gravitational collapse of dense pebble swarms (Youdin and Goodman 2005; Johansen et al. 2014, 2015; Simon et al. 2016; and references therein). The leading physical mechanism for concentrating the pebbles, and thus triggering the gravitational instability in the works above, is the streaming instability (SI). The prerequisites for SI are an enhanced solids/gas density ratio (~several times greater than solar) and suitably low nebular turbulence. There do not appear to be fundamental barriers to increasing the density of nebular solids, either locally through particle drift, piling up at ice lines, in pressure bumps, etc. (e.g., Izidoro et al. 2022), or globally through photoevaporation of nebular gas (Guillot and Hueso

[2] Distinct from the traditional minimum-mass *protosolar* nebula, or MMSN (Hayashi et al. 1985).

[3] Mosqueira and Estrada (2003b) ultimately argue for a circumjovian nebula relatively depleted in gas, perhaps by an order of magnitude.

2006). The turbulence issue is actively debated (e.g., Gole et al. 2020; Umurhan et al. 2020), but there are other potential pebble concentration mechanisms that can operate in turbulence (see Lesur et al. 2022). The success, however, of numerical simulations of SI in explaining the initial size-frequency distributions of asteroids and Kuiper belt objects (e.g., Morbidelli et al. 2009; Li et al. 2019; Kavelaars et al. 2021) and the properties of Kuiper belt binaries (Nesvorný et al. 2019, 2021) does lend some credence to the overall SI planetesimal formation picture.

Accordingly, many recent models for the formation of the Galilean satellites invoke pebble-sized solids, in one form or another, as the initial reservoir for satellite accretion in the context of a circumplanetary disk (Ronnet et al. 2017, 2018; Shibaike et al. 2019; Ronnet and Johansen 2020; Batygin and Morbidelli 2020; Madeira et al. 2021). The ultimate source of the pebbles differs in these various models, however, as will be discussed in Sect. 3.2.2. Once formed by SI or another mechanism, or captured directly from heliocentric orbit, planetesimals (which I now term satellitesimals) can grow further by accreting other satellitesimals (traditional hierarchical coagulation) or by gas-assisted accretion of pebbles (termed *pebble accretion*), the latter being distinct from the formation of any initial satellitesimals *from* pebbles (Johansen and Lambrechts 2017)! Satellitesimals may grow massive enough that pebble accretion becomes dominant (excellent illustrations of this phenomenon, in the context of the protosolar nebula, can be found in Johansen et al. 2015). There is, however, no clear consensus on the importance of pebble accretion per se for the formation of Io and the other Galilean satellites, as will become clear below.

3.1.3 When Did Io Form?

I end this section by discussing the absolute time frame for Io's formation. Nearly all models peg the formation of the Galilean satellites to the end of Jupiter's hydrodynamic collapse phase (Estrada and Mosqueira 2006 excepted). Naturally, Jupiter (and Saturn) must form while the solar nebula exists in order for both to acquire their massive H-He envelopes. Based on the observations of young stars and disks, the lifetime of the solar nebula has been generally taken to lie between a few to 10 million years (Myr) (e.g., Hubickyj et al. 2005; Meyer et al. 2007; Lissauer and Stevenson 2007). Observations of infrared excesses in clusters of young stellar objects indicate that nearly all ($\gtrsim$80%) form with optically thick circumstellar disks, and that the average lifetime of these disks is ~3 Myr (Williams and Cieza 2011) (Fig. 3.2). The observations in Fig. 3.2, compiled by Mamajek (2009), follow the classic plot of Haisch et al. (2001), but are based on either near-infrared excesses (sensitive to hot dust close to the stars in question, within ~0.1 AU), longer wavelength observations from *Spitzer* (sensitive to warm to cold dust out to several 10s of AU from solar-type stars), and/or classical T Tauri Hα emission (indicative of disk accretion onto the central star; Takeuchi et al. 2005). Figure 3.2 is essentially a *cumulative* distribution, but the underlying differential distribution of primordial disk lifetimes is less well constrained.

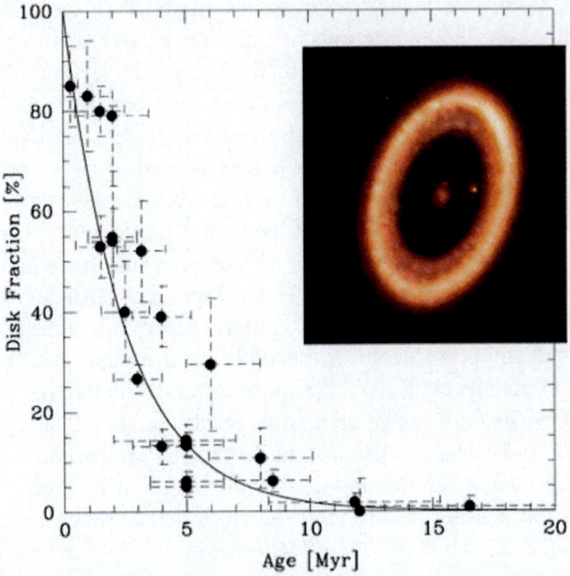

Fig. 3.2 Hα emission or infrared excess diagnostic/disk fraction as a function of mean protostellar cluster or group age (modified from Mamajek 2009). Disk ages depend on the spread of source ages from a single set of pre-main-sequence (PMS) model tracks (mostly for stars under 2 $M_{\odot}$). The uncertainty from different PMS model tracks is indicated, but may be underestimated (see text). The best fit exponential decay curve is plotted with timescale $\tau_{disk} = 2.5$ Myr, but is shown mostly to guide the eye (i.e., it is not a model fit). Inset shows an ALMA continuum dust image of a transition disk around an ~5 Myr old K7 T Tauri star (PDS 70), with an embedded protoplanet and CPD (PDS 70c) visible at the outer edge of the disk cavity (Benisty et al. 2021). Another protoplanet (PDS 70b), lies deeper in the cavity and is not visible in this image. Credit: ALMA (ESO/NAOJ/NRAO)/Benisty et al.

The number of young stellar objects sampled in Fig. 3.2 is large, and age uncertainties are reduced by using stellar cluster averages (especially as not all the stars in a given cluster may have the same exact age), but the stellar birth ages are still subject to systematic uncertainties in the stellar evolutionary tracks used. In this regard, Bell et al. (2013) revised the isochrones for 13 young star-forming regions, and found they were up to twice as old as previously considered. Their age for 50% disk fraction from *Spitzer* infrared excesses is not that different than that shown in Fig. 3.2, but there is a longer-lived tail: ~20% of young late-type stars are seen to retain dust disks at ~10–12 Myr. The current state-of-the-art regarding disk dust mass and temporal evolution is reviewed in Manara et al. (2022).

Of course, data on protostellar disk lifetimes does not tell us when our own protosolar disk dissipated, presumably due to planetesimal and planet formation and T-Tauri driven winds and photoevaporation (e.g., Coleman and Haworth 2022 and references therein). Figure 3.2 provides bounds on what is likely, but more precise constraints have come from the meteorite record (see Burkhardt 2021; Weiss and Bottke 2021). Another substantial advance in the past few years has been the recognition that all the solid bodies from which we have macroscopic samples (Earth, Mars, Moon, asteroids and meteorite parent bodies) come stamped with a subtle but measurable nucleosynthetic isotopic fingerprint. One supergroup (comprised of the Earth, Mars, Moon, ordinary chondrites, the HED clan,[4] most iron meteorites, others), termed non-carbonaceous (NC), is relatively enhanced in slow-neutron capture process nuclides, whereas the carbonaceous (CC) supergroup (carbonaceous chondrites, other irons) is relatively enhanced in rapid-neutron capture process nuclides. These enhancements are subtle, at the 0.1–0.01% level. First pointed out by Warren (2011), the list of elements for which this isotopic dichotomy is seen has grown to include Ti, Cr, Mo, Ca, Ni, Ru, and Zr.

The dichotomy implies spatial and/or temporal separation of the two supergroups during accretion, one presumably in the inner solar system (NCs) and the other farther out (CCs). Kruijer et al. (2017) propose, naturally, that the culprit responsible was Jupiter. Once its core grows enough to reach the so-called "pebble isolation mass," perhaps 10–20 $M_\oplus$ (Earth masses), (proto)Jupiter created a shallow gap in the protosolar gas nebula and a pressure bump outside of its orbit. This prevented most pebbles from drifting inward, and thus throttled Jupiter's core growth (e.g., Alibert et al. 2018; Bitsch et al. 2018; Helled et al. 2022) *and* effectively isolated the inner and outer nebular reservoirs. This isolation mass needs to be reached by no later than 1 Myr after the condensation time of the Solar System's first solids, the calcium-aluminum inclusions,[5] based on thermal evolution models for the early formation of iron cores of both NC and CC iron parent bodies. And this isolation needed to be maintained for at least 3 Myr while the ordinary chondrite and carbonaceous chondrite parent bodies (and their chondrules) accreted (Kruijer et al. 2020).

This range of dates is supported by limited but growing paleomagnetic evidence for the presence or absence of (technically, upper limit on) a protosolar nebular magnetic field from several meteorite groups (LL, CM, CV chondrites, angrites, CR chondrules, and NWA 7325, an ungrouped basaltic achondrite) implying the existence or dispersal of the conducting medium—the protosolar gas nebula—containing the field (Weiss et al. 2021). In their review, these authors conclude that the protosolar nebula dispersed sometime between 1.2 and 3.9 Myr after t_{CAI} in the NC formation region and between 2.5 and 4.9 Myr after t_{CAI} in the CC region. These dates incorporate 2σ limits on meteorite or chondrule formation ages (cf. Weiss and Bottke 2021). The paleomagnetic evidence and accretion timescales of meteorite parent bodies provide evidence for both the loss of nebular gas *and* the cessation of primary planetesimal formation by 4–5 Myr after t_{CAI} (Kruijer et al. 2020). The rapid, runaway acquisition by Jupiter of its massive gaseous envelope (over ~10^5 year; see, e.g., D'Angelo et al. 2010) would also have to have occurred by then, and this sets the timeframe for the formation of Io and the other Galilean satellites.

---

[4] Howardites, eucrites, and diogenites, related meteorites nearly universally regarded as originating from Vesta (e.g., McSween et al. 2013).

[5] t_{CAI}, so defined, is taken to be time zero for the formation of the Solar System.

It should be noted that alternative explanations have been offered for the isotopic dichotomy just described. Lichtenberg et al. (2021) argue that migration of two distinct water snow lines, one early, during the proto-Sun's class-I[6] infall stage, and one later, triggered planetesimal formation in two spatially and temporally distinct waves (as opposed to being temporally concurrent as in Kruijer et al. 2017). The planetesimals formed in both waves are born icy, but the inner (i.e., NC) bodies underwent profound and essentially complete dehydration due to ^{26}Al heating. Mixing of pebbles or other solids between reservoirs is prevented by pressure bumps or other structures in the protosolar disk, but Jupiter's formation or influence is ignored (not cited in the model). The idea that the ordinary and enstatite chondrite parent bodies, as well as Vesta, accreted as icy bodies is petrologically challenging, to say the least. But Lichtenberg et al. (2021) reminds us that concepts such as "the Jupiter barrier" (or for that matter, the streaming instability) are, ultimately, hypotheses to be tested, evaluated, and discarded if found wanting. Subsequently, both Morbidelli et al. (2022) and Izidoro et al. (2022) have offered detailed models whereby rings of planetesimals form at both the silicate vaporization front and the water ice line within the protosolar nebula, providing a natural (and plausible) explanation for the coeval formation of the inner (dry) NC and outer (wet) CC parent body groups.

A final note on the stellar "ages" discussed above: Zero in this case is set at the stellar "birthline," as shown in the Hertzsprung-Russell luminosity-temperature diagram, where the protostar begins its initial gravitational collapse (e.g., Sackmann et al. 1993; Palla and Stahler 1999; Siess et al. 2000). For the range of disk lifetimes discussed above, the formation of Jupiter, and that of Io and the other Galilean satellites, all take place while the Sun is still a PMS star, before hydrogen fusion begins.

3.2 The Protojovian Circumplanetary Disk

As discussed above, Io and Galilean satellites almost certainly formed in a circumjovian accretion (and/or decretion) disk supplied by inflow from the solar nebula after Jupiter formed. Observations of class-II[7] protostellar disks in nearby star-forming regions by the Atacama Large Millimeter Array (ALMA), sensitive to dust in emission, as well as in scattered light with the Very Large Telescope (VLT), clearly (and spectacularly!) show annular gaps, rings, central holes, arcs and spirals, down to the limits of resolution (Andrews 2020; Benisty et al. 2022). The origin of these disk substructures is under intensive investigation; planet formation is thought be at least responsible for some disk gaps (Drążkowska et al. 2022), and at least one candidate circumplanetary disk has been identified in such a gap (Isella

[6] Defined as a protostar and gas-dust disk surrounded by an infalling spherical cloud.

[7] Protostar and gas-dust disk.

et al. 2019; Benisty et al. 2021; Fig. 3.2, inset). Hence, it appears reasonably secure that the Galilean satellites were born in just such a structure. Systems of closely spaced, coplanar, solid exoplanets, some in mean-motion resonances (such as the TRAPPIST-1 and Kepler-223 systems), also imply a disk/nebular origin, so there are likely important parallels in the physical processes responsible (e.g., Kane et al. 2013; chapter by Barr et al.). Below relevant aspects of Jupiter's protosatellite disk structure and evolution are outlined, drawing at first from the now classic work of Canup and Ward (2002, 2006, 2009), though not because their "gas-starved" model is necessarily correct, but because the model is illustrative and serves as springboard to discuss more recent work.

Any model for the formation of the Galilean satellites must satisfy two important constraints: the total mass of satellites (~2×10^{-4} M_J, or Jupiter mass) and the large water ice mass fractions of three of them, especially outermost Ganymede and Callisto. Other constraints, such as the partially differentiated state of Callisto (Schubert et al. 2004), the ~anhydrous state of Io, or the Laplace resonance between Io, Europa, and Ganymede, are more conditional, or possibly due to post-accretional processes, as discussed below.

Figure 3.3 illustrates several key aspects of circumjovian disk models. Solids and gas are delivered to circumplanetary orbit with a range of specific angular momenta such that they achieve orbit in the satellite forming region, out to some radius r_o (Canup and Ward 2002, 2006, 2009). Because the gas closer-in orbits moves faster than more distant gas, shear exists, and if there is viscous coupling in the gas, turbulence and dissipation. This causes mass to flow towards Jupiter, and angular momentum and some mass to flow outward. The model of Canup and Ward (2002, 2006, 2009) is a classic accretion disk model, in which mass is continuously fed into it from the protosolar nebula ($\dot{M}_{\text{disk}}$), and through viscous spreading both inward and outward, achieves a steady-state distribution of surface density and temperature.

Commonly, kinematic disk viscosity (ν) is parameterized by the Shakura-Sunyaev α parameter, according to $\nu = \alpha c H \approx \alpha c^2/\Omega_K$, where c is the sound speed at the nebular midplane, H is the nebular scale height, and Ω_K is the Keplerian orbital frequency. The α model is intended to represent turbulent, eddy viscosity

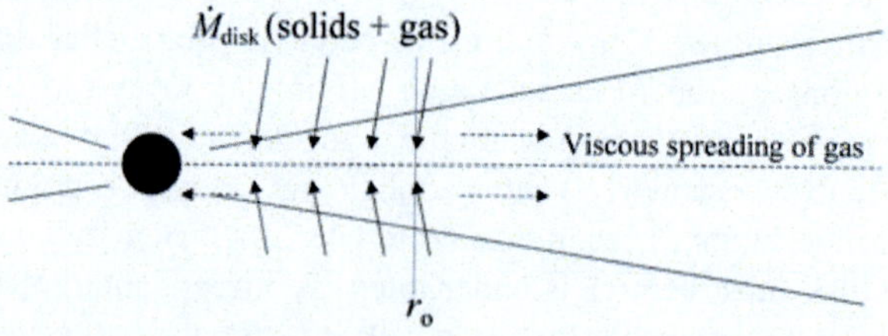

Fig. 3.3 Schematic of a circumjovian accretion disk model. $\dot{M}_{\text{disk}}$ is the flux of gas and solids from the protosolar nebula, which achieves centrifugal balance across the region of satellite formation ($R_J \lesssim r \lesssim r_o$). Nominally, the disk spreads viscously inward and outward, and is both accreted by Jupiter and stripped by other processes (returns to the protosolar nebula) beyond some outer radius (see text). Modified from Canup and Ward (2002)

(molecular viscosity being unimportant by many orders of magnitude), where the sources and strength of the turbulence are all folded into the dimensionless α parameter. Modeling of accretion disks around single protostars has long suggested α in the range between 10^{-4} and 10^{-2} (e.g., Hartmann et al. 1998; Stone et al. 2000; Dullemond et al. 2007; Ward and Canup 2010). Such values are consistent with recent observational constraints, though values at the lower end of the range (10^{-4} to 10^{-3}) are now favored (Manara et al. 2022). Appropriate values for circumplanetary disks are, however, only partially constrained. Potential sources of turbulence include purely hydrodynamic mechanisms such as the (baroclinic) vertical shear instability, or VSI (Nelson et al. 2013), mechanical instabilities driven by infall from the protosolar nebula (e.g., Ward and Canup 2010), and magnetorotational instabilities, or MRI (Balbus and Hawley 1991), when the CPD is thin enough that galactic cosmic rays partially ionize the gas (this requires surface densities $\lesssim 10^2$ g cm^{-2} [Dullemond et al. 2007]) or gas temperatures are high enough for thermal ionization (Batygin 2018).

Temperatures in the protojovian nebula are set by the energy balance between heating due to proto-Jupiter's luminosity, the kinetic energy of infalling matter, viscous dissipation within the disk and illumination from the background protosolar nebula, and radiative cooling from the disk photosphere or surface (e.g., Coradini et al. 1989; Canup and Ward 2002, 2009; Makalkin and Dorofeeva 2014). Viscous dissipation can be dominant in the energy balance, so the choices of α and especially $\dot{M}_{disk}$ are crucial. Figure 3.4 illustrates the midplane temperature and disk surface density profiles for steady-state models of Canup and Ward (2002, 2009). For these models the disk opacity K is set at 1 cm^2 g^{-1} (i.e., dusty and optically thick), so disk

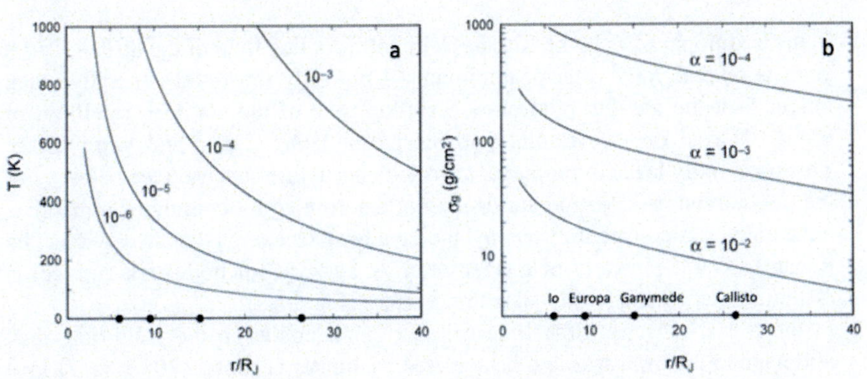

Fig. 3.4 Steady-state (**a**) midplane temperature T as a function of inflow rate from the protosolar nebula, in units of Earth masses ($M_\oplus$) per year, for a specific α of 5×10^{-3}; and (**b**) gas surface density as a function of distance from Jupiter for different values of the α viscosity parameter, for a specific inflow rate of 10^{-5} $M_\oplus$ year^{-1}. For the temperature boundary conditions adopted (500K at Jupiter's surface and a chilly ambient background of 20K), such low inflow rates are necessary to permit ice condensation. The disk is dusty and optically thick, and lower α imply more massive, but still gas-limited, protojovian nebulae. The present positions of Io and the other Galilean satellites are indicated. Modified from Canup and Ward (2009)

photosphere temperatures (both top and bottom) are less than that at the midplane. The infall rate controls the temperatures, while both $\dot{M}_{\rm disk}$ and α determine the surface densities. These are examples of "gas-starved" or at least "gas-limited" CPDs (for lower α). The surface density, σ, when integrated to its outer truncation radius contains much less mass than the MMsN, whose surface density follows a function similar to:

$$\sigma = 1.2 \times 10^6 \ (5.9 \ R_J/r) \ \text{ g cm}^{-2} \tag{3.1}$$

(e.g., Mosqueira and Estrada 2003a), where r is radial distance. A MMsN (~2×10^{-2} M_J) *is* processed through the gas-starved/limited disk in Fig. 3.4b in ~6×10^5 year. However, as long as entrained condensable solids can accrete along the way, and be retained in orbit, Io and the Galilean satellites can (in principle) form on this time scale. It should be noted that at this stage in Solar System history, the solar nebula is probably a few million years old (Sect. 3.1.3), so the ratio of gas to solids in the infalling material is likely to be substantially *non-solar* (discussed further below). If so, the MMsN, when seen as integrated through time, could be larger or smaller than its nominal value.

3.2.1 Advantages of Gas-Starved/Limited Disk Scenarios

The gas-starved disk model solves several long-standing time scale issues with the protojovian nebula (Stevenson et al. 1986; Canup and Ward 2002, 2009; Ward and Canup 2010):

1. In the traditional MMsN the Kelvin-Helmholtz cooling time of the disk, ~$10^3 \sigma K$ year, is several $\times 10^4$ year at minimum (if optically thin) and generally much longer than the gas-free, Safronov accretion time of the Galilean satellites, of order 10^3 year or less (Lunine and Stevenson 1982; Canup and Ward 2002). This potentially leads to problems of rock-rich satellite survival (see below). For the gas-starved disk, temperatures are not determined by cooling of a massive, essentially static structure, but by the steady-state energy balance within the jovian CPD.[8] The extent of condensation or survival (of both rock and ice) is essentially set by the steady-state temperature (Fig. 3.4a).
2. In the MMsN the gas drag timescale for small bodies in the solid-rich, disk midplane to drift inwards and be accreted by Jupiter is short, ~$10^3 \times (R_s/1$ km) year, where R_s is the body (i.e., satellitesimal) radius (Stevenson et al. 1986), and the orbital decay time for individual large satellites is little better, $10^3 \times (R_s/2000$ km) year (Canup and Ward 2002), due to perturbation of the gas by such large bodies. In the slowly cooling MMsN, early (rock-rich) and

[8] For low enough α, the viscous spreading time of the CPD may be commensurate with the time over which the inflow varies, so steady-state conditions may not be realized.

later generations of planetesimals could be lost (accreted to Jupiter). Mosqueira and Estrada (2003a) counter-argued that gas drag actually assists the accretional growth of small bodies (cf. Hayashi et al. 1985) and allows them to reach sizes where gas-drag-induced drift is not important (see next). In the gas-starved disk, however, gas-drag time scales are much increased, and the peril mitigated.

3. Large satellite orbits also migrate inward because of angular momentum transfer by disk tidal torques (type I decay), or if they are sufficiently massive as to open a gap in the circumjovian disk, inward migration follows the viscous spreading of the disk (type II decay). Type I decay in the MMsN is rapid, ~10^2 year for a Galilean satellite (Canup and Ward 2002, 2009). Type II decay is slower, ~$10^3 \times (10^{-3}/\alpha)$ year, but Canup and Ward (2002, 2009) argue that the ability of even Ganymede to open a gap is marginal unless α is low enough, $<10^{-4}$ (though a large enough satellite may be able to open a partial gap; Cilibrasi et al. 2018). Thus unless the MMsN disk viscosity is sufficiently low (facilitating and lengthening type II decay), it is unlikely that early rock-rich satellites could have survived long enough for CPD temperatures to fall and thus ice condensation and accretion (at least for accretion disks that actually accrete onto Jupiter). In contrast, for the gas-starved disk Type I decay times (being $\propto \sigma^{-1}$) are greatly increased (by ~3 orders of magnitude or more), so satellites can potentially survive over much of the gas-starved disk lifetime, and slower, type II decay need not be invoked.

In Fig. 3.4a disk temperatures are much too high for water ice condensation in the region of the Galilean satellites for mass inflow rates $\gtrsim 10^{-4}\ M_{\oplus}$ year^{-1}, and even silicate condensation in the Io region would have been problematic (Coradini et al. 1989; Canup and Ward 2002). Such a fast-inflow accretion disk corresponds to Jupiter at the time of runaway gas accretion or immediately thereafter (Sect. 3.1) (e.g., Coradini et al. 1989; Mosqueira and Estrada 2003a; Bate et al. 2003; D'Angelo et al. 2003a; Tanigawa et al. 2012). Recent three-dimensional radiative-hydrodynamic simulations of circumplanetary disk formation around jovian-mass planets (Szulágyi et al. 2016; Schulik et al. 2020) do clearly show hot disks or rotating envelopes (>1000 K) around the central planet. Such high temperatures are consistent with the luminosity and surface temperature of proto-Jupiter during its major gas accretion (Papaloizou and Nelson 2005; Lissauer et al. 2009).

To condense and accrete the Galilean satellites then, such a thick, hot disk must naturally first cool (e.g., Coradini et al. 1989; Mosqueira and Estrada 2003a; Alibert et al. 2005b), which returns us to the epoch of slow (and diminishing) inflow from the protosolar nebula, during the terminal stage of Jupiter's gas accretion, *or* satellite accretion must occur in cooler disk regions well outside the present positions of the satellites (cf. Fujii et al. 2017; Cilibrasi et al. 2018; Batygin and Morbidelli 2020). To create (or maintain) a circumjovian accretion disk with a mass comparable to the MMsN requires a low α, so that the disk does not spread viscously too rapidly and accrete onto Jupiter. If a MMsN disk is to have "reasonable" temperature structure, i.e., one that predicts ice condensation near Ganymede or Callisto's present positions, then α must be *very low*, ~10^{-6} or less (essentially inviscid and non-dissipative) (Lunine and Stevenson 1982; Canup and Ward 2002). Mosqueira and Estrada (2003a) in fact use such a MMsN model with an assumed temperature profile to constrain α values in the 10^{-6}-to-10^{-5} range in the Io-to-Ganymede region, with the temperatures maintained against radiative losses to space by very weak viscous dissipation.

Such a very-low-viscosity MMsN would necessarily be long-lived, with a viscous lifetime[9] of 10^5 to 10^6 year. From points (2) and (3) above, any satellites then formed would almost certainly have been lost to Jupiter via gas drag or type I decay. In such a case the best hopes for the survival of Io and the other Galilean satellites would either be gap opening and type II decay, which would proceed on the same extended viscous time scales, *or* stalled migration due to a central hole or cavity in the CPD (Sasaki et al. 2010), discussed in greater detail below. The former scenario was advocated by Mosqueira and Estrada (2003b). Canup and Ward (2002, 2009) countered that the presence of Galilean-sized satellites themselves would generate, through density wave interactions, an effective $\alpha \gg 10^{-6}$, and thus a correspondingly more rapid type II orbital decay. But even forming such a low-viscosity (~inviscid) disk is highly problematic, because the gas and solids have to come from somewhere ("the gas must flow"). With such low turbulent viscosities, gas accreted to CPD can neither accrete to Jupiter or flow back to the protosolar nebula, and should simply build up (an MMsN worth of gas every ~10^5 year), possibly to the point of triggering gravitational instabilities in the CPD, not an obvious path for solid satellite formation (Canup and Ward 2009).

Recent radiative, multi-dimensional hydrodynamic calculations of the vertical shear instability within protostellar accretion disks show that VSI can generate sustained turbulence, albeit at a relatively modest level with an effective α between 1 and few $\times 10^{-4}$ (Stoll and Kley 2014, 2016). VSI requires $\partial\Omega/\partial z \neq 0$,where Ω is the angular velocity of the gas and z is the distance from the midplane, and can be triggered as long as there is a temperature variation, either radially or vertically, within the disk. For the protojovian CPD, the latter is guaranteed as long as proto-Jupiter is luminous. The point is that there was likely a minimum level of turbulence within the protojovian CPD, and that very low α values between ~10^{-6} and 10^{-5} are most likely physically unrealizable.

Taken together, all the physical arguments above strongly indicate a preference for something more similar to a gas-starved, or at least gas-limited, accretion disk origin for Io and the Galilean satellites, as opposed to formation in a more massive MMsN. There seems little doubt that an accretion/decretion disk formed about Jupiter after it opened a gap in the protosolar nebula. Jupiter's final growth was likely processed through such a disk, but ultimately, the inflow must have abated as the protosolar nebula "reservoir" was depleted by (1) growth of Jupiter and Saturn; (2) viscous accretion of the inner protosolar nebula onto the Sun; and (3)

[9] Viscous spreading occurs on a timescale $\tau_{\text{visc}} \sim 10^4 \left(10^{-4}/\alpha\right)\left(r/30\,R_J\right)^2$ year (Canup and Ward 2009).

photoevaporation or other T Tauri loss processes. Io and the other Galilean satellites must have formed in this waning stage of *protosolar* nebula evolution. All recent models of satellite formation about Jupiter have adopted the latter point of view.

The nominally partially differentiated structure of Callisto also provides an independent, if contingent, argument for prolonged accretion (>few $\times$ 10^5 year) of the Galilean satellites, which is consistent with the gas-starved disk model (Canup and Ward 2002), and discussed in Sect. 3.3.4. Mosqueira and Estrada (2003a), and later, Batygin and Morbidelli (2020), prefer to argue for independent (and different) accretion scenarios for outermost Callisto, arguments that strike as ad hoc. Given that Ganymede and Callisto are so similar in orbit, density, and mass, it would seem more economical, if not preferable, to seek a common origin for both, and indeed for all four Galilean satellites.

3.2.2 Major Issue: Supply of Solids

In the original work by Canup and Ward (2002), solids were assumed to be delivered to the protojovian disk as small particles, small enough that their gas-drag stopping times were on the order of an orbital period or less. Canup and Ward (2002, 2009) estimated that particles of order 1-m or less would be entrained (for midplane gas densities). Later authors (e.g., Shibaike et al. 2019; Ronnet and Johansen 2020; Batygin and Morbidelli 2020) emphasized that given the vertical nature of the inflow supply to the CPD (e.g., Ayliffe and Bate 2009; Tanigawa et al. 2012), the major supply of gas would come from greater heights in the protosolar nebula, well away from the dust- and pebble-rich midplane, and that the inflow could be correspondingly substantially solids poor.

Figure 3.5 illustrates the geometry and character of such an inflow, from Schulick et al. (2020). The azimuthally averaged flow pattern does show that gas accreting to the central disk comes from higher elevations at radial distances >1 r_H (Hill radius), but the full 3D pattern can be quite complex, and some material may flow in from close to the protosolar midplane (see Fig. 10 in Schulik et al. 2020). The particular calculation in Fig. 3.5 assumes an $\alpha \approx 10^{-2}$ and a dust opacity 100 times less than the interstellar medium (ISM) value. The latter is necessary in order for the gas to cool during the timespan numerically simulated; otherwise Jupiter ends up with an extended, hot, pressure supported, featureless envelope. The gas accretion rates determined are formally in the stage of proto-Jupiter's Kelvin-Helmholtz contraction, but the accretion rates are compatible with those expected during late run-away accretion. Future simulations should give insight into later stages of CPD evolution, when the protosolar gas available for inflow declines.

To this picture we can add the effects of pressure (P) gradients in the protostellar gas nebula. Depending on whether $\partial P/\partial r$ is negative or positive, due to pressure support gas orbital velocities will either be sub- or super-Keplerian, respectively (Weidenschilling 1977). Pebbles and larger solid particles that orbit at Keplerian speeds then feel a "headwind/tailwind" or drag force that causes them to drift

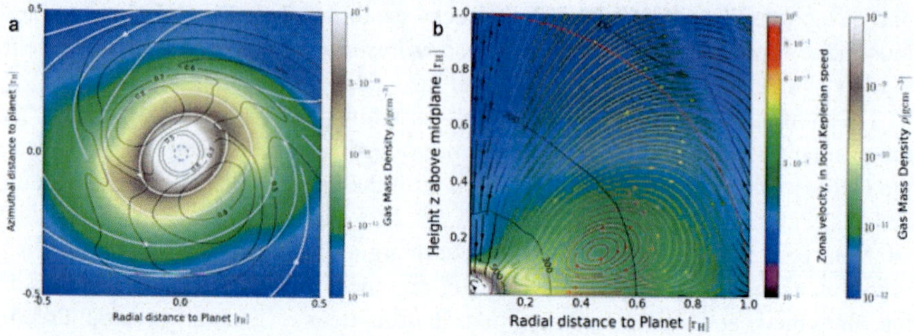

Fig. 3.5 (**a**) Midplane rotational structure for a global 3D radiation-hydrodynamic calculation of a circumplanetary disk forming around a Jupiter-mass planet. Velocity contours are labeled as a fraction of the Kepler speed at Jupiter's distance from the Sun (13.1 km s^{-1}) and distances in terms of Jupiter's Hill radius ($r_H \equiv a_J[M_J/3\ M_\odot]^{1/3} \approx 750\ R_J$, where a_J is Jupiter's semimajor axis). The gravitational smoothing length is marked with the blue, dashed circle, and can be viewed as Jupiter's size (numerically fixed). (**b**) Vertical, azimuthally averaged flow structures are shown with the temperature (in K) as contours. The observed flows are similar to the isothermal results presented in Tanigawa et al. (2012). Modified from Schulik et al. (2020)

radially towards regions of the disk with higher pressure, usually toward the inner disk. As Jupiter forms and opens a gap, however, the *positive* pressure gradient exterior to Jupiter *reverses* the inward radial drift of the pebbles, so in principle pebbles drifting inward from greater distances are trapped at this pressure bump, and cannot reach the protojovian CPD (regardless of the meridional flow pattern, e.g., Fig. 3.5b). On the other hand, smaller particles whose motion is well coupled to the gas[10] can diffuse across the gap, and follow the gas flow as it accretes to proto-Jupiter and its surrounding CPD (Pinilla et al. 2012; cf. Desch et al. 2018). Observations of PDS 70 in particular are consistent with this "dust filtration" picture (Benisty et al. 2021).

Accordingly, focus on solid supply (to form satellitesimals) has shifted from dust and pebble inflow to direct accretion or ablation of solar-orbiting planetesimals as they pass through the protojovian CPD (Ronnet et al. 2018; Shibaike et al. 2019; Ronnet and Johansen 2020). Of course, given that the formation of Io and the Galilean satellites is tied to the final stages of Jupiter's gas accretion has long meant that residual heliocentric planetesimals (those not already accreted to Jupiter or other protoplanets) were available as feedstock for the Galilean satellites (McKinnon and Zolensky 2003; Estrada et al. 2009; Canup and Ward 2009; Fujita et al. 2013; Tanigawa et al. 2014; Suetsugo and Ohtsuki 2017; Kronrod et al. 2020).

Planetesimal accretion is thought to be important for Jupiter itself as well, both as a possible explanation for the heavy element enrichment (over solar values) of its outer gaseous envelope and as a source of accretional energy to sustain

[10] Those whose Stokes numbers, defined as the stopping time divided by the local turbulent eddy timescale, are $\ll$1 (Birnstiel et al. 2010).

an extended gas envelope and delay the Kelvin-Hemholtz contraction phase and runaway gas accretion, possibly for up to 2 Myr so as to maintain the nominally required separation of the NC and CC reservoirs (Alibert et al. 2018; Venturini and Helled 2020). The efficiency of this process for supplying the protojovian CPD with solids (of any composition) is not yet clear. Numerical models of planetesimal dispersion during gap opening may or may not supply Jupiter with the requisite solid (high-Z) material to explain its present-day atmospheric enrichment. Results depend on assumed planetesimal sizes and distribution at gap edges, whether planetesimal formation itself is modeled (the outer pressure bump being a region of possibly substantially enhanced pebble abundance, the greater solid/gas ratio promoting streaming instabilities), and whether giant planet migration, Saturn formation, and planetesimal collisions and fragmentation are included in the modeling (Carter and Stewart 2020; Eriksson et al. 2021, 2022).

Carter and Stewart (2020) find that planetesimal-planetesimal collisions are important when giant planet migration is included whereas Eriksson et al. (2022) in contrast do not. Planetesimal size is not an important determinant of the accretional efficiency of solids for Jupiter and Saturn in Eriksson et al. (2022), due to compensating factors arising from gas drag in the nebula and planetary envelope, but would be important for capture or ablation into the protojovian CPD (as would collisional evolution of the planetesimal population). Although Eriksson et al. (2022) find that the planetesimal accretion efficiency for Jupiter is "low" (10% of the available planetesimals at maximum), the necessary accretional efficiency into the protojovian CPD need not be high to explain, at minimum, a total solid mass of $6 \times 10^{-2}\ M_{\oplus}$ (see Ronnet et al. 2018).

I do note that, for smaller planetesimals ($\lesssim$10 km), Erickson et al. (2022) predict a large implanted mass of outer solar system planetesimals in the inner Solar System (larger than the mass of the terrestrial planets), which is cosmochemically problematic. Possibly, as Jupiter and Saturn opened their gaps, they created an inner cavity in the protosolar gas nebula (reducing implantation efficiency there), or simply, that the formation of larger planetesimals (~100 km, as predicted by SI) was favored. These would have been much less efficiently captured by gas drag, according to Eriksson et al. (2022). Or it may just be that there was, at the time of gap opening and protojovian CPD formation, much less *planetesimal* mass available for such scattering.

3.2.3 *Major Issue: Central Magnetized Cavity?*

In their original works, Canup and Ward (2002, 2006, 2009) and Ward and Canup (2010) emphasized that due to gas drag and type I migration satellites that accrete in giant planet CPDs may ultimately migrate to the planet's surface and either be engulfed or tidally torn apart. Thus as the supply of gas and solids to the CPD declines due to (the eventual) attenuation and loss of the parental protosolar nebula, this inward migration would necessarily slow to a halt, and the surviving satellites would be ones seen today (e.g., the Galileans), subject to any further gas-free accretion and sweep up of smaller satellites. Sasaki et al. (2010) introduced the concept of a magnetically truncated inner cavity, based on the earlier analysis of Takata and Stevenson (1996). The principal idea is that the magnetic field generated in (e.g.) proto-Jupiter due to internal convection would be able to Lorentz-force couple to a partially ionized inner disk, and thus transfer planetary rotational angular momentum outward to the disk, both slowing the planet's rotation and preventing inward accretional disk flow inside the corotation radius.

In this situation, as an inwardly migrating satellite approaches the cavity boundary gas drag ceases or reverses, type I torques reverse, and satellite migration stalls. In particular, the corotation torque opposing the satellite's inward migration strengthens and counteracts the one-sided Lindblad torque pushing the satellite planetward (Liu et al. 2017). As modeling by Sasaki et al. (2010) and Ogihara and Ida (2012) show, exterior satellites can continue to migrate inward until they are captured into mean-motion resonance with the first (if the resonance was not established prior), and entire resonant chains can be created, pinned by the first satellite at the cavity boundary. A negative torque on the tidally coupled outer satellites remains and can in principle push the innermost satellite into the cavity and toward Jupiter,[11] but only if certain dynamic requirements are met.

In *N*-body accretional simulations, Ogihara and Ida (2012) find a strong preference for the formation of Laplace-resonance-like satellite chains, in which *several* satellites are captured into pairwise 2:1 mean-motion resonances. Sasaki et al. (2010) also proposed that the difference between the jovian and saturnian satellite systems is due to the lack of a magnetospheric cavity around Saturn during satellite formation in its CPD. Inner satellites are lost (and indeed, potentially create Saturn's massive icy ring; Canup 2010) with only a single, more distant, Galilean-mass moon (Titan) left standing.

An inner magnetospheric cavity, or an MRI-driven pressure bump (which can also halt inward satellite migration), has been adopted by several subsequent authors (e.g., Ogihara and Ida 2012; Fujii et al. 2017; Shibaike et al. 2019; Ronnet and Johansen 2020; Batygin and Morbidelli 2020). However, the *physical plausibility* of such a cavity was not specifically addressed, other than by appealing to analogy with the rotational properties of young stars and the ostensible necessity of slowing Jupiter's primordial spin from critical rotation, until Batygin (2018). In the latter, it is argued that the most efficient coupling with proto-Jupiter's magnetic field is through thermal ionization of the inner disk. Batygin (2018) adopts 1500 K for surface temperature of Jupiter and the inner disk, based on Lissauer et al. (2009) and the radiative-hydrodynamic calculations of Szulágyi et al. (2016) and Szulágyi (2017). At this temperature silicates are vaporized and alkali metals partially ionized. But this is tricky business. Even at 1500 K, high-temperature oxides can

[11] Technically, once inside the corotation radius, a satellite is also subject to tides raised on Jupiter that will cause the satellite to spiral inward, though generally on timescales much longer than the lifetime of the protojovian CPD (Peale and Canup 2015).

condense as dust, and by 1300 K almost all major rock- and metal-forming elements have condensed, for solar composition (see Table 7 in Lodders 2003). Dust is an effective quencher of gas-phase ionization, though dust itself can get charged, so the question of whether a cavity can form within some truncation radius is, at present, dependent on assumed parameters.

The real issue, however, is not so much whether a magnetized inner cavity can form in the protojovian CPD, but rather, what were its properties? What was the time history of temperature and gas and solids across the inner disk—the ostensible satellite-forming region, including that of Io—and how were the two (inner disk and cavity) connected.

3.2.4 *Emerging Paradigms*

I wrap up this section with a brief tour of promising recent scenarios for satellite formation within the protojovian CPD. A broad variety of approaches continue to be investigated, including formation in massive MMsN-like disks, although as a general rule these fail to produce a Galilean-like system (Miguel and Ida 2016; Moraes et al. 2018). Hence, below the focus is on (1) accretion from pebbles, (2) population syntheses, and (3) a novel decretion disk proposal.

Pebble accretion models. Ronnet and Johansen (2020) have considered in some detail how heliocentric planetesimals can be captured and ablated within the protojovian CPD. In their model, planetesimals are composed of mixed rock and water ice, and frictional heat from passage through the CPD melts or vaporizes any ices, releasing rocky grains that become, or lead to the formation of, pebbles within the CPD. Planetesimal formation from pebbles via SI is not invoked, however; rather, surviving captured planetesimals provide the seeds for pebble accretion. As proto-satellites grow, type I migration rapidly carries them planetward, and they all end up in a resonant chain anchored by the innermost satellite at the edge of the inner magnetospheric cavity. The nebula is assumed to be fairly massive ($\sigma \approx 10^4$ g/cm^2 at 15 R_J; cf. Fig. 3.4b), with an $\alpha = 10^{-4}$. This facilitates the capture of planetesimals seeds, but it also ensures that satellite migration completes well before the satellites can grow to anything resembling the Galilean moons.

Figure 3.6 illustrates the ultimate growth of the satellites, in place, by accreting pebbles that continue to be supplied to and migrate within the CPD. The mass scale is set by the pebble isolation mass,[12] which if Ganymede reaches this threshold first, blocks the inward drift of pebbles exterior to its orbit. This greatly reduces the pebble flux available to the growing Io and Europa, which is a possible explanation for the relatively smaller masses of the two inner moons. Because the satellites are built almost entirely of drifting pebbles, and the inner disk must be hot (for a

[12] Accretional growth stops at the pebble isolation mass—where the accreting body (a satellite in this case) generates a pressure bump in the CPD that traps drifting pebbles outside its orbit.

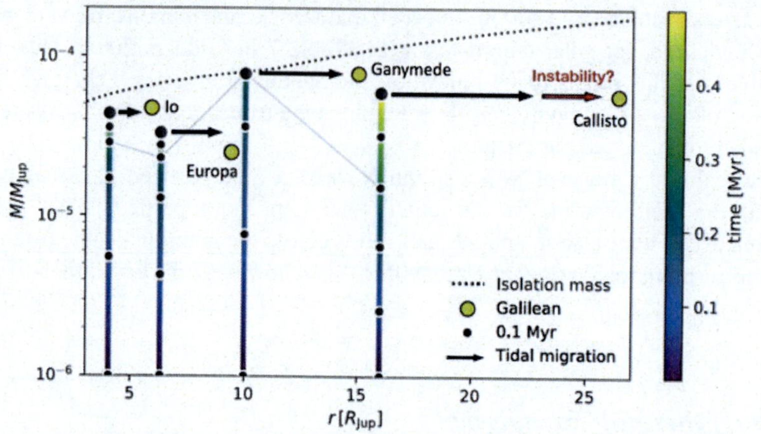

Fig. 3.6 Example pebble accretion growth tracks of the Galilean satellites trapped in their mutual 2:1 mean motion resonances in a protojovian CPD. Color indicates elapsed time, with black dots at intervals of 0.1 Myr, and the larger dots indicate the end state of the system after 0.5 Myr of accretion. The thin gray line connects the system at the time Ganymede reaches the pebble isolation mass (dotted line) at ≈0.3 Myr, and stops growing. Tidal migration after the dissipation of the CPD ultimately brings the orbits of the three inner satellites to their current location; the masses and present positions of the Galilean satellites are indicated. Another mechanism, however, such as a dynamical instability, is required to explain how Callisto ended up outside of the resonance system. Modified from Ronnet and Johansen (2020)

magnetospheric cavity to exist), Io and Europa form mostly (if not entirely in Io's case) from devolatilized pebbles.[13]

The CPD accretion timescale in Ronnet and Johansen (2020) is set at 10 Myr, which is on the long side, though major satellite accretion is complete within 0.5 Myr (Fig. 3.6). A similar study, but invoking a less massive, gas-starved nebula (Shibaike et al. 2019) finds much slower migration and pebble accretion rates, slow enough that the CPD lifetime needs to be ~30 Myr, which is excessive (when compared with the lifetime of the protosolar nebula; Sect. 3.1.3).

Subsequently, Madeira et al. (2021) modeled formation of the Galilean satellites, also incorporating pebble accretion, planetesimal seeds, and the usual satellite migration effects, but also full *N*-body integrations of the satellitesimals/satellites as they migrate, gravitationally interact, and (perfectly) accrete one another. The CPD is based on the gas-starved model of Canup and Ward (2002) with $\dot{M}_{\text{disk}} = 3 \times 10^{-5}$ $M_\oplus$ year^{-1} (declining with an *e*-folding time of 1 Myr), $\alpha = 10^{-3}$, and a relatively cool, and constant, temperature profile (the ice line is fixed at 14.5 R_J), but an inner magnetospheric cavity at 5 R_J is also imposed (note that the last two initial

[13] Cooling of the CPD and inward evolution of the ice line could account for Europa's ice in this scenario.

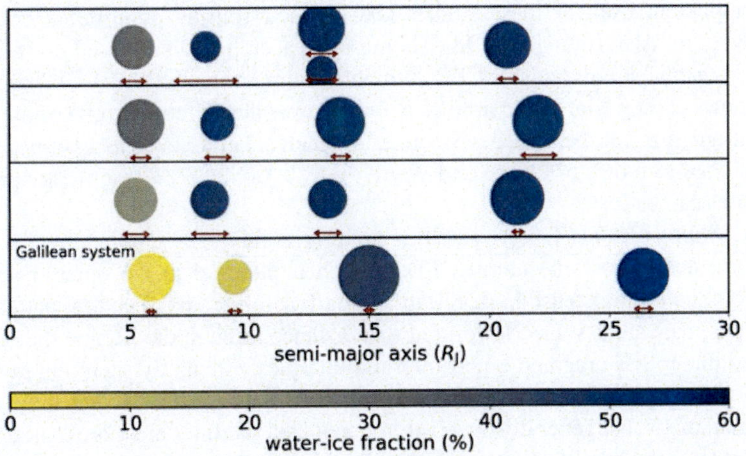

Fig. 3.7 Galilean system analogues at the end of the gas disk phase (2 Myr) in the simulations of Madeira et al. (2021). The top three panels are example model results, while the bottom panel is the observed Galilean system. Satellite sizes scale linearly with satellite mass, and color indicates ice fraction. Orbital eccentricities at this stage are indicated by the variation in jovicentric distance (red arrows), but are subject to later tidal damping. All of the analogue satellites are in pair-wise 2:1 mean-motion resonances. In the top analogue, coorbiting satellites at Ganymede's position could either subsequently merge or the system could undergo a dynamic instability. Modified from Madeira et al. (2021)

conditions are mutually incompatible[14]). The dust supply for pebble formation is radially distributed across a disk that extends to 125 R_J following Shibaike et al. (2019), rather than the ablation-derived distribution in Ronnet and Johansen (2020) above. The simulations typically produce between three and five satellites, with resulting pebble+planetesimal accretion timescales of ~10^6 year.

Four simulations (out of 120, parameters being varied) produced reasonable "matches" to today's Galilean satellites. Three of these are shown in Fig. 3.7. The innermost satellites, the Io and Europa analogues, are relatively or very icy in bulk, respectively, a consequence of the majority of accretion occurring beyond the ice line in the simulations.

Population syntheses. Given increased computational power and advanced data analytics, it is possible to run (and query) many thousands of individual satellite accretion simulations, varying initial conditions such as disk viscosity, dust/gas ratio, gas inflow rate, heliocentric planetesimal flux, etc., conditions or parameters that are generally poorly constrained. Cilibrasi et al. (2018, 2021) have taken

[14] The temperature at their cavity rim (5 R_J) is 380 K, far short of thermal ionization threshold necessary to couple CPD gas to proto-Jupiter's presumed magnetic field (Batygin 2018), but a hotter Jupiter and thus inner disk could be presumed. The gas surface density near the cavity in Madeira et al. (2021) is modest (~10^3 g/cm^2) and decreases with time, so direct ionization by galactic cosmic rays may be (or become) sufficient.

this approach. Both of these studies take as their starting point/initial condition relatively massive (though not MMsN massive) accretion disks based on radiative-hydrodynamic simulations of CPD formation from Szulágyi (2017).[15] These disks start hot, but they cool as the inflows attenuate over their lifetimes (one of the initial conditions that can be varied). No inner magnetospheric cavity is assumed, so as in the classic studies of Canup and Ward (2002, 2006, 2009), earlier generations of satellites can be lost.

In Cilibrasi et al. (2018) satellite-building solids are supplied by vertically infalling dust across the entire CPD, though it piles up in the outer disk (near 100 R_J) due to an ostensible dust trap arising from the radial gas flow pattern. All satellitesimals are assumed to form at that distance through the SI, and then evolve inward due to gas drag and type I migration torques, efficiently accreting dust (not pebbles!) along the way. Multiple satellites are commonly left stranded at the end of the simulations, but generally the resulting, modeled satellites are more massive than the real Io or the other Galilean satellites. The radial positions of the survivors favor locations more distant than the Galilean satellites, and later-forming, outer satellites necessarily accrete less mass (opposite to the pattern actually seen at Jupiter). Given their formation distance and late starts of the survivors, most surviving satellites contain ice. Satellite accretion times vary widely, but most fall between 10^4 and 10^5 year.

In Cilibrasi et al. (2021) satellitesimal seeds, i.e., randomly captured heliocentric planetesimals, are introduced, and full 3D *N*-body integrations of the growing satellites are followed, as the satellites migrate and accrete infalling dust and each other. In this case the surviving satellites skew multiple (>4) and with masses less than Io. Many mean-motion resonant combinations are found, not just 2:1 pairs. Satellite accretion times center on 10^5 year.

While it is possible to view these numerical experiments as a group as unsuccessful, my view is that they illustrate a path forward, one that frees us from the tyranny of the nominal model and its tuned parameters. The question that must be asked, though, is whether searching for an explanation for what may have been a product of a historically contingent set of circumstances (e.g., the Galilean satellites) can reasonably be expected to emerge as "likely" from an ensemble of 2000 simulations (as in Cilibrasi et al. 2021).

Decretion disk. In what is a rather substantial re-conceptualization, Batygin and Morbidelli (2020) have proposed that the Galilean satellites formed in a vertically-fed CPD in which nearly all the mass inflow is channeled close-in to Jupiter, near the magnetic truncation radius (based on arguments in Batygin 2018). Accordingly, the average disk radial velocity υ_r is >0 everywhere, in what they term a decretion disk (Fig. 3.8). The disk model they explore is rather massive ($\sigma \sim 3 \times 10^4$ g/cm^2 at 15 R_J) and low viscosity ($\alpha = 10^{-4}$). Entrained dust particles are the ultimate feedstock for the satellites, though captured and ablated planetesimals could also play a role (perhaps an important one). A key point is that fine dust moves outward

[15] For comparison, the initial surface density at 10 R_J is ~10^4 g/cm^2 (cf. Fig. 3.4b).

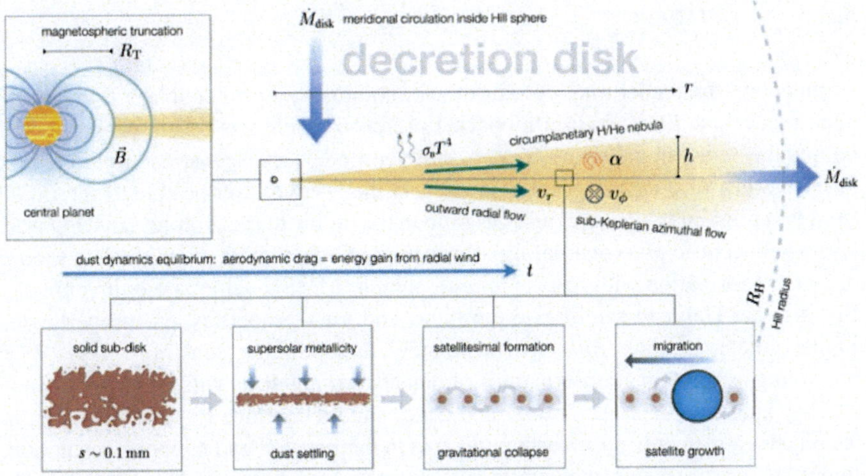

Fig. 3.8 Schematic cross-section of possible processes within a circumplanetary, satellite-forming decretion disk. Material flows in from the protosolar nebula and accretes near the magnetospheric truncation radius R_T, nominally somewhat inward of Io's position today. As gas and dust flow outwards, back towards the protosolar nebula, particles of critical sizes s are trapped (aerodynamic energy loss from the sub-Keplerian headwind balances the energy gain from the radial wind), and settle toward the CPD midplane. With sufficient mass density, the particle sheet gravitationally fragments into a swarm of 100-km-scale satellitesimals. Satellites form from these "satellitesimals," migrate toward Jupiter, and link up in a Laplace-like resonant chain (see text). The satellites that form are all icy, at least initially. $\vec{B}$ is proto-Jupiter's powerful magnetic field, v_r and v_ϕ are the radial and azimuthal gas velocities, and $\sigma_B T^4$ is the blackbody radiation flux from the disk surface (both top and bottom) due to turbulent gas dissipation and re-radiated protojovian luminosity (σ_B is the Stefan-Boltzmann constant). Modified from Batygin and Morbidelli (2020)

with the decretion flow, large objects spiral inward due to gas drag (and type I drag if large enough), but large grains or small pebbles may stall in the outer disk (beyond ~75 R_J). If sufficient dust is "trapped" and settles to the mid-plane, this dust disk can undergo classic gravitational fragmentation to form satellitesimals (Goldreich and Ward 1973). Batygin and Morbidelli (2020) argue that satellite accretion ensues, mainly by oligarchic, binary coagulation of these satellitesimals.

Growth through satellitesimal collisions is brisk, with formation timescales of only a few $\times 10^3$ years. Satellites emerge from their icy nursery in sequence and then exit inward via type I migration. Depending on capture and ablation efficiency, the satellites should continue to accrete pebbles and other heliocentric planetesimal debris as they migrate inward, though this is not part of the decretion disk scenario (Fig. 3.8). The first satellite out of the gate (Io) eventually stalls at the magnetospheric cavity radius, and subsequent satellites migrate inward and are captured, one by one, in 2:1 mean-motion resonances. The process halts when the CPD dissipates by T-Tauri driven photoevaporation. An exploration of decretion disk parameter space, and the effects on outcome, were not intended in Batygin and Morbidelli (2020). But the concepts therein are sufficiently novel that further scrutiny is anticipated.

3.2.5 A Perspective

At this point the reader may despair of ever coming to a firm conclusion regarding giant planet satellite accretion models. I prefer to emphasize recent progress, and remind the reader that the earlier era of simpler models and scenarios offered at best false comfort. As a case in point, it is unlikely that we will ever be able to determine from first principles what the dust/gas ratio in the inflow to the protojovian CPD was, or the heliocentric planetesimal flux through the CPD, though given the late timing of satellite formation with respect to t_{CAI}, solar or MMSN values are quite unlikely. But it is the Galilean satellites themselves and their properties, the properties of Jupiter, and meteorites, that will ultimately determine or constrain these. Ever improving astronomical observations of planet-forming disks will provide guidance and key analogue data. And numerical codes should continue to grow in power, in the physics and chemistry incorporated, and in their spatial and temporal resolution. Important matters requiring continued, rigorous attention are the inflow rate to the circumjovian CPD, its geometry, strength, and time history, the origin of disk viscosity, and the myriad details of disk structure and satellitesimal formation and aggregation. But given the importance of detailed knowledge of the internal structures and compositions of Io and the other Galilean satellites to constraining the origin scenarios above, future telescopic and spacecraft measurements will surely play a critical role, points I will return to below.

3.3 Accretion of Io

3.3.1 Composition

I turn now specifically to the composition of the "building blocks" that originally made up Io. I do this in the context of the protojovian CPD models above. A fundamental point of agreement from most of these models is that the solids that ultimately built Io were fed into the circumjovian accretion disk from solar orbit, as opposed to, say, forming from condensation in a hot but cooling MMsN (Prinn and Fegley 1981; Mosqueira and Estrada 2003a). Thus the initial chemistry of these solids is, nominally, characteristic of planetesimals near the orbit of proto-Jupiter, rather than materials from the terrestrial planet zone, the inner asteroid belt, or the more distant comets, for example. Logically, the closest analogues to, or survivors from, this primordial planetesimal population might be assumed to be the dark, reddish asteroids of the outer asteroid belt and Trojan clouds, specifically the P- and D-type asteroids (Gradie et al. 1989; DeMeo and Carry 2014). The P- and D-type asteroids are, however, now thought to have largely been dynamically injected into cisjovian orbits at a later epoch in Solar System history (Morbidelli et al. 2005; Levison et al. 2009; Nesvorný 2018). The current view is that asteroids, or more generally, dust, pebbles and planetesimals, near Jupiter at the end of its gas accretion (and thus during the satellite building epoch) could represent both the NC and CC compositional supergroups, inward and outward of the jovian gap, respectively (see Sect. 3.1.3).

In principle, both NC and CC materials could have contributed to the growing Galilean satellites. For example, in the elaborated model of Desch et al. (2018) the ureilite parent body (an NC) accreted closest to Jupiter on the inward flank of the gap while a variety of carbonaceous chondrite (CC) parent body types accreted outside the gap. But because CC parent body accretion, from chondrule ages and thermal models, was prolonged compared with that of NC parent bodies (Kruijer et al. 2017, 2020; Desch et al. 2018; Scott et al. 2018), it is logical to assume that the Galilean satellites (at least their rock portions) were dominated by CC materials. But this does not preclude some contribution from inner, NC materials, or for that matter, contributions (fragments perhaps) from early-formed, highly thermally processed NC and CC bodies (e.g., ureilites, CC irons).[16] To this we can add surviving or re-generated dust and pebbles, or even new planetesimals formed from such pebbles (Eriksson et al. 2021, 2022).

Carbonaceous chondrites represent the most primitive class of chondritic materials, and the CI chondrites in particular contain the full complement of elements at solar abundance levels (except H, C, O, N, and the noble gases) (Lodders 2003, 2021; Palme et al. 2014). Desch et al. (2018) argue that CI chondrites are sufficiently enhanced in water (e.g., phyllosilicates) and organics yet lack chondrules and CAIs, and so likely accreted far from the Sun, in particular beyond proto-Saturn's nebular gap. Now, as proto-Saturn's core grew, it would have increasingly perturbed the orbits of any extant, proximate planetesimals inside and outside of its orbit (Raymond and Izidoro 2017; Ronnet et al. 2018). Such planetesimals would have been scattered on increasingly wide orbits as proto-Saturn gained mass, as the protosolar nebular gas declined (reducing the effects of gas drag); if Saturn migrated, as in the Grand Tack scenario (Walsh et al. 2011), scattering would have been even greater (Raymond and Izidoro 2017). Ronnet et al. (2018) in fact argue that scattered Saturn-zone planetesimals are the *dominant* contributor to the solids acquired by the protojovian CPD. Whether this is true or not, the relevance of primitive CI-like materials (or other primitive carbonaceous CI- or CM-like materials, such as represented by the Tagish Lake, Sutter's Mill, Aguas Zarcas, and Tarda meteorite falls; see Brown et al. 2000, Zolensky et al. 2002, Kerraouch et al. 2021, Marrocchi et al. 2021) for the compositions of Io and the other Galileans cannot be discounted.

In sum, Io likely accreted from a near-solar mixture of refractory solids, metal, and silicates, materials whose state of hydration, oxidation, sulfidization, etc.,

[16] Determining which flavor of "ice cream cone" the Galilean satellites represent, or whether they are a "swirl" (a mixture of NC and CC) will probably require the return of a volcanic rock sample from Io itself to terrestrial laboratories. Rocky materials from, say, the icy surfaces of Ganymede and Callisto are likely largely or entirely composed of later exogenous, meteoritic debris (McKinnon and Parmentier 1986; Bottke et al. 2013), and would be less relevant in this regard.

depended on specifics of a given dust particle, pebble, or planetesimal's history. At proto-Jupiter's distance from the Sun, pebbles formed from dust particles (themselves condensed as anhydrous grains), along with water and other ices and organic matter of various volatilities, as well as inherited pre-solar matter (e.g., Sephton 2005; Pizzarello et al. 2006). Thermal evolution in planetesimals led to melting of ices, and hydration and hydroxylation, carbonation, and sulfidization of non-ice components, as well as alteration of organics (Alexander et al. 2017; Suttle et al. 2021), to various degrees. If meter-scale and smaller boulders, pebbles, and dust of such compositions followed the gas inflow into the circumjovian accretion disk (Sect. 3.2) (Canup and Ward 2002, 2006, 2009), they should have encountered shocks (e.g., D'Angelo et al. 2002, 2003a; Schulik et al. 2020). The effects of such *circumplanetary* shocks on entrained solids have not been studied specifically to my knowledge, but heating if not mechanical disruption is likely.

Accreting onto the circumjovian CPD at or near the position of Io as in the model of Canup and Ward (2002) would have subjected these solids to ambient temperatures sufficient to vaporize any ices and dehydrate most hydrated silicates (e.g., serpentine breaks down, i.e., dehydrates, above 250–300 K at the CPD midplane pressures near Io's position in such models [Fegley 1999]). Most organics would also be lost (Nakano et al. 2003), in particular the more volatile, so-called soluble organics, along with associated volatile sulfur compounds. If, however, much higher temperatures (>1000 K) were encountered, as noted above was likely close to Jupiter (Batygin 2018), then all (or nearly all) of the solar nebula solids encountering such temperatures should have vaporized and subsequently recondensed as the protojovian CPD cooled.

What solids should condense from a hot circumjovian nebula can be judged from the classic papers of Prinn and Fegley (1981, 1989), in which thermochemical equilibrium calculations in a solar-composition gas were combined with kinetic or rate considerations. Further elucidation of gas-grain reactions can be found in Fegley (1999). It turns out that even the gas-starved disks exemplified by Fig. 3.4 are denser, higher pressure disks than classic *protosolar* nebula models (e.g., Lewis 1974), meaning that pressures are higher at a given temperature (by up to 2 orders of magnitude). They are, however, lower pressure environments than the minimum mass giant planet subnebulae examined in detail in Prinn and Fegley (1981, 1989). For gas-starved disks, the corrosion of Fe alloy to troilite (FeS) in the presence of subnebular H_2S gas remains facile at the FeS stability temperature of ≈710 K (Fegley 1999). Oxidation of remaining iron to magnetite (Fe_3O_4) at ≈370 K is also kinetically favored. What is likely to be kinetically inhibited, and in contrast to the MMsN prediction in Prinn and Fegley (1989), is the hydration of mafic silicates to form minerals such as serpentine at still lower temperatures (see Figs. 2 and 3 in Fegley 1999).

In contrast, if Io accreted mainly from satellitesimals that themselves derived from substantial heliocentric planetesimals captured by gas drag (Estrada et al. 2009; Fujita et al. 2013; Tanigawa et al. 2014; Suetsugu and Ohtsuki 2017), then except for ablated material, these satellitesimals should have carried the petrological and cosmochemical signatures of their heliocentric formation regions, ices and all. Even ablated and (possibly) recondensed materials, should reflect their primordial, isotopic inheritance. Planetesimal capture (and ablation) are weighted towards the inner portion of the protojovian CPD, especially for larger (~100-km scale) planetesimals (Ronnet and Johansen 2020). So in this case, some contribution of ice and other volatiles to Io's original composition would seem inevitable. As for the ablated (and fragmented) material, in general it should have been of small enough scale (pebble-sized?) to be able to thermochemically reequilibrate with the protojovian CPD, if subnebular temperatures and pressures were sufficiently high, though this has yet to be modeled in any detail. Planetesimal scattering, as examined in the *N*-body accretion model of Dwyer et al. (2013), would also have promoted the accretion of icy satellitesimals from the outer CPD into Io or any other satellites growing in the inner part of the disk.

3.3.2 Did Io Accrete Wet or Dry?

At this point it is useful to walk through what the various accretion scenarios described earlier predict for Io's original volatile budget. This is an an important point, not only for Io but for Europa as well, and has been highlighted by studies that address the possibility of Io and Europa having lost bulk ice due to accretionary or evolutionary processes, and whether the densities (and iciness) of Io and Europa actually provide constraints on accretion models (Dwyer et al. 2013; Bierson and Nimmo 2020). I address these processes below, but first discuss the issue of volatile accretion *sensu stricto*.

In the model of Madeira et al. (2021), simulations that produce Galilean satellite analogues predict a relatively icy Io (Fig. 3.7), because Io accretes copious pebbles and satellitesimals that themselves formed in or were supplied to more distant, colder regions of the protojovian CPD (beyond the ice line). In the models of Cilibrasi et al. (2018) and Batygin and Morbidelli (2020) all the Galilean satellites accrete at a relatively great distance from Jupiter (compared with the radial range of the satellites today) and migrate inward due to type I torques. By definition, any Io analogue formed this way must have been quite icy, presumably as icy as Ganymede is today. In Cilibrasi et al. (2021), as in the original Canup and Ward (2002, 2006, 2009) models, infalling dust and heliocentric planetesimals are captured across the entire CPD, so at least in principle Io could have largely accreted from dry or devolatilized materials, but there would likely be a non-trivial icy component. Detailed modeling of the ice/rock ratio in such scenarios have not been presented, however.

In Dwyer et al. (2013), the compositional and provenance implications of the Ogihara and Ida (2012) *N*-body study were studied through post-processing the collisional outcomes. Though limited in dynamical realism, as admitted by the authors, even making the extreme assumptions that all satellitesimals are differentiated (i.e., possess rock cores and icy mantles) and that all impact ejected ice is lost from the system (when in reality most ultimately reaccretes), they find that due to dynamical mixing the only way to recover a strong radial compositional gradient in the final satellite system is to impose it from the beginning on the initial satellitesimals as they are added to the CPD. Ogihara and Ida (2012) adopted the gas-starved model of Canup and Ward (2002) for the gas disk, so greater gas surface densities (Fig. 3.4a) could further sharpen any such compositional gradient by suppressing the dynamical mixing of smaller satellitesimals. But this would likely be obviated by more realistic models of heliocentric planetesimal capture (Fujita et al. 2013; Tanigawa et al. 2014; Ronnet and Johansen 2020).

Finally, in the pebble accretion model of Ronnet and Johansen (2020), Io necessarily accretes from high temperature materials (Fig. 3.6). In fact, given the Io analogue's stalled position near the edge of the magnetospheric cavity, it is questionable whether Io can continue to accrete at all, as even inward drifting rocky pebbles should largely vaporize. What would actually happen then? Possibly, because the other satellites locked into resonance with the Io analogue would continue to accrete pebbles and grow, while "Io" could not, eventually the linked type I torques would push the innermost satellite deep into the cavity and towards its doom. The entire resonant chain would then migrate inward and restabilize with a new, more massive Io analogue. This cycle would repeat (if necessary) until either the innermost satellite was massive enough, or the protojovian CPD attenuated sufficiently. In principle, the Io that ultimately survives could be relatively dry, but as above, such a scenario would need to be modeled in detail.

So the answer to the question posed by this subsection ("Did Io accrete wet or dry?") is "it depends." While this may appear less than satisfactory, at least some implications for Io's bulk composition are relatively clear. Refractory oxides, metal and silicates, whether delivered directly from solar orbit or condensed from a hot circumjovian disk,[17] should have existed in *essentially solar proportions*.

Sulfur obviously plays a big role in Io's geology in general and volcanology is particular, and is responsible for the satellite's distinctive appearance. Io's total sulfur abundance, as sulfide, could at least approach solar, but this depends on whether Io accreted icy (or "wet"), or if not, how much of the original sulfur input was sulfide or sulfidic (e.g., tochilinite; Suttle et al. 2021), how much was altered by aqueous activity on planetesimal parent bodies, and ultimately, on the original sulfur abundance of the planetesimal parent bodies. CM and CI meteorites show evidence of aqueous alteration, which can produce elemental sulfur, volatile organic sulfur species, and likely some sulfate (e.g., Bullock et al. 2010).

Now, meteoritic sulfate may be pre-terrestrial in origin, or may be the result of terrestrial alteration (Gounelle and Zolensky 2001; Airieau et al. 2005), or both (see McKinnon and Zolensky 2003 for discussion of this issue). The four recent CM or ungrouped CM-like falls noted above (and for which terrestrial alteration

[17] It is notable, in the scenario of Batygin and Morbidelli (2020), that all inflowing protosolar gas and dust is processed through (vaporized in) the hot inner portion of their decretion disk. It would be of interest to model the thermochemical consequences of condensation within the cooling decretion flow.

is limited) have little if any identified sulfate (e.g., Zolensky et al. 2002).[18] Pristine samples of the asteroid Ryugu, returned by the *Hayabusa 2* mission, are similar to CI meteorites, and contain negligible sulfate (Yokoyama et al. 2022). If Io accreted from pebbles equilibrated at higher temperatures, such volatile sulfur species as sulfate would have likely been largely driven off beforehand regardless. Notably, CM chondrites are depleted in volatile elements such as sulfur and alkalis with respect to CIs,[19] by 25–45% (Lodders and Fegley 1998; Lodders 2021), so Io may reflect such bulk abundances.

A similar line of argument would apply to Io's carbon abundance. Io is likely to have been initially carbon-depleted compared with solar abundance, if only because accretion of gas-phase carbon (CO_2, and especially CH_4 and CO) would nominally have been quite difficult under protojovian CPD temperature conditions (see next sections). Retention of refractory organics and/or graphite is likely, however (Prinn and Fegley 1989; Nakano et al. 2003; Lodders 2003, 2004), whereas the accretion of more volatile, soluble organic materials (characteristic of the more primitive carbonaceous chondrites) and carbonates depends, as with volatile sulfur compounds, on whether Io accreted icy or wet.

Some of Io's iron metal may have accreted as magnetite—an oxide—which is predicted in thermochemical models of the protojovian CPD (Prinn and Fegley 1981), and is recognized in carbonaceous chondrites (in the great majority of examples) as a product of low-temperature, aqueous alteration (e.g., Zolensky et al. 2002; Sutton et al. 2017; Suttle et al. 2021). This magnetite could have formed in solar-orbiting planetesimals that were fed (or whose collisional or ablated fragments were fed) into the CPD at the time of Io's accretion. Such low-temperature aqueous alteration depends on relatively late ice melting (late in the context of the great isotopic dichotomy), which would have been driven by heating due to the decay of ^{26}Al (e.g., Grimm and McSween 1989) and/or planetesimal impacts (Suttle et al. 2021). Mn-Cr ages and models estimate aqueous alteration times in carbonaceous asteroids between 3 and at least 5 Myr after t_{CAI} (e.g., Kruijer et al. 2020; Yokoyama et al. 2022).

3.3.3 Disk Cooling

Here I briefly consider the end state of the protojovian CPD's temperature evolution, and what it might imply for Io once formed. The blackbody temperature (T_e) of a

[18] Izawa et al. (2010) detected sulfate anions in aqueous leaching experiments involving three Tagish Lake samples, two of which had been previously exposed to terrestrial water. All three samples were leached in open air and likely underwent prompt (if not prior) oxidation of previously identified Tagish Lake sulfides to form the sulfates (M. Zolensky, pers. comm., 2010).

[19] The difference in sulfur, alkali, and halogen abundances between CM and CI carbonaceous chondrites, both relatively primitive, carbon and volatile-rich meteorite types, may reflect greater incorporation of volatiles such as H_2S ice at the formation distance of the CI parent body or bodies (Pasek et al. 2005), or alternatively, volatile loss from CM and CM-like parent bodies.

flat disk (adequate for the argument that follows) in equilibrium with proto-Jupiter's luminosity is given, following Chiang and Goldreich (1997), by

$$T_e \approx \left(\frac{2}{3\pi}\right)^{\frac{1}{4}} \left(\frac{R_{PJ}}{r}\right)^{\frac{3}{4}} T_{PJ}, \tag{3.2}$$

where R_{PJ} and T_{PJ} are the protojovian radius and blackbody temperature, respectively. For $r = 5.9\ R_J$ (Io's present position), $R_{PJ} \approx 2\ R_J$, and $T_{PJ} \approx 1500$ K (from Lissauer et al. 2009), $T_e \approx 450$ K, more than adequate to prevent ice condensation near Io's position, though not enough for silicate vaporization or ionization without additional viscous and/or infall heating. Now, as accretion of protosolar gas ended for Jupiter, its radius would have shrunk (to ~1.6 R_J) and its surface temperature would stabilize near 600 K (Lissauer et al. 2009). This yields $T_e \approx 150$ K at $r = 5.9\ R_J$, sufficient to stabilize water ice (noting that the snow line temperature at the higher pressures of the protojovian CPD, compared with the protosolar nebula, may exceed 200 K). Therefore, regardless of whether Io accreted hot or merely warm (see next subsection), minor ice accretion onto Io appears possible. A late, final "frosting" of ice in the Io region could also account for the inferred but otherwise enigmatic iciness of Amalthea (Takato et al. 2004; Anderson et al. 2005).

These calculations also make clear that the long-held idea that Jupiter's early luminosity was responsible for the compositional gradient of the Galilean satellites (Kuiper 1952; Pollack and Reynolds 1974) is not necessarily incorrect. Viscous dissipation and infall kinetic energy are, however, more important than jovian insolation in determining disk radial temperature structure (Canup and Ward 2009; Makalkin and Dorofeeva 2014). Figure 3.4a illustrates this point.

In this overall context, it is also worth noting that the temperature of the Io-forming region of the protojovian CPD probably cannot be supported by the background solar nebula radiation bath either. While classic analytical solar nebula models (e.g., Lewis 1974) propose temperatures near 150 K, close to the present-day solar insolation temperature at Jupiter (~120 K), modern accretion disk models obtain lower temperatures near 5 AU (see, e.g., Wood 2000). An extreme example is the solar nebula model of Hersant et al. (2001), where midplane temperatures evolve with time and drop below 20 K at 5 AU after 5 Myr of viscous evolution, and the effect of a similar nebular boundary condition can be seen in Fig. 3.4a for low α (cf. Bell et al. 1997).

More detailed protoplanetary disk models, which incorporate heating by dissipation *and* the central star (missing in Hersant et al. 2001), dust evolution, and vertical *and* radial radiative transport, support a more nuanced view (D'Alessio et al. 1999, 2001). The latter models, for 0.5-M_{sun} T Tauri stars, consistently show midplane temperatures near ~60 K at Jupiter's present position, even for very low accretion rates onto the star. Similar models for a 1-M_{sun} T Tauri star would no doubt lead to modestly higher temperatures, and for comparison, the Desch et al. (2018) model posits protosolar nebular temperatures near 90 K at Jupiter's position late in the nebular lifetime.

The position of the circumjovian accretion disk within the solar nebula is also important. Nestled deep within the gap in the solar nebula opened by Jupiter, the satellite-forming disk would find itself plunged into shadow; no direct solar radiation would reach the disk (D'Angelo et al. 2003a), though the illuminated tops of the far wall of the gap would reradiate in the infrared into the gap (Turner et al. 2012). But the background protosolar nebula would no longer fill 4π steradians, and the circumjovian disk would be exposed to space (and whatever the birth environment of the Sun was; see Vacher et al. 2021) and to ionizing cosmic radiation. The outer boundary of the CPD may have been no colder than ~100 K (Turner et al. 2012; cf. Canup and Ward 2002; Fig. 3.4a), and this may ultimately play an important role in the formulation of realistic, time-dependent, thermochemical accretion/decretion disk models for Io and the Galilean satellites.

3.3.4 Initial Thermal State of Io

Io should have accreted solid material as such was supplied to its "feeding zone" within the protojovian nebula, either by infall, condensation, or inward drift, and as Io itself migrated. Sweep-up of local, small bodies would have been rapid (even if not perfectly efficient) and faster than overall subnebular evolution times in either the gas-starved (Canup and Ward 2002, 2006, 2009) or MMsN models (Mosqueira and Estrada 2003a). In the context of the gas-starved subnebula, total accretion times would be long enough (set by the declining infall rate) and accreting bodies small enough (possibly m-scale or less) that accretional heating may be severely limited by radiative losses to space or the CPD (Stevenson et al. 1986). This can in principle provide the necessary cold or lukewarm start to Callisto, consistent with its present inferred state of partial differentiation (Schubert et al. 2004), and obviously also implies limited accretional heating for Io as well. But because the hydrostatic interpretation upon which the interpretation of Callisto's state of differentiation rests is questioned (McKinnon 1997, 2003; Gao and Stevenson 2013), this accretional constraint cannot be taken for granted.

In the formation scenarios outlined in Sect. 3.2.4 above, accretion timescales for Io range from as long as 10^6 year (gas-starved accretion in Madeira et al. 2021) or several $\times 10^5$ year (pebble accretion in Ronnet and Johansen 2020) to as little as a few $\times 10^3$ year (oligarchic growth in Batygin and Morbidelli 2020). This full range will be considered below. In addition, I note that in the gas-starved model, there is insufficient subnebular gas to form an optically thick, convective envelope (Lunine and Stevenson 1982) about Io as it accretes, but this would not necessarily be the case for denser (more massive) protojovian CPDs.

The characteristic length scale for thermal conduction in an accreting satellite is κ/u, where κ is the thermal diffusivity of the accreted material (10^{-6} m^2 s^{-1} is typical for solid rock) and u is the radial growth rate (McKinnon 2002). For Io, and assuming a constant rate of *mass* accretion,

$$\frac{\kappa}{u} \sim 0.5\,m \times \left(\frac{\kappa}{10^{-7}\ \mathrm{m^2 s^{-1}}}\right) \times \left(\frac{\tau_{\mathrm{Io}}}{10^5\ \mathrm{yr}}\right), \tag{3.3}$$

where τ_{Io} is Io's accretion time and a reduced κ ~ 10^{-7} m^2 s^{-1} is assumed appropriate for *porous* rock+metal. Heat buried greater than this depth is *not* in good conductive communication with the accreting surface, and cannot be efficiently radiated to space as Io accretes. This estimate implies that for Io to remain cool it must accrete m-sized satellitesimals on time scales $\gg 10^5$ year or that the accreting particles are $\ll$1 m in scale.

How cool might this be? Following Stevenson et al. (1986), and maximizing radiative losses,

$$T(R) = \left[\frac{1}{4\pi R^2 \sigma_{\mathrm{B}}}\left(\frac{GM(R)}{R} + \frac{\langle \upsilon \rangle^2}{2}\right)\frac{\mathrm{d}m}{\mathrm{d}t} + T_0^4\right]^{1/4}, \tag{3.4}$$

where $T(R)$ is the accretional temperature profile, $M(R)$ the mass contained within a radius R, G the gravitational constant, $\langle \upsilon \rangle$ the mean encounter velocity, dm/dt the mass accretion rate, and T_0 the background radiative equilibrium temperature (which for an optically thin protojovian nebula is less than the physical gas temperature; e.g., Canup and Ward 2009). Figure 3.9 shows T as a function of T_0 for Io for accretion times between 10^3 and 10^6 year. Note that for dm/dt = constant and $\langle \upsilon \rangle = 0$, the accretional temperature increase is constant throughout the satellite (i.e., $T(R) = T$). These temperature increases are minimized here for Io, because heat is neither buried nor is impactor pre-encounter kinetic energy included, but heat capacity effects are ignored as compensation. For accretion times $\lesssim 10^5$ year, accretional energy dominates the background term in Eq. (3.4) over the T_0 range illustrated. For example, for $\tau_{\mathrm{Io}} = 10^5$ year, Io's initial interior temperature is ~500 K for CPD background temperatures $\lesssim$500 K. In contrast, for $\tau_{\mathrm{Io}} = 3 \times 10^3$ year, temperature increases are much greater, insufficient to melt Io's rock, but more than adequate for dehydration and/or ice melting.[20] If Io accreted from icy satellitesimals, as in Batygin and Morbidelli 2020), rapid accretion would have led to formations of a deep surface ocean. Whether this amount of accretional energy would have yielded a steam-H_2 atmosphere in this case as well is discussed below.

If Io accreted from bodies that were larger than the limit implied by Eq. (3.3), which need be no greater than 10-to-100-m in diameter, then some satellitesimal kinetic energy must be trapped as heat. The fraction trapped in a symmetrically accreting uniform satellite, as a function of depth, is $hGM(R)/R$, where $0 \le h \le 1$

[20] Accounting for heat capacity and latent heat in Eq. (3.4) would reduce $T(R)$ and radiative losses, and thus *increase* the total accretional energy retained.

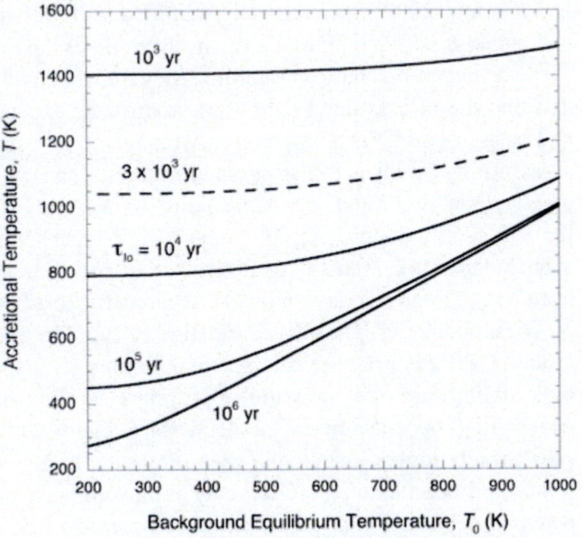

Fig. 3.9 Temperature within Io in the limit of small satellitesimal accretion (Eq. 3.4), limited by radiative loss and as a function of background radiative equilibrium temperature. The contribution of satellitesimal encounter velocity is ignored, and the mass accretion rate is assumed constant over the accretion time τ_{Io}

and I have again neglected the mean encounter velocity[21] (e.g., Schubert et al. 1981; Stevenson et al. 1986). For a characteristic rock+metal STP heat capacity of 920 J kg^{-1} K^{-1} (Kirk and Stevenson 1987), accretional temperature increases in Io's outer layers could have reached ≈350 K × (h/0.1). The factor h is empirical, and can in principle be determined by detailed calculation of the deposition and redistribution of impact energy (e.g., Coradini et al. 1982; Squyres et al. 1988), plus radiative and, possibly, atmospheric convective cooling (Lunine and Stevenson 1982). Large accretionary impactors bury their kinetic energy as heat, so $h \gtrsim 0.5$ is expected. In this circumstance Io's outer layers should melt, if Io accreted from rock+metal. If Io accreted from rock+ice, the latent heat of melting and vaporization would buffer the temperature increase, but ice from rock differentiation and formation of a surface water ocean is expected (Lunine and Stevenson 1982; Bierson and Nimmo 2020).

Io today is an anhydrous body, or at least there are no confirmed volcanic emissions of hydrogen-bearing compounds (Dalton et al. 2010), and ionian magmas must be significantly drier compared with terrestrial counterparts (Zolotov and Fegley 2000b). This does not mean Io has always been anhydrous. Water released from dehydrated minerals or melted ice could have been vented early in Io's evolution and lost to space (Lewis 1982; Zolotov and Fegley 1999). The question, however, of whether Io could have once possessed a massive water ocean, thus making it more similar to Ganymede in overall structure, deserves careful evaluation. Bierson

[21] Mostly for convenience, but this neglect is justifiable when the eccentricities of the accreting satellitesimal swarm are low, which obtains for very small bodies orbiting in the presence of CPD gas.

and Nimmo (2020) have recently examined such, in which they, adapting the work of Lehmer et al. (2017), model hydrodynamic escape of a water vapor atmosphere above an accreting Galilean satellite. They find that hydrodynamic escape is possible and that the effects maybe far from nontrivial, depending on accretion timescale and background CPD temperature. Although certain simplifications are made, such as assuming small ice-bearing satellitesimals can themselves have temperatures greater than the water ice triple point temperature (273.16 K), they show that for sufficiently long accretion times ($\gtrsim 10^5$ year) and high enough background subnebular temperatures ($\gtrsim$300 K), that an entire, massive water ocean may have been lost to hydrodynamic blowoff (technically, lost before it can form).

These results are, however, sensitive to satellite mass and, especially, to background CPD gas pressure. Bierson and Nimmo (2020) assume either zero pressure or a static, very low pressure appropriate to the gas-starved nebula of Canup and Ward (2006). Some of the accretion scenarios discussed in this review have substantially higher subnebular pressures even if they are not full MMsN models, so this hydrodynamic possibility may or may not have obtained. The exact degree of potential dehydration of an accreting Io depends on several physical processes and the sequence for them to occur, which requires further study. The model of Bierson and Nimmo (2020) is testable, however, not for Io in particular, but by detection of the predicted isotopic shifts (in D/H, $^{18}O/^{16}O$, and $^{17}O/^{16}O$) between the water ice on Ganymede and Callisto and that retained on Europa (if the latter lost ice by the same mechanism).

3.3.5 Core Formation and Io's Oxidation State

Although the total gravitational potential energy of Io's assembly ($3GM_I/5R_I$ = 1950 kJ kg^{-1}, where M_I and R_I are the satellite's present-day mass and radius) is just enough to completely melt it, pebble accretion could have limited its initial interior temperatures to $\lesssim$1000 K irrespective of accretion timescale, as long as the latter was not too short (Fig. 3.9). The shortest τ_{Io} discussed above pertains to the decretion disk model of Batygin and Morbidelli (2020), but their model posits purely hierarchical coagulation, rendering pebble accretion moot. Temperatures approaching 1000 K are non-trivial (e.g., serpentine breaks down at pressure below this temperature; McKinnon and Zolensky 2003), but fall short of the Fe-FeS euctectic melting temperature ($\approx$1250 K), at which point downward percolation of iron-sulfide melt occurs, and formation of a metallic core begins.

The presence and characteristics of Io's core deserve some comment here. The density (3530 kg m^{-3}) and normalized moment-of-inertia (0.378) determined by *Galileo* imply the existence of a substantial metallic core within Io (see the review by Schubert et al. (2004) and the chapter by Keane et al.). Internal structural models can further constrain the properties of this core, and the surrounding mantle, subject to (presently unavoidable) assumptions of composition, chemistry, and temperature. For example, by modeling Io as a pure olivine mantle surrounding a solid Fe-FeS core, Sohl et al. (2002) found Io's bulk Fe/Si mass ratio to lie between 1 and 1.25 for solid mantles and between 1.25 and 1.5 for partially molten ones. Kuskov and Kronrod (2001) found, by basing mantle chemistries on chondritic meteorites and assuming plausible but uniform core densities, that Io's Fe/Si ratio probably lies within the 0.8–1.2 range, and that Io in bulk is most compatible with an L or LL ordinary chondrite composition. Obviously, these two works do not agree on the important point of iron content. Subtle (and not so subtle) differences in modeling assumptions underlie this difference, but both conclude that Io's Fe/Si mass ratio is less than the solar value of ≈1.7 (Lodders 2003, 2021; Asplund et al. 2009).

There is, however, no obvious way to fractionate iron from rock in the context of gas-starved, gas-limited, or even MMsN accretion/decretion disk models (at least to my knowledge), given the continuous input of ~solar-composition (i.e., carbonaceous) "feedstock." Temperature is obviously important for strongly tidally heated Io, and if one allows for a molten core and a full solar abundance of S (as in CI chondrites), then it is possible to construct internal models of Io with solar Fe/Si (McKinnon and Desai 2003). In this case Io's core would be relatively large (~1000-km in radius) and S (and possibly O) rich to account for Io's relatively high moment-of-inertia. Such a large, fluid interior would serve to maximize tidal flexing and dissipation in the mantle, for a given orbital eccentricity (e.g., Cassen et al. 1982). While such a model cannot be proven by gravity data alone, it is cosmochemically compatible with the oxidation state of Io's mantle (similar to that of the Earth's upper mantle; Zolotov and Fegley 2000a), and consistent with the cosmochemical argument of Lewis (1982) that Io could not have formed from a metal-bearing (as distinct from sulfide-bearing) chondritic assemblage—such as L or LL ordinary chondrites (see McKinnon 2004).

Lewis (1982) argued in particular that the copious sulfur and sulfur dioxide at Io's surface precluded direct derivation of Io's crust/lithosphere from a metal-bearing source (i.e., from a metal-bearing mantle). He favored CM (or metal-free CV) carbonaceous chondrites as the most plausible chondritic antecedents. If the oxidation state of Io's early rock interior were not high enough, then total extraction of Io's sulfur to its core would have ultimately occurred. These CC types have less total sulfur (i.e., lower S/Si and S/Fe ratios than CI chondrites; Lodders and Fegley 1998; Lodders 2021), but are also plausible analogues for Io's initial bulk composition, in that Io's density and moment-of-inertia can be matched (given variation in the Mg# of mantle silicates and internal temperature structure). Lewis (1982) further argued that Io would have lost the bulk of its accreted carbon by venting of CO_2/CO gas. This does put constraints on the amount of refractory carbon (insoluble organic matter and graphite) accreted by Io, for if it were too much, it would have constrained the initial oxygen fugacity of Io's rock interior to the CCO (graphite–CO-CO_2) oxygen buffer, which is more reducing than FMQ (fayalite-magnetite-quartz), and incompatible with elemental sulfur or pyrite (FeS_2), the latter of which can disproportionate to yield sulfur.

3.3.6 Spin State

I end this section with a note on Io's rotational history. Peale (1977) estimated that Io would have been despun by tides raised by Jupiter from an initially rapid rotation to its present synchronous period of 1.77 days in only ~5000 × (Q/100) year, where Q is Io's specific dissipation factor at rotational tidal frequencies. Even with modern parameter values (density, moment-of-inertia, etc.), this time scale is so short for plausible Q values (generally taken to be of order 100) that it is likely that Io accreted in synchronous lock, subject to periodic unlocking from any relatively large accretionary impacts. The despinning time goes as a_{I}^{6}, however, where a_{I} is Io's semimajor axis, so if proto-Io first formed in a more distant orbit and migrated inward (as discussed earlier), it may have despun while it was migrating and accreting.

3.4 Early Dynamical Evolution

It is fitting, in this penultimate section, to briefly discuss one of the more interesting applications of the slow-inflow accretion disk model for the formation of Io and the Galilean satellites. Io's special place in the geophysical pantheon is owed to the tidal heating that follows from the satellite's resonant orbital configuration with Europa and Ganymede—the Laplace resonance. A subject of much research, the Laplace relation has long been thought to have been a by-product of the outward tidal evolution of Jupiter's satellites. As discussed in the comprehensive reviews of Peale (1999) and Peale and Canup (2015), Io post-formation should evolve outward under the action of jovian tides more rapidly than Europa, and both more rapidly than Ganymede. As Io does so, it first captures Europa into the 2:1 mean-motion resonance, and then the coevolving pair capture Ganymede into the 2:1 with Europa.

Difficulties, at least as then understood, with the required extent of orbital evolution by gravitational tides alone as well as with maintaining Io's present volcanic power prompted Greenberg (1982, 1987) to offer that Io, Europa, and Ganymede were actually evolving out of a deep, primordial resonance. There was not, at that time, a concrete mechanism to account for such a primordial resonance. As has been emphasized above, assembly of the satellites within a gaseous protojovian disk predicts inward type I migration as the satellites grow to larger and larger sizes (Canup and Ward 2002 and following). Ganymede is by far the most massive of the three, and as type I drift is proportional to satellite mass and nebula surface density, Ganymede can in principle migrate faster. Peale and Lee (2002) demonstrated numerically that in doing so Ganymede can capture Europa into the 2:1 mean-motion resonance, and then the pair can migrate fast enough to capture Io as well (Fig. 3.10). Given Io's greater mass than Europa, and the arguably greater disk surface density closer to Jupiter, Io's capture into the 2:1 with Europa

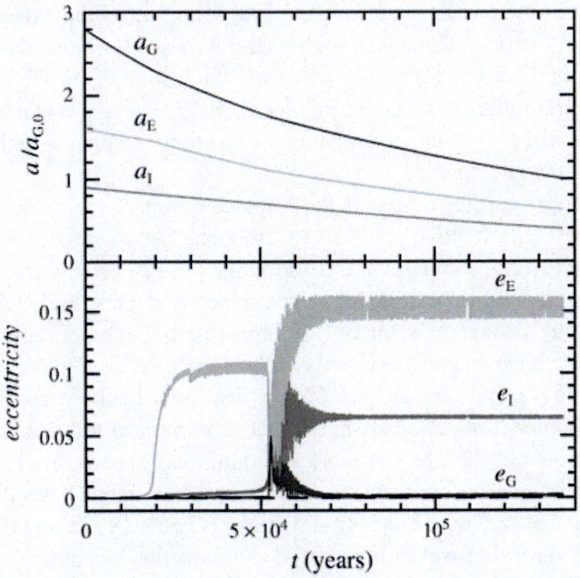

Fig. 3.10 Nebula-induced evolution of the Galilean satellites into the Laplace resonance. The semimajor axes of Io (a_I), Europa (a_E), and Ganymede (a_G) are normalized to Ganymede's current distance from Jupiter, $a_{G,0}$. e_I, e_E, and e_G are the eccentricities of Io, Europa, and Ganymede, respectively. Inward migration and eccentricity damping time scales are imposed (10^5 year for the former); and the initial orbits are circular and coplanar. After circumjovian nebula dispersal, eccentricities are damped by solid-body tides within each satellite. Modified from Peale (2003)

depends on it not evolving faster than Europa after Europa is captured into resonance with Ganymede. Thus, Io can run, but it cannot hide.

Io's eccentricity in Fig. 3.10, as well as that of Europa, is well above current values. They are due to tidal interactions with the spiral density waves launched in the protojovian CPD by the satellites. As the disk dissipates (i.e., as infall from the solar nebula abates), these eccentricities should begin to decline due to tidal dissipation within the satellites. Peale and Lee (2002) show that the entire system (Io-Europa-Ganymede) naturally relaxes to its current Laplace configuration, and in detail.

There are, however, caveats to this picture. The relative migration rates of the satellites in Fig. 3.10 were chosen to be consistent with the gas-starved model of Canup and Ward (2002), but are nevertheless ad hoc, as discussed by Peale and Canup (2015). Different satellite accretion scenarios, described in this review, would predict different outcomes. A more serious challenge comes from Batygin and Morbidelli (2020), who argue that this particular Laplace resonance assembly sequence (Ganymede capturing Europa followed by the pair capturing Io) is, actually, not dynamically stable. They argue that the Ganymede-Europa 2:1 resonance is "overstable" due to the large mass ratio between these two satellites (close to 3-to-1), and that stable assembly of the Laplace resonance requires establishment of the Io-Europa resonance first, followed by Ganymede locking into resonance with the pair. This inside-out assembly is seen in several of the accretion models described earlier, especially those in which Io stalls at the outer edge of a disk cavity (e.g., Sasaki et al. 2010; Fujii et al. 2017; Batygin and Morbidelli 2020; Madeira et al. 2021), but the discrepancy in assembly sequence with that of Peale and Lee (2002) is unresolved.

To some, a primordial origin of the Laplace resonance may be distressing, in that it removes an attractive solution to the Ganymede-Callisto dichotomy from the playing field (e.g., Showman and Malhotra 1997). Yet a primordial, outside-in or inside-out assembly of the Laplace resonance is elegant, and makes interesting predictions. One is that Io's history of high internal temperatures and extreme volcanism may have begun very early as well (Peale 2003). As an illustration, the decay of Io's primordial $e_{\mathrm{I}} \approx 0.063$ in Fig. 3.10 to the present-day value of 0.004 would imply dissipation within Io and a minimum temperature increase of $GM_{\mathrm{J}}e_{\mathrm{I}}^2/2C_{\mathrm{P}}a_{\mathrm{I}} \approx 600$ K. This estimate is a minimum because it (1) assumes Io at its present semimajor axis, whereas Io, Europa and Ganymede have almost certainly tidally evolved outward over Solar System history (Peale 1999); and (2) neglects any torque from Jupiter that would have acted to retard the eccentricity decay. A heat impulse of this magnitude, when added to Io's likely warm to hot initial accretional state (Sect. 3.2.4), implies that sulfide or Fe-FeS eutectic melting and core formation started soon after the protojovian CPD disappeared and accretion ended (within a few Myr), if not before. With this, Io would have entered the realm of enhanced tidal flexing and dissipation, both being necessary to explain its prodigious volcanic output (e.g., Schubert et al. 1981; chapters by Barr et al. and Keane et al.).

Other accretion models/scenarios predict different post-CPD eccentricities for Io, but values between 0.05 and 0.1 are not uncommon (e.g., Ogihara and Ida 2012; Madeira et al. 2021). If such eccentricities could be sustained, for example, when and if Io reached the magnetospheric cavity, at the inner edge of the CPD, the tidal power dissipated might have been far above today's observed rate of ~2.5 W m^{-2} (e.g., Veeder et al. 1994).

3.5 Summary and Prospects

Our understanding of the origin of Io and the Galilean satellites has evolved from early, protosolar nebula derived or analogue discussions, based on the revelations of the *Voyager* missions (e.g., Pollack and Fanale 1982), into modern, quantitative analyses of accretion disk models around Jupiter informed by *Galileo* data, starting with Canup and Ward (2002) and propagating forward to the present. Much newer work was and is driven by ever-improving observations of extrasolar planet systems and protoplanetary disks, and concomitant theoretical and numerical advances in understanding star and giant planet formation.

While many details remain to be clarified, it seems clear that formation of Jupiter in the core accretion–gas capture model inevitably ends with an accretion and/or decretion disk around the planet, and as Jupiter at some point must stop accreting gas and solids, that protojovian CPD must become, at some point, one of the slower inflow, gas-limited, and ultimately, gas-starved variety. Numerous and varied alternative models for satellite formation have been proposed, ranging from more massive, cooling MMsN models (Mosqueira and Estrada 2003a, b; Estrada et al. 2009), to pebble and gas-drag captured heliocentric planetesimal accretion (e.g., Ronnet et al. 2017, 2018; Ronnet and Johansen 2020) to post-CPD, purely planetesimal collisional-capture accretion (Estrada and Mosqueira 2006), to cite only a subset of the literature. Accretion models range from semi-analytical to full *N*-body, and adopt different parameters for disk structure (surface density, temperature, opacity, viscosity), but almost all now attempt to follow the time evolution of the CPD. It goes without saying that all these models and scenarios cannot be correct, and maybe none are, but it is more likely that elements of many will turn out to be important to a more comprehensive understanding. But all should be tied to models of Jupiter's formation in a direct and compelling way, so that they account for the physical and compositional characteristics of the satellites themselves. The Galilean satellites are a well-ordered system, so the processes that created them should yield that order.

All circumjovian disk models rely on inferences and assumptions of mass and angular momentum inflow to Jupiter from the protosolar nebula. This work has advanced (from, e.g., Lubow et al. 1999 to Schulik et al. 2020), and at each stage has profoundly influenced satellite formation models. Early, 2D numerical gap opening and inflow calculations provided the basis for the gas-starved model of Canup and Ward (2002). Later 3D radiative-hydrodynamic inflow calculations prompted consideration (or renewed consideration) of heliocentric planetesimal capture as the dominant source of satellite-building solids. These later inflow calculations have also led to very different hypotheses of dust and small pebble deposition, ranging from distant infall (e.g., Cilibrasi et al. 2018) to close-in infall and formation of an outflowing, decretion disk (Batygin and Morbidelli 2020). Continued advances in 3D, thermohydrodynamic giant planet formation models, especially ones that can resolve circumjovian disk or envelope structure within the inner 10% of Jupiter's Hill sphere (~75 R_J), should prove telling. Uncertainties in disk physics (both protosolar and protoplanetary), especially the sources and strengths of disk turbulence, and their evolution through time, will likely remain for some time, but constraints from protostellar disk observations and theoretical modeling may yet lead to breakthroughs. These will propagate forward to satellite formation concepts and models, as will observations (at the limits of resolution) of CPDs associated with nearby protostars (e.g., Benisty et al. 2021).

If sufficient mass was processed through the protojovian CPD, then multiple generations of satellites may have been formed and lost (Canup and Ward 2006). The Galilean satellites would then be the last generation, formed from the inflow of at least an MMsN worth of solid matter. Or maybe not, if a magnetospheric cavity opens up close to Jupiter and the innermost satellite, destined to become Io stalls there (e.g., Sasaki et al. 2010; Batygin 2018). In accordance with abundant evidence from meteorites, the non-volatile solids that accreted to form Io and the other Galilean moons should have been essentially solar in composition (though not necessarily in every element), and plausibly dominated by the CC isotopic supergroup (See Sect. 3.1.3). The thermal environment at Io's formation distance, or range of formation distances, was determined mostly by the subnebula itself, except close to Jupiter, and could have been hot enough to largely devolatilize any small, accreting satellitesimals, including the ablated fragments of heliocentric planetesimals that encounter the CPD (Ronnet and Johansen 2020)—some of which may have come from as far away as the orbit of Saturn (Ronnet et al. 2018).

Thermochemical processing of at least some of the solids that built Io is possible (e.g., Fegley 1999), depending on the pressure and temperature conditions in the CPD, and the residence timescales of these disk solids, but this aspect has received very little attention in recent years. But the accretion of hydrated rock (of protosolar origin) or minor or even major ice, is far from precluded. Some models call for accretion of Io in the cold outer disk followed by inward type I migration (e.g., Batygin and Morbidelli 2020; Cilibrasi et al. 2021). Needless to say, removal of major amounts of ice, corresponding to >10% of Io's total mass, is a non-trivial requirement in such scenarios. The one mechanism that shows promise, hydrodynamic blowoff of an H_2O atmosphere (Bierson and Nimmo 2020) does not work in all circumstances, and should be subject to further scrutiny. It is not sufficient to state that a hydrodynamic wind removed proto-Io's ice for one set of boundary conditions and assume it might work for another. The compositional gradient in the Galilean satellite system is a fundamental constraint that must always be satisfied.

Regardless, at the very end of the circumjovian disk's lifetime, temperatures in the disk must inevitably have fallen, which would have had two consequences. One is that, even if composed in bulk of higher-temperature, non-icy solids, very late accretion of some hydrated minerals and/or ice is possible, even if Io accreted relatively close to Jupiter. The second is that maintenance of the magnetospheric cavity may end if there is insufficient thermal ionization (even for a sufficiently powerful protojovian magnetic field), and so inward migration of a stalled Io may resume. But perhaps the overall CPD gas density was low enough at that stage so that Io's overall migration was limited, or perhaps galactic cosmic rays were be able to "pick up the slack" in terms of ionization. These details remain to be assessed in a rigorous way.

Io traveled a great deal in its youth, at least in most of the models presented in this review. Io probably began accreting well outside its present position, and continued accreting while migrating inward due to disk tidal torques. In the scenario of Peale and Lee (2002), its migration sped up once Io was captured into the Laplace resonance with Europa and Ganymede, which had been migrating inward as well, and more rapidly. Or as noted in the majority of the newer works cited here, the Laplace resonance was assembled from the inside out, either during Io's migration or subsequent to its arrival at the magnetospheric cavity boundary. In either circumstance, orbital decay probably brought Io closer to Jupiter than it is today while the protojovian CPD evanesced. Subsequently the weaker action of jovian tidal torques slowly pushed all three satellites back outward, to their present positions, over Solar System history (Fuller et al. 2016; chapter by Keane).

The combination of CPD conductive and radiant heating of inflowing or gas-drag captured solar dust, pebbles and planetesimals, the kinetic energy of satellitesimal accretion onto Io, and tidal heating during and consequent to the protojovian epoch probably pushed Io past the threshold for metallic core formation. That is, unless it also accreted copious ice, which would have melted to form a surface ocean and buffered proto-Io's internal temperatures. In this case Io would have formed with an undifferentiated but likely hydrated rock+metal core (e.g., McKinnon and Zolensky 2003). But in the former, non-ice-rich case, once such a relatively deformable, dissipative internal structure (i.e., a liquid metallic core) was created, the conditions for vigorous volcanism from long-term tidal heating were assured. This early heating would have been more than sufficient to drive any and all accreted water out of Io, if minor, but the oxidation of rock and metal by water, with loss of hydrogen, was likely very important in establishing Io's overall chemical and oxidation state (e.g., Lewis 1982; Zolotov and Fegley 1999).

Continued work on satellite formation in a gas-starved or gas-limited accretion/decretion disk is likely to prove fruitful. This is perhaps one of the more subtle legacies of the *Galileo* mission (e.g., Coradini and Magni 1997): a reasonably successful, quantitative model or model path for satellite formation around gas giant planets, and one which has provided a springboard for much elaboration, variation, and revision.

3.5.1 Future Spacecraft Measurements

Satellite accretion modeling is now being driven forward largely by other observational, theoretical, and numerical advances, but new observations of Io and the Galilean satellites should also prove telling. Close passes of Io by *Juno* in 2023 and 2024 should permit at least some refinement of its degree-2 gravity field and therefore yield tighter constraints on its moment-of-inertia and thus internal structure. Infrared spectral measurements of silicate volcanic units may provide constraints on internal composition and petrology. Looking farther into the future, ESA's *Jupiter Icy Moons Explorer* and NASA's *Europa Clipper* should be able to determine Callisto's state of hydrodynamic equilibrium and whether it is actually partially differentiated. If so, it is a powerful constraint on the timescales of Galilean satellite formation. And as previously noted, measured isotopic differences (D/H, $\delta^{18}O$, $\delta^{17}O$) between Europa and Ganymede or Callisto, from in situ mass spectrometry, should provide tracers of accretionary or evolutionary processes that may have affected Io as well. But the most necessary information to determine Io's present state and thus constrain its formation and evolution requires a dedicated spacecraft mission, as described elsewhere in this volume.

Of these future data, isotopic measurements are some of the most powerful yet least heralded in terms of constraining the origin and evolution of these worlds.

Europa Clipper will carry two types of mass spectrometers, the MAss Spectrometer for PlanetaryExploration, or MASPEX (Waite et al. 2019), to analyze gases, and the Europa SUrface Dust Analyzer, or SUDA (Kempf et al. 2019), to analyze the dust grains ejected by micrometeorites striking Europa. MASPEX's measurements will concentrate on the exosphere and, if active, plumes of Europa. The instrument has sufficient dynamical range and mass resolution to determine the D/H ratio from H_2 and H_2O, as well as oxygen isotopic abundances in H_2O, CO_2, and other oxygen-bearing species, even if at low abundance, through use of a cryotrap on each individual close pass to the surface of Europa (Miller et al. 2022). As noted in Sect. 3.3.4, such measurements are most telling in context, both in the context of the other Galilean satellites and in the context of the outer Solar System generally. As such it is critically important that MASPEX and SUDA data be acquired during *Europa Clipper*'s close passes of Ganymede and Callisto during the pump-down phase of *Clipper*'s orbital tour, if for no other reason than calibration. *JUICE* carries the Neutral Ion Mass Spectrometer (NIMS; Föhn et al. 2021), and a Submillimetre Wave Instrument (or SWI; Wirström et al. 2020), a microwave spectrometer that will characterize the composition and dominant isotopic abundances of the thin atmospheres and exospheres of Ganymede, Callisto, and to a more limited extent, Europa. These latter measurements will complement the in-situ mass spectrometry above. And both *Europa Clipper* and *JUICE* carry imaging infrared spectrometers that should be able to constrain the D/H ratio in surface water ice through detection of the subtle 4.13 μm O-D stretch (Clark et al. 2019).

None of these future isotopic measurements are directed at Io specifically, but as stated, to the degree that they pertain to all the Galilean satellites, they can constrain Io's origin and evolution. Moreover, SUDA should measure the compositions of dust particles sourced from Io, which are likely dominated by S and SO_2. Thus there is the possibility, in principle, of determining major isotopic ratios for Io's volcanoes "in situ," such as $^{34}S/^{32}S$. I note a tentative detection of Io's $^{34}S/^{32}S$ by ALMA indicates a possible enhancement over the terrestrial value (Moullet et al. 2013).[22] Moreover, even from Ganymede's distance, SWI should be able to resolve Io and its limb for favorable viewing geometries, which may allow for determinations of $^{34}S/^{32}S$, $^{18}O/^{16}O$, and other important ratios in Io's atmosphere (if suitable rotational lines exist in the two SWI bandpasses). Ultimately, however, it will be necessary to carry out in situ compositional and isotopic measurements of gaseous and particulate material vented in Io's plumes and ejected or otherwise "liberated" from its surface. Looking even deeper into the future, a sample return mission from Io or its environs could truly unlock its cosmochemical secrets (and by

[22] de Pater et al. (2020) point out that interpretation of this measurement requires better knowledge of the temperature structure of Io's atmosphere. Such may be forthcoming from planned *JWST* observations of Io (which may also detect isotopologues of SO_2 in surface ices directly [cf. Howell et al. 1989]). Moreover, if Iogenic sulfur is "heavy," e.g., from a long history of volcanic S fractionation and loss, then it may provide a way to discriminate Iogenic S from sulfur native to Europa (and its ocean).

inference, those of the other Galilean satellites), given the atom-level measurement capabilities of terrestrial laboratories.

Acknowledgements This work was originally supported by grants from the NASA Planetary Geology and Geophysics and Outer Planets Research Programs. The revised version was completed while on sabbatical from Washington University during spring 2022. It has benefited greatly from close reads by James Tuttle Keane, Ryan Ogliore, and Kevin Zahnle.

References

Airieau, S.A., Farquhar, J., Thiemens, M.H., Leshin, L.A., Bao, H., Young, E.: Planetesimal sulfate and aqueous alteration in CM and CI carbonaceous chondrites. Geochim. Cosmochim. Acta. **69**, 4167–4172 (2005)

Alexander, C.M., Cody, G.D., De Gregorio, B.T., Nittler, L.R., Stroud, R.M.: The nature, origin and modification of insoluble organic matter in chondrites, the major source of Earth's C and N. Chem. Erde. **77**, 227–256 (2017)

Alibert, Y., Mordasini, C., Benz, W., Winisdoerffer, C.: Models of giant planet formation with migration and disc evolution. Astron. Astrophys. **434**, 343–353 (2005a)

Alibert, Y., O. Mousis, and W. Benz, Modeling the jovian subnebula: I – Thermodynamical conditions and migration of proto-satellites, Astron. Astrophys., 439, 1205–1213, 2005b.

Alibert, Y., Venturini, J., Helled, R., Ataiee, S., Burn, R., Senecal, L., Benz, W., Mayer, L., Mordasini, C., Quanz, S., Schönbächler, M.: The formation of Jupiter by hybrid pebble–planetesimal accretion. Nature Astron. **2**, 873–877 (2018)

Anderson, J.D., Johnson, T.V., Schubert, G., Asmar, S., Jacobson, R.A., Johnson, D., Lau, E.L., Lewis, G., Moore, W.B., Taylor, A., Thomas, P.C., Weinwurm, G.: Almathea's density is less than that of water. Science. **308**, 1291–1293 (2005)

Andrews, S.M.: Observations of protoplanetary disk structures. Annu. Rev. Astron. Astrophys. **58**, 483–528 (2020)

Asplund, M., Grevesse, N., Sauval, A.J., Scott, P.: The chemical composition of the Sun. Annu. Rev. Astron. Astrophys. **47**, 481–522 (2009)

Ayliffe, B.A., Bate, M.R.: Circumplanetary disc properties obtained from radiation hydrodynamical simulations of gas accretion by protoplanets. Mon. Not. R. Astron. Soc. **397**, 657–665 (2009)

Ayliffe, B., Bate, M.R.: The growth and hydrodynamic collapse of a protoplanet envelope. Mon. Not. R. Astron. Soc. **427**, 2597–2612 (2012)

Balbus, S.A., Hawley, J.F.: A powerful local shear instability in weakly magnetized disks. I. Linear analysis. Astrophys. J. **376**, 214–233 (1991)

Baruteau, C., Masset, F.: Recent developments in planet migration theory. In: Souchay, J., Mathis, S., Tokieda, T. (eds.) Tides in Astronomy and Astrophysics Lecture Notes in Physics, vol. 861, pp. 201–253. Springer, Berlin (2013)

Bate, M.R., Lubow, S.H., Ogilvie, G.I., Miller, K.A.: Three-dimensional calculations of high- and low-mass planets embedded in protoplanetary disks. Mon. Not. Roy. Astron. Soc. **341**, 213–229 (2003)

Batygin, K.: On the terminal rotation rates of giant planets. Astrophys. J. **155**, 178 (2018)

Batygin, K., Morbidelli, A.: Formation of giant planet satellites. Astrophys. J. **894**, 143 (2020)

Bell, K.R., Cassen, P.M., Klahr, H.H., Henning, T.: The structure and appearance of protostellar accretion disks: limits on disk flaring. Astrophys. J. **486**, 372–387 (1997)

Bell, C.P.M., Naylor, T., Mayne, N.J., Jeffries, R.D., Littlefair, S.P.: Pre-main-sequence 5176 isochrones – II. Revising star and planet formation time-scales. Month. Not. R. Astron. Soc. **434**, 806–831 (2013)

Benisty, M., Bae, J., Facchini, S., Keppler, M., Teague, R., Isella, A., Kurtovic, N.T., Pérez, L.M., Sierra, A., Andrews, S.M., et al.: A circumplanetary disk around PDS70c. Astrophys. J. Lett. **916**, L2 (2021)

Benisty, M., Dominik, C., Follette, K., Garufi, A., Ginski, C., Hashimoto, J., Keppler, M., Kley, W., Monnier, J.: Optical and near-infrared view of planet-forming disks and protoplanets. In: Inutsuka, S., Aikawa, Y., Muto, T., Tomida, K., Tamura, M. (eds.) Protostars and Planets VII. University of Arizona Press, Tucson, AZ (2022)

Bierson, C.J., Nimmo, F.: Explaining the Galilean satellites' density gradient by hydrodynamic escape. Astrophys. J. **897**, L43 (2020)

Birnstiel, T., Dullemond, C.P., Brauer, F.: Gas- and dust evolution in protoplanetary disks. Astron. Astrophys. **513**, A79 (2010)

Bitsch, B., Morbidelli, A., Johansen, A., Lega, E., Lambrechts, M., Crida, A.: Pebble-isolation mass: scaling law and implications for the formation of super-Earths and giant planets. Astron. Astrophys. **612**, A30 (2018)

Bodenheimer, P., Pollack, J.B.: Calculations of the accretion and evolution of giant planets: the effects of solid cores. Icarus. **67**, 391–408 (1986)

Boley, A.C.: The two modes of gas giant planet formation. Astrophys. J. **695**, L53–L57 (2009)

Boss, A.P.: Formation of gas and ice giant planets. Earth Planet. Sci. Lett. **202**, 513–523 (2002)

Bottke, W.F., Vokrouhlický, D., Nesvorný, D., Moore, J.M.: Black rain: the burial of the Galilean satellites in irregular satellite debris. Icarus. **223**, 775–795 (2013)

Brown, P.G., Hildebrand, A.R., Zolensky, M.E., Grady, M., Clayton, R.N., Mayeda, T.K., Tagliaferri, E., Spalding, R., MacRae, N.D., Hoffman, E.L., et al.: The fall, recovery, orbit, and composition of the Tagish Lake meteorite: a new type of carbonaceous chondrite. Science. **290**, 320–325 (2000)

Bryden, G., Chen, X., Lin, D.N.C., Nelson, R.P., Papaloizou, J.C.B.: Tidally induced gap formation in protostellar disks: gap clearing and suppression of protoplanetary growth. Astrophys. J. **514**, 344–367 (1999)

Bullock, E.S., Mckeegan, K.D., Gounelle, M., Grady, M.M., Russell, S.S.: Sulfur isotopic composition of Fe-Ni sulfide grains in CI and CM carbonaceous chondrites. Meteor. Planet. Sci. **45**, 885–898 (2010)

Burkhardt, C.: Planetary geneology. Elements. **17**, 395–400 (2021)

Cameron, A.G.W.: Physics of the primitive solar accretion disk. Moon Planets. **18**, 5–40 (1978)

Canup, R.M.: Origin of Saturn's rings and inner moons by mass removal from a lost Titan-sized satellite. Nature. **468**, 943–946 (2010)

Canup, R.M., Ward, W.R.: Formation of the Galilean satellites: conditions of accretion. Astron. J. **124**, 3404–3423 (2002)

Canup, R.M., Ward, W.R.: Formation of large satellites around gas planets. Nature. **441**, 834–839 (2006)

Canup, R.M., Ward, W.R.: Origin of Europa and the Galilean satellites. In: Pappalardo, R.T., McKinnon, W.B., Khurana, K. (eds.) Europa, pp. 59–83. University of Arizona Press, Tucson, AZ (2009)

Carter, P.J., Stewart, S.T.: Colliding in the shadows of giants: planetesimal collisions during the growth and migration of gas giants. Planet. Sci. J. **1**, 45 (2020)

Cassen, P.M., Peale, S.J., Reynolds, R.T.: Structure and thermal evolution of the Galilean satellites. In: Morrison, D. (ed.) Satellites of Jupiter, pp. 93–128. University of Arizona Press, Tucson, AZ (1982)

Chiang, E.L., Goldreich, P.: Spectral energy distributions of T Tauri stars with passive circumstellar disks. Astrophys. J. **490**, 368–376 (1997)

Cilibrasi, M., Szulágyi, J., Mayer, L., Drążkowska, J., Miguel, Y., Inderbitzi, P.: Satellites form fast & late: a population synthesis for the Galilean moons. Mon. Not. R. Astron. Soc. **480**, 4355–4368 (2018)

Cilibrasi, M., Szulágyi, J., Grimm, S.L., Mayer, L.: An N-body population synthesis framework for the formation of moons around Jupiter-like planets. Mon. Not. R. Astron. Soc. **504**, 5455–5474 (2021)

Clark, R.N., Brown, R.H., Cruikshank, D.P., Swayze, G.A.: Isotopic ratios of Saturn's rings and satellites: implications for the origin of water and Phoebe. Icarus. **321**, 791–802 (2019)

Coleman, G.A.L., Haworth, T.J.: Dispersal of protoplanetary discs: how stellar properties and the local environment determine the pathway of evolution. Mon. Not. R. Astron. Soc. **514**, 2315–2332 (2022)

Coradini, A., Magni, G.: The formation of Jupiter's satellites: relation of present appearance with past history. In: Barbieri, C., Rahe, J.H., Johnson, T.V., Sohus, A.M. (eds.) The Three Galileos: The Man, The Spacecraft, The Telescope, pp. 177–190. Kluwer Academic Publishers, the Netherlands (1997)

Coradini, A., Federico, C., Lanciano, P.: Ganymede and Callisto: accumulation heat content. In: Coradini, A., Fulchignoni, M. (eds.) Comparative Study of the Planets, pp. 61–70. Reidel, Dordrecht (1982)

Coradini, A., Cerroni, P., Magni, G., Federico, C.: Formation of the satellites of the outer solar system: sources of their atmospheres. In: Atreya, S.K., Pollack, J.B., Matthews, M.S. (eds.) Origin and Evolution of Planetary and Satellite Atmospheres, pp. 723–762. University of Arizona Press, Tucson, AZ (1989)

Coradini, A., Federico, C., Forni, O., Magni, G.: Origin and thermal evolution of icy satellites. Surv. Geophys. **16**, 533–591 (1995)

Coradini, A., Magni, G., Turrini, D.: From gas to satellitesimals: disk formation and evolution. Space Sci. Rev. **153**, 411–429 (2010)

D'Alessio, P., Calvet, N., Hartmann, L., Lizano, S., Canto, J.: Accretion disks around young objects. II. Tests of well-mixed models with ISM dust. Astrophys. J. **527**, 893–909 (1999)

D'Alessio, P., Calvet, N., Hartmann, L.: Accretion disks around young objects. III. Grain growth. Astrophys. J. **553**, 321–334 (2001)

D'Angelo, G., Henning, T., Kley, W.: Nested-grid calculations of disk-planet interaction. Astron. Astrophys. **385**, 647–670 (2002)

D'Angelo, G., Henning, T., Kley, W.: Thermohydrodynamics of circumstellar disks with high-mass planets. Astrophys. J. **599**, 548–576 (2003a)

D'Angelo, G., Kley, W., Henning, T.: Orbital migration and mass accretion of protoplanets in three-dimensional global computations with nested grids. Astrophys. J. **586**, 540–561 (2003b)

Dalton, J.B., Cruikshank, D.P., Stephan, K., McCord, T.B., Coustenis, A., Carlson, R.W., Coradini, A.: Chemical composition of icy satellite surfaces. Space. Sci. Rev. **153**, 113–154 (2010)

D'Angelo, G., Durisen, R.H., Lissauer, J.J.: Giant planet formation. In: Seager, S. (ed.) Exoplanets, pp. 319–346. University of Arizona Press, Tucson, AZ (2010)

de Pater, I., Keane, J.T., de Kleer, K., Davies, A.G.: A 2020 observational perspective of Io. Annu. Rev. Earth Planet. Sci. **49**, 643–678 (2020)

DeMeo, F.E., Carry, B.: Solar system evolution from compositional mapping of the asteroid belt. Nature. **505**, 629–634 (2014)

Desch, S.J., Kalyaan, A., Alexander, C.M.: The effect of Jupiter's formation on the distribution of refractory elements and inclusions in meteorites. Astrophys. J. Suppl. **238**, 11 (2018)

Drążkowska, J., Bitsch, B., Lambrechts, M., Mulders, G.D., Harsono, D., Vazan, A., Liu, B., Ormel, C.W., Kretke, K., Morbidelli, A.: Planet formation theory in the era of ALMA and Kepler: from pebbles to exoplanets. In: Inutsuka, S., Aikawa, Y., Muto, T., Tomida, K., Tamura, M. (eds.) Protostars and Planets VII. University of Arizona Press, Tucson, AZ (2022)

Dullemond, C.P., Hollenbach, D., Kamp, I., D'Alessio, P.: Models of the structure and evolution of protoplanetary disks. In: Reipurth, B., Jewitt, D., Keil, K. (eds.) Protostars and Planets V, pp. 555–572. University of Arizona Press, Tucson, AZ (2007)

Durisen, R.H., Boss, A.P., Mayer, L., Nelson, A.F., Quinn, T., Rice, W.K.M.: Gravitational instabilities in gaseous protoplanetary disks and implications for giant planet formation. In: Reipurth, B., Jewitt, D., Keil, K. (eds.) Protostars and Planets V, pp. 607–622. University of Arizona Press, Tucson, AZ (2007)

Dwyer, C.A., Nimmo, F., Ogihara, M., Ida, S.: The influence of imperfect accretion and radial mixing on ice:rock ratios in the Galilean system. Icarus. **225**, 390–402 (2013)

Eriksson, L.E.J., Ronnet, T., Johansen, A.: The fate of planetesimals formed at planetary gap edges. Astron. Astrophys. **648**, A112 (2021)
Eriksson, L.E.J., Ronnet, T., Johansen, A., Helled, R., Valletta, C., Petit, A.C.: A low accretion efficiency of planetesimals formed at planetary gap edges. Astron. Astrophys. **661**, A73 (2022)
Estrada, P.R., Mosqueira, I.: A gas-poor planetesimal capture model for the formation of giant planet satellite systems. Icarus. **181**, 486–509 (2006)
Estrada, P.R., Mosqueira, I., Lissauer, J.J., D'Angelo, G., Cruikshank, D.P.: Formation of Jupiter and conditions for accretion of the Galilean satellites. In: Pappalardo, R.T., McKinnon, W.B., Khurana, K. (eds.) Europa, pp. 27–58. University of Arizona Press, Tucson, AZ (2009)
Fegley, B.: Kinetics of gas-grain reactions in the solar nebula. Space Sci. Rev. **92**, 177–200 (1999)
Fischer, D.A., Valenti, J.: The planet-metallicity correlation. Astrophys. J. **622**, 1102–1117 (2005)
Föhn, M., Galli, A., Vorburger, A., Tulej, M., Lasi, D., et al.: Description of the mass spectrometer for the Jupiter Icy Moons Explorer mission. In: Proceedings of the 42nd IEEE Aerospace Conference, pp. 1–14. (2021)
Fujii, Y.I., Kobayashi, H., Takahashi, S.Z., Gressel, O.: Orbital evolution of moons in weakly accreting circumplanetary disks. Astron. J. **153**, 194 (2017)
Fujita, T., Ohtsuki, K., Tanigawa, T., Suetsugu, R.: Capture of planetesimals by gas drag from circumplanetary disks. Astrophys. J. **146**, 140 (2013)
Fuller, J., Luan, J., Quataert, E.: Resonance locking as the source of rapid tidal migration in the Jupiter and Saturn moon systems. Mon. Not. R. Astron. Soc. **458**, 3867 (2016)
Gao, P., Stevenson, D.J.: Nonhydrostatic effects and the determination of icy satellites' moment of inertia. Icarus. **226**, 1185–1191 (2013)
Goldreich, P., Ward, W.R.: The formation of planetesimals. Astrophys. J. **183**, 1051–1062 (1973)
Gole, D.A., Simon, J.B., Li, R., Youdin, A.N., Armitage, P.J.: Turbulence regulates the rate of planetesimal formation via gravitational collapse. Astrophys. J. **904**, 132 (2020)
Gounelle, M., Zolensky, M.E.: A terrestrial origin for sulfate veins in CI1 chondrites. Meteor. Planet. Sci. **36**, 1321–1329 (2001)
Gradie, J.C., Chapman, C.R., Tedesco, E.F.: Distribution of taxonomic classes and the compositional structure of the asteroid belt. In: Binzel, R.P., Gehrels, T., Matthews, M.S. (eds.) Asteroids II, pp. 316–335. University of Arizona Press, Tucson, AZ (1989)
Greenberg, R.: Orbital evolution of the Galilean satellites. In: Morrison, D. (ed.) Satellites of Jupiter, pp. 65–92. University of Arizona Press, Tucson, AZ (1982)
Greenberg, R.: Galilean satellites: evolutionary paths in deep resonance. Icarus. **70**, 334–347 (1987)
Grimm, R.E., McSween, H.Y.: Water and the thermal evolution of carbonaceous chondrite parent bodies. Icarus. **82**, 244–280 (1989)
Guillot, T., Hueso, R.: The composition of Jupiter: sign of a (relatively) late formation in a chemically evolved protosolar disc. Mon. Not. R. Astron. Soc. **367**, L47–L51 (2006)
Haisch, K.E., Lada, E.A., Lada, C.J.: Disk frequencies and lifetimes in young clusters. Astrophys. J., L153–L156 (2001)
Hartmann, L., Calvet, N., Gullbring, E., D'Alessio, P.: Accretion and the evolution of T Tauri disks. Astrophys. J. **495**, 385–400 (1998)
Hayashi, C., Nakazawa, K., Y. Nakagawa Y.: Formation of the solar system. In: Black, D.C., Matthews, M.S. (eds.) Protostars and Planets II, pp. 1100–1153. University of Arizona Press, Tucson, AZ (1985)
Helled, R., Morbidelli, A.: Planet formation. In: Madhusudhan, N. (ed.) ExoFrontiers, pp. 12-1–12-15. IOP Publishing, Bristol (2021)
Helled, R., Bodenheimer, P., Podolak, M., Boley, A., Meru, F., Nayakshin, S., Fortney, J.J., Mayer, L., Alibert, Y., Boss, A.P.: Giant planet formation, evolution, and internal structure. In: Beuther, H., Klessen, R.S., Dullemond, C.P., Henning, T. (eds.) Protostars and Planets VI, pp. 643–665. University of Arizona Press, Tucson, AZ (2014)
Helled, R., Stevenson, D.J., Lunine, J.I., Bolton, S.J., Nettelmann, N., Atreya, S., Guillot, T., Militzer, B., Miguel, Y., Hubbard, W.B.: Revelations on Jupiter's formation, evolution and interior: challenges from Juno results. Icarus. **378**, 114937 (2022)

Hersant, F., Gautier, D., Huré, J.-M.: A two-dimensional model for the primordial nebula constrained by D/H measurements in the solar system: implications for the formation of giant planets. Astrophys. J. **554**, 391–407 (2001)

Howell, R.R., Nash, D.B., Geballe, T.R., Cruikshank, D.P.: High-resolution infrared spectroscopy of Io and possible surface materials. Icarus. **78**, 27–37 (1989)

Hubickyj, O., Bodenheimer, P., Lissauer, J.J.: Accretion of the gaseous envelope of Jupiter around a 5–10 Earth-mass core. Icarus. **179**, 415–431 (2005)

Inaba, S., Wetherill, G.W., Ikoma, M.: Formation of gas giant planets: core accretion models with fragmentation and planetary envelope. Icarus. **166**, 46–62 (2003)

Inderbitzi, C., Szulágyi, J., Cilibrasi, M., Mayer, L.: Formation of satellites in circumplanetary discs generated by disc instability. Mon Not. Roy. Astron. Soc. **499**, 1023–1036 (2020)

Isella, A., Benisty, M., Teague, R., Bae, J., Keppler, M., Facchini, S., Pérez, L.: Detection of continuum submillimeter emission associated with candidate protoplanets. Astrophys. J. Lett. **879**, L25 (2019)

Izawa, M.R.M., Nesbitt, H.W., Macrae, N.D., Hoffman, E.L.: Composition and evolution of the early oceans: evidence from the Tagish Lake meteorite. Earth Planet. Sci. Lett. **298**, 443–449 (2010)

Izidoro, A., Dasgupta, R., Raymond, S.N., Deienno, R., Bitsch, B., Isella, A.: Planetesimal rings as the cause of the Solar System's planetary architecture. Nature Astron. **6**, 357–366 (2022)

Johansen, A., Lambrechts, M.: Forming planets via pebble accretion. Annu. Rev. Earth Planet. Sci. **45**, 359–387 (2017)

Johansen, A., Blum, J., Tanaka, H., Ormel, C., Bizzarro, M., Rickman, H.: The multifaceted planetesimal formation process. In: Beuther, H., Klessen, R.S., Dullemond, C.P., Henning, T. (eds.) Protostars and Planets VI, pp. 547–570. University of Arizona Press, Tucson, AZ (2014)

Johansen, A., Mac Low, M.-M., Lacerda, P., Bizzaro, M.: Growth of asteroids, planetary embryos, and Kuiper belt objects by chondrule accretion. Sci. Adv. **1**, e1500109 (2015)

Kane, S.R., Hinkel, N.R., Raymond, S.N.: Solar system moons as analogs for compact exoplanetary systems. Astron. J. **146**, 122 (2013)

Kavelaars, J.J., Petit, J.-M., Gladman, B., Bannister, M.T., Alexandersen, M., Chen, Y.-T., Gwyn, S.D.J., Volk, K.: OSSOS finds an exponential cutoff in the size distribution of the cold classical Kuiper Belt. Astrophys. J. **920**, L28 (2021)

Kempf, S., Sternovsky, Z., Horanyi, M., Hand, K.P., Srama, R., Postberg, F., Altobelli, N., Gruen, E., Gudipati, M.S., Schmidt, J., et al.: The SUrface Dust Analyzer (SUDA): Compositional Mapping of Europa's Surface. American Geophysical Union, Fall Meeting 2019, abstract #P53D-3500 (2019)

Kerraouch, I., Bischoff, A., Zolensky, M.E., Pack, A., Patzek, M., Hanna, R., Fries, M.D., Harries, D., Kebukawa, Y., Le, L., et al.: The polymict carbonaceous breccia Aguas Zarcas: a potential analog to samples being returned by the OSIRIS-REx and Hayabusa2 missions. Meteor. Planet. Sci. **56**, 277–310 (2021)

Kirk, R.L., Stevenson, D.J.: Thermal evolution of a differentiated Ganymede and implications for surface features. Icarus. **69**, 91–134 (1987)

Klahr, H., Bodenheimer, P.: Formation of giant planets by concurrent accretion of solids and gas inside an anticyclonic vortex. Astrophys. J. **639**, 432–440 (2006)

Korycansky, D.G., Pollack, J.B., Bodenheimer, P.: Numerical models of giant planet formation with rotation. Icarus. **92**, 234–251 (1991)

Kronrod, V.A., Makalkin, A.B., Kronrod, E.V.: Fragmentation and ablation of the planetesimals in the protoplanetary disks of Jupiter and Saturn. J. Phys. Conf. Ser. **1705**, 012003 (2020)

Kruijer, T.S., Burkhardt, C., Budde, G., Kleine, T.: Age of Jupiter inferred from the distinct genetics and formation times of meteorites. Proc. Natl. Acad. Sci. USA. **114**, 6712–6716 (2017)

Kruijer, T.S., Klein, T., Borg, L.E.: The great isotopic dichotomy of the early solar system. Nature Astron. **4**, 32–40 (2020)

Kuiper, G.P.: Planetary atmospheres and their origins. In: Kuiper, G.P. (ed.) The Atmospheres of the Earth and Planets, Revised Ed. University of Chicago Press, Chicago, IL (1952)

Kuskov, O.L., Kronrod, V.A.: L- and LL-chondritic models of the chemical composition of Io. Solar Syst. Res. **35**, 198–208 (2001)

Lehmer, O.R., Catling, D.C., Zahnle, K.J.: The longevity of water ice on Ganymedes and Europas around migrated giant planets. Astrophys. J. **839**, 32 (2017)

Lesur, G., Ercolano, B., Flock, M., Lin, M.-K., Yang, C.-C., Barranco, J.A., Benitez-Llambay, P., Goodman, J., Johansen, A., Klahr, H., et al.: Hydro-, magnetohydro-, and dust-gas dynamics of protoplanetary disks. In: Inutsuka, S., Aikawa, Y., Muto, T., Tomida, K., Tamura, M. (eds.) Protostars and Planets VII. University of Arizona Press, Tucson, AZ (2022)

Levison, H.F., Bottke, W.F., Gounelle, M., Morbidelli, A., Nesvorný, D., Tsiganis, K.: Contamination of the asteroid belt by primordial trans-Neptunian objects. Nature. **460**, 364–366 (2009)

Lewis, J.: The temperature gradient in the solar nebula. Science. **186**, 440–443 (1974)

Lewis, J.S.: Io: geochemistry of sulfur. Icarus. **50**, 105–114 (1982)

Li, R., Youdin, A.N., Simon, J.B.: Demographics of planetesimals formed by the streaming instability. Astrophys. J. **885**, 69 (2019)

Lichtenberg, T., Drążkowska, J., Schönbächler, M., Golabek, G.J., Hands, T.O.: Bifurcation of planetary building blocks during solar system formation. Science. **371**, 365–370 (2021)

Lissauer, J.J., Stevenson, D.J.: Formation of giant planets. In: Reipurth, B., Jewitt, D., Keil, K. (eds.) Protostars and Planets V, pp. 591–606. University of Arizona Press, Tucson, AZ (2007)

Lissauer, J.J., Hubickyj, O., D'Angelo, G., Bodenheimer, P.: Models of Jupiter's growth incorporating thermal and hydrodynamic constraints. Icarus. **199**, 338–350 (2009)

Liu, B., Ormel, C.W., Lin, D.N.C.: Dynamical rearrangement of super-Earths during disk dispersal I. Outline of the magnetospheric rebound model. Astron. Astrophys. **601**, A15 (2017)

Liu, S.-F., Hori, Y., Müller, S., Zheng, X., Helled, R., Lin, D., Isella, A.: The formation of Jupiter's diluted core by a giant impact. Nature. **572**, 355–360 (2019)

Lodders, K.: Solar system abundances and condensation temperatures of the elements. Astrophys. J. **591**, 1220–1247 (2003)

Lodders, K.: Jupiter formed with more tar than ice. Astrophys. J. **611**, 587–597 (2004)

Lodders, K.: Relative atomic solar system abundances, mass fractions, and atomic masses of the elements and their isotopes, composition of the solar photosphere, and compositions of the major chondritic meteorite groups. Space Sci. Rev. **217**, 44 (2021)

Lodders, K., Fegley Jr., B.: The Planetary Scientist's Companion. Oxford University Press, New York (1998)

Lubow, S.H., Seibert, M., Artymowicz, P.: Disk accretion onto high-mass planets. Astrophys. J. **526**, 1001–1012 (1999)

Lunine, J.I., Stevenson, D.J.: Formation of the Galilean satellites in a gaseous nebula. Icarus. **52**, 14–39 (1982)

Lunine, J.I., Coradini, A., Gautier, D., Owen, T.C., Wuchterl, G.: The origin of Jupiter. In: Bagenal, F., Dowling, T.E., McKinnon, W.B. (eds.) Jupiter – The Planet, Satellites and Magnetosphere, pp. 19–34. Cambridge University Press, Cambridge (2004)

Madeira, G., Izidoro, A., Giuliatti Winter, S.M.: Building the Galilean moons system via pebble accretion and migration: a primordial resonant chain. Mon. Not. R. Astron. Soc. **504**, 1854–1872 (2021)

Magni, G., Coradini, A.: Formation of Jupiter by nucleated instability. Planet. Space. Sci. **52**, 343–360 (2004)

Makalkin, A.B., Dorofeeva, V.A.: Accretion disks around Jupiter and Saturn at the stage of regular satellite formation. Solar Syst. Res. **48**, 62–78 (2014)

Mamajek, E.E.: Initial conditions of planet formation: lifetimes of primordial disks. In: Exoplanets and Disks: Their Formation and Diversity. Proceedings of the International Conference. AIP Conference Proceedings, vol. 1158, pp. 3–10 (2009)

Manara, C.F., Ansdell, M., Rosotti, G.P., Hughes, A.M., Armitage, P.J., Lodato, G., Williams, J.P.: Demographics of young stars and their protoplanetary disks: lessons learned on disk evolution and its connection to planet formation. In: Inutsuka, S., Aikawa, Y., Muto, T., Tomida, K., Tamura, M. (eds.) Protostars and Planets VII. University of Arizona Press, Tucson, AZ (2022) in press

Mankovich, C.R., Fuller, J.: A diffuse core in Saturn revealed by ring seismology. Nature Astron. **5**, 1103–1109 (2021)

Marrocchi, Y., Avice, G., Barrat, J.-A.: The Tarda meteorite: a window into the formation of D-type asteroids. Astrophys. J. Lett. **913**, L9 (2021)

Mayer, L., Quinn, T., Wadsley, J., Staedel, J.: The evolution of gravitationally unstable protoplanetary disks: fragmentation and possible giant planet formation. Astrophys. J. **609**, 1045–1064 (2004)

McKinnon, W.B.: Mystery of Callisto: is it undifferentiated? Icarus. **130**, 540–543 (1997)

McKinnon, W.B.: On the initial thermal evolution of Kuiper Belt Objects. In: Proc. Asteroids, Comets, Meteors (ACM 2002), ESA SP-500, pp. 29–38 (2002)

McKinnon, W.B.: Internal structures of the Galilean satellites: unfinished business and the road ahead. In: Forum on Concepts and Approaches for Jupiter Icy Moons Orbiter. LPI Contribution 1163, p. 53. Lunar and Planetary Institute, Houston, TX (2003)

McKinnon, W.B.: On the oxidation states of the Galilean satellites: implications for internal structures, ocean chemistry, and magnetic fields. Lunar Planet. Sci. **XXXV, #2137** (2004)

McKinnon, W.B., Desai, S.S.: Internal structures of the Galilean satellites: what can we really tell? Lunar Planet. Sci. Conf. **XXXIV, #2104** (2003)

McKinnon, W.B., Parmentier, E.M.: Ganymede and Callisto. In: Burns, J.A. (ed.) Satellites, pp. 718–763. University of Arizona Press, Tucson, AZ (1986)

McKinnon, W.B., Zolensky, M.E.: Sulfate content of Europa's ocean and shell: evolutionary considerations and geological and astrobiological implications. Astrobiology. **3**, 879–897 (2003)

McSween, H.Y., Binzel, R.P., De Sanctis, M.C., Ammannito, E., Prettyman, T.H., Beck, A.W., Reddy, V., Le Corre, L., Gaffey, M.J., McCord, T.B., Raymond, C.A., Russell, C.T., the Dawn Science Team: Dawn; the Vesta–HED connection; and the geologic context for eucrites, diogenites, and howardites. Meteor. Planet. Sci. **48**, 2090–2104 (2013)

Meyer, M.R., Backman, D.E., Weinberger, A.J., Wyatt, M.C.: Evolution of circumstellar disks around normal stars: placing our Solar System in context. In: Reipurth, B., Jewitt, D., Keil, K. (eds.) Protostars and Planets V, pp. 573–588. University of Arizona Press, Tucson, AZ (2007)

Miguel, Y., Ida, S.: A semi-analytical model for exploring Galilean satellites formation from a massive disk. Icarus. **266**, 1–14 (2016)

Miller, K.E., Miller, G.P., Franke, K., Hoeper, P., Waite, Jr., J.H., Brockwell, T., Perryman, R.S., Dunn, G., Hanley, J., Magee, B., et al.: An update on the MASPEX instrument: new capabilities for planetary chemistry and habitability. In: 53rd Lunar Planet. Sci. Conf., abstract #2664 (2022)

Mizuno, H.: Formation of the giant planets. Prog. Theor. Phys. **64**, 544–557 (1980)

Moraes, R.A., Kley, W., Neto, E.V.: Growth and evolution of satellites in a Jovian massive disc. Mon. Not. R. Astron. Soc. **475**, 1347–1362 (2018)

Morbidelli, A., Levison, H.F., Tsiganis, K., Gomes, R.: Chaotic capture of Jupiter's Trojan asteroids in the early Solar System. Nature. **435**, 462–465 (2005)

Morbidelli, A., Bottke, W.F., Nesvorn, D., Levison, H.F.: Asteroids were born big. Icarus. **204**, 558–573 (2009)

Morbidelli, A., Baillié, K., Batygin, K., Charnoz, S., Guillot, T., Rubie, D.C., Kleine, T.: Contemporary formation of early Solar System planetesimals at two distinct radial locations. Nature Astron. **6**, 72–79 (2022)

Mosqueira, I., Estrada, P.R.: Formation of the regular satellites of giant planets in an extended gaseous nebula I: Subnebula model and accretion of satellites. Icarus. **163**, 198–231 (2003a)

Mosqueira, I., Estrada, P.R.: Formation of the regular satellites of giant planets in an extended gaseous nebula II: Satellite migration and survival. Icarus. **163**, 232–255 (2003b)

Moullet, A., Lellouch, E., Moreno, R., Gurwell, M.A., Black, J.H., Butler, B.: Exploring Io's atmospheric composition with APEX: first measurement of $^{34}SO_2$ and tentative detection of KCl. Astrophys. J. **776**, 32 (2013)

Nakano, H., Kouchi, A., Tachibana, S., Tsuchiyama, A.: Evaporation of interstellar organic materials in the solar nebula. Astrophys. J. **592**, 1252–1262 (2003)

Nelson, R.P., Gressel, O., Umurhan, O.M.: Linear and non-linear evolution of the vertical shear instability in accretion discs. Mon. Not. R. Astron. Soc. **435**, 2610–2632 (2013)
Nesvorný, D.: Dynamical evolution of the early Solar System. Annu. Rev. Astron. Astrophys. **56**, 137–174 (2018)
Nesvorný, D., Li, R., Youdin, A.N., Simon, J.B., Grundy, W.M.: Trans-Neptunian binaries as evidence for planetesimal formation by the streaming instability. Nature Astron. **3**, 808–812 (2019)
Nesvorný, D., Li, R., Simon, J.B., Youdin, A.N., Richardson, D.C., Marschall, R., Grundy, W.M.: Binary planetesimal formation from gravitationally collapsing pebble clouds. Planet. Sci. J. **3**, 808–812 (2021)
Ogihara, M., Ida, S.: N-body simulations of satellite formation around giant planets: origin of orbital configuration of the Galilean moons. Astrophys. J. **753**, 60–77 (2012)
Palla, F., Stahler, S.W.: Star formation in the Orion nebula cluster. Astrophys. J. **525**, 772–783 (1999)
Palme, H., Lodders, K., Jones, A.: Solar system abundances of the elements. In: Turekian, K. (ed.) Treatise on Geochemistry Planets, Asteroids, Comets and the Solar System, vol. 2, 2nd edn, pp. 15–36. Elsevier, Amsterdam (2014)
Papaloizou, J.C.B., Nelson, R.P.: Models of accreting gas giant protoplanets in protostellar disks. Astron. Astrophys. **433**, 247–265 (2005)
Pasek, M.A., Milsom, J.A., Ciesla, F.J., Lauretta, D.S., Sharp, C.M., Lunine, J.I.: Sulfur chemistry with time-varying oxygen abundance during Solar System formation. Icarus. **175**, 1–14 (2005)
Peale, S.J.: Rotation histories of the natural satellites. In: Burns, J.A., Matthews, M.S. (eds.) Planetary Satellites, pp. 87–111. University of Arizona Press, Tucson, AZ (1977)
Peale, S.J.: Origin and evolution of the natural satellites. Annu. Rev. Astron. Astrophys. **37**, 533–602 (1999)
Peale, S.J.: Tidally induced volcanism. Celest. Mech. Dynam. Astron. **87**, 129–155 (2003)
Peale, S.J., Canup, R.M.: The origin of the natural satellites. In: Schubert, G. (ed.) Treatise on Geophysics Planets and Moons, vol. 10, 2nd edn, pp. 559–604. Elsevier, Boston, MA (2015)
Peale, S.J., Lee, M.H.: A primordial origin of the Laplace relation among the Galilean satellites. Science. **298**, 593–597 (2002)
Pinilla, P., Birnstiel, T., Ricci, L., Dullemond, C.P., Uribe, A.L., Testi, L., Natta, A.: Trapping dust particles in the outer regions of protoplanetary disks. Astron. Astrophys. **538**, A114 (2012)
Pizzarello, S., Cooper, G.W., Flynn, G.J.: The nature and distribution of the organic material in carbonaceous chondrites and interplanetary dust particles. In: Lauretta, D.S., McSween, H.Y. (eds.) Meteorites and the Early Solar System II, pp. 625–651. University of Arizona Press, Tucson, AZ (2006)
Pollack, J.B., Fanale, F.: Origin and evolution of the Jupiter satellite system. In: Morrison, D. (ed.) Satellites of Jupiter, pp. 872–891. University of Arizona Press, Tucson, AZ (1982)
Pollack, J.B., Reynolds, R.T.: Implications of Jupiter's early contraction history for the composition of the Galilean satellites. Icarus. **21**, 248–253 (1974)
Pollack, J.B., Lunine, J.I., Tittemore, W.C.: Origin of the Uranian satellites. In: Bergstralh, J.T., Miner, E.D., Matthews, M.S. (eds.) Uranus, pp. 469–512. University of Arizona Press, Tucson, AZ (1991)
Pollack, J.B., Hubickyj, O., Bodenheimer, P., Lissauer, J.J., Podolak, M., Greenzweig, Y.: Formation of the giant planets by concurrent accretion of solids and gases. Icarus. **124**, 62–85 (1996)
Prinn, R.G., Fegley, B.: Kinetic inhibition of CO and N_2 reduction in circumplanetary nebulae: implications for satellite composition. Astrophys. J. **249**, 308–317 (1981)
Prinn, R.G., Fegley, B.: Solar nebula chemistry: origin of planetary, satellite, and cometary volatiles. In: Atreya, S.K., Pollack, J.B., Matthews, M.S. (eds.) Origin and Evolution of Planetary and Satellite Atmospheres, pp. 8–136. University of Arizona Press, Tucson, AZ (1989)
Raymond, S.N., Izidoro, A.: Origin of water in the inner solar system: planetesimals scattered inward during Jupiter and Saturn's rapid gas accretion. Icarus. **297**, 134–148 (2017)

Ronnet, T., Johansen, A.: Formation of moon systems around giant planets. Capture and ablation of planetesimals as foundation for a pebble accretion scenario. Astron. Astrophys. **633**, 93 (2020)
Ronnet, T., Mousis, O., Vernazza, P.: Pebble accretion at the origin of water in Europa. Astrophys. J. **845**, 92 (2017)
Ronnet, T., Mousis, O., Vernazza, P., Lunine, J.I., Crida, A.: Saturn's formation and early evolution at the origin of Jupiter's massive moons. Astron. J. **155**, 224 (2018)
Sackmann, I.-J., Boothroyd, A.I., Kraemer, K.E.: Our Sun. III. Present and future. Astrophys. J. **418**, 457–468 (1993)
Safronov, V.S., Pechernikova, G.V., Ruskol, E.L., Vitiazev, A.V.: Protosatellite swarms. In: Burns, J.A., Mathews, M.S. (eds.) Satellites, pp. 86–116. University of Arizona Press, Tucson, AZ (1986)
Sasaki, T., Stewart, G.R., Ida, S.: Origin of the different architectures of the Jovian and Saturnian satellite systems. Astrophys. J. **714**, 1052–1064 (2010)
Schubert, G., Stevenson, D.J., Ellsworth, K.: Internal structures of the Galilean satellites. Icarus. **47**, 46–59 (1981)
Schubert, G., Anderson, J.D., Spohn, T., McKinnon, W.B.: Interior composition, structure and dynamics of the Galilean satellites. In: Bagenal, F., Dowling, T.E., McKinnon, W.B. (eds.) Jupiter – The Planet, Satellites and Magnetosphere, pp. 281–306. Cambridge University Press, Cambridge (2004)
Schulik, M., Johansen, A., Bitsch, B., Bertram, E.L., Lambrechts, M.: On the structure and mass delivery towards circumplanetary discs. Astron. Astrophys. **642**, A187 (2020)
Scott, E.R.D., Krot, A.N., Sanders, I.S.: Isotopic dichotomy among meteorites and its bearing on the protoplanetary disk. Astrophys. J. **854**, 164 (2018)
Sephton, M.: Organic matter in carbonaceous meteorites: past, present, and future research. Phil. Trans. R. Soc. A. **363**, 2729–2742 (2005)
Shibaike, Y., Ormel, C.W., Ida, S., Okuzumi, S., Sasaki, T.: The Galilean satellites formed slowly from pebbles. Astrophys. J. **885**, 79 (2019)
Showman, A.P., Malhotra, R.: Tidal evolution into the Laplace resonance and the resurfacing of Ganymede. Icarus. **127**, 93–111 (1997)
Siess, L., Dufour, E., Forestini, M.: An internet server for pre-main sequence tracks of low- and intermediate-mass stars. Astron. Astrophys. **358**, 593–599 (2000)
Simon, J.B., Armitage, P.J., Li, R., Youdin, A.N.: The mass and size distribution of planetesimals formed by the streaming instability. I. The role of self-gravity. Astrophys. J. **822**, 55 (2016)
Sohl, F., Spohn, T., Breuer, D., Nagel, K.: Implications from Galileo observations on the interior structure and chemistry of the Galilean satellites. Icarus. **157**, 104–119 (2002)
Squyres, S.W., Reynolds, R.T., Summers, A.L., Shung, F.: Accretional heating of the satellites of Saturn and Uranus. J. Geophys. Res. **93**, 8779–8794 (1988)
Stevenson, D.J.: Formation of the giant planets. Planet. Space Sci. **30**, 755–764 (1982)
Stevenson, D.J.: Jupiter and its moons. Science. **294**, 71–72 (2001)
Stevenson, D.J., Harris, A.W., Lunine, J.I.: Origins of satellites. In: Burns, J.A., Matthews, M.S. (eds.) Satellites, pp. 39–88. University of Arizona Press, Tucson, AZ (1986)
Stoll, M.H.R., Kley, W.: Vertical shear instability in accretion disc models with radiation transport. Astron. Astrophys. **572**, A77 (2014)
Stoll, M.H.R., Kley, W.: Particle dynamics in discs with turbulence generated by the vertical shear instability. Astron. Astrophys. **594**, A57 (2016)
Stone, J.M., Gammie, C.F., Balbus, S.A., Hawley, J.F.: Transport processes in protostellar disks. In: Mannings, V., Boss, A.P., Russell, S.S. (eds.) Protostars and Planets IV, pp. 589–613. University of Arizona Press, Tucson, AZ (2000)
Suetsugu, R., Ohtsuki, K.: Distribution of captured planetesimals in circumplanetary gas disks and implications for accretion of regular satellites. Astrophys. J. **839**, 66 (2017)
Suttle, M.D., King, A.J., Schofield, P.F., Bates, H., Russell, S.S.: The aqueous alteration of CM chondrites, a review. Geochim. Cosmochim. Acta. **299**, 219–256 (2021)

Sutton, S., Alexander, C.M., Bryant, A., Lanzirotti, A., Newville, M., Cloutis, E.A.: The bulk valence state of Fe and the origin of water in chondrites. Geochim. Cosmochim. Acta. **211**, 115–132 (2017)

Szulágyi, J.: Effects of the planetary temperature on the circumplanetary disk and on the gap. Astrophys. J. **842**, 103 (2017)

Szulágyi, J., Masset, F., Legal, E., Crida, A., Morbidelli, A., Guillot, T.: Circumplanetary disc or circumplanetary envelope? Mon. Not. R. Astron. Soc. **460**, 2853–2861 (2016)

Takata, T., Stevenson, D.J.: Despin mechanism for protogiant planets and ionization state of protogiant planetary disks. Icarus. **123**, 404–421 (1996)

Takato, N., Bus, S.J., Terada, H., Pyo, T.-S., Kobayashi, N.: Detection of a deep 3-μm absorption feature in the spectrum of Amalthea (JV). Science. **306**, 2224–2227 (2004)

Takeuchi, T., Clarke, C.J., Lin, D.N.C.: The differential lifetimes of protostellar gas and dust disks. Astrophys. J. **627**, 286–292 (2005)

Tanigawa, T., Ohtsuki, K., Machid, M.N.: Distribution of accreting gas and angular momentum onto circumplanetary disks. Astrophys. J. **747**, 47–63 (2012)

Tanigawa, T., Maruta, A., Machida, M.: Accretion of solid materials onto circumplanetary disks from protoplanetary disks. Astrophys. J. **784**, 109 (2014)

Turner, N.J., Choukroun, M., Castillo-Rogez, J., Bryden, G.: A hot gap around Jupiter's orbit in the solar nebula. Astrophys. J. **748**, 92 (2012)

Umurhan, O.M., Estrada, P.R., Cuzzi, J.N.: Streaming instability in turbulent protoplanetary disks. Astrophys. J. **895**, 4 (2020)

Vacher, L.G., Ogliore, R.C., Jones, C., Liu, N., Fike, D.A.: Cosmic symplectite recorded irradiation by nearby massive stars in the solar system's parent molecular cloud. Geochim. Cosmochim. Acta. **309**, 135–150 (2021)

Veeder, G.J., Matson, D.L., Johnson, T.V., Blaney, D.L., Goguen, J.D.: Io's heat flow from infrared radiometry: 1983–1993. J. Geophys. Res. **99**, 17095–17162 (1994)

Venturini, J., Helled, R.: Jupiter's heavy-element enrichment expected from formation models. Astron. Astrophys. **634**, A31 (2020)

Wahl, S.M., Hubbard, W.B., Militzer, B., Guillot, T., Miguel, Y., Movshovitz, N., Kaspi, Y., Helled, R., Reese, D., Galanti, E., et al.: Comparing Jupiter interior structure models to Juno gravity measurements and the role of a dilute core. Geophys. Res. Lett. **44**, 4649–4659 (2017)

Waite, J.H., Brockwell, T., Glein, C., Perry, R., Bolton, S., McGrath, M., McKinnon, W., Miller, K., Mousis, O., Ray, C., et al.: MASPEX-Europa aboard Clipper: a mass spectrometer for investigating the habitability of Europa. EPSC Abs. **13**, EPSC-DPS2019-559-1 (2019)

Walsh, K.J., Morbidelli, A., Raymond, S.N., O'Brien, D.P., Mandell, A.M.: A low mass for Mars from Jupiter's early gas-driven migration. Nature. **475**, 206–209 (2011)

Ward, W.R., Canup, R.M.: Circumplanetary disk formation. Astron. J. **140**, 1168–1193 (2010)

Warren, P.H.: Stable-isotopic anomalies and the accretionary assemblage of the Earth and Mars: a subordinate role for carbonaceous chondrites. Earth Planet. Sci. Lett. **311**, 93–100 (2011)

Weidenschilling, S.J.: Aerodynamics of solid bodies in the solar nebula. Month. Not. R. Astron. Soc. **180**, 57–70 (1977)

Weiss, B.P., Bottke, W.F.: What can meteorites tell us about the formation of Jupiter? AGU Adv. **2**, e2020AV000376 (2021)

Weiss, B.P., Bai, X.-N., Fu, R.R.: History of the solar nebula from meteorite paleomagnetism. Sci. Adv. **7**, eaba5967 (2021)

Williams, J.P., Cieza, L.A.: Protoplanetary disks and their evolution. Annu. Rev. Astron. Astrophys. **49**, 67–117 (2011)

Wirström, E.S., Bjerkeli, P., Rezac, L., Brinch, C., Hartogh, P.: Effect of the 3D distribution on water observations made with the SWI. Astron. Astrophys. **637**, A90 (2020)

Wood, J.A.: Pressure and temperature profiles in the solar nebula. Space Sci. Rev. **92**, 87–93 (2000)

Yokoyama, T., Nagashima, K., Nakai, I., Yound, E.D., Abe, Y., Aléon, J., Alexander, C.M., Amari, S., Amelin, Y., Bajo, K., et al.: Samples returned from the asteroid Ryugu are similar to Ivuna-type carbonaceous asteroids. Science. (2022). https://doi.org/10.1126/science.abn7850

Youdin, A.N., Goodman, J.: Streaming instabilities in protoplanetary disks. Astrophys. J. **620**, 459–469 (2005)

Zolensky, M.E., Nakamura, K., Gounelle, M., Mikouchi, T., Kasama, T., Tachikawa, O., Tonoi, E.: Mineralogy of Tagish Lake: an ungrouped type 2 carbonaceous chondrite. Meteor. Planet. Sci. **37**, 737–761 (2002)

Zolotov, M.Y., Fegley, B.: Eruption conditions of Pele volcano on Io inferred from chemistry of its volcanic plume. Icarus. **141**, 40–52 (1999)

Zolotov, M.Y., Fegley, B.: Eruption conditions of Pele volcano on Io inferred from chemistry of its volcanic plume. Geophys. Res. Lett. **27**, 2789–2792 (2000a)

Zolotov, M.Yu., Fegley, B.: Volcanic degassing of hydrogen compounds on Io. Lunar Planet Sci. XXXI, abstract #1186 (2000b)

Chapter 4
Tidal Heating and the Interior Structure of Io

James Tuttle Keane, Isamu Matsuyama, Carver J. Bierson, and Antony Trinh

Abstract Io is the most tidally heated world in the Solar System. In this chapter, we review the basics of how Io (and other satellites) responds to tidal forces, and how those forces shape Io's interior structure and geologic activity. We also summarize the current state of knowledge of Io's interior, largely gleaned from the *Voyager* and *Galileo* missions, along with ground and space-based observations. Despite decades of work, there are still many important, unanswered questions about tidal heating and the interior structure of Io, motivating continued research and exploration.

4.1 Introduction

Jupiter's moon, Io, is the most volcanically active world in the Solar System. Io's rampant activity is powered by tidal heating, where the periodic forces from Io's neighboring moons, Europa and Ganymede, preserve the eccentricity of its orbit, subjecting Io to extreme tides from massive Jupiter. These tides continuously reshape Io, dissipating enormous amounts of heat within its interior—so much so that it may melt Io's interior, creating a subsurface magma ocean.

Since the *Galileo* era, there have been major advances in our understanding of Io's interior structure and evolution, and the fundamental process of tidal heating. These geophysical advances provide the context for interpreting Earth-based observations, which are often designed to interrogate Io's interior structure and evolution from afar. In this chapter, we summarize the most notable developments in our

J. T. Keane (✉)
Jet Propulsion Laboratory, California Institute of Technology, Pasadena, CA, USA
e-mail: James.t.keane@jpl.nasa.gov

I. Matsuyama · A. Trinh
University of Arizona, Tucson, AZ, USA
e-mail: isa@lpl.arizona.edu; atrinh@lpl.arizona.edu

C. J. Bierson
Arizona State University, Phoenix, AZ, USA
e-mail: CBierson@asu.edu

R. M. C. Lopes et al. (eds.), *Io: A New View of Jupiter's Moon*, Astrophysics and Space Science Library 468, https://doi.org/10.1007/978-3-031-25670-7_4

understanding of Io's geophysics since the close of the *Galileo* mission. We build upon past review documents, particularly Schubert et al. (2004), Moore et al. (2007), and de Kleer et al. (2019b).

This chapter is organized in three sections. In Sect. 4.2, we review the theory for the shape and interior structure of a tidally-deformed world. In Sect. 4.3, we review the basics of tidal heating. In Sect. 4.4 we review the relevant observations of Io that inform our understanding of its interior structure and tidal heating. Finally, in Sect. 4.5, we synthesize the results and identify major outstanding problems.

Throughout this chapter, we will present maps of different observed and predicted geologic quantities. Figure 4.1 shows the global image mosaic of Io for context.

4.2 Theory, Part 1: Satellite Figures

In planetary science, "figure" is a catch-all term used to characterize some aspect of the shape or dynamical character of a planetary body. The simplest use of "figure" is to describe the actual shape of a planetary body, usually characterized in terms of a radius, flattening, triaxial shape, or more esoteric metrics (e.g., spherical harmonics). While planetary shape is intuitive, "figure" can also be used when describing far more abstract quantities, like an object's gravity field or moments of inertia. All three of these quantities (shape, gravity, moments of inertia) are measured in different ways. Planetary shape can be measured from stellar occultation, limb profiles, stereophotogrammetry, laser or radar ranging, etc. Planetary gravity fields are measured by carefully tracking the accelerations of spacecraft (or other test particles) as they orbit, or fly past, planetary bodies. Planetary moments of inertia are measured by observing how planets react to external torques. Despite the myriad of physical parameters, measurement types, and nuances therein—these quantities can all be lumped under the umbrella of planetary "figures" because they are intimately related to the interior structure of the object and the forces acting on it. Measuring and understanding planetary figures is one of the basic ways that we determine the interior structures and histories of planetary bodies.

In this section, we will describe how planetary figures are quantified, and how planetary figures are controlled by rotational and tidal deformation.

4.2.1 Shape

The most convenient way to describe a planetary body's shape or gravity field is to expand it in spherical harmonics. While mathematically daunting, spherical harmonics are advantageous because they form a natural, complete basis of orthogonal

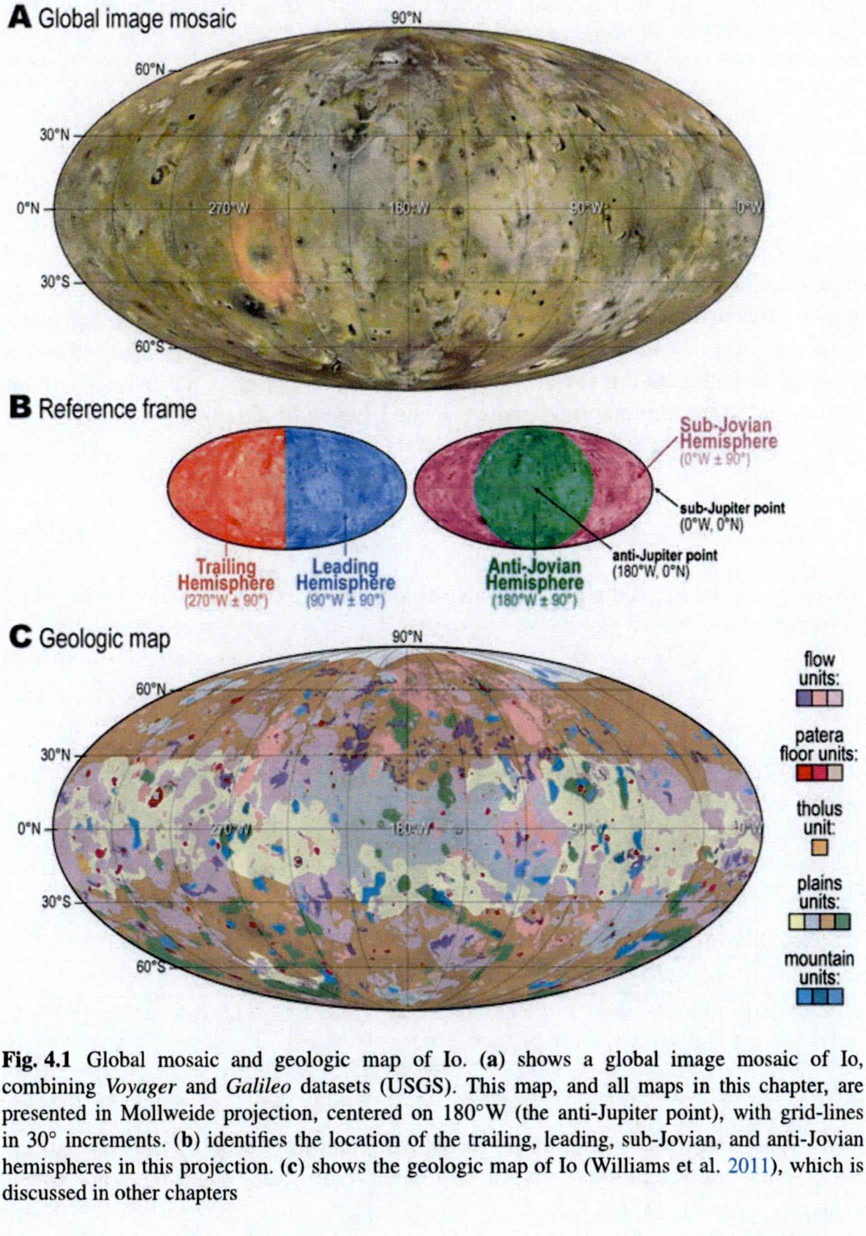

Fig. 4.1 Global mosaic and geologic map of Io. (**a**) shows a global image mosaic of Io, combining *Voyager* and *Galileo* datasets (USGS). This map, and all maps in this chapter, are presented in Mollweide projection, centered on 180°W (the anti-Jupiter point), with grid-lines in 30° increments. (**b**) identifies the location of the trailing, leading, sub-Jovian, and anti-Jovian hemispheres in this projection. (**c**) shows the geologic map of Io (Williams et al. 2011), which is discussed in other chapters

functions on a sphere, which simplifies mathematical manipulations in spherical geometries. Even the global-scale forces acting on planetary bodies as a result of rotation and tides can be written compactly in spherical harmonics as the gradient of a scalar potential field.

The shape of a planetary body, defined as the radial distance r of the free surface, can be expanded in spherical harmonics as:

$$r(\theta,\phi) = R\sum_{l=0}^{\infty}\sum_{m=0}^{l}\left(c_{\ell,m}\mathrm{P}_{\ell,m}(\cos\theta)\cos(m\phi) + s_{\ell,m}\mathrm{P}_{\ell,m}(\cos\theta)\sin(m\phi)\right), \tag{4.1}$$

where R is the mean radius of the planetary body; θ and ϕ are the co-latitude[1] and eastward longitude at the location where the radius is being evaluated; $c_{\ell,m}$ and $s_{\ell,m}$ are the (dimensionless) unnormalized spherical harmonic coefficients of the planet's shape, of degree ℓ and order m. For clarity, we use a comma to separate the indices in subscripts—although this is rarely done in the literature (e.g., $c_{2,0}$ is usually written as c_{20}). $\mathrm{P}_{\ell,m}$ are the unnormalized associated Legendre functions, defined for an arbitrary argument x by (e.g. Arfken and Weber 1995):

$$\mathrm{P}_{\ell,m}(x) = (1-x^2)^{m/2}\frac{d^m}{dx^m}\mathrm{P}_{\ell}(x), \tag{4.2}$$

where P_{ℓ} are the Legendre polynomials, which can be expressed using Rodrigues's formula:

$$\mathrm{P}_{\ell}(x) = \frac{1}{2^{\ell}\ell!}\frac{d^m}{dx^m}(x^2-1)^{\ell}, \tag{4.3}$$

For spherical harmonics, x is replaced with $\cos\theta$ (e.g., Eq. 4.1). The first few associated Legendre functions are:

Associated Legendre functions, $\mathrm{P}_{\ell,m}$			
	$m = 0$	$m = 1$	$m = 2$
$\ell = 0$	$\mathrm{P}_{0,0}(\cos\theta) = 1$	Not used here ($m > l$)	Not used here ($m > l$)
$\ell = 1$	$\mathrm{P}_{1,0}(\cos\theta) = \cos\theta$	$\mathrm{P}_{1,1}(\cos\theta) = \sin\theta$	Not used here ($m > l$)
$\ell = 2$	$\mathrm{P}_{2,0}(\cos\theta) = 1/2\,(3\cos^2\theta) - 1/2$	$\mathrm{P}_{2,1}(\cos\theta) = 3\cos\theta\sin\theta$	$\mathrm{P}_{2,2}(\cos\theta) = 3\sin^2\theta$

(4.4)

where functions with $m > \ell$ are not used. Note that the Legendre polynomials of degree ℓ are equivalent to the associated Legendre functions of degree ℓ and order 0 (i.e., $\mathrm{P}_{\ell}(x) = \mathrm{P}_{\ell,0}(x)$).

[1] Co-latitude is the angular distance from the north pole: co-latitude = 90° − latitude.

The expansion coefficients, $c_{\ell,m}$ and $s_{\ell,m}$, can be found by integrating:

$$c_{\ell,m} = \frac{1}{R}\frac{1}{1+\delta_{m,0}}\frac{2l+1}{2\pi}\frac{(\ell-m)!}{(\ell+m)!}\int_0^{\pi}\int_0^{2\pi} r(\theta,\phi)\left[\mathrm{P}_{\ell,m}(\cos\theta)\cos(m\phi)\right] \times \sin(\theta)d\theta d\phi \tag{4.5}$$

$$s_{\ell,m} = \frac{1}{R}(1-\delta_{m,0})\frac{2l+1}{2\pi}\frac{(\ell-m)!}{(\ell+m)!}\int_0^{\pi}\int_0^{2\pi} r(\theta,\phi)\left[\mathrm{P}_{\ell,m}(\cos\theta)\sin(m\phi)\right] \times \sin(\theta)d\theta d\phi\,, \tag{4.6}$$

where $\delta_{i,j}$ is the Kronecker delta ($\delta_{i,j} = 0$ if $i \neq j$; $\delta_{i,j} = 1$ if $i = j$).

Some simple relationships between spherical harmonic degree (ℓ) and order (m) can help visualize the geometry of the spherical harmonics. Spherical harmonics possess $2m$ zero crossings in the longitudinal direction, and $\ell - m$ zero crossings in the latitudinal direction (excluding the poles). If $m = 0$, the zero crossings (and the spherical harmonics themselves indeed) only depend on latitude and the harmonics are called *zonal*. If $m = \ell$, the zero crossings only depend on longitude and the harmonics are called *sectoral*. If $0 < m < \ell$, the harmonics are called *tesseral*. Additionally, for a given spherical harmonic degree, the equivalent wavelength, λ, is given by the Jeans relation (e.g. Wieczorek 2015):

$$\lambda = \frac{2\pi R}{\sqrt{\ell(\ell+1)}}\,. \tag{4.7}$$

This relationship shows that spherical harmonics of higher degree and order provide information at smaller spatial scales.

4.2.2 *Gravity*

The gravitational potential of a planetary body can be similarly expanded in spherical harmonics (Kaula 1966, 1968; Wieczorek 2015):

$$U(r,\theta,\phi) = \frac{GM}{r}\sum_{\ell=0}^{\infty}\sum_{m=0}^{\ell}\left(\frac{\mathscr{R}}{r}\right)^{\ell} \times \left(C_{\ell,m}\mathrm{P}_{\ell,m}(\cos\theta)\cos(m\phi) + S_{\ell,m}\mathrm{P}_{\ell,m}(\cos\theta)\sin(m\phi)\right), \tag{4.8}$$

where r, θ, and ϕ are the radius, co-latitude, and longitude at which the potential field is evaluated; G is the gravitational constant; M is the mass of the body; $\mathscr{R}$ is the reference radius; and $C_{\ell,m}$ and $S_{\ell,m}$ are the (dimensionless) unnormalized spherical harmonic coefficients of the body's gravitational potential, of degree ℓ and order m.

Throughout this work, we follow the geodesy sign convention that the gravitational field is the positive gradient of the gravitational potential. The gravitational potential is in units of ($m^2 s^{-2}$), or equivalently, ($J\,kg^{-1}$). Note that the gravitational potential is often expressed relative to a prescribed reference radius $\mathscr{R}$, which can be different than the average radius R, in which case we assume that the coefficients $C_{\ell,m}$ and $S_{\ell,m}$ in Eq. 4.8 have been appropriately rescaled.

Spherical harmonic expansions of the gravity field are only explicitly correct if evaluated above the largest radial distance displayed by the free surface (the so-called Brillouin sphere). If the spherical harmonic expansion is evaluated below the largest equatorial radius, a process called "downward continuation", the expansion *can* diverge and generate spurious values for the gravitational potential. This is particularly problematic for smaller, irregular objects (e.g., asteroids and comets), which is why gravity models of small bodies often use other methods for calculating gravity fields (e.g., Scheeres 2012). For larger planetary bodies (e.g., terrestrial planets, dwarf planets, and the mid/large-sized regular satellites of the gas giants), this effect is small, and often ignored when evaluating the gravity field on the surface of the object even if it is slightly below the largest equatorial radius. Above the largest equatorial radius, the term $(\mathscr{R}/r)^\ell$ reveals how short-wavelength (large ℓ) gravity becomes attenuated with altitude—and thus why spacecraft missions that want to measure the gravity field must have close encounters with their target.

In planetary science, it is common for zonal harmonics to be written as J_ℓ, where $J_\ell \equiv -C_{\ell,0}$. For example, $J_2 = -C_{2,0}$.

4.2.3 *Inertia*

Gravity can be related to the inertia tensor. Using the orthogonality conditions of the spherical harmonics, it is possible to express the inertia tensor, $\boldsymbol{I}$, as a function of solely the degree-two spherical harmonic coefficients of the gravitational potential (Lambeck 1980):

$$\boldsymbol{I} = \begin{bmatrix} I_{xx} & I_{xy} & I_{xz} \\ I_{yx} & I_{yy} & I_{yz} \\ I_{zx} & I_{zy} & I_{zz} \end{bmatrix} = I_0 \begin{bmatrix} 1 & 0 & 0 \\ 0 & 1 & 0 \\ 0 & 0 & 1 \end{bmatrix} + MR^2 \begin{bmatrix} \frac{1}{3}C_{2,0} - 2C_{2,2} & -2S_{2,2} & -C_{2,1} \\ -2S_{2,2} & \frac{1}{3}C_{2,0} + 2C_{2,2} & -S_{2,1} \\ -C_{2,1} & -S_{2,1} & -\frac{2}{3}C_{2,0} \end{bmatrix}, \tag{4.9}$$

where I_0 is the spherically symmetric contribution to the inertia tensor. Since I_0 is spherically symmetric it does not affect the orientation of the principal axes, the ordering between the principal axes, or the relative differences between the principal moments.

The inertia tensor can be diagonalized and therefore simplified in the set of principal axes:

$$\boldsymbol{I}^{\backslash} = \begin{bmatrix} A & 0 & 0 \\ 0 & B & 0 \\ 0 & 0 & C \end{bmatrix}, \tag{4.10}$$

where the backslash (\) indicates that we are in the reference frame aligned with the principal axes, and $C > B > A$ are the principal moments of inertia. These can be related to two degree-2 spherical harmonic coefficients of the gravitational potential:

$$J_2^{\backslash} = -C_{2,0}^{\backslash} = \frac{C - (A + B)/2}{MR^2}, \tag{4.11}$$

$$C_{2,2}^{\backslash} = \frac{B - A}{4MR^2}. \tag{4.12}$$

These relationships are commonly used to relate the spherical harmonic gravity coefficients (which can be measured by tracking accelerations of a spacecraft in proximity to the body of interest), to the *differences* between principal moments of inertia. It is important to note that it is not possible to directly relate gravity coefficients to individual moments of inertia without additional assumptions. This is troublesome, since usually we are most interested in the moments themselves, as they are more uniquely related to the radial density structure (and, for example, the size/state of a core).

4.2.4 Rotational Deformation

Consider a quasi-spherical world rotating at a constant angular rate, Ω (where $\Omega = 2\pi/P$, and P is the object's spin period). In a reference frame where this rotating world seems to be at rest, it is customary to invoke a fictitious centrifugal acceleration field $\vec{a}_C$, directed perpendicularly away from the spin axis, to cancel out the combination of gravitational forces, pressures, and tractions that allows the body to rotate steadily like a rigid body. The magnitude of this acceleration is given by:

$$a_C = \Omega^2 r_\perp, \tag{4.13}$$

where $r_\perp$ is the perpendicular distance between the rotation axis and the point of interest. Note that the centrifugal acceleration increases as you move away from the rotation axis; thus, along the surface of a quasi-spherical world, it is zero at the north and south poles where $r_\perp = 0$ ($\theta = 0°$ or $\theta = 180°$), and greatest at the equator ($\theta = 90°$). The centrifugal force produced by this latitudinally-varying acceleration field acts like an additional external force applied on the non-rotating body and drives the formation of a rotational bulge (hereafter referred to as 'rotational deformation').

To relate centrifugal acceleration (Eq. 4.13) to planetary figures, it is useful to transform these relationships into spherical harmonics (e.g., Kaula 1966, 1968; Murray and Dermott 2000). The centrifugal acceleration can be rewritten as the gradient of a centrifugal potential (i.e., $\vec{a}_C = \nabla\Phi^{\mathcal{R}}$), which, in spherical coordinates, writes:

$$\Phi^{\mathcal{R}}(r,\theta) = -\frac{1}{2}\Omega^2 r^2 \sin^2\theta\,, \tag{4.14}$$

where r is radial distance and θ is co-latitude. Note that $\Phi^{\mathcal{R}}(r,\theta)$ does not depend on longitude, ϕ, because the rotational potential is symmetric about the spin axis. Equation 4.14 can be rewritten in terms of associated Legendre functions by combining it with Eq. 4.4 and the trigonometric identity $\sin^2\theta + \cos^2\theta = 1$

$$\Phi^{\mathcal{R}}(r,\theta) = \frac{1}{3}\Omega^2 r^2 - \frac{1}{3}\Omega^2 r^2 \mathrm{P}_{2,0}(\cos\theta)\,. \tag{4.15}$$

Finally, we can rewrite the rotational potential Eq. 4.15 in a way similar to Eq. 4.8 in terms of two spherical harmonic coefficients:

$$\Phi^{\mathcal{R}}(r,\theta) = \frac{GM}{R}\left(\frac{r}{R}\right)^2 \left(C^{\mathcal{R}}_{0,0} + C^{\mathcal{R}}_{2,0}\mathrm{P}_{2,0}(\cos\theta)\right), \tag{4.16}$$

with coefficients:

$$-C^{\mathcal{R}}_{0,0} = C^{\mathcal{R}}_{2,0} = -\frac{\Omega^2 R^3}{3GM}\,. \tag{4.17}$$

Note that the superscript in these equations, $\mathcal{R}$, is not an exponent, but rather a notation indicating that this coefficient relates to rotation. We only write this as a superscript to clearly separate it from the subscripts that indicate degree and order. Despite all of the subscripts and superscripts in Eq. 4.17 (and similar equations throughout), $C^{\mathcal{R}}_{2,0}$ is a (dimensionless) scalar (i.e., a single number). We use this style for all subsequent equations related to rotation ($\mathcal{R}$), rotational deformation ($\mathcal{RD}$), tides ($\mathcal{T}$), and tidal deformation ($\mathcal{TD}$). The first term in Eq. 4.15, corresponding to the coefficient $C^{\mathcal{R}}_{0,0}$ in Eq. 4.16, is spherically symmetric (i.e., has no latitude or longitude dependence), and does not contribute to the aspherical shape or tidal heating; therefore, it will not be discussed further hereafter.

If our hypothetical quasi-spherical, rotating world was infinitely rigid, Eqs. 4.15–4.17 would completely describe the potential across the surface. However, planetary bodies are never completely rigid; they have finite strength, and tend to deform in response to applied potentials. For a rotating world, this drives the formation of an equatorial, rotational bulge. This deformation contributes to the total potential because the equatorial bulge has mass that perturbs the gravity field. The most common way to account for this contribution is to assume that the body deforms linearly with the applied potential. With this approach, the gravitational potential arising from the rotational deformation of the planet, $\Phi^{\mathcal{RD}}$, is:

$$\Phi^{\mathcal{RD}}(r,\theta) = \frac{GM}{r}\left(\frac{R}{r}\right)^2 \left(C_{0,0}^{\mathcal{RD}} + C_{2,0}^{\mathcal{RD}} \mathrm{P}_{2,0}(\cos\theta)\right), \tag{4.18}$$

where

$$C_{2,0}^{\mathcal{RD}} = k_2^{\infty} C_{2,0}^{\mathcal{R}} = -k_2^{\infty} \frac{\Omega^2 R^3}{3GM}, \tag{4.19}$$

and similarly in terms of shape spherical harmonic coefficients, $c^{\mathcal{RD}}$:

$$c_{2,0}^{\mathcal{RD}} = h_2^{\infty} C_{2,0}^{\mathcal{R}} = -h_2^{\infty} \frac{\Omega^2 R^3}{3GM}, \tag{4.20}$$

where k_2^{∞} and h_2^{∞} are the so-called *long-term* "Love numbers" governing the linear relationship between the applied potential and the resulting *long-term* deformation. k_2^{∞} is the degree-2 potential Love number, and h_2^{∞} is the degree-2 radial displacement Love number.

For a homogeneous, elastic body, k_2^{∞} can be derived analytically (Love 1944):

$$k_2^{\infty} = \frac{3}{2}\left(1 + \frac{19\mu}{2\rho g R}\right)^{-1}, \tag{4.21}$$

where μ is the shear modulus of the body, g is the surface gravity, and R is the body radius. The long-term Love numbers are maximum, $k_2^{\infty} = 3/2$ and $h_2^{\infty} = 5/2$, for a homogeneous body without long-term elasticity (i.e. behaving like a fluid with no shear strength on long timescales). Note that the equality $1 + k_2^{\infty} = h_2^{\infty}$ still holds for a non-homogeneous body without long-term elasticity. Lower values, $k_2^{\infty} < 3/2$ and $h_2^{\infty} < 5/2$, indicate either that the body possesses some shear strength, and/or is not entirely homogeneous, i.e. is differentiated (e.g. worlds with cores).

Rotational deformation produces flattened worlds shaped like oblate spheroids, defined by two unequal radii: $a > c$, where a is the equatorial radius and c is the polar radius. Substituting Eq. 4.20 into Eq. 4.1, we can solve for these two axes:

$$r(\theta, \phi) = R\left[1 + c_{2,0}\left(\frac{3\cos^2\theta - 1}{2}\right)\right], \tag{4.22}$$

$$a = r(90°, 0°) = R\left[1 - \frac{1}{2}c_{2,0}\right], \tag{4.23}$$

$$c = r(0°, 0°) = R\left[1 + c_{2,0}\right]. \tag{4.24}$$

To simplify this further, we often quantify the extent of polar flattening by defining the oblateness or flattening of the object as:

$$f = \frac{a - c}{a}. \tag{4.25}$$

Combining Eq. 4.20–4.25 reveals:

$$f = \frac{3h_2^{\infty}q}{6 + h_2^{\infty}q} \simeq \frac{1}{2}h_2^{\infty}q, \tag{4.26}$$

where we define the dimensionless parameter, q:

$$q = \frac{\Omega^2 R^3}{GM}, \tag{4.27}$$

which corresponds roughly to the ratio of the centrifugal acceleration to the gravitational acceleration at the equator of the body.

4.2.5 Tidal Deformation

In addition to rotational deformation (Sect. 4.2.4), many planetary bodies experience tidal deformation. Tides are the result of differential gravitational accelerations across the surface of a body due to the presence of a nearby companion—be it a planet orbiting a star (e.g., tidal deformation of Mercury due to the Sun), or a satellite orbiting a planet (e.g., tidal deformation of Io due to Jupiter). In this section, we focus on tidal deformation of planetary satellites like Io, which are synchronously rotating and have small orbital eccentricity.

It is useful to decompose the total tidal gravitational potential acting on Io ($\Phi^{\mathcal{T}}_{\rm tot}$) into static ($\Phi^{\mathcal{T}}$) and time-varying tides ($\Phi^{\mathcal{T}}_e$) because the shape of Io is determined by the former and tidal heating is driven by the latter. In both cases, the potential (and associated deformation) can be expanded in spherical harmonics in a similar fashion as the rotational potential (Sect. 4.2.4).

4.2.5.1 Static Tides

First, let us consider static tides. Static tides are the time-averaged tidal perturbation arising from orbiting (or being orbited by) a planetary body.

Consider a quasi-spherical world with a tide-raising object located at a distance along the +z-axis. This axis connecting the center of the primary body and the tide-raising body is referred to as the "tidal axis." Note that this is not the standard set of axes for most planetary bodies; usually, the spin axis is nearly orthogonal to the tidal axis, and the rotational axis defines the +z-axis (as in Sect. 4.2.4), and the tidal axis defines the +x-axis. For now, we use this atypical coordinate system because it simplifies the derivation of tidal deformation, and we will shortly convert back to the standard coordinate system. In this set of axes, the tidal potential is given by (e.g. Murray and Dermott 2000):

$$\Phi^{\mathcal{T}'}(r,\theta) = \frac{GM_{\mathcal{T}}r^2}{a^3}\mathrm{P}_{2,0}(\cos\theta)\,, \tag{4.28}$$

where $M_{\mathcal{T}}$ is the mass of the tide-raising body, and a is the semimajor axis of the tide-raising body's orbit. The prime ($'$) indicates that we are in our atypical set of axes, with the tidal axis directed along the +z-axis. We can rewrite the tidal potential Eq. 4.28 in a way similar to Eq. 4.8 in terms of a spherical harmonic coefficient for when the tidal axis is aligned with the +z axis:

$$\Phi^{\mathcal{T}'}(r,\theta) = \frac{GM}{R}\left(\frac{r}{R}\right)^2 C^{\mathcal{T}'}_{2,0}\mathrm{P}_{2,0}(\cos\theta)\,, \tag{4.29}$$

with coefficient $C^{\mathcal{T}'}_{2,0}$:

$$C^{\mathcal{T}'}_{2,0} = \frac{R}{GM}\frac{GM_{\mathcal{T}}R^2}{a^3}\,. \tag{4.30}$$

As in Sect. 4.2.4, we assume that the object deforms in response to this forcing potential, and use long-term Love numbers to write out the gravitational potential and shape of the tidally deformed world,

$$\Phi^{\mathcal{TD}'}(r,\theta) = \frac{GM}{r}\left(\frac{R}{r}\right)^2 C^{\mathcal{TD}'}_{2,0}\mathrm{P}_{2,0}(\cos\theta)\,, \tag{4.31}$$

where

$$C^{\mathcal{TD}'}_{2,0} = k_2^{\infty}C^{\mathcal{T}'}_{2,0}\,, \tag{4.32}$$

$$c^{\mathcal{TD}'}_{2,0} = h_2^{\infty}C^{\mathcal{T}'}_{2,0}\,, \tag{4.33}$$

where k_2^∞ is the degree-2 potential Love number, and h_2^∞ is the degree-2 radial displacement Love number. For the case of a synchronously rotating moon like Io (where the moon's spin period is equal to its orbital period), these relationships can be further simplified with Kepler's third law: $\Omega^2 a^3 = G(M + M_\mathcal{T})$:

$$C_{2,0}^{\mathcal{TD}'} = k_2^\infty \left(\frac{M_\mathcal{T}}{M_\mathcal{T} + M}\right)\left(\frac{\Omega^2 R^3}{GM}\right), \tag{4.34}$$

$$c_{2,0}^{\mathcal{TD}'} = h_2^\infty \left(\frac{M_\mathcal{T}}{M_\mathcal{T} + M}\right)\left(\frac{\Omega^2 R^3}{GM}\right). \tag{4.35}$$

In Eq. 4.28–4.35 we used an atypical coordinate system, with the tidal axis aligned with the +z direction. To convert to the standard set of axes (with the tidal axis aligned with the +x direction), we make use of the spherical harmonic addition theorem (e.g. Arfken and Weber 1995), which allows us to rotate spherical harmonics:

$$\begin{bmatrix} C_{\ell,m}^{\mathcal{TD}} \\ S_{\ell,m}^{\mathcal{TD}} \end{bmatrix} = C_{2,0}^{\mathcal{TD}'} (2 - \delta_{m,0}) \frac{(l-m)!}{(l+m)!} \mathrm{P}_{\ell,m}(\cos\theta) \begin{bmatrix} \cos(m\phi) \\ \sin(m\phi) \end{bmatrix}, \tag{4.36}$$

where θ and ϕ are the co-latitude and longitude we wish the tidal axis to be aligned with. While Eq. 4.36 is framed around rotating $C_{2,0}^{\mathcal{TD}'}$, this theorem is incredibly general and can be used to rotate any set of spherical harmonic coefficients. Using the standard set of axes, with the tidal axis aligned with the +x-axis ($\theta = 90°$, $\phi = 0°$), and performing the transform, yields two degree-two spherical harmonic coefficients for the gravitational potential from tidal deformation:

$$C_{2,0}^{\mathcal{TD}} = -\frac{1}{2} C_{2,0}^{\mathcal{TD}'} = -\frac{1}{2} k_2^\infty \left(\frac{M_\mathcal{T}}{M_\mathcal{T} + M}\right)\left(\frac{\Omega^2 R^3}{GM}\right), \tag{4.37}$$

$$C_{2,2}^{\mathcal{TD}} = \frac{1}{4} C_{2,0}^{\mathcal{TD}'} = \frac{1}{4} k_2^\infty \left(\frac{M_\mathcal{T}}{M_\mathcal{T} + M}\right)\left(\frac{\Omega^2 R^3}{GM}\right). \tag{4.38}$$

And similarly for the shape arising from the tidal deformation (Eq. 4.35):

$$c_{2,0}^{\mathcal{TD}} = -\frac{1}{2} c_{2,0}^{\mathcal{TD}'} = -\frac{1}{2} h_2^\infty \left(\frac{M_\mathcal{T}}{M_\mathcal{T} + M}\right)\left(\frac{\Omega^2 R^3}{GM}\right), \tag{4.39}$$

$$c_{2,2}^{\mathcal{TD}} = \frac{1}{4} c_{2,0}^{\mathcal{TD}'} = \frac{1}{4} h_2^\infty \left(\frac{M_\mathcal{T}}{M_\mathcal{T} + M}\right)\left(\frac{\Omega^2 R^3}{GM}\right). \tag{4.40}$$

Combining our relationships for rotational deformation (Eq. 4.19–4.20) and tidal deformation (Eq. 4.37–4.40) we finally arrive at the spherical harmonic expression for the gravity field of a tidally-deformed, synchronously rotating world:

$$C_{2,0} = C_{2,0}^{\mathcal{TD}} + C_{2,0}^{\mathcal{RD}} = -\frac{1}{3}k_2^{\infty}\left(\frac{\Omega^2 R^3}{GM}\right) - \frac{1}{2}k_2^{\infty}\left(\frac{M_{\mathcal{T}}}{M_{\mathcal{T}}+M}\right)\left(\frac{\Omega^2 R^3}{GM}\right), \tag{4.41}$$

$$C_{2,2} = C_{2,2}^{\mathcal{TD}} = \frac{1}{4}k_2^{\infty}\left(\frac{M_{\mathcal{T}}}{M_{\mathcal{T}}+M}\right)\left(\frac{\Omega^2 R^3}{GM}\right). \tag{4.42}$$

We can perform the similar operations to determine the spherical harmonic expression for the shape of similarly for the shape of a tidally-deformed, synchronously rotating world:

$$c_{2,0} = c_{2,0}^{\mathcal{TD}} + c_{2,0}^{\mathcal{RD}} = -\frac{1}{3}h_2^{\infty}\left(\frac{\Omega^2 R^3}{GM}\right) - \frac{1}{2}h_2^{\infty}\left(\frac{M_{\mathcal{T}}}{M_{\mathcal{T}}+M}\right)\left(\frac{\Omega^2 R^3}{GM}\right), \tag{4.43}$$

$$c_{2,2} = c_{2,2}^{\mathcal{TD}} = \frac{1}{4}h_2^{\infty}\left(\frac{M_{\mathcal{T}}}{M_{\mathcal{T}}+M}\right)\left(\frac{\Omega^2 R^3}{GM}\right). \tag{4.44}$$

The combination of tidal and rotational deformation yields worlds shaped similar to a triaxial ellipsoid, with three unequal axes: $a > b > c$. Substituting Eq. 4.43–4.44 into Eq. 4.1, we can solve for these axes:

$$r(\theta,\phi) = R\left[1 + c_{2,0}\left(\frac{3\cos^2\theta - 1}{2}\right) + c_{2,2}\left(3\sin^2\theta\right)\cos(2\phi)\right], \tag{4.45}$$

$$a = r(90°, 0°) = R\left[1 - \frac{1}{2}c_{2,0} + 3c_{2,2}\right], \tag{4.46}$$

$$b = r(90°, 90°) = R\left[1 - \frac{1}{2}c_{2,0} - 3c_{2,2}\right], \tag{4.47}$$

$$c = r(0°, 0°) = R\left[1 + c_{2,0}\right]. \tag{4.48}$$

These expressions for the gravity field and shape of a tidally-deformed, synchronously rotating world (Eqs. 4.41–4.42 and 4.43–4.44) assume that the body is *hydrostatic* at degree/order-2, meaning that it conforms to the applied tidal and rotational potential. Other geologic processes can effectively "contaminate" this long-wavelength signal, although it is generally assumed that non-hydrostatic components to long-wavelength gravity and topography are small. In some cases, this contamination can be quantified, like for the Moon (Keane and Matsuyama 2014) or Enceladus (Hemingway et al. 2018), although these methods usually require knowledge of short-wavelength gravity or shape—which are currently absent for Io.

Neglecting non-hydrostatic effects, it can be informative to look at the ratios between degree-2 coefficients of the gravity field and shape. For example, taking the ratio of Eqs. 4.41 and 4.42, and assuming that the tide-raising body (Jupiter) has a mass much larger than the satellite (Io), it can be shown that $C_{2,0}/C_{2,2} = -10/3$. This ratio is, in principal, independent of the interior radial structure (which factors into the Love numbers). Because of this, some gravity solutions force $C_{2,0}/C_{2,2}$ to be equal to the "hydrostatic" value of $-10/3$ when data is sparse. We will return to this assumption in Sect. 4.4.3.

4.2.5.2 Time-Dependent Tides

The Laplace resonance between Io, Europa, and Ganymede produces periodic gravitational pulls between the satellites, generating non-circular orbits. This results in periodic variations in the Io–Jupiter distance $\mathcal{R}$ and the direction of Jupiter as seen in a reference frame fixed to Io and rotating at the synchronous rotation rate. Therefore, in addition to the permanent tidal bulge discussed above, Io experiences time-varying tides driven by Jupiter's gravitational forcing. Io's deformation associated with these time-dependent tides generates frictional energy that heats the interior.

The total tidal potential at a point with spherical coordinates $(r,\ \theta,\ \phi)$, including static and time-dependent components, can be written:

$$\Phi^{\mathcal{T}}_{\text{tot}}(r,\theta,\phi) = \frac{GM_{\mathcal{T}}}{\mathcal{R}} \sum_{\ell=2}^{\infty} \left(\frac{r}{\mathcal{R}}\right)^{\ell} \sum_{m=0}^{\ell} (2-\delta_{m0}) \frac{(\ell-m)!}{(\ell+m)!}$$
$$\times\, \mathrm{P}_{\ell,m}(\cos\theta)\mathrm{P}_{\ell,m}(\cos\theta_{\mathcal{T}})\cos(m(\phi-\phi_{\mathcal{T}}))\,, \qquad (4.49)$$

where we use the spherical harmonics addition theorem, and $(\theta_{\mathcal{T}},\ \phi_{\mathcal{T}})$ are the spherical coordinates of the planet. In a reference frame fixed to the satellite and rotating at the synchronous rotation rate with the z-axis aligned with the rotation axis, $\theta_{\mathcal{T}} = 90°$ and $\phi_{\mathcal{T}} = f - \Omega t = f - \Omega t$, where f is the true anomaly, t is time with $t = 0$ corresponding to the time of pericenter passage, Ω is the mean motion, and we assume zero obliquity and zero orbit inclination. To first order in orbital eccentricity e, $\mathcal{R} = a[1 - e\cos(\Omega t)]$, and $\phi_{\mathcal{T}} = 2e\sin(\Omega t)$ (Murray and Dermott 2000), and the dominant $\ell = 2$ term of the time-dependent component of the tidal potential can be written:

$$\Phi^{\mathcal{T}}_{e}(r,\theta,\phi) = \frac{1}{4}\frac{GM_{\mathcal{T}}}{a}\left(\frac{r}{a}\right)^2 e\Big[-6\mathrm{P}_{2,0}(\cos\theta)\cos(\Omega\, t)$$
$$+\,\mathrm{P}_{2,2}(\cos\theta)\big(3\cos(2\phi)\cos(\Omega\, t) + 4\sin(2\phi)\sin(\Omega\, t)\big)\Big]. \qquad (4.50)$$

We can rewrite the tidal potential Eq. 4.50 in a way similar to Eq. 4.8 in terms of three spherical harmonic coefficients

$$\Phi_e^{\mathcal{T}}(r,\theta,\phi) = \frac{GM}{R}\left(\frac{r}{R}\right)^2 \Big[C_e^{\mathcal{T}}{}_{2,0}\mathrm{P}_{2,0}(\cos\theta) + \mathrm{P}_{2,2}(\cos\theta)(C_e^{\mathcal{T}}{}_{2,2}\cos(2\phi) + S_e^{\mathcal{T}}{}_{2,2}\sin(2\phi)) \Big]. \tag{4.51}$$

The (complex) Fourier expansion reads

$$C_e^{\mathcal{T}}{}_{2,0} = C_e^{\mathcal{T}*}{}_{2,0}(-\Omega)\mathrm{e}^{-\mathrm{i}\Omega t} + C_e^{\mathcal{T}*}{}_{2,0}(\Omega)\mathrm{e}^{\mathrm{i}\Omega t} = -\frac{3M_{\mathcal{T}}}{2M}e\left(\frac{R}{a}\right)^3 \cos(\Omega t) \tag{4.52}$$

$$C_e^{\mathcal{T}}{}_{2,2} = C_e^{\mathcal{T}*}{}_{2,2}(-\Omega)\mathrm{e}^{-\mathrm{i}\Omega t} + C_e^{\mathcal{T}*}{}_{2,2}(\Omega)\mathrm{e}^{\mathrm{i}\Omega t} = \frac{3M_{\mathcal{T}}}{4M}e\left(\frac{R}{a}\right)^3 \cos(\Omega t) \tag{4.53}$$

$$S_e^{\mathcal{T}}{}_{2,2} = S_e^{\mathcal{T}*}{}_{2,2}(-\Omega)\mathrm{e}^{-\mathrm{i}\Omega t} + S_e^{\mathcal{T}*}{}_{2,2}(\Omega)\mathrm{e}^{\mathrm{i}\Omega t} = \frac{M_{\mathcal{T}}}{M}e\left(\frac{R}{a}\right)^3 \sin(\Omega t) \tag{4.54}$$

where

$$C_e^{\mathcal{T}*}{}_{2,0}(\Omega) = -\frac{3M_{\mathcal{T}}}{4M}e\left(\frac{R}{a}\right)^3 \tag{4.55}$$

$$C_e^{\mathcal{T}*}{}_{2,2}(\Omega) = \frac{3M_{\mathcal{T}}}{8M}e\left(\frac{R}{a}\right)^3 \tag{4.56}$$

$$S_e^{\mathcal{T}*}{}_{2,2}(\Omega) = -\frac{\mathrm{i}M_{\mathcal{T}}}{2M}e\left(\frac{R}{a}\right)^3 \tag{4.57}$$

and $C_e^{\mathcal{T}*}{}_{2,0}(-\Omega)$, $C_e^{\mathcal{T}*}{}_{2,2}(-\Omega)$, $S_e^{\mathcal{T}*}{}_{2,2}(-\Omega)$ are their respective complex conjugates. Higher order expansions can be obtained by converting the spherical coordinates of the planet to Keplerian elements (Kaula 1964; Chyba et al. 1989).

The time-dependent gravitational potential arising from the satellite deformation in response to the time-dependent forcing potential is given by

$$\Phi_e^{\mathcal{T}\mathcal{D}}(r,\theta,\phi) = \frac{GM}{r}\left(\frac{R}{r}\right)^2 \Big[C_e^{\mathcal{T}\mathcal{D}}{}_{2,0}\mathrm{P}_{2,0}(\cos\theta) + \mathrm{P}_{2,2}(\cos\theta)(C_e^{\mathcal{T}\mathcal{D}}{}_{2,2}\cos(2\phi) + S_e^{\mathcal{T}\mathcal{D}}{}_{2,2}\sin(2\phi)) \Big]. \tag{4.59}$$

$$\begin{aligned} C_e^{\mathcal{TD}}{}_{2,0} &= C_e^{\mathcal{TD}*}{}_{2,0}(-\Omega)\mathrm{e}^{-\mathrm{i}\Omega t} + C_e^{\mathcal{TD}*}{}_{2,0}(\Omega)\mathrm{e}^{\mathrm{i}\Omega t} \\ &= -\frac{3M_{\mathcal{T}}}{2M} e k_2 \left(\frac{R}{a}\right)^3 \cos(\Omega t - \delta) \end{aligned} \tag{4.60}$$

$$\begin{aligned} C_e^{\mathcal{TD}}{}_{2,2} &= C_e^{\mathcal{TD}*}{}_{2,2}(-\Omega)\mathrm{e}^{-\mathrm{i}\Omega t} + C_e^{\mathcal{TD}*}{}_{2,2}(\Omega)\mathrm{e}^{\mathrm{i}\Omega t} \\ &= \frac{3M_{\mathcal{T}}}{4M} e k_2 \left(\frac{R}{a}\right)^3 \cos(\Omega t - \delta) \end{aligned} \tag{4.61}$$

$$\begin{aligned} S_e^{\mathcal{TD}}{}_{2,2} &= S_e^{\mathcal{TD}*}{}_{2,2}(-\Omega)\mathrm{e}^{-\mathrm{i}\Omega t} + S_e^{\mathcal{TD}*}{}_{2,2}(\Omega)\mathrm{e}^{\mathrm{i}\Omega t} \\ &= \frac{M_{\mathcal{T}}}{M} e k_2 \left(\frac{R}{a}\right)^3 \sin(\Omega t - \delta) \end{aligned} \tag{4.62}$$

$$C_e^{\mathcal{TD}*}{}_{2,0}(\pm\Omega) = k_2^*(\pm\Omega) C_e^{\mathcal{T}*}{}_{2,0}(\pm\Omega) \tag{4.63}$$

$$C_e^{\mathcal{TD}*}{}_{2,2}(\pm\Omega) = k_2^*(\pm\Omega) C_e^{\mathcal{T}*}{}_{2,2}(\pm\Omega) \tag{4.64}$$

$$S_e^{\mathcal{TD}*}{}_{2,2}(\pm\Omega) = k_2^*(\pm\Omega) S_e^{\mathcal{T}*}{}_{2,2}(\pm\Omega) \tag{4.65}$$

where $k_2^*(\Omega) = k_2\mathrm{e}^{-\mathrm{i}\delta}$ is the complex potential Love number at the tidal forcing frequency (equal to the rotation rate Ω assuming synchronous rotation) and $k_2^*(-\Omega) = k_2\mathrm{e}^{\mathrm{i}\delta}$ is its complex conjugate (note that the star (*) does not denote complex conjugation). This Love number differs from the long-term potential Love number k_2^∞ used to describe static tides above in two important ways. First, k_2^∞ describes the permanent deformation in response to the constant part of the rotational and tidal forcing (Eqs. 4.13 and 4.28) while $k_2^*(\Omega)$ describes the periodic deformation of the satellite in response to the time-depending tidal forcing (Eq. 4.50). Second, k_2^∞ is real while $k_2^*(\Omega)$ is complex to characterize the lag between the satellite deformation and the forcing tidal potential. This lag arises because of the anelastic deformation of the satellite (Sect. 4.3). The complex number $k_2^*(\Omega)$ can be described by its module

$$k_2 = |k_2^*(\Omega)| = \sqrt{\mathrm{Re}(k_2^*(\Omega))^2 + \mathrm{Im}(k_2^*(\Omega)^2} \tag{4.66}$$

and its argument

$$-\delta = \arg k_2^*(\Omega) \tag{4.67}$$

which implies

$$\tan\delta = -\frac{\mathrm{Im}(k_2^*(\Omega))}{\mathrm{Re}(k_2^*(\Omega))}. \tag{4.68}$$

4.2.6 *Libration*

Another consequence of Io's triaxial figure and orbital eccentricity is a small periodic fluctuation of its sidereal rotation rate, known as the diurnal libration. It is the result of the variable gravitational torque exerted by Jupiter as Io moves along its orbit. To the first order in the orbital eccentricity, Io's equatorial bulge remains aligned with the empty focus of its elliptical orbit, rather than with the focus occupied by Jupiter. The torque resulting from this misalignment is counterclockwise from perijove to apojove, and clockwise from apojove to perijove. Consequently, Io's equatorial bulge departs from steady rotation, lagging slightly behind from perijove to apojove, and leading slightly ahead from apojove to perijove. The amplitude of the diurnal libration is typically inversely proportional to the polar moment of inertia. However, in the presence of a global subsurface magma ocean, the solid lithosphere and the solid interior (here meant as the combined core and mantle) become somewhat decoupled from each other, and the libration amplitude of the outermost layer can be resonantly amplified. Elastic deformations tend to consolidate the lithosphere-interior coupling, so that the resonant amplification is not as strong as it would be were the lithosphere completely rigid.

Physically, the lithosphere-interior decoupling allowed by an intermediate layer of inviscid fluid endows the system with a free mode where the lithosphere and the interior librate out of phase, in addition to the more usual free mode, also possessed by oceanless bodies, where the lithosphere and the interior librate in phase. The periodically-forced response of a physical system can be depicted as a superposition of its free modes oscillating at the frequency of the forcing. Resonant amplification of the forced response occurs when one of the eigenfrequency approaches the forcing frequency. For Io, resonant amplification of the forced diurnal libration is possible because the frequency of the out-of-phase free libration, which depends on the lithospheric thickness, approaches the orbital frequency (Van Hoolst et al. 2020).

Measurements of the diurnal libration amplitude, as deduced from a reconstruction of the time-dependent orientation of a control-point network against the background of distant stars, could settle the presence or absence of a global subsurface magma ocean in Io's interior, and, if Io is indeed an ocean world, provide additional constraints on the thickness and rigidity of the overlying lithosphere.

Mathematically, the librations forced by Jupiter (indexed 'Jup') can be calculated from a linearized balance of vertical angular momentum in an inertial frame,

$$\frac{\mathrm{d}}{\mathrm{d}t}\left[C_0\frac{\mathrm{d}\gamma}{\mathrm{d}t}+\delta C\Omega_0\right]=\Gamma_{/\mathrm{Jup}} \qquad \text{for an oceanless Io} \qquad (4.69)$$

or, for an Io consisting of a solid lithosphere (indexed 'lit'), a fluid magma ocean (indexed 'mag'), and a solid interior (indexed 'int'),

$$\frac{\mathrm{d}}{\mathrm{d}t}\left[C_{\mathrm{lit}0}\frac{\mathrm{d}\gamma_{\mathrm{lit}}}{\mathrm{d}t}+\delta C_{\mathrm{lit}}\Omega_0\right]=\Gamma_{\mathrm{lit/Jup}}+\Gamma_{\mathrm{lit/mag}}$$
$$+\Pi_{\mathrm{lit/mag}}+\Gamma_{\mathrm{lit/int}} \qquad \text{for the lithosphere} \qquad (4.70)$$

$$\frac{\mathrm{d}}{\mathrm{d}t}\left[C_{\mathrm{int}0}\frac{\mathrm{d}\gamma_{\mathrm{int}}}{\mathrm{d}t} + \delta C_{\mathrm{int}}\Omega_0\right] = \Gamma_{\mathrm{int/Jup}} + \Gamma_{\mathrm{int/mag}} + \Pi_{\mathrm{int/mag}} + \Gamma_{\mathrm{int/lit}} \quad \text{for the interior} \tag{4.71}$$

where the bracketed quantities are incremental vertical angular momenta relative to the steadily rotating frame, C_{i0} and δC_i are the equilibrium and incremental polar principal moment of inertia of i, γ_i is the libration angle (measured from the steadily rotating frame) of i, $\Gamma_{i/j}$ is the gravitational torque exerted on i by j, and $\Pi_{i/j}$ is the pressure torque exerted on i along the boundary shared with j. The quantities $\delta C_i, \Gamma_{i/j}, \Pi_{i/j}$ can be expressed in terms of γ_k to the first order in the perturbations; $\Gamma_{i/\mathrm{Jup}}$ additionally contains a forcing term

$$\iiint_{V_{i0}} \rho(\boldsymbol{r}')\left(x'\frac{\partial \Phi_e^{\mathrm{T}}}{\partial y}(\boldsymbol{r}') - y'\frac{\partial \Phi_e^{\mathrm{T}}}{\partial x}(\boldsymbol{r}')\right) \mathrm{d}V(\boldsymbol{r}') \tag{4.72}$$

independent of γ_k and proportional to $\sin M$, where M is Io's mean anomaly. The system of equations, of the form

$$\frac{\mathrm{d}^2}{\mathrm{d}t^2}\begin{bmatrix}\gamma_{\mathrm{lit}}\\ \gamma_{\mathrm{int}}\end{bmatrix} + \begin{bmatrix}\bullet & \bullet\\ \bullet & \bullet\end{bmatrix}\cdot\begin{bmatrix}\gamma_{\mathrm{lit}}\\ \gamma_{\mathrm{int}}\end{bmatrix} = \begin{bmatrix}\bullet\\ \bullet\end{bmatrix}\sin M \tag{4.73}$$

can then be solved for the observable libration angle γ_{lit} as a function of time.

Figure 4.2 shows how libration (Fig. 4.2a) can be combined with other measurements of Io's tidal deformation (k_2 and h_2, in Fig. 4.2b and c, respectively) to constrain the interior structure of Io. Any precise measurement of one geophysical quantity only places weak bounds on the interior structure of Io. However, when multiple geophysical measurements are combined (e.g., Fig. 4.2d), it is possible to tightly constrain the state of Io's interior structure—including characterizing the properties of the lithosphere and testing for a magma ocean.

4.3 Theory, Part 2: Tidal Heating

Io's deformation in response to the strong, periodic gravitational tides from Jupiter generates internal friction that heats the interior, a process commonly referred to as tidal heating. Peale et al. (1979) predicted that the orbital eccentricity generated by the Laplace resonance would produce strong tidal heating, melting a large fraction of the interior and likely causing widespread surface volcanism. This prediction was confirmed by *Voyager 1* observations of active volcanic plumes and widespread volcanic features (Morabito et al. 1979; Smith et al. 1979).

The deformation in response to Jupiter's forcing contains elastic and anelastic components. The latter is responsible for tidal heating and is determined by the relaxation of the materials in the interior layers, with a larger anelastic response

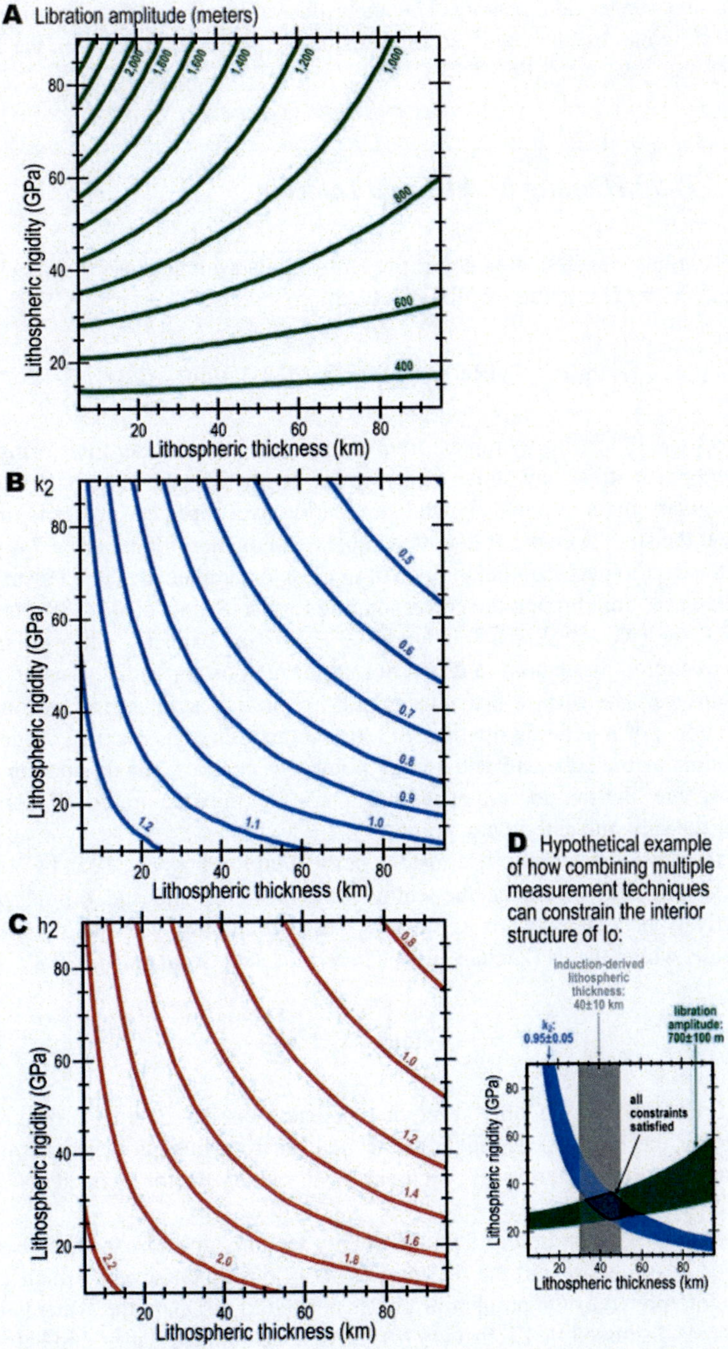

Fig. 4.2 Forward models of Io's predicted libration amplitude (**a**), and tidal Love numbers $k_2^*(\Omega)$ (**b**) and $h_2^*(\Omega)$ (**c**) The combination of multiple geophysical measurements can provide tight constraints on Io's interior structure, as shown in a notional example in (**d**). Induction is described in Sect. 4.4.4

generating stronger tidal heating. The anelastic deformation and the corresponding tidal heating can be characterized in terms of the phase lag δ between the forcing and response potentials (Eq. 4.68).

4.3.1 Tidal Heating in the Solid Interior

In solid regions, the dissipated power per unit volume averaged over the tidal forcing period $T = 2\pi/\Omega$ is given by (Tobie et al. 2005)

$$\dot{E}_V = \frac{1}{2}\Omega\Big[\mathrm{Im}(\sigma_{ij}^*(\Omega))\mathrm{Re}(\epsilon_{ij}^*(\Omega)) - \mathrm{Re}(\sigma_{ij}^*(\Omega))\mathrm{Im}(\epsilon_{ij}^*(\Omega))\Big], \tag{4.74}$$

where σ_{ij}^* and ϵ_{ij}^* are the (complex, frequency-dependent) Fourier transforms of the time-dependent stress and strain tensors respectively, Re and Im denote their real and imaginary parts respectively, and summation over repeated indices is implied. Note that the star (*) does not denote complex conjugation. Equation (4.74) can be integrated over a specific radial interval of interest to compute the lateral distribution of surface heat flux through the corresponding region (Segatz et al. 1988; Hamilton et al. 2013b; Tyler et al. 2015). Figure 4.3 shows the expected surface heat flux pattern assuming dissipation in different interior regions and radial transfer of heat to the surface. The surface heat flux in solid regions is symmetrical in longitude and latitude. For mantle dissipation, the surface heat flux has maxima at the poles and minima at the sub- and anti-Jovian points. In contrast, for dissipation in the asthenosphere, the minima are at the poles, and the maxima are at 30° north and south of the sub- and anti-Jovian points.

The total dissipated power averaged over the tidal forcing period can be obtained by integrating Eq. 4.74 over the entire volume of Io and the orbital period. Alternatively, using energy conservation, it can be found from the rate of work done by the tide, which yields (Zschau 1978; Platzman 1984; Segatz et al. 1988):

$$\dot{E} = -\frac{21}{2}\mathrm{Im}(k_2^*(\Omega))\frac{\Omega^5 R^5}{G}e^2, \tag{4.75}$$

where $\mathrm{Im}(k_2^*(\Omega))$ is the only term that depends on the internal structure and rheology of the satellite. Throughout the tidal heating literature there is another common notation in terms of a global tidal quality factor defined as $Q = -\left|k_2^*(\Omega)\right|/\mathrm{Im}(k_2^*(\Omega))$.

The traditional definition of the tidal quality factor is given by the ratio between the stored peak energy and the dissipated energy during one cycle, which can be roughly interpreted as the number of cycles needed to dissipate the system energy. However, as discussed by Zschau (1978), there is no simple relationship between a tidal quality factor defined this way and the imaginary and real parts of the $k_2^*(\Omega)$ Love number which determine the phase lag of the tidal bulge gravitational potential

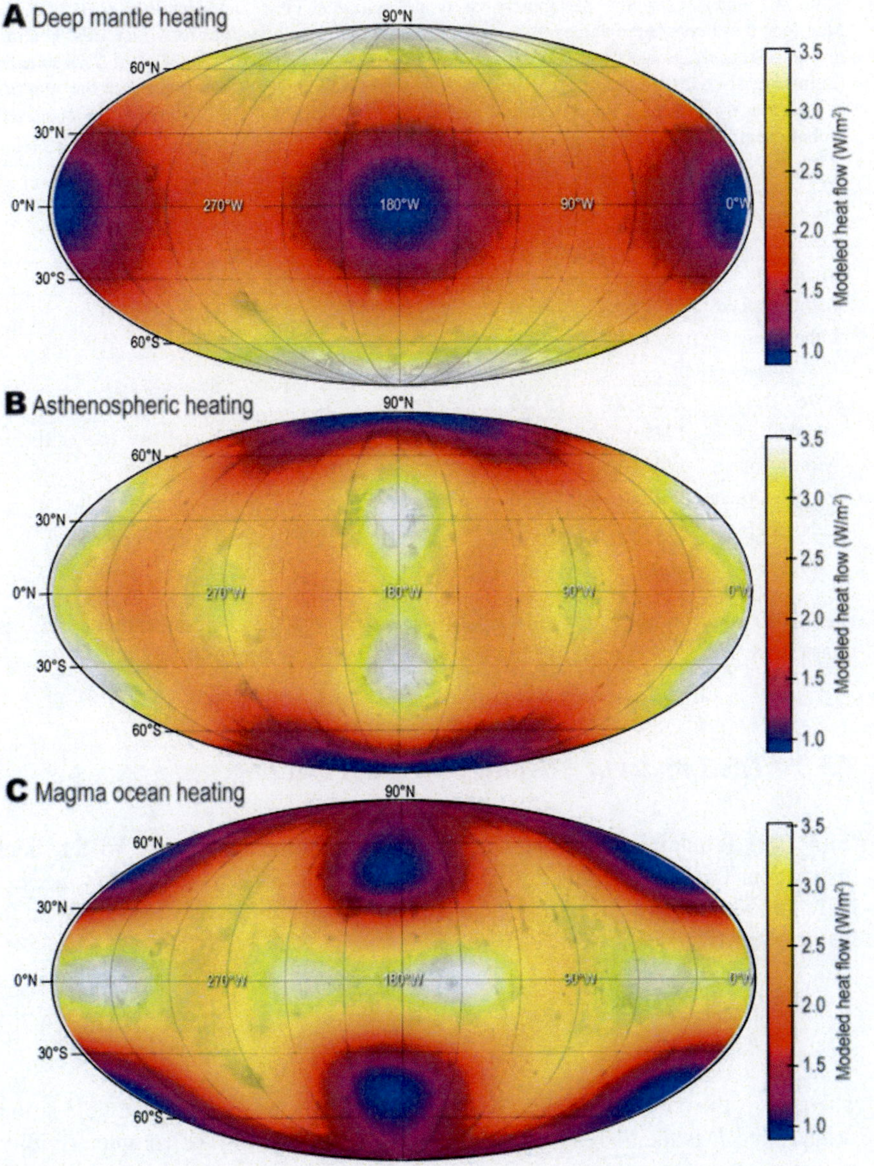

Fig. 4.3 Maps of predicted tidal dissipation within Io as a function of latitude and longitude for three end-member hypotheses: (**a**) tidal heating concentrated entirely in the deep mantle, (**b**) tidal heating concentrated entirely in the (shallow) asthenosphere, and (**c**) tidal heating concentrated in a magma ocean. The first two assume that there is no magma ocean (or that it has no impact on tidal heating). The associated radius, density, and other properties of these models are detailed in Table 4.1. Data adapted from de Kleer et al. (2019b). Maps are in Mollweide projection, centered on 180°W (the anti-Jupiter point), with grid-lines in 30° increments—as described in Fig. 4.1

Table 4.1 Interior structure parameters based on Segatz et al. (1988) and Hamilton et al. (2013b). Mantle and asthenosphere values in parentheses correspond to the values for which tidal heating in one of these regions alone can explain the observed global mean heat flux of 2.24 W m^{-2} (Lainey et al. 2009) assuming that heat is transferred radially to the surface. For the magma ocean case, we assume a linear friction coefficient $\alpha = 2.3 \times 10^{-5}$ s^{-1} to match the observed global mean heat flux

Layer	Radius (km)	Density (kg/m^3)	Shear modulus, μ (Pa)	Viscosity (Pa s)
Without magma ocean				
Core	980	5150	–	–
Mantle	1741.6	3200	6×10^{10} (3.5×10^9)	10^{20} (6.9×10^{14})
Asthenosphere	1791.6	3000	6×10^{10} (5×10^4)	10^{20} (1.3×10^{14})
Lithosphere	1821.6	2700	6.5×10^{10}	10^{23}
With magma ocean				
Core	980	5150	–	–
Mantle	1741.6	3200	6×10^{10} (3.5×10^9)	10^{20} (6.9×10^{14})
Asthenosphere	1766.6	3000	6×10^{10}	10^{20}
Magma Ocean	1791.6	2900	–	–
Lithosphere	1821.6	2700	6.5×10^{10}	10^{23}

(Eq. 4.68) and average energy dissipation rate (Eq. 4.75). For the remainder of this chapter we will continue to use the complex k_2 notation shown in Eq. 4.75.

4.3.2 The Impact of Rheology on Tidal Heating

$\mathrm{Im}(k_2^*(\Omega))$ is an integrated quantity relating the internal structure to the tidal dissipation. To build intuition it is useful to consider the case of a uniform spherical body, for which the complex potential Love number can be analytically related to the rheology by

$$k_2^*(\Omega) = \frac{3/2}{1 + \frac{19}{2\rho g R}\mu^*(\Omega)} \tag{4.76}$$

where μ^* is the complex rigidity, ρ is the density, g is the gravitational field at the surface, and R is the body radius (Ross and Schubert 1986). The complex rigidity μ^* can also be expressed in terms of the complex compliance, j^* (Findley et al. 1976). These quantities are related by

$$\mu^*(\Omega) = 1/j^*(\Omega) \tag{4.77}$$

$$j^*(\Omega) = J_1(\Omega) - i J_2(\Omega) \tag{4.78}$$

where $i^2 = -1$.

Different rheological models relate the material properties to the complex compliance. The simplest rheological model is the Maxwell rheology. The Maxwell model is based on the viscoelastic response of an elastic spring, capturing the instantaneous recoverable strain through a rigidity μ_0, and a dashpot, capturing the irrecoverable time-dependent strain through a viscosity η_0, connected in series. For a Maxwell rheology the complex compliance is given by

$$J_1(\Omega) = 1/\mu_0 \tag{4.79}$$

$$J_2(\Omega) = 1/(\Omega\eta_0) \tag{4.80}$$

Because of its simplicity, and clear relationship between well-known material properties (η_0, μ_0), the Maxwell model has been used extensively to describe Io's tidal dissipation (Moore 2003; Hussmann and Spohn 2004; Steinke et al. 2020a; Spencer et al. 2020). In laboratory studies of cyclically stressed materials the Maxwell model does a very poor job of characterizing the dissipation as you move away from the dissipation peak (Faul and Jackson 2015). Alternative rheological models exist which add more unknown parameters (and therefore complexity) to better replicate the behavior of real materials. Importantly these models capture the anelastic response (recoverable time-dependent strain) of the material. The most prominent of these are the Burgers, Sundberg-Cooper, and Andrade models. Of these the Andrade model (Andrade 1910; Jackson et al. 2004; Jackson and Faul 2010) has been most commonly applied to Io (Bierson and Nimmo 2016; Renaud and Henning 2018) and is based on the complex compliance:

$$J_1(\Omega) = 1/\mu_0 + \beta\Omega^{-n}\Gamma(n+1)\cos\left(\frac{n\pi}{2}\right) \tag{4.81}$$

$$J_2(\Omega) = \beta\Omega^{-n}\Gamma(n+1)\sin\left(\frac{n\pi}{2}\right) + \frac{1}{\Omega\eta_0} \tag{4.82}$$

Here R_g is the ideal gas constant, Γ is the gamma function (Findley et al. 1976). The parameters β and n describe the magnitude and time decay of anelastic deformation. Note that in the limit where $\beta \to 0$ the Andrade model reverts to the Maxwell approximation.

A comparison of these different rheological models applied to Io is shown in Fig. 4.4. Using a Maxwell rheology the high tidal dissipation of Io can only be explained if the mantle has a particularly low viscosity ($\sim 10^{15}$ Pa s). This is significantly lower than estimates for the viscosity of Earth's mantle which generally range between 10^{18} Pa s and 10^{22} Pa s (Mitrovica and Forte 2004; Bürgmann and Dresen 2008; Lambeck et al. 2017). Even if Io's upper mantle has a high melt fraction, such a low viscosity is unlikely. As melt fraction increases in a system it abruptly transitions between the viscosity of the solid matrix ($> 10^{15}$ Pa s) to the viscosity of the liquid ($\ll 10^{15}$ Pa s) at the disaggregation limit (around 30% melt fraction) (Costa et al. 2009). The difference between the Maxwell rheology and other rheological models grows the farther you get from the Maxwell dissipation

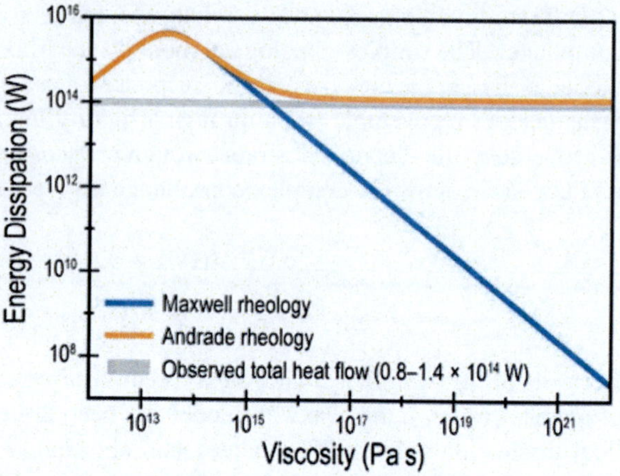

Fig. 4.4 Andrade rheology predicts larger dissipation at higher viscosity due to anelastic dissipation, not captured by the Maxwell model. Dissipation is calculated for a homogeneous body (Eq. 4.76). This example uses $\mu_0 = 60\,\text{GPa}$, $\beta = 10^{-12}\ \text{Pa}^{-1}\ \text{s}^{-n}$, and $n = 1/3$. Observed total heat flux spans $0.8 \times 10^{14} - 1.2 \times 10^{14}$ W

peak. Laboratory studies find higher dissipation at larger viscosity due to anelastic dissipation (Faul and Jackson 2015). The result is that Io's dissipation can be matched with a mantle viscosity similar to that of the Earth. For this reason the Andrade model has been increasingly applied to understanding Io's dissipation (Bierson and Nimmo 2016; Renaud and Henning 2018)

4.3.3 The Influence of Internal Structure

The magnitude and spatial pattern of tidal dissipation also depends on the internal structure of Io. In a uniform body tidal heating is higher towards the body center due to higher stresses (Kaula 1964; Peale and Cassen 1978). Because of this, if Io (or any satellite) has a homogeneous interior, heating will be concentrated in the deep mantle. However, if the upper mantle has a lower viscosity, potentially due to higher temperatures or partial melt, heating can be concentrated in that region. In general, if the tidal dissipation is concentrated deeper in the interior, it will also be concentrated at the poles, while if dissipation is higher in the upper mantle, it will also be higher in equatorial regions (Segatz et al. 1988; Ross et al. 1990; Beuthe 2013). These two contrasting cases are usually described as a "Deep mantle heating" and "Asthenospheric heating" (shown in Fig. 4.3). It is expected that Io itself is somewhere in between these two extreme end-member cases (Tackley et al. 2001).

It is important to note that, while these two end-members are appropriate for Io, additional modes of tidal heating within a solid body can be important for other worlds. For example, in icy satellites with thin outer shells, the dissipation may exhibit a pattern different than the "thick" shell, "asthenospheric heating" case appropriate for Io (e.g., Ojakangas and Stevenson 1989; Beuthe 2019).

4.3.4 Tidal Heating in a Magma Ocean

If Io contains a magma ocean, tidal heating can also occur in this fluid region driven by the same gravitational forces responsible for tidal heating in solid regions. The response to gravitational forcing and the corresponding tidal heating in fluid regions can differ significantly from those in solid regions due to large horizontal fluid motions and their damping mechanisms. On Earth, tidal heating is dominated by energy dissipation in the oceans (Egbert and Ray 2001), suggesting the possibility of strong tidal heating in a magma ocean.

In fluid regions, tidal heating occurs due to viscous dissipation associated with the excitation and damping of waves and turbulence. A magma ocean may be in a laminar or turbulent flow regime depending on the degree of partial melt, for which there is only a lower bound (Khurana et al. 2011). In the laminar flow regime, heat is generated in the fluid by viscous dissipation (Ross and Schubert 1985). In the turbulent flow regime, eddies and turbulence lead to dissipation via the cascade of energy from large scales to small scales where viscous dissipation occurs. The damping of fluid motions at solid-fluid interfaces is commonly referred to as bottom friction for Earth's oceans and modeled as a quadratic term in velocity. Friction can also arise in the turbulent flow regime between adjacent fluid parcels moving relative to each other; such an interaction is commonly referred to as Rayleigh or linear drag due to the linear dependence on velocity.

Tidal heating in a turbulent magma ocean can be found by solving the Laplace tidal equations (Tyler et al. 2015). These equations describing mass and momentum conservation in a thin fluid layer, but do not include the effect of a solid shell overlying a magma ocean, although they can be modified to include this effect using pressure Love numbers (Beuthe 2016; Matsuyama et al. 2018). Figure 4.3 shows the expected surface heat flux pattern assuming linear drag in a magma ocean. Similar to the patterns due to tidal heating in solid regions, the surface heat flux is symmetrical in latitude. However, unlike the patterns for tidal heating in the solid regions, the surface heat flux pattern is not symmetrical in longitude, which has been suggested as a possible explanation for the observed eastward shift in volcanic features (Hamilton et al. 2013b; Tyler et al. 2015). Tidal heating in a magma ocean can reach values that are consistent with the observed heat flux for a wide range of model parameters including the magma ocean thickness and linear drag coefficient (Tyler et al. 2015). However, these parameters are poorly constrained.

4.3.5 Heat Transport

Connecting the tidal heat dissipation to surface observations requires understanding the processes transporting heat in the interior. Within Io's mantle the two relevant processes are thermal convection and melt advection. Important to both of these is the way a material changes as the partial melt increases. For small melt fractions, the melt acts as a fluid moving through the solid matrix. In this regime, the solid itself is largely unchanged in terms of bulk viscosity, although the shear strength does begin to drop. Above some critical melt fraction there is an abrupt transition from melt within a porous matrix to solid crystals suspended in a liquid melt. This transition usually occurs at melt fractions around 30% and is referred to as the breakdown temperature. At this point the viscosity rapidly decides and begins to approach the viscosity of the melt phase (Costa et al. 2009).

Tying this material behavior to these heat transport processes, the amount of heat that can be transported by convection is most sensitive to the mantle viscosity (Moore 2003). Because of this, the total heat transport is only weakly dependent on the temperature of the mantle until it hits the breakdown temperature. At this point the convective heat flux rapidly increases as the viscosity approaches that of a fluid. Direct melt advection can very efficiently transfer large amounts of heat, but only within a narrow regime. Melt must both be present and the solid matrix must still be intact to facilitate porous flow (Moore 2001).

The amount of heat dissipated within the mantle also depends on the internal temperature. As interior temperatures rise the viscosity drops, generally allowing for more tidal dissipation (See Fig. 4.4). However once a significant amount of melt develops (T_c) the shear modulus begins to drop, lowering dissipation (Berckhemer et al. 1982).

Figure 4.5 compares the behavior of these different heat sources and heat transport mechanisms for Io's mantle. For Io, two stable equilibrium states are possible. The high temperature equilibrium is a balance between internal convection and tidal heating. This possibility was found to be stable but inconsistent with the high surface heat flow (Moore 2003). Thus, it is generally thought that the high tidal dissipation is accommodated by melt advecting through a porous matrix (Moore 2001). As Io's eccentricity evolves over time the tidal heating curve will fall and rise adjusting the stable melt fraction in the interior.

The final stage in Io's heat transport is through the cold lithosphere. Io's surface has mountains many kilometers high requiring that Io has a stiff lithosphere capable of supporting large loads (Schenk and Bulmer 1998; White et al. 2014). If heat was transported conductively through the lithosphere, it would be less than 2 km thick (O'Reilly and Davies 1981). From this it has been inferred that the dominant heat loss mechanism in the lithosphere is also melt transport, sometimes called a "heat pipe" model (O'Reilly and Davies 1981; Moore 2001). The central premise of the "heat pipe" model is that tidal dissipation within Io causes melting (Moore 2001). That melt is buoyant and rises to the surface where it cools after erupting or becoming emplaced within the cold lithosphere. Unlike models for conductive or convective heat transport, there is no clear relationship between the heat transport and lithospheric thickness. Modeling work that coupled the tidal dissipation, melt segregation, and lithospheric thickness have found that the ratio of intrusive to extrusive eruptions is the main control on the lithosphere thickness (Spencer et al. 2020).

Given the clear differences between the predicted surface heat flux patterns for tidal heating in a mantle, asthenosphere, or a magma ocean (Fig. 4.3), an observation

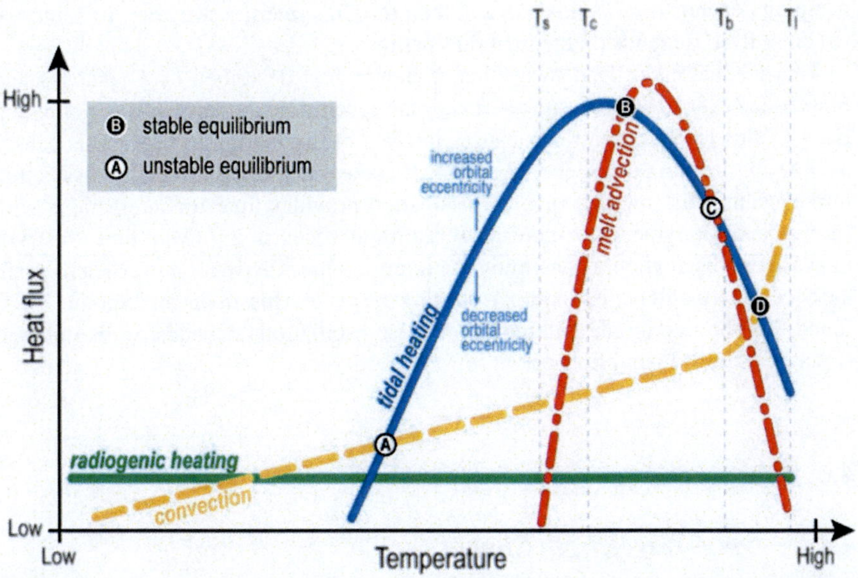

Fig. 4.5 Schematic figure showing the possible equilibrium between heat production (solid lines) and heat transport (dashed lines). Filled circles are equilibrium that are stable to temperature perturbations. The four temperatures of interest are the solidus (T_s), critical temperature where shear modulus rapidly drops (T_c), breakdown temperature where the material starts to act as a fluid (T_b), and liquidus (T_l)

of the heat flux pattern could be used to determine the dominant region of tidal heating. Because "heat pipe" transport is driven by buoyancy it would be expected that, at least of large scales, heat dissipation in the interior would be well correlated with its surface expression. The most common assumption is that surface expression is either volcano density or activity (which have different spatial patterns discussed in the following section). Complicating this picture, modeling work has found that even without the presence of a magma ocean, silicate convection could laterally redistribute approximately 20% of Io's heat flow (Steinke et al. 2020a,b). If Io does have a magma ocean it could redistribute far more heat, dramatically changing the spatial heat flux pattern (Tyler et al. 2015; Steinke et al. 2020b).

4.3.6 Dissipation-Orbit Coupling

As Io tidally dissipates energy, that energy is extracted from the eccentricity of its orbit (Murray and Dermott 2000). This causes the eccentricity to dampen towards a circular orbit. However counteracting this are periodic perturbations on Io's orbit due to the Laplace resonance with Europa and Ganymede. The rate of eccentricity damping depends on Io's tidal k_2, and therefore Io's internal structure. Io's internal structure itself depends on the tidal dissipation.

This feedback can lead to cyclical behavior first described by Ojakangas and Stevenson (1986). As Io dissipates energy the eccentricity begins to drop. Following Eq. 4.75 this lower eccentricity causes the rate of dissipation to drop. This causes the interior of Io to cool. Generally this will make the interior stiffer and dissipation lower again. This low rate of tidal dissipation provides time for interactions with Europa and Ganymede to increase Io's orbital eccentricity. Over time this will increase the dissipation, re-warming the interior. Once dissipation is efficient again the eccentricity will begin to drop, restarting the cycle. Hussmann and Spohn (2004) found that the magnitude and period of these oscillations depends on the internal structure of both Io and Europa.

4.4 Observations

4.4.1 Long-Wavelength Shape

The long-wavelength shape of Io is set by the combination of rotational and tidal forces (see Sects. 4.2.1 and 4.2.5.1). The combination of these two forces is expected to produce an equilibrium shape described by a triaxial ellipsoid. The long axis (a) is aligned with Jupiter (the "tidal axis" that defines 0°W and 180°W in Io's longitude system); the intermediate axis (b) is aligned with Io's orbital velocity vector (90°W and 270°W); and the short axis (c) is aligned with Io's spin pole (defining 90°N and 90°S). The exact values of a, b, and c are related to the rotational and tidal forces acting on Io—which we know a priori from Io's rotation and orbital parameters (e.g., spin and orbital period)—and to the specific but unknown properties of Io's interior structure.

At present, Io's triaxial shape is moderately-well constrained from *Voyager* and *Galileo* images (Table 4.2, Fig. 4.6a) using limb profiles (Thomas et al. 1998) and control point networks (Oberst and Schuster 2004). Long-wavelength shape can be useful for inferring interior structure (e.g., Schubert et al. 2004; Oberst and Schuster 2004), although for the case of Io, the constraints from gravity field measurements tend to be stronger (see Sect. 4.4.3).

4.4.2 Short-Wavelength Shape

At local scales, Io's topography is affected by an array of geological and geophysical processes (Fig. 4.6b-c). White et al. (2014) present the most complete topography model for Io derived from a stereophotogrammetry analysis of overlapping *Voyager* and *Galileo* images. This topography model covers approximately ~75% of Io,

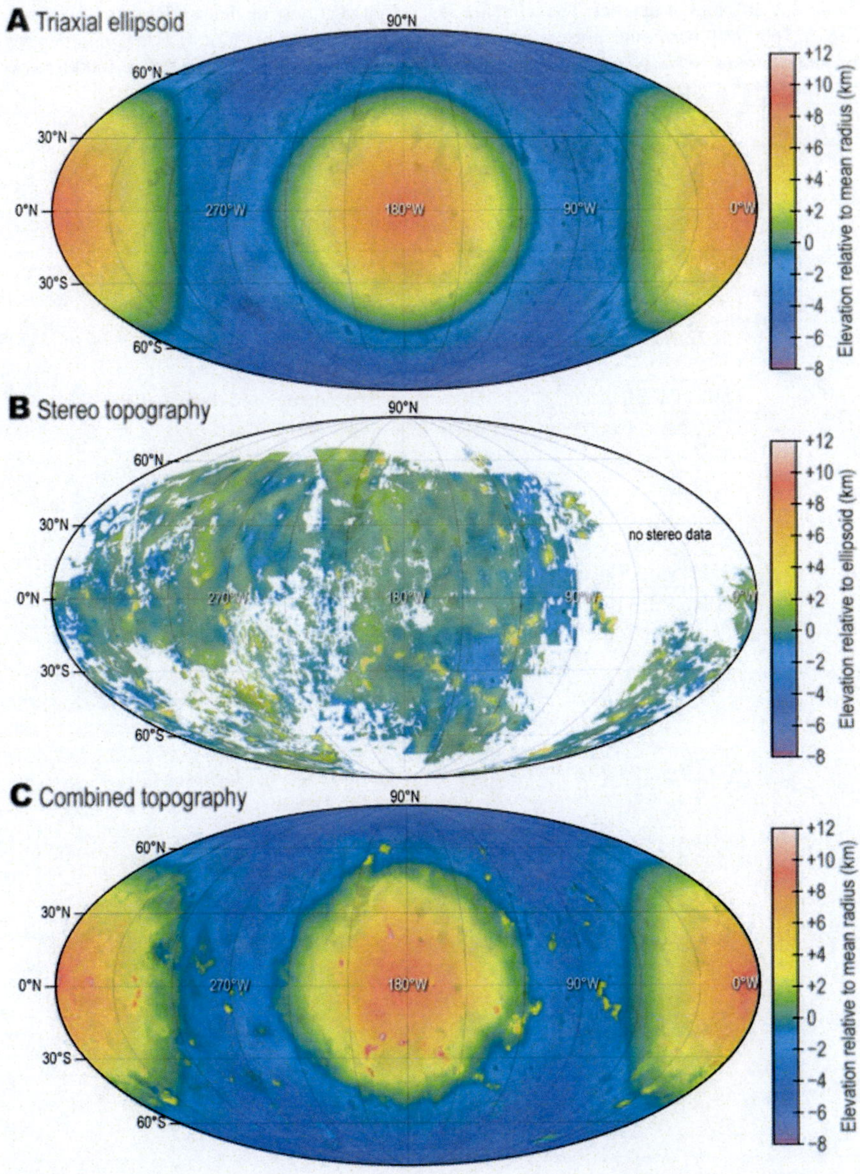

Fig. 4.6 The shape of Io. (**a**) The triaxial ellipsoid shape for Io, from Oberst and Schuster (2004). Elevations are referenced to the mean radius, 1822.7 km (Table 4.2). (**b**) Stereo topography, from White et al. (2014). Elevations are referenced to the triaxial ellipsoid, as described in (White et al. 2014). Regions in white do not have stereo topography data. (**c**) The combined topographic model of Io, which is the sum of the best-fit triaxial ellipsoid (**a**) and the stereo topography (**b**). Elevations are referenced to the mean radius, as in (**a**). Note that while this is shown as a global dataset, this map also suffers the same data gaps as in the previous two figures. Maps are in Mollweide projection, centered on 180°W (the anti-Jupiter point), with grid-lines in 30° increments—as described in Fig. 4.1

Table 4.2 Io's bulk properties. The reference radius for the gravity model is 1821.6 km (Jacobson 2013). Spherical harmonic normalization follows Schubert et al. (2004). Uncertainties are 1σ uncertainties. The Newtonian constant of gravitation is $G = (6.67430 \pm 0.00015) \times 10^{-11}\ \mathrm{m^3\ kg^{-1}\ s^{-2}}$ (per the National Institute of Standards and Technology)

Parameter	Value	Description	Reference
a	1831.1 ± 0.3 km	Triaxial figure, longest axis (tidal axis)	Oberst and Schuster (2004)
b	1820.4 ± 0.5 km	Triaxial figure, intermediate axis	Oberst and Schuster (2004)
c	1816.6 ± 0.5 km	Triaxial figure, shortest axis (spin axis)	Oberst and Schuster (2004)
R_{mean}	1822.7 ± 0.2 km	Mean radius, $R_{\mathrm{mean}} = (abc)^{1/3}$	–
GM	$5959.91 \pm 0.02\ \mathrm{km^3\ s^{-2}}$	Gravitational parameter	Schubert et al. (2004)
M	$(8.9296 \pm 0.0002) \times 10^{22}$ kg	Mass, $M = GM/G$	–
g	$1.79\ \mathrm{m/s^2}$	Surface gravitational acceleration $g = GM/R_{mean}^2$	–
ρ	$3527.5 \pm 2.9\ \mathrm{kg\ m^{-3}}$	Density	Schubert et al. (2004)
$J_2 = -C_{2,0}$	$(1846.7 \pm 3.6) \times 10^{-6}$	Spherical harmonic coefficient of the gravity field	Jacobson (2013)
$C_{2,1}$	$(4.4 \pm 0.9) \times 10^{-6}$	Spherical harmonic coefficient of the gravity field	Jacobson (2013)
$S_{2,1}$	$(-2.9 \pm 2.0) \times 10^{-6}$	Spherical harmonic coefficient of the gravity field	Jacobson (2013)
$C_{2,2}$	$(556.4 \pm 0.6) \times 10^{-6}$	Spherical harmonic coefficient of the gravity field	Jacobson (2013)
$S_{2,2}$	$(0.6 \pm 0.6) \times 10^{-6}$	Spherical harmonic coefficient of the gravity field	Jacobson (2013)
k_2^{∞}	1.3043 ± 0.0019	Degree-2 long-term potential Love number	Schubert et al. (2004)
h_2^{∞}	2.242 ± 0.167	Degree-2 long-term radial displacement Love number	Moore et al. (2007)

albeit with varying vertical and horizontal resolution (100's of meters to several kilometers).

One of the most widely debated features in Io's short-wavelength topography is the identification of putative basins and swells. In early control point analyses, Gaskell et al. (1988) identified broad (~500 km diameter) low-amplitude (±1 km) alternating patterns of basins and swells separated by 90° of arc. If these features exist, they may reflect tidal heating patterns at depth, and/or the processes by which heat escapes Io (e.g. Ross et al. 1990). While basins and swells have motivated substantial theoretical work, subsequent topography analyses have not consistently identified basins and swells, or identified basins and swells in different regions. The White et al. (2014) model does identify several longitudinally-arranged basins and swells that appear correlated with the spatial distribution of mountains and volcanoes, respectively.

Beyond putative basins/swells, the White et al. (2014) topography data reveals a number of specific topographic features–including mountain arcs, a broad depression encompassing Loki Patera, and moderately high-resolution topographic data of a handful of specific volcanic features (e.g., Ra Patera, Tvashtar Paterae). While Io is home to some of the tallest mountains in the solar system (e.g. Turtle et al. 2007), outside of these localized, high-relief features, Io's topography is muted–with a standard deviation of 0.61 km across Io's plains (with respect to the triaxial figure; Sect. 4.4.1).

4.4.3 *Gravity Field*

Gravity measurements, acquired by precise Doppler tracking of spacecraft accelerations during flybys, provide direct constraints on the distribution of mass within a planetary body. The current state of knowledge of Io's gravity field is limited due to the small number of close flybys, constrained flyby geometry, and overall data quality from past missions to the Jupiter system. Only the *Galileo* spacecraft has performed close flybys of Io (within ∼2 Io radii), and not every flyby was suitable for gravity science analysis. The immediate, post-*Galileo* gravity analysis retrieved a partial, long-wavelength gravity field of Io (Anderson et al. 2001; Schubert et al. 2004), described as a spherical harmonic degree/order-2 gravity field, consisting of J_2 (the gravity-equivalent of polar oblateness or flattening) and $C_{2,2}$ (which quantifies how Io is stretched towards Jupiter). However, in these early analyses, these coefficients were not truly independent, since the models included an a priori constraint that Io be in hydrostatic equilibrium—forcing $J_2/C_{2,2} = 10/3$ (Sect. 4.2.5.1). More recent analyses of the *Galileo* flyby data have yielded a complete degree-2 gravity field for Io (Table 4.2; Jacobson 2013). The available data is insufficient to constrain Io's gravity field at shorter wavelength—inhibiting the ability to use gravity to probe regional or local-scale phenomenon. For comparison, the current state of knowledge of Io's gravity field is comparable to our knowledge of the Moon's before the *Apollo* era. In general, it is advisable to view the gravity field of Io (and the other Galileans satellites) with a healthy amount of skepticism. For example, for neighboring Europa, recent, independent analyses of the *Galileo* data have yielded different gravity solutions (with much larger error bars) than earlier analyses (Gomez Casajus et al. 2021).

Figure 4.7 shows the measured gravity field of Io. The measured gravity field of Io is largely dominated by the contribution from tides and rotation, producing

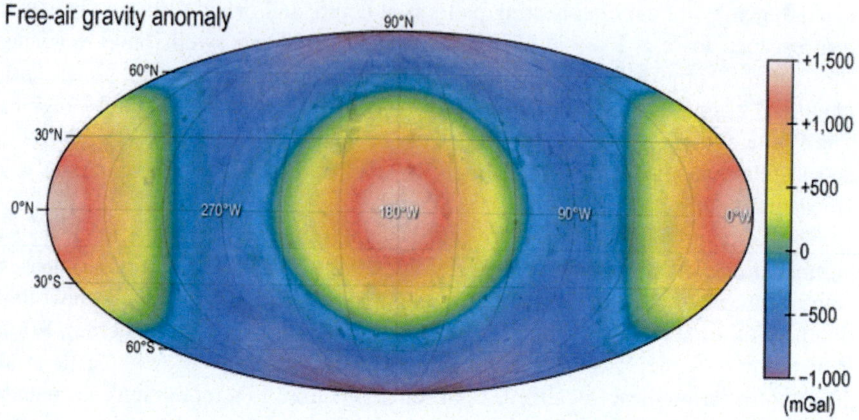

Fig. 4.7 The observed gravity field of Io. This gravity map is only complete to spherical harmonic degree/order 2, and there is likely shorter wavelength structure that is unresolved due to the lack of measurements. This map is in Mollweide projection, centered on 180°W (the anti-Jupiter point), with grid-lines in 30° increments—as described in Fig. 4.1

large values of J_2 and $C_{2,2}$. The ratio of $J_2/C_{2,2}$ is close to the hydrostatic value of 10/3, but not exactly equal to it, differing from 10/3 by 0.014 ± 0.007 (i.e., a 2σ difference). There are additional hints for non-hydrostatic effects from the existence of other non-zero degree-2 gravity coefficients, particularly $C_{2,1}$, which is inconsistent with the hydrostatic value of $C_{2,1} = 0$ at the $\sim 5\sigma$ level.

From the observed coefficients, it is possible to determine the long-term degree-2 potential Love number, k_2^∞ (Table 4.2), which quantifies how a body's gravitational potential responds to an external forcing potential, and is an important measurement for constraining the interior structure and bulk characteristics of a body (Sect. 4.2.5.1). Io's k_2^∞ is significantly lower than 1.5, indicating that it possesses some shear strength and/or is differentiated.

First, let us consider k_2^∞ in the case of elasticity. From Eq. 4.21, and the observed value of k_2^∞ (Table 4.2), we can place an upper bound on Io's average shear modulus of $\mu < 2 \times 10^8$ Pa. This is several orders of magnitude lower than values for solids, which are typically between 10^{10}–10^{11} Pa (Moore et al. 2007), indicating that Io's interior cannot be entirely solid.

Second, let us consider k_2^∞ in the case that Io is not entirely homogeneous. For entirely hydrostatic bodies, it is possible to relate k_2^∞ to the body's moment of inertia via the Radau-Darwin relationship (Darwin 1899; Murray and Dermott 2000; Hemingway et al. 2018), which is given by:

$$\frac{C}{MR^2} = \frac{2}{3}\left(1 - \frac{2}{5}\sqrt{\frac{4 - k_2^\infty}{1 + k_2^\infty}}\right), \tag{4.83}$$

where C is the maximum principal moment of inertia (Sect. 4.2.3), and M and R are the body's mass and radius, respectively. Using this relationship, we can infer Io's maximum principal moment of inertia, $C/MR^2 = 0.3782 \pm 0.0002$. This moment of inertia is lower than for a homogeneous sphere ($C/MR^2 = 0.4$), indicating that Io is centrally condensed.

With a Io's moment of inertia inferred from the Radau-Darwin relationship, it is possible to consider more detailed interior structure models. There is extensive literature for determining the moments of inertia for multi-layered hydrostatic bodies deformed by tides and rotation (for a review, see Tricarico 2014). The simplest approach is to consider a two-layer, spherical world consisting of a core and a mantle is given by:

$$\frac{C}{MR^2} = \frac{2}{5}\left[\frac{\rho_m}{\rho} + \left(1 - \frac{\rho_m}{\rho}\right)\left(\frac{r_c}{R}\right)^2\right], \tag{4.84}$$

where ρ is the density of the entire body, ρ_m is the density of the mantle, and r_c is the radius of the core. While this relationship (which can be derived by simply adding and subtracting moments of inertia for spheres, $C = 0.4MR^2$) omits many of the processes affecting planetary bodies, it is sufficient for the analysis of many worlds—including Io (e.g. Schubert et al. 2004; Moore et al. 2007). Beyond these simple models, there are semi-analytical and numerical solutions to determine the moment of inertia for worlds with arbitrary number of interior layers, and accounting for how tidal and rotational deformation distort the body and affect the moments of inertia (e.g., Tricarico 2014).

Figure 4.8a shows a family of two-layer, hydrostatic models for Io that are consistent with the moment of inertia derived from Io's observed k_2^∞, based on the recursive solutions derived by Tricarico (2014), which account for Io's fast rotation. Despite the added complexity, these new models are consistent with previous work (Schubert et al. 2004; Moore et al. 2007)—only differing by a few percent. Figure 4.8a shows that there is a trade-off between core size, core density, and mantle density. This two-layer, hydrostatic model can be further constrained by making assumptions about the composition of the core. It is generally assumed that Io's core has an iron or iron-sulfide rich composition (Moore et al. 2007), which would allow a density ranging from 8000 kg/m^3 for pure metallic iron down to 5150 kg/m^3 for an iron-iron sulfide eutectic composition (Usselman 1975a,b). These density end-members would result in core radii between $0.35R$ and $0.50R$, respectively. These core sizes and density would require mantle densities between 3335 and 3300 kg/m^3. For comparison, these densities are very similar to the inferred densities of the Moon's core and mantle (e.g., Matsuyama et al. 2016).

Io's volcanic activity may process the uppermost layers of Io, generating a compositionally distinct crust with a different, lower density compared to the mantle (e.g., Keszthelyi and McEwen 1997). Some researchers have attempted to develop corresponding three-layer models (Anderson et al. 2001; Moore et al. 2007). While insightful, the inversion is sufficiently degenerate that not much additional information can be gleaned without the use of major assumptions.

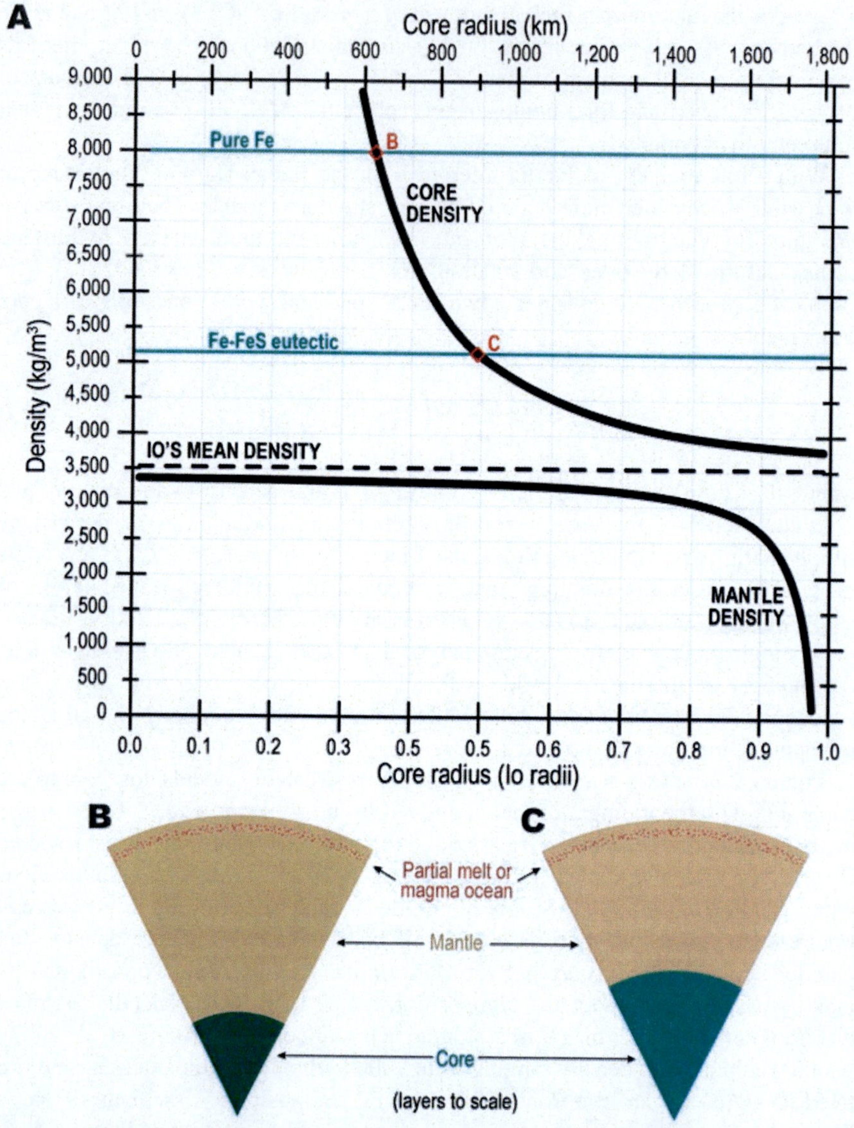

Fig. 4.8 Two-layer, hydrostatic models of Io consistent with the measured k_2 and mean density of Io. (**a**) shows the full parameter space. Io's mean density (3527.5 kg/m^3) is shown as the dashed horizontal line. The density of Io's core is the black line above the mean density (since the core is denser than the mean), and the density of Io's mantle is the black line below the mean density (since the mantle is less dense than the mean). Horizontal green lines indicate end-member compositions for the core: pure iron, and iron-iron sulfide eutectic. (**b**) and (**c**) show cross-sections of Io's interior for these two end-members, with layers to scale. In addition to the core and mantle, these cross-sections also show the location and thickness of Io's partial melt or magma ocean layer, as constrained by magnetic induction measurements (Khurana et al. 2011). The melt fraction and thickness of this layer is not well constrained (in particular, it could be much thicker). The presence of partial melt or a magma ocean is *not* included in the two-layer, hydrostatic models of Io

It should be noted that, while Io's core is believed to be largely metallic iron or iron-sulfide, the thermal state of the core is uncertain. It is unclear whether Io's core is entirely molten, partially molten, or entirely solid. We may expect Io to have a partially molten core based on analogy to its neighbor Ganymede. Gravity measurements indicate that Gravity has a rocky interior comparable to the size of Io, and a metallic core roughly the same size as the one inferred within Io (Schubert et al. 2004). Ganymede, however, has a well-measured intrinsic magnetic field, likely driven by a core dynamo (Kivelson et al. 1996, 2004). In any case, more work and measurements are required to determine the state of Io's core.

Thus far, all gravity measurements have been of Io's "static" gravity field—i.e., the unchanging, time-averaged gravity field. However, time-varying gravity would be a powerful tool for probing Io's interior structure (See Sect. 4.2.5.2). Io's degree-2 potential Love number, k_2, quantifies how a body's gravitational potential responds to an external, time-varying forcing potential. We distinguish between the *long-term* Love number, k_2^{∞}, and the *tidal* Love number, $k_2^*(\Omega)$. k_2^{∞} describes Io's response to the long-term, time-averaged, forcing potential and is captured in the static degree-2 gravity field (Table 4.2). It is related to the interior density distribution and (under certain assumptions) to the body's moment of inertia (the Radau-Darwin approximation; Darwin 1899; Murray and Dermott 2000). In contrast, $k_2^*(\Omega)$, describes how Io responds on a tidal timescale (i.e., Io's frequency, $\Omega = 2\pi/(42 \text{ hours})$), and would yield information about Io's viscoelastic response to tides, the thickness and rigidity of the lithosphere, and presence of a magma ocean. For example, if Io would be completely solid, $k_2^*(\Omega) \sim 0.09$, while $k_2^*(\Omega) \sim 0.5$ if it has a fluid magma ocean (Bierson and Nimmo 2016; de Kleer et al. 2019b). Figure 4.2 shows how these quantities vary for different interior structure models.

Since planetary materials do not respond perfectly elastically, we expect a phase lag between the tidal forcing potential and Io's response. This phase lag is directly related to Io's rigidity, viscosity, and the energy dissipation rate within Io expressed as k_2/Q, where Q is the tidal dissipation quality factor. Unfortunately, this phase lag has not been directly measured, although it has been estimated using different methods—including assuming that Io's tidal dissipation is in equilibrium with Io's orbital evolution (Sect. 4.4.5), or that Io's tidal dissipation can be inferred from Io's thermal output (Sect. 4.4.7).

While we have focused on the potential Love number, k_2, it is worth noting that it is also possible to measure how Io's shape responds to tides with the radial and lateral surface displacement Love numbers, h_2 and l_2, respectively. These surface displacement Love numbers can be divided in the same way as described above, and like for k_2^{∞}, only Io's h_2^{∞} is measured (Table 4.2; Moore et al. 2007). k_2, h_2, l_2 have similar—but not identical—dependencies on Io's interior structure and rheology. Like k_2, for a solid Io $h_2 \sim 0.1$, and $h_2 \sim 0.5$–1.0 for an Io with a fluid magma ocean (de Kleer et al. 2019b).

4.4.4 Magnetic Induction

As Io orbits Jupiter, it is subjected to a time-varying external magnetic field (since Jupiter's magnetic field is inclined with respect to Io's orbital plane). In the reference frame of Io, the applied magnetic field oscillates at the synodic period, which is the period it takes for Io to return to the same phase with respect to Jupiter's magnetosphere. For Io's orbital period, $P_{Io} = 42.459$ hours, and Jupiter's rotational period, $P_{Jupiter} = 9.925$ hours, the synodic period is $1/(1/P_{Jupiter} - 1/P_{Io}) = 12.95$ hours. This time-varying magnetic field generates eddy currents within Io's conductive layers, producing an induced magnetic field. Figure 4.9 shows a schematic illustration of this process. The strength and geometry of this induced field varies as the applied external field varies, and is sensitive to the electrical properties within Io. (This is distinct from a core dynamo, which would be largely insensitive to these external perturbations.) By using laboratory measurements of analogues, it is possible to interpret these magnetic field anomalies in terms of melt fraction, depth distribution, composition, and temperature (e.g. Khan et al. 2014; Pommier et al. 2015). Magnetic induction (or magnetic sounding) has been well-demonstrated at the Earth and Moon (Hood et al. 1982; Constable and Constable 2004), and used to detect salty (i.e., conductive) subsurface water oceans on Europa (Khurana et al. 1998; Kivelson et al. 2000), Ganymede (Kivelson et al. 2002), and Callisto (Khurana et al. 1998; Zimmer et al. 2000). While magnetic induction is traditionally deduced from in situ magnetic field and plasma measurements, there have been recent attempts to use Earth-based observations of aurora to sound the interiors of the Galilean satellites (e.g. Ganymede; Saur et al. 2015). Aurora respond to both the external and induced field, and provide complementary insights, although the technique lacks the precision of in situ measurements.

Analysis of *Galileo* magnetometer data revealed the presence of an induced magnetic field at Io (Khurana et al. 2011). The induced field is global, dipolar, time-varying, and out-of-phase with the applied field—confirming that it is not a permanent dynamo. By modeling the induced magnetic field, Khurana et al. (2011) showed that the induced signal was inconsistent with a completely solid mantle, and was best fit by a $\gtrsim$50 km thick shell of $\gtrsim$20% partial melt located ~50 km beneath the surface—consistent with the presence of a subsurface magma ocean (Sect. 4.5). At present, this is the best evidence for an extant magma ocean within Io, although this has been subject of extensive debate within the community. Roth et al. (2017) found that *Hubble Space Telescope (HST)* observations of Io's auroral spots (in conjunction with magnetohydrodynamic (MHD) simulations) were inconsistent with the presence of a magma ocean, and instead favored either atmosphere-plasma interactions, or induction in Io's metallic core. The same group later modeled the *Galileo* magnetometer data with a more complex 3-D MHD model and found that plasma interactions with Io's asymmetric atmosphere could equally explain the observed magnetic field signal (Blöcker et al. 2018). A third group, Šebek et al. (2019) subsequently performed independent MHD models and strongly favored the presence of an induced magnetic field from a magma ocean. In short, the nature of

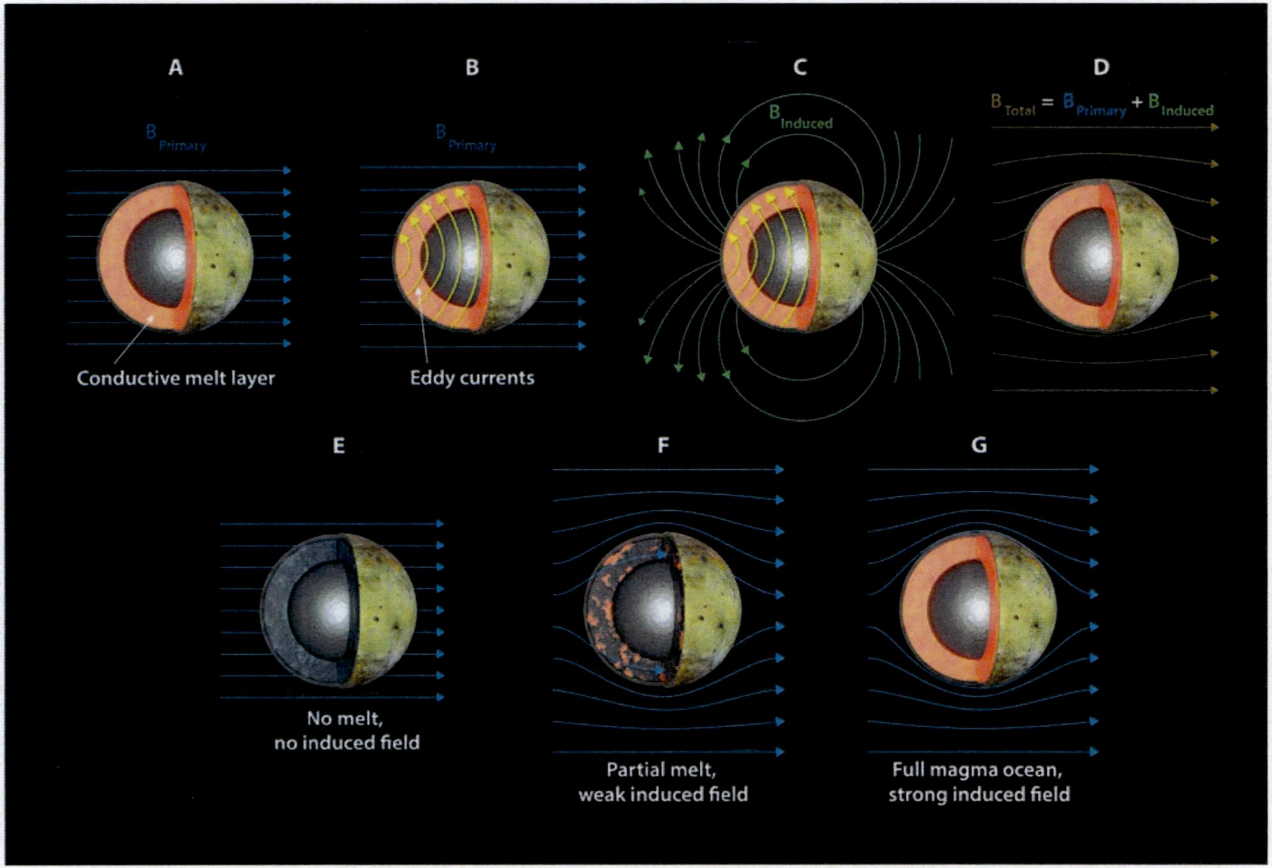

Fig. 4.9 Schematic illustration of electromagnetic sounding of Io's interior from de Kleer et al. (2019b). (**a**) Io experiences a time-varying magnetic field, $B_{primary}$, as it orbits through Jupiter's magnetosphere. (**b**) This time-varying magnetic field drives eddy currents in Io's conducting layers. (**c**) The eddy currents produce an induced magnetic field, $B_{induced}$. (**d**) The observed magnetic field, B_{total} is the combination of these different processes. (**e–g**) The magnitude of the induced magnetic field depends on the physical and electromagnetic properties of Io's interior structure. Image credit: Chuck Carter, James Tuttle Keane, Keck Institute for Space Studies (KISS; de Kleer et al. 2019b)

Io's induced magnetic field and the presence of a subsurface magma ocean is still uncertain.

Thus far, most induction studies have focused on looking for induced fields at the synodic period (12.95 hours). While this period corresponds to the strongest amplitude signal (~850 nT), there are other periodicities in the time-varying magnetic field applied to Io. For example, the amplitude at Io's orbital period (42.459 hours) is also strong (~50 nT; de Kleer et al. 2019b). Future in situ measurements—either from an orbiter or multi-flyby spacecraft—could exploit these other periodicities to uniquely probe the interior structure of Io. Such measurements would also require detailed characterization of the surrounding plasma environment.

4.4.5 The Laplace Resonance and Astrometry

Io's activity is maintained by the orbital resonance between Io, Europa, and Ganymede. In this configuration, first described by Pierre Simon de Laplace in 1771, Io completes four orbits for every two orbits of Europa, and one orbit of Ganymede, i.e., a 4:2:1 resonance (also known as the Laplace resonance). This resonance can be described by the following relationships, (e.g., Murray and Dermott 2000):

$$\Omega_{Io} - 3\Omega_{Europa} + 2\Omega_{Ganymede} = 0 \tag{4.85}$$

$$\phi_L = \lambda_{Io} - 3\lambda_{Europa} + 2\lambda_{Ganymede} \approx 180^\circ \tag{4.86}$$

where ϕ_L is the resonant argument, λ is the mean longitude, and Ω are mean motions ($\Omega = 2\pi/P$, where P is the orbital period). Equation 4.85 shows that the mean motions of the three Galilean are related by a simple integer ratio, while Eq. 4.86 describes the relative geometry between the three satellites at any given time. For an object to truly be in a stable mean-motion resonance, it must not only have an orbital period in a small-integer ratio with another object, but the system must be stable against small perturbations. For the Galilean satellites, the resonant argument librates around 180° with an amplitude of 0.064° and a period of 2071 days (Murray and Dermott 2000). Since Galilean satellites librate about 180°, this means that there can never be a triple conjunction between Io, Europa, and Ganymede (i.e., the three never "line-up" on one side of Jupiter). The full dynamics of the Laplace resonance is quite complicated, and we refer the reader to Murray and Dermott (2000) and Yoder and Peale (1981) for more thorough discussion.

Curiously, Io–Europa–Ganymede is one of only two three-body mean-motion resonance in the Solar System, which is otherwise dominated by two-body mean motion resonances (e.g., Neptune–Pluto, Mimas–Tethys, Enceladus–Dione, Titan–Hyperion). The only other Laplace-like resonance in the Solar System is a resonance between Pluto's small moons, Styx, Nix, and Hydra (Showalter and Hamilton 2015). However, the deluge of exoplanet discoveries in the past two decades has revealed numerous exoplanetary systems in Laplace, or Laplace-like resonances. Planets c, b, and e in Gliese 876 (an M-dwarf star) are in Laplace resonance, with an orbital period ratio of 4:2:1—although unlike the Io–Europa–Ganymede system, these Jupiter-mass planets librate about 0°, meaning that triple conjuctions are possible (Rivera et al. 2010). In some instances, chains of Laplace-like resonances have been found: the five super-Earth and mini-Neptunes around TOI-178 are in a chain of *two* Laplace-like resonances (period ratio: 18:9:6:4:3; Leleu et al. 2021); the seven Earth-mass planets around TRAPPIST-1 are in a chain of *three* Laplace-like resonances (period ratio: 24:15:9:6:4:3:2; Luger et al. 2017).

While the Laplace resonance was known for hundreds of years, its importance was not realized until Peale et al. (1979) noted that the Laplace resonance creates a substantial forced eccentricity in Io and Europa. Peale et al. (1979) predicted that this forced eccentricity would result in substantial tidal heating—plausibly melting Io's interior and driving volcanic activity (see Sect. 4.3). This prediction was spectacularly confirmed by the *Voyager 1* flyby of Io only a few months later, which discovered active volcanic plumes (Morabito et al. 1979; Morabito 2012).

Despite the fundamental role of the Laplace resonance in driving the tidal heating of Io and the Galilean satellites, it is unclear if this configuration has been constant throughout Solar System history, or varied with time (e.g., de Kleer et al. 2019b). For example, Io could be in orbital equilibrium—where the total heat output is constant with time, and Io's small forced eccentricity is constant, resulting in a constant rate at which tidal heat is delivered via Jupiter. Alternatively, Io could be oscillating about equilibrium in some cyclical or episodic fashion. For example, if Io is currently migrating away from the exact resonance with Europa (Lainey et al. 2009), its eccentricity will decrease, resulting in decreased tidal heating, and net cooling of the interior. As this happens, this less dissipative Io will then migrate back outwards in response to Jupiter's tides, ultimately increasing its eccentricity and tidal heating again. This cyclic behavior has been proposed for Io, Europa, and several other tidally-heated worlds, and would occur over timescales of $\sim$100 Myr (Ojakangas and Stevenson 1986; Hussmann and Spohn 2004; Shoji and Kurita 2014). A recent complication to this story has been the development of the "resonant locking" hypothesis (Fuller et al. 2016), which more accurately accounts for how the evolving interior structure of gas giants couples into the tidal evolution of their satellites.

The orbital evolution of the Galilean satellites can be directly measured via astrometry—the precise observations of satellite positions over time—which enables derivation of very accurate models of orbital motion. Lainey et al. (2009) performed an extensive analysis of astrometric observations, spanning from 1891 to 2007, and were able to directly measure the secular drift of Io, Europa, and Ganymede (Table 4.3). These changes are quantified either in terms of changes of the satellite's semimajor axis (a) or mean motion ($\Omega = 2\pi/P = (G(M+m)/a^3)^{1/2}$, where P is the satellite's orbital period, M is the mass of Jupiter, m is the mass of the satellite, and G is the gravitational constant).

At present, Io is believed to be migrating inward, towards Jupiter (indicated by the positive $\dot{\Omega}/\Omega$ in Table 4.3), while Europa and Ganymede are migrating outward. From astrometric observations, it is possible to directly infer the tidal dissipation

Table 4.3 Astrometric data for Io, Europa, and Ganymede (Lainey et al. 2009; de Kleer et al. 2019b)

	Io	Europa	Ganymede
da/dt	-1.934 m/yr	$+13.632$ m/yr	$+3.720$ m/yr
$\dot{\Omega}/\Omega$	$(+0.14 \pm 0.01) \times 10^{-10}$ 1/yr	$(-0.43 \pm 0.10) \times 10^{-10}$ 1/yr	$(-1.57 \pm 0.27) \times 10^{-10}$ 1/yr

rate (and specifically, k_2/Q) in the various bodies. Lainey et al. (2009) constrained k_2/Q for both Jupiter ($k_2/Q = (1.102 \pm 0.203) \times 10^{-5}$) and Io ($k_2/Q = 0.015 \pm 0.003$), and neglected dissipation within Europa and Ganymede. Combining this measurement with Eq. 4.75, allowed Lainey et al. (2009) to estimate the heating rate within Io to be: $\dot{E} = (9.33 \pm 1.87) \times 10^{13}$ W, equivalent to 2.24 ± 0.45 W/m^2 when converted to a heat flux out of Io's surface.

4.4.6 Hotspots, Volcanoes, and Mountains

Io's extreme tidal heating drives its extreme geology—so it is only natural to try to use Io's geology to investigate Io's interior structure and tidal heating. However, this is complicated by the sheer complexity of Io's geology, coupled with the lack of complete data. In this section, we focus on how Io's geologic features—particularly hotspots, volcanoes, and mountains—contribute to our understanding of Io's interior structure and tidal heating. For more details about Io's geology, we refer the reader to other chapters in this book.

As shown in Fig. 4.3, tidal heating models often predict specific patterns of tidal heating. Many researchers have attempted to test these models by comparing to the spatial distribution of Io's geologic features (e.g., Carr et al. 1998; Lopes-Gautier et al. 1999; Tackley et al. 2001; Radebaugh et al. 2001; Schenk et al. 2001; McEwen et al. 2004; Veeder et al. 2011; Veeder et al. 2012; Kirchoff et al. 2011; Hamilton et al. 2013b; Davies et al. 2015; de Kleer and de Pater 2016; Shoji and Hussmann 2016; Keane et al. 2018; Rathbun et al. 2018; de Kleer et al. 2019b). Figure 4.10 shows some example datasets and analyses. Most analyses focus on three types of data: (1) *Hotspots*: regions of enhanced thermal output, typically observed in ground or spacebased infrared observations, corresponding to sites of active volcanism. (2) *Paterae*: the predominant volcanic landform on Io, which are caldera-like volcano-tectonic depressions. Paterae are usually mapped in visible imagery. Paterae may or may not be volcanically active. (3) *Mountains*, which are typically observed in visible imagery. Many patterns have been repeatedly identified:

1. Io's paterae are globally distributed, but have a definitive, non-random, degree-2 pattern (Fig. 4.10b). More volcanoes are located at low-latitude, with the highest spatial density in two regions near the sub/anti-Jupiter points—but shifted roughly 30° westward (e.g., Carr et al. 1998; Kirchoff et al. 2011; Hamilton et al. 2013b; Davies et al. 2015; Keane et al. 2018).

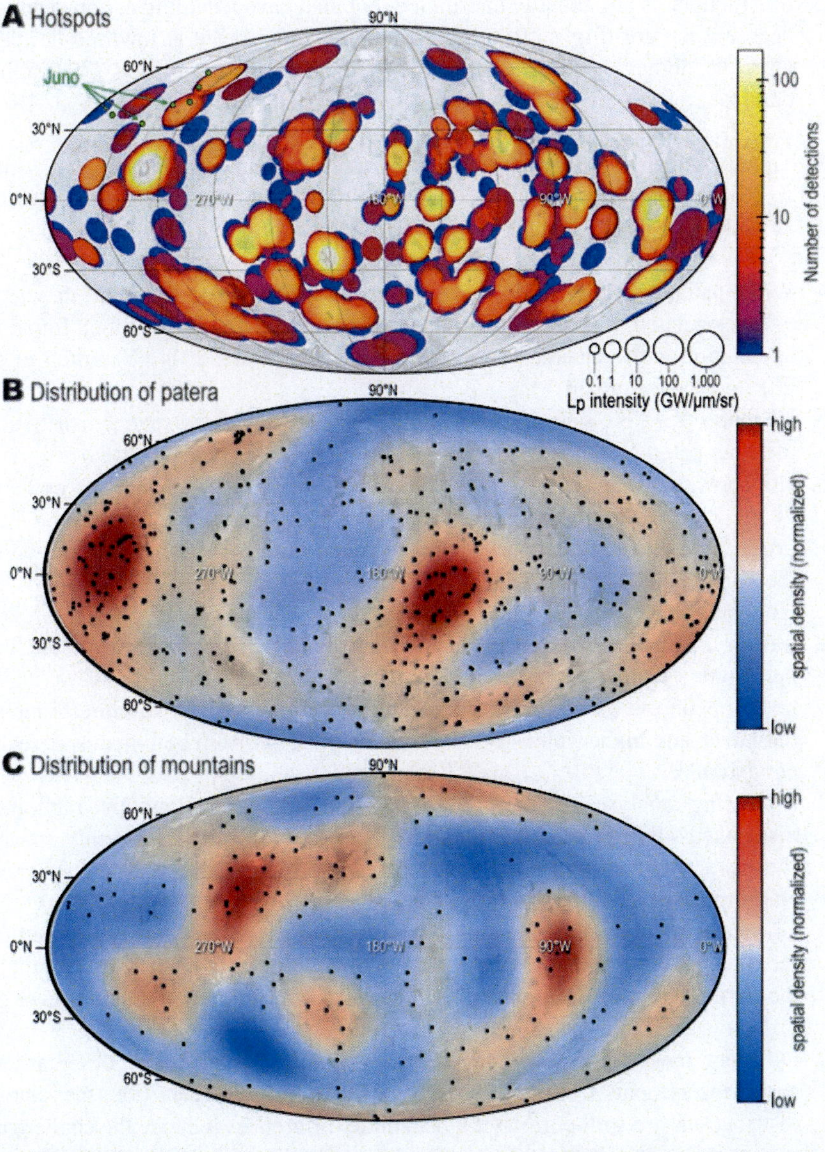

Fig. 4.10 Spatial distribution of hotspots, volcanoes, and mountains on Io. (**a**) Hotspots as observed from Earth (adapted from de Pater et al. 2021). In this plot, the size of each point corresponds to the brightness, and the color corresponds to the number of times emission was detected in a certain region. Green points indicate new hot spots observed by *Juno*. (**b**) The spatial distribution of Io's patera, expanded to spherical harmonic degree/order 6 (Hamilton et al. 2013b; Keane et al. 2018). Black points indicate individual patera, while colors indicate the spatial density calculated in Keane et al. (2018). (**c**) The spatial distribution of Io's mountains, expanded to spherical harmonic degree/order 6 (Hamilton et al. 2013b; Keane et al. 2018). Black points indicate individual mountains, while colors indicate the spatial density calculated in Keane et al. (2018). Maps are in Mollweide projection, centered on 180°W (the anti-Jupiter point), with grid-lines in 30° increments—as described in Fig. 4.1

2. Io's mountains are globally distributed, but also have a definitive, non-random, degree-2 pattern (Fig. 4.10c). More mountains are locate at low/mid-latitude, with the highest spatial density in two regions on the leading and trailing hemispheres (e.g., Carr et al. 1998; Schenk et al. 2001; Kirchoff et al. 2011; Hamilton et al. 2013b; Keane et al. 2018).
3. The spatial distributions of Io's paterae and mountains are statistically anti-correlated at long wavelengths (e.g., Fig. 4.10b and c) (e.g., Carr et al. 1998; Schenk et al. 2001; Kirchoff et al. 2011; Hamilton et al. 2013b; Keane et al. 2018).
4. While paterae and mountains are anticorrelated at long wavelengths, they tend to be correlated at short-wavelengths. Individual mountains are often found in association with volcanic features (e.g., Kirchoff et al. 2011; Hamilton et al. 2013a; Keane et al. 2018).
5. Hot spots (e.g., Fig. 4.10a) are not highly correlated with the spatial distribution of either paterae or mountains (Fig. 4.10b–c) (e.g., Keane et al. 2018).
6. Hot spots (e.g., Fig. 4.10a), show latitudinal trends (e.g., de Kleer and de Pater 2016). Persistent hot spots tend to be located within ±30° of the equator, while bright, transient outbursts tend to be located between 40° and 65° latitude in the north and south hemisphere.
7. Hot spots (e.g., Fig. 4.10a), show strong differences between Io's leading and trailing hemisphere (de Kleer and de Pater 2016). Hot spots on the leading hemisphere tend to be more numerous, persistent, but lower intensity; while hot spots on the trailing hemisphere tend to be less numerous, more prone to outbursts, and higher intensity. The overall emission from both hemispheres is comparable.
8. Clustering analyses indicate that Io's hot spots are globally randomly distributed—except near the equator where they are more uniformly spaced (i.e., more widely spaced than expected from randomly located features; Hamilton et al. 2013a).
9. Clustering analyses indicate that Io's individual patera tend to be clustered on local scales. At global scales, patera tend to be uniformly distributed, except in the northern regions where the distribution is closer to random (Hamilton et al. 2013a).
10. Different measurements of heat flow, including ground-based observations and measurements from *Galileo*, are generally in agreement (i.e., the spatial distributions are well-correlated), but subtle differences indicate the challenges in measuring and interpreting the data (e.g. Keane et al. 2018). Strong biases affect both types of data.

While many of these patterns are compelling (particularly items 1–4), confidence in all of these patterns are hampered by the incompleteness in the underlying datasets. All ground- and Earth-based observations (e.g., Fig. 4.3a), have limited visibility of Io's poles. *Voyager*, *Galileo*, and *New Horizons* observations are not globally uniform, with major gaps towards the poles and in select longitude bands. There are many datasets where global analyses are infeasible, like tectonics (although regional analyses are feasible and informative, e.g., Ahern et al. 2017).

Despite the challenges with analyzing global patterns, Io's geology has placed many important constraints on the interior structure of Io. In particular, Io's extremely tall mountains—occasionally exceeding 17 km above the surrounding terrain—place constraints on the thickness of the lithosphere. Io's mountains are believed to form from the accumulation of stress as Io's surface is progressively buried, which ultimately drives the formation of deep-seated thrust faults and the creation of mountains (e.g., Schenk et al. 2001; Bland and McKinnon 2016). The combination of geophysical inferences (e.g., mountain heights Schenk et al. 2001; Turtle et al. 2007), modeling of crustal stresses Bland and McKinnon (2016), and induction measurements Khurana et al. (2011), suggest lithospheric thicknesses between 20 and 50 km (Moore et al. 2007). Independent analyses of the magma ascent process (Spencer et al. 2020) and lateral variations in tidal heating (Steinke et al. 2020a) suggest similar mean values, although they indicate that there may be substantial lateral variations in crustal thickness.

In the end, we are left with many evocative patterns in Io's geology (Fig. 4.10) which do not nicely correlate with theoretical predictions of tidal heating (Fig. 4.3). Some researchers attempt to solve this with admixtures of different basic models (e.g., half deep mantle heating and half asthenospheric heating), or using other processes to match observations (e.g., "blurring" tidal heating models by assuming convective processes, or shifting tidal heating models by assuming the action of an ocean or non-synchronous rotation of Io's lithosphere). Many of these approaches may be overly simplistic, and it is likely that many geologic processes interact in non-trivial ways (e.g., tidal heating can affect interior structure, which in turn affects where tidal heating occurs, and so on; Steinke et al. 2020a).

In addition to *spatial* patterns, many researchers have attempted to find *temporal* patterns in Io's volcanic activity. These investigations are partly motivated by observations of Saturn's tidally heated and cryovolcanically active moon: Enceladus. Enceladus's erupted water plume is known to vary in intensity cyclically over the period of moon's 33-hour long orbit (Hedman et al. 2013). This periodicity is believed to be related to the tectonic opening and closing of the faults on Enceladus's south pole (Hurford et al. 2007, 2012; Nimmo et al. 2014; Kite and Rubin 2016; Běhounková et al. 2017). In addition to modulation on the tidal cycle, there is evidence Enceladus's plume activity may also be modulated on longer time periods due to decade-long variations in Enceladus's orbital eccentricity (Ingersoll and Ewald 2017; Ingersoll et al. 2020). While Io has the distinction of being both the most tidally-heated and volcanically active world in the Solar System, there is almost no discernable tidal modulation of Io's volcanic activity. The one exception is Loki Patera, which is one of the most consistently active and observable volcanoes on Io (de Kleer et al. 2019a). Loki appears to exhibit quasi-periodic behavior on timescales comparable to the timescales of Io's eccentricity and semimajor axes variations (~480 and ~460 days, respectively). It is unclear why Loki's activity appears to be more sensitive to these long-period variations (which are generally weaker), and not the short-period variations that occur over a single tidal cycle (which are generally stronger). de Kleer et al. (2019a) hypothesized that some geological processes may act as a low-pass filter, including poroelastic flow of magma. Longer baseline, high-cadence monitoring of Io's volcanoes are critical to test these hypotheses.

4.4.7 *Total Heat Flow*

One of the most important observational constraints for tidal dissipation models is the globally integrated heat flow. Estimates for Io's total heat flow have been derived from three sources, Earth based observations, Galileo thermal data,and astrometry. Observations from the NASA Infrared Telescope Facility (IRTF) provide the longest record of Io's heat flow. From observations spanning 4.8–20 μm, Veeder et al. (1994) placed a lower limit on Io's heat flow of $(1.05 \pm 0.21) \times 10^{14}$ W. Using *Galileo*'s photopolarimeter–radiometer (PPR) instrument Rathbun et al. (2004) estimated Io's total global heat flow to be $(0.96 \pm 0.12) \times 10^{14}$ W. These long-wavelength observation are important, as they constrains the total output of Io. Most other investigations focus on the thermal output of individual volcanoes. From astrometric observations Io's tidal $\mathrm{Im}(k_2^*)$ has been estimated to be 0.015 ± 0.003 which translates to a heat dissipation of $(0.933 \pm 0.187) \times 10^{14}$ W (Section 4.4.5; Lainey et al. 2009). All these independent constraints generally agree placing Io's total dissipation between $(0.8 - 1.2) \times 10^{14}$ W.

While different measures of global heat flow are generally consistent, there is a problem when trying to identify the specific mechanisms by which this heat is lost. Multiple studies have estimated the thermal emission from all of Io's individual, active (or recently active) volcanoes, and the total heat flow measurements always comes up short—0.56×10^{14} W, or only half of the total heat flow (Veeder et al. 2011; Veeder et al. 2012; Veeder et al. 2015; Davies et al. 2015; de Kleer et al. 2019b). The cause of this discrepancy is unknown. It may be that Io loses a substantial amount of heat from unresolved volcanic features, or via conduction through the lithosphere. Alternatively, there may be errors in our knowledge of Io's surface properties (e.g., albedo, thermal emissivity), which are critical to correctly separate endogenic heat flow from re-radiated solar illumination.

4.5 Synthesis and Open Questions

In this chapter, we have synthesized the theory (Sects. 4.2 and 4.3) and observations (Sect. 4.4) pertinent to understanding the interior structure of Io. The current state of knowledge of Io's interior structure can be synthesized as such: From gravity measurements, we believe Io is likely differentiated, with an iron-rich core, overlain by a rocky mantle (Sect. 4.2.2). *Galileo* magnetic induction measurements imply the presence of a partial melt layer, at least $\gtrsim$50 km thick, at a depth of ∼50 km, with a melt fraction of $\gtrsim$20%. This partial melt layer may be consistent with the presence of a magma ocean, although this interpretation is debated (Sect. 4.4.4). The lithosphere is believed to be 20–50 km thick, based on analyses of Io's mountains and tectonic features (Sect. 4.4.6). The total heat output is believed to be $\sim 1 \times 10^{14}$ W, based on a combination of thermal and astrometric observations—although there is substantial uncertainty in this analysis, how the heat is transported, and whether Io is in steady-state (Sects. 4.4.5 and 4.4.7).

Despite the substantial advances in the past decade, many key questions remain about Io's interior structure and the nature of tidal heating (e.g., de Kleer et al. 2019b; Keane et al. 2021a,b):

1. **What is the interior structure of Io, and does Io have a magma ocean?** Despite decades of research, the basic interior structure of Io still remains a mystery. Four end-member hypotheses are shown in Fig. 4.11. Perhaps the most important outstanding question for Io is whether it has a magma ocean (Fig. 4.11c), a predominantly solid mantle (Fig. 4.11a–b), or something different altogether—like "magma sponge" (Fig. 4.11d). (A magma sponge is a high-melt region where all of the melt is interconnected through an structurally coherent

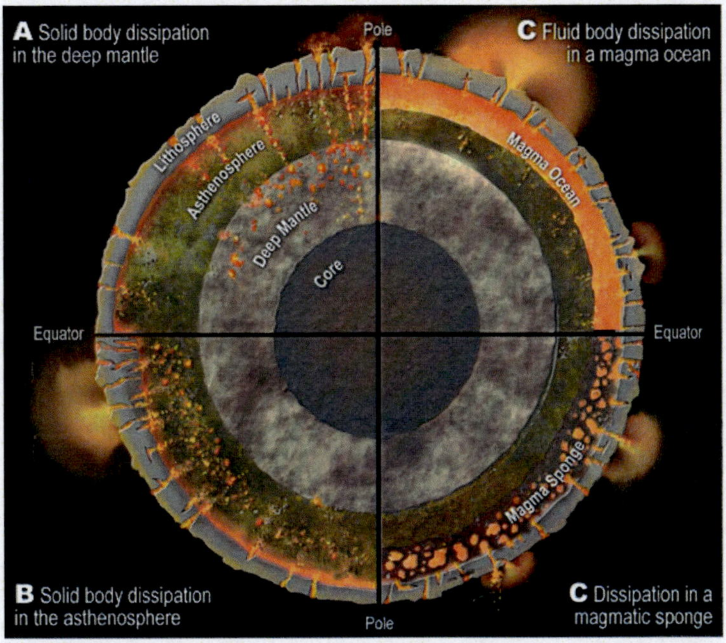

Fig. 4.11 Four end-member models for the interior structure of Io and where tidal heat is dissipated (adapted from de Kleer et al. 2019b). Each of these different interior structures would have different implications for geophysical measurements (Table 4.4). Figure create by Chuck Carter and James Tuttle Keane for the Keck Institute for Space Studies. Layer thicknesses and other features are not to scale

Table 4.4 The relationship between four end-member models for the interior structure of Io (Fig. 4.11) and geophysical measurements (de Kleer et al. 2019b)

	A Solid Io, with dissipation in the deep mantle	**B** Solid Io, with dissipation in the asthenosphere	**C** Io with a magma ocean	**D** Io with a magma "sponge"
Tidal deformation (e.g., k_2)	Low	Low	High	Low
Libration amplitude	Small	Small	Large	Small
Magnetic induction	Weak	Weak	Strong	Strong
Lava temperature	High-temperature basaltic	Basaltic	Very high-temperature ultramafic	Very high-temperature ultramafic
Heat flux distribution	More polar	More equatorial	More equatorial or uniform	More equatorial or uniform
Crustal thickness variations	Thin at equator	Thin at pole	Uniform	Uniform

sponge-like matrix (de Kleer et al. 2019b). It is unclear how such a structure would form, and how it would affect tidal dissipation. There has been some work investigating tidal dissipation in porous media, particularly the porous, fluid-filled core of Enceladus (Rovira-Navarro et al. 2022), although little work about analogous processes within Io.) At present, the best evidence for the magma ocean comes from *Galileo* induction measurements, although the interpretation of this data is debated (Sect. 4.4.4). While models of the internal structure of Io can provide some useful insight, our understanding of the basic nature of Io's interior will likely remain uncertain until new measurements are made in situ. In particular, a combination of geophysical measurements—including measuring tidal Love numbers (Sect. 4.4.3), libration amplitude (Sect. 4.2.6), magnetic induction (particularly at multiple frequencies, along with the plasma measurements necessary to interpret them; Sect. 4.4.4), global shape (Sects. 4.4.1 and 4.4.2), and better constraints on Io's lava composition and temperature—would be able to test these different end-member hypotheses (Table 4.4; de Kleer et al. 2019b).

2. **Where and how is tidal heat being dissipated in the interior of Io?** Io's activity is driven by extreme tidal heating, yet it is unclear where and how that heat is dissipated within Io. This question is intimately coupled with Io's uncertain interior structure—for example, the presence or absence of a magma ocean changes the available dissipation mechanisms, and how dissipation at depth is manifested at the surface (i.e., circulation or motion in the magma ocean may blur or erase tidal heating patterns originating at depth). Even if Io is entirely solid, it is unclear if dissipation occurs predominantly in the deep mantle (Hypothesis A in Fig. 4.11 and Table 4.4), or in the shallower

asthenosphere (Hypothesis B in Fig. 4.11 and Table 4.4), or in some combination thereof. Models for dissipation in each of these layers yields different predicted patterns for Io's heat flux (Fig. 4.3), although it is unclear if these patterns would be manifest in Io's observed heat flux or geology (Fig. 4.10), since geologic processes and feedbacks may act to obfuscate the pattern (e.g., Steinke et al. 2020a). More fundamentally, the precise dissipation mechanisms within Io are unknown. Dissipation depends on the rheology of the material, and there are currently insufficient laboratory analyses of material at Io-relevant conditions (composition, melt fraction, forcing frequency, etc.), and advanced theoretical models, to understand this process (de Kleer et al. 2019b). Similarly, the mechanisms for dissipation in a fluid magma ocean are uncertain.

3. **How does Io lose its internal heat, what is the balance between different heat loss mechanisms (conduction, heat-pipe volcanism, intrusions, convection, etc.), and how do those processes affect Io's internal structure?** Io produces an enormous amount of heat (1.5–4.0 W/m^2 Moore et al. 2007), far exceeding the heat output of the Earth (0.09 W/m^2 Turcotte and Schubert 2002). It is believed that this heat is released via the advection of heat by rising magma and erupted lava—so-called "heat pipe" volcanism (O'Reilly and Davies 1981). Continuous volcanic eruptions bury old and cold lavas, which are subsequently re-melted and mixed back into the system. While heat-pipe volcanism is the canonical mechanism for Io's heat loss, there is growing evidence that this hypothesis is incomplete. For example, magmatic intrusions may play a major role in Io's heat loss (Spencer et al. 2020). Io's volcanism may be quite complicated, and it is unclear how volcanic eruptions and their characteristics (temperature, composition, spatial distribution, variability, etc.) relate to Io's deep interior and the nature of tidal heating.
4. **Is Io (and the Laplace resonance between Io, Europa, Ganymede) in equilibrium?** We do not know if Io's present-day, extreme activity is representative of Io's entire history. Io's total heat flow appears to be broadly consistent with the astrometric measurements, but there are substantial uncertainties with both measurements (Sects. 4.4.5 and 4.4.7). More stringent measurements of Io's total heat flow and astrometric observations are necessary to make definitive conclusions—like testing if Io's activity is continuous or episodic. In addition to the equilibrium between input/output energy, there is a related question about whether Io's actual internal structure (i.e., the thickness of the lithosphere, the existence and thickness of the magma ocean, etc.) is in equilibrium.
5. **How does tectonism and magmatism interact on local and global scales on Io?** Io exhibits peculiar patterns in tectonism and magmatism (Sect. 4.4.6). While there has been some work explaining aspects of these correlations and anti-correlations (e.g., Kirchoff et al. 2011; Bland and McKinnon 2016), and tides likely play a critical role, there is no coherent hypothesis capable of explaining these patterns.

Acknowledgments The authors acknowledges substantial input from the Tidal Heating—Lessons from Io and the Jovian System workshop organized by the W.M. Keck Institute for Space Studies in 2018 (de Kleer et al. 2019b). A portion of this work was carried out at the Jet Propulsion Laboratory, California Institute of Technology, under a contract with the National Aeronautics and Space Administration (80NM0018D0004). The authors thank Francis Nimmo for thoughtful edits which enhanced this chapter.

References

Ahern, A.A., Radebaugh, J., Christiansen, E.H., Harris, R.A., Tass, E.S.: Icarus **297**, 14 (2017). https://doi.org/10.1016/j.icarus.2017.06.004

Anderson, J.D., Jacobson, R.A., Lau, E.L., Moore, W.B., Schubert, G.: J. Geophys. Res. **106**, 32963 (2001). https://doi.org/10.1029/2000JE001367

Andrade, E.N.d.C.: Proc. R. Soc. Lond. A, Containing Papers of a Mathematical and Physical Character **84**, 1 (1910). http://www.jstor.org/stable/92859

Arfken, G.B., Weber, H.J.: Mathematical Methods for Physicists, 4th edn. Academic Press, San Diego (1995)

Běhounková, M., Souček, O., Hron, J., Čadek, O.: Astrobiology **17**, 941 (2017). https://doi.org/10.1089/ast.2016.1629

Berckhemer, H., Kampfmann, W., Aulbach, E., Schmeling, H.: Phys. Earth Planet. Interiors **29**, 30 (1982). https://doi.org/10.1016/0031-9201(82)90135-2

Beuthe, M.: Icarus **223**, 308 (2013)

Beuthe, M.: Icarus **280**, 278 (2016)

Beuthe, M.: Icarus **332**, 66 (2019). https://doi.org/10.1016/j.icarus.2019.05.035

Bierson, C.J., Nimmo, F.: J. Geophys. Res. Planets **121**, 2211 (2016). https://doi.org/10.1002/2016JE005005

Bland, M.T., McKinnon, W.B.: Nat. Geosci **9**, 429 (2016). https://doi.org/10.1038/ngeo2711

Blöcker, A., Saur, J., Roth, L., Strobel, D.F.: J. Geophys. Res. (Space Phys.) **123**, 9286. https://doi.org/10.1029/2018JA025747

Bürgmann, R., Dresen, G.: Annu. Rev. Earth Planet. Sci. **36**, 531 (2008)

Carr, M.H., McEwen, A.S., Howard, K.A., et al. Icarus **135**, 146 (1998). https://doi.org/10.1006/icar.1998.5979

Chyba, C.F., Jankowski, D.G., Nicholson, P.D.: Astron. Astrophys. **219**, L23 (1989). http://adsabs.harvard.edu/abs/1989A&A...219L..23C

Constable, S., Constable, C.: Geochem. Geophys. Geosyst. **5**, Q01006 (2004). https://doi.org/10.1029/2003GC000634

Costa, A., Caricchi, L., Bagdassarov, N.: Geochem. Geophys. Geosyst. **10**, n/a (2009). https://doi.org/10.1029/2008GC002138

Darwin, G.H.: MNRAS **60**, 82 (1899). https://doi.org/10.1093/mnras/60.2.82

Davies, A.G., Veeder, G.J., Matson, D.L., Johnson, T.V.: Icarus **262**, 67 (2015). https://doi.org/10.1016/j.icarus.2015.08.003

de Kleer, K., de Pater, I.: Icarus **280**, 405 (2016). https://doi.org/10.1016/j.icarus.2016.06.018

de Kleer, K., Nimmo, F., Kite, E.: Geophys. Res. Lett. **46**, 6327 (2019a). https://doi.org/10.1029/2019GL082691

de Kleer, K., McEwen, A.S., Park, R.S., et al.: Final Report for the Keck Institute for Space Studies (KISS) (2019b). https://www.kiss.caltech.edu/final_reports/Tidal_Heating_final_report.pdf

de Pater, I., Keane, J.T., de Kleer, K., Davies, A.G.: Annu. Rev. Earth Planet. Sci. **49**, 643 (2021). https://doi.org/10.1146/annurev-earth-082420-095244

Egbert, G.D., Ray, R.D.: J. Geophys. Res. Oceans **106**, 22475 (2001). https://doi.org/10.1029/2000jc000699

Faul, U., Jackson, I.: Annu. Rev. Earth Planet. Sci. **43**, 541 (2015)

Findley, W.N., Lai, J.S., Onaran, K.: Creep and Relaxation of Nonlinear Viscoelastic Materials With an Introduction to Linear Viscoelasticity. Dover (1976). https://doi.org/10.1115/1.3424077

Fuller, J., Luan, J., Quataert, E.: MNRAS **458**, 3867 (2016). https://doi.org/10.1093/mnras/stw609

Gaskell, R.W., Synnott, S.P., McEwen, A.S., Schaber, G.G.: Geophys. Res. Lett. **15**, 581 (1988). https://doi.org/10.1029/GL015i006p00581

Gomez Casajus, L., Zannoni, M., Modenini, D., et al.: Icarus **358**, 114187 (2021). https://doi.org/10.1016/j.icarus.2020.114187

Hamilton, C.W., Beggan, C.D., Still, S., et al.: Earth Planet. Sci. Lett. **361**, 272 (2013a)

Hamilton, C.W., Beggan, C.D., Still, S., et al.: Earth Planet. Sci. Lett. **361**, 272 (2013b)

Hedman, M.M., Gosmeyer, C.M., Nicholson, P.D., et al.: Nature **500**, 182 (2013). https://doi.org/10.1038/nature12371

Hemingway, D., Iess, L., Tajeddine, R., Tobie, G.: The interior of enceladus. In: Schenk, P.M., Clark, R.N., Howett, C.J.A., Verbiscer, A.J., Waite, J.H. (eds.) Enceladus and the Icy Moons of Satur, vol. 57 (2018). https://doi.org/10.2458/azu_uapress_9780816537075-ch004

Hood, L.L., Herbert, F., Sonett, C.P.: J. Geophys. Res. **87**, 5311 (1982). https://doi.org/10.1029/JB087iB07p05311

Hurford, T.A., Helfenstein, P., Hoppa, G.V., Greenberg, R., Bills, B.G.: Nature **447**, 292 (2007). https://doi.org/10.1038/nature05821

Hurford, T.A., Helfenstein, P., Spitale, J.N.: Icarus **220**, 896 (2012). https://doi.org/10.1016/j.icarus.2012.06.022

Hussmann, H., Spohn, T.: Icarus **171**, 391 (2004). https://doi.org/10.1016/j.icarus.2004.05.020

Ingersoll, A.P., Ewald, S.P.: Icarus **282**, 260 (2017). https://doi.org/10.1016/j.icarus.2016.09.018

Ingersoll, A.P., Ewald, S.P., Trumbo, S.K.: Icarus **344**, 113345 (2020). https://doi.org/10.1016/j.icarus.2019.06.006

Jackson, I., Faul, U.H.: Phys. Earth Planet. Interiors **183**, 151 (2010). https://doi.org/10.1016/j.pepi.2010.09.005

Jackson, I., Faul, U.H., Fitz Gerald, J.D., Tan, B.H.: J. Geophys. Res. B Solid Earth **109**, 1 (2004). https://doi.org/10.1029/2003JB002406

Jacobson, R.A.: AAS/Division for Planetary Sciences Meeting Abstracts, vol. 45, AAS/Division for Planetary Sciences Meeting Abstracts #45, 418.06 (2013)

Kaula, W.M.: Rev. Geophys. Space Phys. **2**, 661 (1964)

Kaula, W.M.: Theory of Satellite Geodesy. Applications of Satellites to Geodesy. Blaisdell, Waltham (1966)

Kaula, W.M.: An Introduction to Planetary Physics - The Terrestrial Planets. Space Science Text Series. Wiley, New York (1968)

Keane, J.T., Matsuyama, I.: Geophys. Res. Lett. **41**, 6610 (2014). https://doi.org/10.1002/2014GL061195

Keane, J.T., de Kleer, K., Rathbun, J., Ahern, A., Radebaugh, J.: AGU Fall Meeting Abstracts, vol. 2018, P53C–2983 (2018)

Keane, J.T., Ahern, A.A., Bagenal, F., et al.: Bull. Am. Astron. Soc. **53**, 178 (2021a). https://doi.org/10.3847/25c2cfeb.f844ca0e

Keane, J.T., Ahern, A.A., Bagenal, F., et al.: Bull. Am. Astron. Soc. **53**, 179 (2021b). https://doi.org/10.3847/25c2cfeb.3de45b59

Keszthelyi, L., McEwen, A.: Icarus **130**, 437 (1997). https://doi.org/10.1006/icar.1997.5837

Khan, A., Connolly, J.A.D., Pommier, A., Noir, J.: J. Geophys. Res. (Planets) **119**, 2197 (2014). http://doi.org/10.1002/2014JE004661

Khurana, K.K., Kivelson, M.G., Stevenson, D.J., et al.: Nature **395**, 777 (1998). https://doi.org/10.1038/27394

Khurana, K.K., Jia, X., Kivelson, M.G., et al.: Science **332**, 1186 (2011). https://doi.org/10.1126/science.1201425

Kirchoff, M.R., McKinnon, W.B., Schenk, P.M.: Earth Planet. Sci. Lett. **301**, 22 (2011). https://doi.org/10.1016/j.epsl.2010.11.018

Kite, E.S., Rubin, A.M.: Proc. Natl. Acad. Sci. **113**, 3972 (2016). https://doi.org/10.1073/pnas.1520507113

Kivelson, M.G., Khurana, K.K., Russell, C.T., et al.: Nature **384**, 537 (1996). https://doi.org/10.1038/384537a0

Kivelson, M.G., Khurana, K.K., Russell, C.T., et al.: Science **289**, 1340 (2000). https://doi.org/10.1126/science.289.5483.1340

Kivelson, M.G., Khurana, K.K., Volwerk, M.: Icarus **157**, 507 (2002). https://doi.org/10.1006/icar.2002.6834

Kivelson, M.G., Bagenal, F., Kurth, W.S., et al.: In: Bagenal, F., Dowling, T.E., McKinnon, W.B. (eds.) Jupiter. The Planet, Satellites and Magnetosphere, vol. 1, pp. 513–536 (2004)

Lainey, V., Arlot, J.-E., Karatekin, Ö., Van Hoolst, T.: Nature **459**, 957 (2009)

Lambeck, K.: The earth's Variable Rotation: Geophysical Causes and Consequences. Cambridge University Press (1980)

Lambeck, K., Purcell, A., Zhao, S.: Q. Sci. Rev. **158**, 172 (2017)

Leleu, A., Alibert, Y., Hara, N.C., et al.: Astron. Astrophys. **649**, A26 (2021). https://doi.org/10.1051/0004-6361/202039767

Lopes-Gautier, R., McEwen, A.S., Smythe, W.B., et al.: Icarus **140**, 243 (1999). https://doi.org/10.1006/icar.1999.6129

Love, A.E.H.: A Treatise on the Mathematical Theory of Elasticity, Dover Books on Advanced Mathematics, 4th edn. Dover, New York (1944)

Luger, R., Sestovic, M., Kruse, E., et al.: Nat. Astron. **1**, 0129 (2017). https://doi.org/10.1038/s41550-017-0129

Matsuyama, I., Nimmo, F., Keane, J.T., et al.: Geophys. Res. Lett. **43**, 8365 (2016). https://doi.org/10.1002/2016GL069952

Matsuyama, I., Beuthe, M., Hay, H.C.F.C., Nimmo, F., Kamata, S.: Icarus **312**, 208 (2018)

McEwen, A.S., Keszthelyi, L.P., Lopes, R., Schenk, P.M., Spencer, J.R.: The lithosphere and surface of Io. In: Bagenal, F., Dowling, T.E., McKinnon, W.B. (eds.) Jupiter: The Planet, Satellites and Magnetosphere, vol. 1, pp. 307–328 (2004)

Mitrovica, J., Forte, A.: Earth Planet. Sci. Lett. **225**, 177 (2004)

Moore, W.B.: Icarus **154**, 548 (2001). https://doi.org/10.1006/icar.2001.6739

Moore, W.B.: J. Geophys. Res. **108**, 1 (2003). https://doi.org/10.1029/2002JE001943

Moore, W.B., Schubert, G., Anderson, J.D., Spencer, J.R.: The interior of Io. In: Lopes, R.M.C., Spencer, J.R. (eds.) Io After Galileo, vol. 89 (2007). https://doi.org/10.1007/978-3-540-48841-5_5

Morabito, L.A.: arXiv e-prints, arXiv:1211.2554 (2012). https://arxiv.org/abs/1211.2554

Morabito, L.A., Synnott, S.P., Kupferman, P.N., Collins, S.A.: Science **204**, 972 (1979). https://doi.org/10.1126/science.204.4396.972.a

Murray, C.D., Dermott, S.F.: Solar System Dynamics. Cambridge University Press. (2000). https://doi.org/10.1017/CBO9781139174817

Nimmo, F., Porco, C., Mitchell, C.: Astron. J. **148**, 46 (2014). https://doi.org/10.1088/0004-6256/148/3/46

O'Reilly, T.C., Davies, G.F.: Geophys. Res. Lett. **8**, 313 (1981). https://doi.org/10.1029/GL008i004p00313

Oberst, J., Schuster, P.: J. Geophys. Res. (Planets) **109**, E04003 (2004). https://doi.org/10.1029/2003JE002159

Ojakangas, G., Stevenson, D.: Icarus, **66**, 341 (1986). https://doi.org/10.1016/0019-1035(86)90163-6

Ojakangas, G.W., Stevenson, D.J.: Icarus **81**, 220 (1989). https://doi.org/10.1016/0019-1035(89)90052-3

Peale, S., Cassen, P.: Icarus **36**, 245 (1978)

Peale, S.J., Cassen, P., Reynolds, R.T.: Science **203**, 892 (1979)

Platzman, G.W.: Rev. Geophys. Space Phys. **22**, 73 (1984)

Pommier, A., Leinenweber, K., Kohlstedt, D.L., et al.: Nature **522**, 202 (2015). https://doi.org/10.1038/nature14502

Radebaugh, J., Keszthelyi, L.P., McEwen, A.S., et al.: J. Geophys. Res. **106**, 33005 (2001). https://doi.org/10.1029/2000JE001406

Rathbun, J., Spencer, J., Tamppari, L., et al.: Icarus **169**, 127 (2004)

Rathbun, J.A., Lopes, R. M.C., Spencer, J.R.: AJ **156**, 207 (2018). https://doi.org/10.3847/1538-3881/aae370

Renaud, J.P., Henning, W.G.: Astrophys. J. **857**, 98 (2018)

Rivera, E.J., Laughlin, G., Butler, R.P., et al.: ApJ **719**, 890 (2010). https://doi.org/10.1088/0004-637X/719/1/890

Ross, M.N., Schubert, G.: Icarus **64**, 391 (1985)

Ross, M., Schubert, G.: J. Geophys. Res. Solid (1986). https://doi.org/10.1029/JB091iB04p0D447/full

Ross, M.N., Schubert, G., Spohn, T., Gaskell, R.W.: Icarus **85**, 309 (1990). https://doi.org/10.1016/0019-1035(90)90119-T

Roth, L., Saur, J., Retherford, K.D., et al.: J. Geophys. Res. (Space Phys.) **122**, 1903 (2017). https://doi.org/10.1002/2016JA023701

Rovira-Navarro, M., Katz, R.F., Liao, Y., van der Wal, W., Nimmo, F.: J. Geophys. Res. (Planets) **127**, e07117 (2022). https://doi.org/10.1029/2021JE007117

Saur, J., Duling, S., Roth, L., et al.: J. Geophys. Res. (Space Phys.) **120**, 1715 (2015). https://doi.org/10.1002/2014JA020778

Scheeres, D.J.: Orbital Motion in Strongly Perturbed Environments (2012)

Schenk, P.M., Bulmer, M.H.: Science **279**, 1514 (1998). https://doi.org/10.1126/science.279.5356.1514

Schenk, P., Hargitai, H., Wilson, R., McEwen, A., Thomas, P.: J. Geophys. Res. **106**, 33201 (2001). https://doi.org/10.1029/2000JE001408

Schubert, G., Anderson, J.D., Spohn, T., McKinnon, W.B.: Interior composition, structure and dynamics of the Galilean satellites. In: Bagenal, F., Dowling, T.E., McKinnon, W.B. (eds.) Jupiter: The Planet, Satellites and Magnetosphere, vol. 1, pp. 281–306 (2004)

Šebek, O., Trávníček, P.M., Walker, R.J., Hellinger, P.: J. Geophys. Res. (Space Phys.) **124**, 313 (2019). https://doi.org/10.1029/2018JA026153

Segatz, M., Spohn, T., Ross, M.N., Schubert, G.: Icarus **75**, 187 (1988)

Shoji, D., Hussmann, H.: J. Geophys. Res. (Planets) **121**, 2055 (2016). https://doi.org/10.1002/2016JE005053

Shoji, D., Kurita, K.: ApJ **789**, 3 (2014). https://doi.org/10.1088/0004-637X/789/1/3

Showalter, M.R., Hamilton, D.P.: Nature **522**, 45 (2015). https://doi.org/10.1038/nature14469

Smith, B.A., Soderblom, L.A., Johnson, T.V., et al.: Science **204**, 951 (1979)

Spencer, D.C., Katz, R.F., Hewitt, I.J.: J. Geophys. Res. Planets **125**, e2020JE006443 (2020)

Steinke, T., Hu, H., Höning, D., Van der Wal, W., Vermeersen, B.: Icarus **335**, 113299 (2020a)

Steinke, T., van Sliedregt, D., Vilella, K., van der Wal, W., Vermeersen, B.: J. Geophys. Res. Planets **125**, e2020JE006521 (2020b)

Tackley, P.J., Schubert, G., Glatzmaier, G.A., et al.: Icarus **149**, 79 (2001). https://doi.org/10.1006/icar.2000.6536

Thomas, P.C., Davies, M.E., Colvin, T.R., et al.: Icarus **135**, 175 (1998). https://doi.org/10.1006/icar.1998.5987

Tobie, G., Mocquet, A., Sotin, C.: Icarus **177**, 534 (2005)

Tricarico, P.: ApJ **782**, 99 (2014). https://doi.org/10.1088/0004-637X/782/2/99

Turcotte, D.L., Schubert, G.: Geodynamics, 2nd edn. (2002). https://doi.org/10.2277/0521661862

Turtle, E.P., Jaeger, W.L., Schenk, P.M.: Ionian mountains and tectonics: insights into what lies beneath Io's lofty peaks. In: Lopes, R.M.C., Spencer, J.R. (eds.) Io After Galileo, p. 109 (2007). https://doi.org/10.1007/978-3-540-48841-5_6

Tyler, R.H., Henning, W.G., Hamilton, C.W.: Astrophys. J. Suppl. Ser. **218**, 22 (2015)

Usselman, T.M.: Am. J. Sci. **275**, 291 (1975a). https://doi.org/10.2475/ajs.275.3.291

Usselman, T.M.: Am. J. Sci. **275**, 278 (1975b). https://doi.org/10.2475/ajs.275.3.278

Van Hoolst, T., Baland, R.-M., Trinh, A., Yseboodt, M., Nimmo, F.: J. Geophys. Res. (Planets) **125**, e06473 (2020). https://doi.org/10.1029/2020JE006473

Veeder, G.J., Matson, D.L., Johnson, T.V., Blaney, D.L., Goguen, J.D.: J. Geophys. Res. Planets **99**, 17095 (1994)
Veeder, G.J., Davies, A.G., Williams, D.A., et al.: Icarus **212**, 236 (2011). https://doi.org/10.1016/j.icarus.2010.09.026
Veeder, G.J., Davies, A.G., Matson, D.L., et al.: Icarus **219**, 701 (2012)
Veeder, G.J., Davies, A.G., Matson, D.L., et al.: Icarus **245**, 379 (2015). https://doi.org/10.1016/j.icarus.2014.07.028
White, O.L., Schenk, P.M., Nimmo, F., Hoogenboom, T.: J. Geophys. Res. (Planets) **119**, 1276 (2014). https://doi.org/10.1002/2013JE004591
Wieczorek, M.: Gravity and topography of the terrestrial planets, pp. 153–193 (2015). https://doi.org/10.1016/B978-0-444-53802-4.00169-X
Williams, D.A., Keszthelyi, L.P., Crown, D.A., et al.: Icarus **214**, 91 (2011). https://doi.org/10.1016/j.icarus.2011.05.007
Yoder, C.F., Peale, S.J.: Icarus **47**, 1 (1981). https://doi.org/10.1016/0019-1035(81)90088-9
Zimmer, C., Khurana, K.K., Kivelson, M.G.: Icarus **147**, 329 (2000). https://doi.org/10.1006/icar.2000.6456
Zschau, J.: Tidal Friction and the Earth's Rotation. Springer, Berlin, Heidelberg, pp. 62–94 (1978)

Chapter 5
Geology of Io

David A. Williams, Paul M. Schenk, and Jani Radebaugh

Abstract Io is the most volcanically active object in the Solar System beyond Earth. This tidally-heated moon manifests its activity via >400 volcanic vents (many of which have been active since spacecraft observations began in 1979), a surface covered in dark and bright lava flow fields and colorful, sulfur-bearing pyroclastic deposits. There are also isolated fault-block mountains up to 14 km high, and explosive plumes that jet gas and dust hundreds of kilometers above its surface. In this chapter we provide an overview of the variety of geologic features found on Io's surface. We discuss the variations in Io's five major geologic feature types (plains, lava flows, paterae (volcano-tectonic depressions), mountains, and diffuse (pyroclastic) deposits), and the variety of structural and geologic features found on its surface. We also review what is known about Io's global and long-wavelength topography, based on limb and stereo-based imaging. We conclude the chapter with a discussion of the necessary spacecraft observations required to improve our understanding of Io's geology, and how those observations have informed the design of recent Io-dedicated mission concepts. Finally, we provide a download link to the proto-Io Planetary Spatial Data Infrastructure (PSDI), the Io GIS Database version 1.0.

D. A. Williams (✉)
Arizona State University, Tempe, AZ, USA
e-mail: David.Williams@asu.edu

P. M. Schenk
Lunar and Planetary Institute, Houston, TX, USA
e-mail: schenk@lpi.usra.edu

J. Radebaugh
Brigham Young University, Provo, UT, USA
e-mail: jani.radebaugh@byu.edu

R. M. C. Lopes et al. (eds.), *Io: A New View of Jupiter's Moon*, Astrophysics and Space Science Library 468, https://doi.org/10.1007/978-3-031-25670-7_5

5.1 Introduction

The surface of Io is unique in the Solar System, in that its surface geology is almost completely dominated by volcanic and tectonic features. These features result from tidally-induced, extreme volcanic activity on a body with a tenuous, volcano-produced atmosphere. This extreme volcanism, causing a resurfacing rate of ~1 cm/year (Johnson et al. 1979), has effectively erased all impact craters from Io's surface. In fact, Io is the only object in the Solar System on which there have been zero impact craters identified in any image at any available resolution. Io is thus an end member for planetary surfaces, and is perhaps an ideal analog for "lava exoplanets" now being discovered in other solar systems (Chao et al. 2021).

In this chapter, we will review the geology of Io's surface, as recognized primarily from images returned by NASA's *Voyager* and *Galileo* missions. Both explosive and effusive volcanic activity manifest themselves in Io's geologic features, as well as the tectonic response to all of this activity (i.e., Io's unique mountains). We will review Io's surface feature types and extents as revealed through global geologic mapping and topographic analyses. Finally, we will conclude with a statement about what data are needed to better understand Io's geology through future missions.

5.2 Background

We began to learn about the geologic structures and features of Io thanks to the images obtained by the NASA *Voyager* spacecraft during their 1979 flybys (Smith et al. 1979a, b). Analysis of the *Voyager* Imaging Science Subsystem (ISS) images first revealed the nature of this volcanically active world. Through comparison to aerial photographs of terrestrial volcanic features, and to the images of volcanic features on the Moon obtained by *Lunar Orbiter* and *Apollo* spacecraft and of Mars obtained by the *Mariner 9* and *Viking* orbiters, the initial characterization of Io's geology was summarized in Schaber (1980, 1982). In particular Schaber (1980, 1982) recognized the five primary types of geologic features on Io: plains, paterae (caldera-like depressions), lava flows, mountains, and diffuse deposits (pyroclastic materials and condensed gases). This generalized classification has held up over time and through analysis of high-resolution images obtained by multiple flybys from NASA's *Galileo* orbiter.

Global syntheses of Io's geology were presented through two global geologic maps, in the *Voyager* era by Crown et al. (1992), and the post-*Galileo* era by Williams et al. (2011a). The Crown et al. (1992) map synthesized four regional geologic maps (Moore 1987; Greeley et al. 1988; Whitford-Stark et al. 1991; Schaber et al. 1989) covering much of the subjovian hemisphere and south polar region, but with large portions of the surface unmapped. The Williams et al. (2011a, b) map (Figs. 5.1 and 5.2), based upon a set of combined *Galileo-Voyager* mosaics produced by the U.S. Geological Survey (Becker and Geissler 2005), provided the

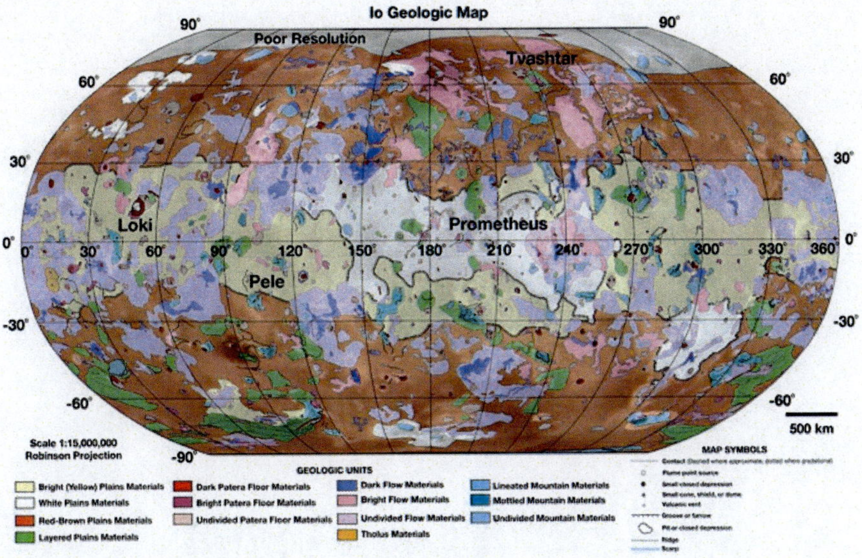

Fig. 5.1 Global geologic map of Io, modified from Williams et al. (2011a)

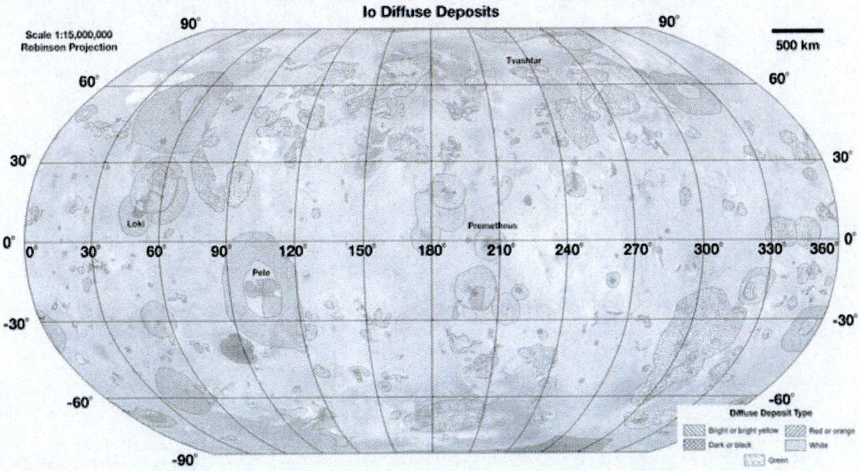

Fig. 5.2 Global map of Io's diffuse deposits, modified from Williams et al. (2011a)

first nearly complete global accounting of Io's geologic features, and the areal distribution of these major feature types is provided in Table 5.1. This global mapping exercise took advantage of lessons learned through a series of regional geologic mapping projects based on *Galileo* SSI images (Williams et al. 2002, 2004, 2005, 2007; Bunte et al. 2008, 2010; Leone et al. 2009).

Table 5.1 Geologic units of Io, expressed as fraction of surface area. After Williams et al. (2011b)

Material unit	Area (km^2)	Area (%)	Material unit	Area (km^2)	Area (%)
Plains	**2.74E7**	**65.75**	**Patera floors**	**1.03E6**	**2.47**
Red-brown plains	1.38E7	33.15	Bright patera floors	1.72E5	0.41
Bright (yellow) plains	7.47E6	17.92	Dark patera floors	2.14E5	0.51
White plains	3.73E6	8.94	Undivided patera floors	6.44E5	1.54
Layered plains	1.98E6	4.74			
			Lava flows	**1.19E7**	**28.52**
Region of poor resolution	**4.11E5**	**0.99**	Bright lava flows	1.97E6	4.71
			Dark lava flows	1.51E6	3.62
Mountains	**1.36E6**	**3.27**	Undivided lava flows	8.42E6	20.19
Lineated mountains	6.62E5	1.59			
Mottled mountains	8.02E4	0.19	**Diffuse deposits**	**7.66E6**	**18.2**
Undivided mountains	5.64E5	1.35	Bright (yellow) diffuse dep.	8.76E5	2.1
Tholi (domes)	5.57E4	0.13	White diffuse deposits	2.90E6	6.9
			Red diffuse deposits	3.61E6	8.6
			Dark diffuse deposits	2.68E5	0.6
			Green diffuse deposits	4.09E3	0.01

Note: Diffuse deposits are superposed on all other materials, and cover 18.2% of Io's surface

Geologic features on Io are named after gods and goddesses of fire, war, or thunder, befitting a volcanic planet. The greatest percentage of Io's surface (~66%) is covered in **plains**, composed of accumulations of lava flows, pyroclastic deposits, and condensed gases from Io's many volcanic eruptions. The equatorial region of Io is dominated by **yellow plains**, which are enriched in sulfur-bearing pyroclastic materials, whereas **white plains** have enhanced SO_2-signatures consistent with cold traps containing condensed SO_2 gases. The mid-latitude to polar regions are dominated by **red-brown plains**, which are thought to consist of a combination of radiation-altered yellow and white plains, and thick concentrations of S_2-rich condensates and short-chain sulfur allotropes (S_3, S_4) produced in volcanic eruptions (Spencer et al. 2000).

The second greatest percentage of Io's surface (~29%) is composed of **lava flows**, fed by either fissure vents or point source vents in and around Io's paterae. Lava flows are classified by color as a proxy for composition: **Dark flows** are thought to be silicate in nature, perhaps akin to terrestrial basaltic or komatiitic flows (McEwen et al. 1998a, b; Williams et al. 2000), whereas **bright flows** are thought to be sulfurous in nature, composed of sulfur or more rarely SO_2 effusions (Williams et al. 2001a). On Io over time, radiation exposure and mantling by pyroclastic materials cause dark flows to brighten and bright flows to darken, producing flows of indeterminate age that have been mapped as **undivided flows**.

Although active hot spots have been detected in Io's lava flow fields (~31% of all hot spots), the majority of hot spots (~66%) are located in Io's paterae (sing. patera). **Paterae** are caldera-like volcano-tectonic depressions in Io's crust that appear to host lava flows and lava lakes on their floors (Radebaugh et al. 2001; Lopes et al. 2004), although many flows break through paterae walls and flow out onto the plains. Patera floors make up ~2.5% of Io's surface. Williams et al. (2011a, b) mapped 425 individual paterae, and there are likely more in areas of poor resolution.

Io's tectonism is exemplified by **mountains**, where 125 mapped structures cover ~3.3% of the surface. Io's mountains range from 1 to 17 km in height (Schenk et al. 2001; Turtle et al. 2001), and ~40% of mountains are in close proximity (within 25 km) to paterae (Jaeger et al. 2003). Only a small number of positive-relief volcanic constructs such as domes, called **tholi** (sing. tholus) have been identified in *Voyager* and *Galileo* images, and these features make about one-tenth of 1% of Io's surface. Likely there are many more present, beyond the limit of current image spatial resolution.

Diffuse deposits are the result of explosive eruptions on Io, and it was the detection of a plume on Io's limb in a *Voyager* navigation image that confirmed Io as an active volcanic world (Morabito et al. 1979). Io's explosive eruptions at central vents are dominated by umbrella-shaped plumes, which deposit both pyroclastic ash and condensed gases in colorful rings around many paterae. Plumes also form at the fronts of lava flows (Kieffer et al. 2000), vaporizing solid ices that jet away perpendicular to the flow fronts. Thanks to color imaging by the *Galileo* SSI, five types of colored plume deposits were identified and mapped.

In the following sections we go into greater detail and show examples of each of Io's geologic features. Following, we discuss the variety of structural and tectonic features observed.

5.3 Plains

Plains cover ~66% of Io's surface, and are subdivided into red-brown (~33%), yellow (~18%), and white (~9%) color subunits (Williams et al. 2011b). Layered plains is a fourth subunit covering ~5% of the surface, and includes plains of any color that are morphologically separated from underlying units by a bounding scarp, indicative of erosional processes (Moore et al. 2001). Morphologically, plains appear smooth in lower resolution, global to regional images, but in highest resolution *Galileo* SSI images they have a distinct hilly texture (Figs. 5.3 and 5.4), possibly due to repeated sublimation and deposition of volatile condensates (e.g., Milazzo et al. 2001).

Plains are thought to be composed of a variety of buried volcanic materials (lava flows, pyroclastic deposits, condensed gases in the form of frosts) from multiple volcanic sources, superposed by fresher deposits traceable to specific active

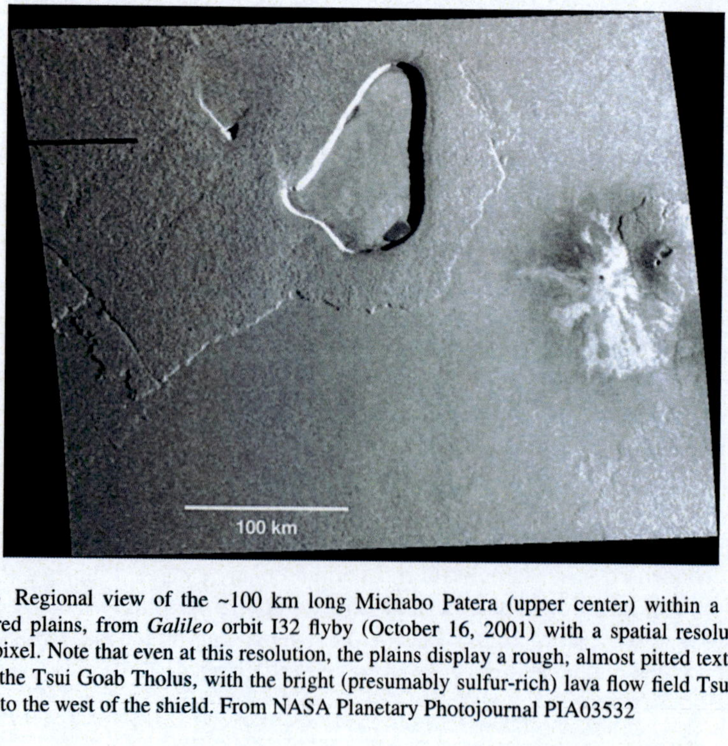

Fig. 5.3 Regional view of the ~100 km long Michabo Patera (upper center) within a section of layered plains, from *Galileo* orbit I32 flyby (October 16, 2001) with a spatial resolution of 330 m/pixel. Note that even at this resolution, the plains display a rough, almost pitted texture. At right is the Tsui Goab Tholus, with the bright (presumably sulfur-rich) lava flow field Tsui Goab Fluctus to the west of the shield. From NASA Planetary Photojournal PIA03532

Fig. 5.4 Close up of margin of Prometheus lava flow field and surrounding white plains, from *Galileo* orbit I27 flyby (February 22, 2000) with a spatial resolution of 12 m/pixel. Note the hilly texture of the plains, some of which are covered by SO_2 frosts redeposited after being vaporized at the lava flow fronts (Kieffer et al. 2000; Milazzo et al. 2001). From NASA Planetary Photojournal PIA02557

volcanic centers (Smythe et al. 1979; Williams et al. 2002). The overriding color of plains units is caused by the dominant material composing them: White plains consist of coarse- to moderate-sized grains of SO_2 frosts and contaminants (Pearl et al. 1979; Carlson et al. 1997; Douté et al. 2001, 2002, 2004); Yellow plains consist of sulfur-rich compounds (S_8, $\pm$ S_nO and S_2O) and likely mixed with SO_2 (Hapke 1989; Geissler et al. 1999); Red-brown plains consist of deposits containing short-chain sulfur (S_3, S_4) recrystallized from condensed S_2 gas (Spencer et al. 2000), possibly mixed with salts such as sulfur chlorides (Lellouch et al. 2003;

Schmidt and Rodriguez 2003). Additional red-brown plains may form at higher latitudes from alteration by radiation exposure (Johnson 1997; Geissler et al. 1999). Global geologic mapping determined that most white plains are geographically concentrated between 30°N and 30°S and between ~90° and 230°E in the antijovian hemisphere, suggestive of a regional cold trap (Williams et al. 2011a, b).

While plains are generally flat, there are elevated, plateau-like regions that exhibit erosional textures at their margins in the form of lobate, straight-walled scarps (Moore et al. 1996). These may be formed by sublimation-related erosion, which removes SO_2 and other volatiles from the walls of the plateaus, weakening the walls and causing collapse (Moore et al. 2001). More traditional landslides are also found near tall mountains, evidencing the presence of stresses in the lithosphere or weakening of mountain structures through volatile layers that can act as failure surfaces (Schenk and Bulmer 1998; Schenk et al. 2001; Turtle et al. 2001; Ahern et al. 2017).

5.4 Paterae and Paterae Floors

Global geologic mapping (Williams et al. 2011a, b; Radebaugh et al. 2001) of the combined *Galileo-Voyager* mosaics (Becker and Geissler 2005) has identified 425 paterae. More are likely present, but have not been resolved in available images, and most likely buried or inactive paterae at the time of the *Galile*o mission (1996–2003) have become reactivated, such that additional paterae will be observed during the next Io observing mission obtaining high spatial resolution images. Paterae manifest themselves as closed topographic depressions (Figs. 5.3 and 5.5), with steep walls (up to several km; Radebaugh et al. 2001) over relatively flat floors, and with margins that vary from circular to polygonal to irregular (Dundas 2017). They are interpreted to be caldera-like, volcano-tectonic depressions (McEwen et al. 1985), and resurfacing of their floors by lava flows and/or lava lakes (Fig. 5.6) was observed to occur during repeated flybys of the *Galileo* mission (Radebaugh et al. 2001; Lopes et al. 2004).

Paterae may form like traditional collapse calderas on Earth, through removal of subsurface materials and subsequent piston-like or trap-door collapse (Radebaugh et al. 2001). However, many paterae have irregular shapes, including some with long, straight margins suggestive of formation related to faulting (Fig. 5.3). Paterae with these shapes, along with the location of many paterae close to regions of inferred orogenic faults, suggest that at least some paterae are pull-apart basins related to mountain formation (Fig. 5.7) (McEwen et al. 2000; Radebaugh et al. 2001; Jaeger et al. 2003). The presence of SO_2 frosts on Io's plains, and the observance of sulfur-rich fluids in paterae (e.g., Fig. 5.5) indicates that volatiles may be an important component in patera formation. Perhaps interlayering of basaltic lavas with SO_2 frosts could lead to a preponderance of volatiles in the crust that can be liberated by hot, rising magma (Radebaugh et al. 2004). This would cause collapse of the cold, brittle crust and subsumption into an underlying, near-surface magma chamber

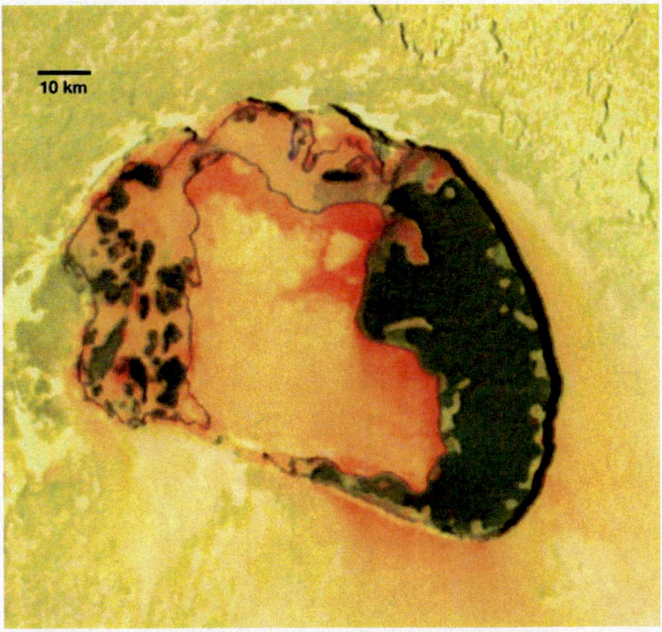

Fig. 5.5 Tupan Patera as imaged during the *Galileo* orbit I32 flyby (October 16, 2001) with a spatial resolution of 135 m/pixel. Shadows indicate a wall height of 1 km. This is the highest resolution color view of an Ionian patera. Note the fresh, dark lava lake on the right, and the diffuse rind of red, short chain sulfur deposits derived from S_2 gas around the margins of the depression

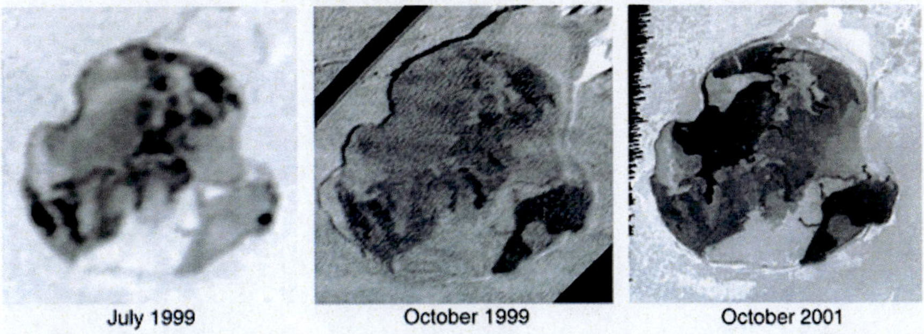

Fig. 5.6 Gish Bar Patera (106.3 × 115 km), observed during three different *Galileo* spacecraft flybys. Note the fresh, dark lava flows that have resurfaced the northwest and southeast sections of the patera floor. From NASA Photojournal PIA03884

(Keszthelyi et al. 2004). After collapse, the pit fills with lava accessed from near the surface, forming lava lakes visible from spacecraft and telescopes (Lopes et al. 2004). In many ways Io's paterae can thus be thought of as unroofed intrusions (Keszthelyi et al. 2007). Paterae across Io have some patterns of distribution; more paterae are found at the sub- and anti-jovian quadrants, while fewer and larger

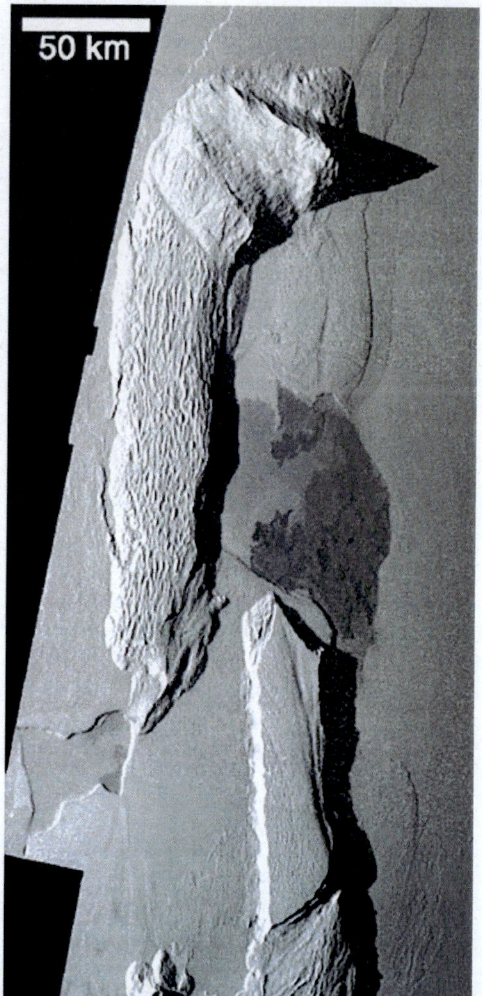

Fig. 5.7 Hi'iaka Montes and Patera, imaged during the *Galileo* orbit I25 flyby (November 25, 1999) with a spatial resolution of 260 m/pixel. Faulting and subsequent formation of Hi'iaka Patera was suggested to pull apart the two segments of Hi'iaka Montes (Jaeger et al. 2003)

paterae are found near the poles (Radebaugh et al. 2001). This distribution correlates with an asthenospheric tidal heating model (Tackley et al. 2001; Kirchoff et al. 2011) in which tidal heating is directed along the Io-Jupiter axis, leading to crustal thinning, while thicker crust is present at polar regions, leading to a barrier to the formation of all but the largest paterae (Radebaugh et al. 2001). On a finer examination, paterae appear to repel, according to Monte Carlo models of their distributions (Hamilton et al. 2013). This may indicate there is a necessary volume of lava required for patera formation, and thus paterae cannot cluster together. While many paterae are found adjacent to mountains (Radebaugh et al. 2001), there is also a global anti-correlation of mountains and paterae (Kirchoff et al. 2011; McGovern et al. 2016; Bland and McKinnon 2016), indicating that processes that lead to formation of one feature does not allow for formation of the other.

Paterae floors compose ~2.5% of Io's surface, and are subdivided into three geologic units based on albedo and color: Dark patera floors are dominated by silicate (basaltic to ultramafic) lava flows and/or lava lakes; bright patera floors are dominated by sulfur-rich materials, including sulfur lava flows/lava lakes and/or concentrated SO_2 deposits in very white floors; undivided patera floors have intermediate albedos between dark and bright, and likely represent older materials that have been mantled by pyroclastics and condensed gases. Although they cover only a small fraction of Io's surface, paterae host a majority (~66%) of Io's thermal emission anomalies, or "hot spots" detected by Earth-based telescopes and passing spacecraft (Williams et al. 2011a, b). These hot spots were found to represent fresh eruptions of lava (Keszthelyi et al. 2001; Milazzo et al. 2005; Gregg and Lopes 2008), or overturn of crusts on lava lakes (Rathbun and Spencer 2006; Lopes et al. 2004).

5.5 Lava Flows

Lava flows manifest as generally elongated (lengths >> widths), lobate features with sharp, curvilinear to crenulate margins, and sharp contacts with surrounding units (Figs. 5.3, 5.4, 5.8 and 5.9). They make up ~28.5% of Io's surface (Williams et al. 2011a, b). Like patera floors, flows are subdivided based on albedo and color into three subunits: Dark flows (~3.6% of surface) are dominantly silicate (basaltic to ultramafic) lava flows (Figs. 5.4 and 5.8); bright flows (~4.7% of surface, Figs. 5.3 and 5.9) are either dominantly sulfur or sulfur-bearing lava flows (Williams et al. 2001a; rare white flows may be SO_2 flows) or silicate flows mantled by bright sulfur-rich pyroclastic materials (Kieffer 1982; McEwen et al. 1998a); and undivided flows (~20.2% of surface) are older flows that could be either silicate or sulfur in origin, that have been mantled by pyroclastics and condensed gases such that their original composition cannot be determined.

Albedo variations suggest age of flows: the freshest flows are the darkest or brightest, and over time dark flows brighten and bright flows darken to an intermediate albedo (McEwen et al. 1998a; Williams et al. 2002). Radiation exposure appears to alter silicate flows with thin mantles of sulfur-rich materials into a gray-green color in SSI images (Geissler et al. 1999). It has been suggested that all bright flows are just dark silicate flows mantled by bright S/SO_2-rich frosts, but distinctive, bright flows exist that appear to lack thick mantles (e.g., Fig. 5.9), more likely indicative of fresh sulfur flows (Williams et al. 2001a, 2002).

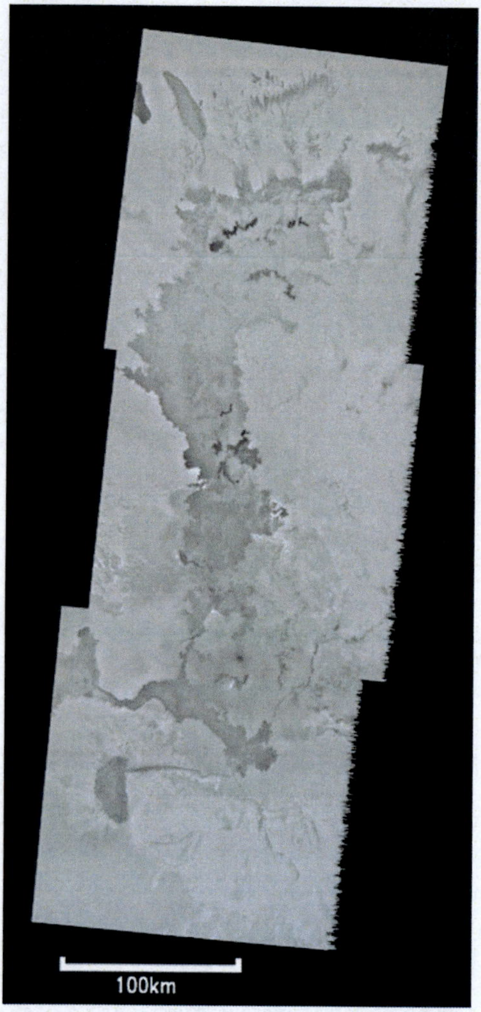

Fig. 5.8 *Galileo* SSI mosaic of the Amirani lava flow field, obtained during the I27 flyby (February 22, 2000) with a spatial resolution of 210 m/pixel. Lava tubes or channels are thought to deliver silicate lava from the interior to the distal margins of the flow, some 300 km distant (Keszthelyi et al. 2001)

Comparison of the highest resolution SSI images of Io's flows (e.g., Fig. 5.4) to aerial photos of terrestrial lava flows suggest that Io's lava flows most closely resemble terrestrial compound pahoehoe flow fields or platy ridged lava flows as found in Hawaii and Iceland (Keszthelyi et al. 2001; Williams et al. 2001b). Lava channels have been seen in some images (Keszthelyi et al. 2001; Schenk and Williams 2004), and presumably lava tubes also exist on Io to deliver low-viscosity silicate and sulfur lava to edges of flow fields, up to several hundred kilometers. Lava flows are distributed at all latitudes and longitudes, but with a slight drop off toward the poles possibly reflecting a pole-ward change in crustal thickness, magma distribution, or heat flow (Williams et al. 2011b). Interestingly, there are 30% more bright flows than dark flows, although most active hot spots in flow fields

Fig. 5.9 *Galileo* SSI mosaic showing the Sobo Fluctus lava flow field, imaged during the I27 flyby (February 22, 2000) with a spatial resolution of 186 m/pixel. Note the smaller, "running man"-shaped darker flows north of a larger bright, presumably sulfur-rich flow field. Melting of sulfur-rich country rock by nearby hot silicate lavas could induce secondary sulfur volcanism (Greeley et al. 1984) to produce adjacent sulfur flow fields (Williams et al. 2002)

coincide with dark flows. This discrepancy might suggest a role for secondary sulfur volcanism (Greeley et al. 1984), or a greater role for recent primary sulfur volcanism than previously thought (Williams et al. 2011b).

5.6 Mountains and Tholi

There has been a total of 125 mountains (not counting layered plains) identified and mapped on Io (Williams et al. 2011b), compared to ~150 mountains and

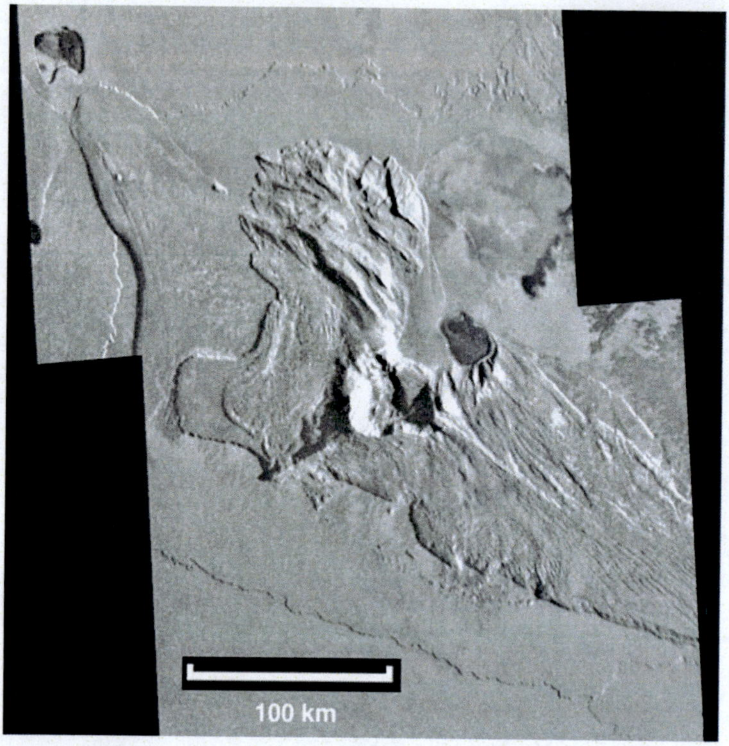

Fig. 5.10 Tohil Mons, as imaged during the *Galileo* I32 flyby (October 16, 2001) with a spatial resolution of 330 m/pixel. This mountain is 5.4 km (18,000 ft) in height above the surrounding plains, and contains both lineated (fractured) and mottled (displaced by mass movement) units

layered plains identified by prior work (Carr et al. 1998; Schenk and Bulmer 1998; McKinnon et al. 2001; Schenk et al. 2001; Turtle et al. 2001; Jaeger et al. 2003). The difference may be one of interpretation of features in images with a variety of illumination conditions. Mountains (Figs. 5.7, 5.10 and 5.11) are mapped as positive relief topographic features, but are only visible in low-sun or stereo images that highlight scarps and shadows, for which there is incomplete global coverage. Mountains make up ~3.3% of Io's surface, and can be hard to identify because their albedos and colors tend to match surrounding plains.

Mountains have been subdivided into three geologic subunits, plus ~3–4% that are tholi (discussed shortly): lineated (~48% of mountains) contain well-defined grooves, ridges, and peaks, and are interpreted as tectonically uplifted autochthonous crustal blocks; mottled (~6% of mountains) contain dome-like hills and lobes, and are interpreted as materials displaced by mass movement or other forms of erosion; and undivided (~42% of mountains) lack distinguishing features (due in part to low image quality) such that they cannot be classified as either lineated or mottled. Global mapping found that ~35% of mapped mountains (not

Fig. 5.11 Closeup of Tohil Mons, including the peak, the mottled south flank, and Radegast Patera on the northern margin of the mountain. This *Galileo* SSI mosaic was also taken during the I32 flyby (October 16, 2001) with a spatial resolution of 50 m/pixel

counting layered plains) are within 25 km of paterae, and ~17% of mapped mountains are in contact with paterae (Williams et al. 2011b), consistent with other research showing associations of mountains to paterae (Radebaugh et al. 2001; Jaeger et al. 2003).

Mass wasting is evident on Io both in the scarps in layered plains, and in the mottled mountain units. Mass wasting appears as alcove-carving slopes and slides (Moore et al. 2001). These features are most likely the result of block release and brittle slope failure (Moore et al. 2001). Mountains are likely large-scale crustal blocks that have undergone modification through local responses of subsurface structures to local and regional stresses (Ahern et al. 2017).

The presence of mountains on Io has had important implications for understanding its crustal structure and interior dynamics. Current thinking (Keszthelyi et al. 2004, 2007) builds on past theories that accumulation of materials on the surface from volcanic eruptions puts the surface under compression, until stress

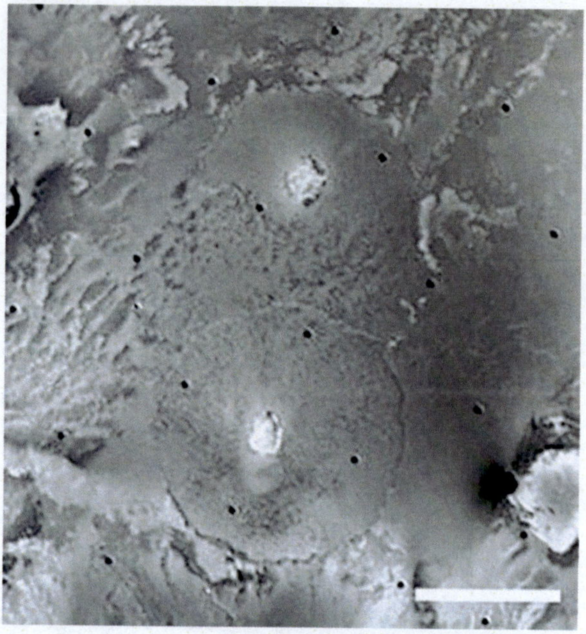

Fig. 5.12 Apis Tholus (top) and Inachus Tholus (bottom) are the only two large shield volcanoes identified on Io. *Voyager* image FDS 1639034. Scale bar is 100 km

is released by tectonic faulting and crustal blocks are raised as lineated mountains, that erode to mottled mountains over geologic time (e.g., Carr et al. 1998; Schenk and Bulmer 1998; McKinnon et al. 2001; Schenk et al. 2001; Turtle et al. 2001; Jaeger et al. 2003). Recent studies indicate Io's mountains may form by a variety of stresses in the upper lithosphere (Ahern et al. 2017), with modeling that corroborates the combination of deep-seated thrusting that leads to near-surface tensile stresses that also allow for magma propagation upward (Bland and McKinnon 2016). An examination of the nature and distribution of volcanic centers and mountains led to the proposed model for heat flow from Io's interior of a heat pipe, wherein magma ascends and cool crust descends, vertically and spaced across the body (Moore 2001). Similar models for heat flow were proposed for the Earth and other terrestrial bodies in their early stages; this is another way in which Io is a model for the early Earth and superheated exoplanets (Moore et al. 2017).

Composite cones or stratovolcanoes have not yet been identified on Io, but two shield volcanoes, termed tholi (sing. tholus), are 1–3 km in height and a few hundred km in width have been observed (Figs. 5.3 and 5.12). Their low slopes are consistent with relatively low viscosity, sulfur and/or mafic to ultramafic silicate lavas rather than more evolved lava compositions (Schenk et al. 2004b). They make up ~0.1% of Io's surface and ~3% of Io's mountains.

5.7 Diffuse Deposits

Explosive volcanic eruptions on Io produce both umbrella-shaped plumes (Fig. 5.13) of gas and dust (Morabito et al. 1979; Smith et al. 1979a, b) that emanate from central vents like paterae or fissures, and irregular flow front plumes (Fig. 5.4) that jet vaporized frosts perpendicular to lava flow fronts (Kieffer et al. 2000). Io's surface geology can be mantled by thin to thick deposits of these fine-grained materials expelled from Io's many active volcanoes, including both pyroclastic ash and condensed gases. We refer to these as diffuse deposits.

Diffuse deposits typically occur on, near, or around active volcanic centers. The umbrella-shaped plumes that are sourced from central vents manifest most typically as ring deposits around the vents. The most famous examples of this is the continuously-replenished, ~1200-diameter ring around the volcano Pele, which is thought to be composed of short chain sulfur (S_3–S_4) that is recrystallized from S_2 gas emanated from Pele (Spencer et al. 2000). In contrast, flow front plumes produce asymmetric, irregular deposits on the sides of a vent where flows are active. Based on global mapping (Fig. 5.2), at any given time diffuse deposits cover ~18% of Io's surface, overlying all other features (Williams et al. 2011a, b).

Fig. 5.13 Gas and dust plume erupting at the Tvashtar volcano, as imaged by the NASA *New Horizon*'s Long Range Reconnaissance Imager (LORRI) on February 28, 2007 at a distance of 2.3 million km. From NASA Planetary Photojournal PIA 09250. See also Spencer et al. (2007)

Galileo SSI images, with their superior color data over *Voyager*, enabled identification of five distinct colors of diffuse deposits. These colors are thought to be related to the composition or state of their constituents: Red (short-chain sulfur $\pm$ sulfur chlorides, ~8.6% of surface), white ($SO_2 \pm$ colorless contaminants, ~6.9% of surface), yellow (sulfur $\pm$ contaminants, ~2.1% of surface), black (silicate ash, ~0.6% of surface), and green (products of silicate-sulfur alteration, ~0.01% of surface).

Diffuse deposits are ephemeral, and if produced in a single eruption event, they will be altered by radiation and fade away in weeks to months (Geissler et al. 1999). This was observed, for example, during the Summer 1997 eruption at Pillan Patera, which emplaced an ~400 km diameter dark ash deposit around its vent (Davies et al. 2001; Williams et al. 2001b). For recurrent or continuous eruptions, diffuse deposits are continuously visible over months to years, as was noted of the Pele red ring and the Prometheus white ring deposits. The long-term accumulation of red diffuse, yellow diffuse and white diffuse deposits over years to decades could lead to the formation of red-brown plains, yellow plains, and white plains materials, respectively (Williams et al. 2011b).

In terms of abundance, red (47% of all diffuse deposits) and white (38% of all diffuse deposits) are the two most commonly occurring diffuse deposits types on Io. This observation suggests that S_2 and SO_2 gases are the two most abundant volatiles that are released during explosive eruption of Io's volcanoes. Red diffuse deposits occur mostly as ring-like features around active vents, including Pele, Loki, Tvashtar, and others (Geissler et al. 2004). The Prometheus white ring deposit is the best example of a continuously emplaced white diffuse deposit. White diffuse deposits mostly manifest as irregularly-shaped, flow front plumes surrounding lava flow margins, resulting from vaporization, condensation, and reaccumulation of SO_2 around warm flow margins (Kieffer et al. 2000; Milazzo et al. 2001). Yellow diffuse deposits (11.5% of all diffuse deposits) are thought to be composed of polysulfur oxides and S_2O (Hapke 1989), or sulfur particles that recrystallize to cyclo-S_8 (Geissler et al. 1999). Black diffuse deposits (0.6% of all diffuse deposits) are thought to be composed of silicate ash (Geissler et al. 1999). Green diffuse deposits are typically too small to be mapped, and exist on some old patera floors. They are thought to be an alteration product of sulfur particulates from plumes and warm silicate lavas (Geissler et al. 1999). The dominance of gas-derived diffuse deposits (red + white, 85% of all deposits), compared to presumably pyroclast-bearing diffuse deposits (dark (silicate) + yellow (sulfur), 15% of all deposits), may indicate that the gas and pyroclasts typically do not travel together in Io's plumes (Williams et al. 2011b).

5.8 Structural and Tectonic Features

A wide range of structural features are visible on Io's surface, mostly in regional (<250 m/pixel) or better resolution images, including, ridges, scarps, grooves, mountains, mesas, pits, graben, and lineaments. A few clearly recognizable lava channels, such as Tawhaki Vallis (Schenk and Williams 2004) and the channels at Emakong (Williams et al. 2001a) are known, but because there is a lack of high-resolution (i.e., meters to tens of meters/pixel) images of Io's flow fields, their prevalence and role in flow field formation is unknown at present.

In low-sun images, scarps are visible that enable recognition of patera rims, layered plains, and the margins of mountains. Such images have been key to determine the depths of paterae, the heights of scarps and mountains, and other aspects of morphometry. In the plains, scarps appear to be mostly degradational features marking sites of sublimation erosion (Moore et al. 2001). Does Io's tidal flexing (~100 m every 1.7 days: Peale et al. 1979) manifest itself in Io's surface geology? Bart et al. (2004) noted grooves and ridges in plains that they suggested could represent tectonic fractures derived from tidal flexing, or possibly nonsynchronous rotation (Ross et al. 1990). Do we not see Europa-like ridges or ridged plains on Io, or even large impact craters, because they are quickly buried by accumulated volcanic deposits, or because the silicate crust of Io prevents their formation (as opposed to the icy crust of Europa)? The resurfacing rate of ~0.03–0.07 cm/year suggests that craters 5–20 km in diameter should be erased in 300,000–2.3 million years (Schenk et al. 2004a, b).

The limitations on imaging coverage at the proper sun angle and resolution, along with burial by volcanic deposits, inhibits the detection of normal or thrust faults, although detailed study has revealed some examples (see Jaeger et al. 2003). Ahern et al. (2017) identified 353 structural (non-patera) lineations on Io, dominantly oriented at 75°–100° and 165°–180°, indicating regional or global stresses are present.

5.9 Large-Scale Topography

The global topographic characteristics of Io may reveal aspects of internal dynamics such as heat flow distribution or asthenospheric convection (e.g., Ross et al. 1990; Tackley et al. 2001). Efforts to map global topographic variations are limited to image analysis of limb relief along approximately 25 north-south ground tracks (e.g., Thomas et al. 1998) and stereo image topography (White et al. 2014). While these maps cover approximately 75% of the surface, they are incomplete and in some areas are of poor quality because of limitations in image resolution or coverage. Despite these issues, the topographic data reveal a surface dominated by lava plains of low relief of only 1–2 km, relative to the best fit global triaxial ellipsoid of Thomas et al. (1998). This result is consistent with the lack of steep sided or high viscosity volcanism generally (e.g., Schenk et al. 2004a, b), but it is not yet sensitive enough to resolve between different heat flow models. Io features some of the highest mountains in the solar system with more than a few towering >10 km above the plains and one reaching ~17 km (Schenk et al. 2001). That these rise abruptly from level plains probably relates to burial of these thrust blocks by ongoing volcanism (Jaeger et al. 2003; Keszthelyi et al. 2007). Stereo photoclinometry (shape-from-shading) also reveals some aspects of smaller scale features, such as relief on mountains (White et al. 2014) or flow channels (Schenk and Williams 2004), but such data are limited to roughly 10% of the global surface.

5.10 Summary and Future Work

5.10.1 Summary

Images and other data of Io were obtained by NASA's *Voyager* spacecraft flybys (1979) and from NASA's *Galileo* orbiter (1996–2003), *Cassini* (2000) and *New Horizon*'s (2007) flybys. The geology of Io has been assessed through study of these data, including global geological mapping (Crown et al. 1992; Williams et al. 2011a, b). Io has five primary types of geological features: plains, lava flows, paterae, mountains, and diffuse deposits. These features are representative of mostly volcanic processes, although tectonism and gradation also have a role in their formation or modification. No impact craters have been observed in any image of Io, indicative of a high volcanic resurfacing rate.

All of Io's feature types are mappable as geologic units at global to regional scales, and can be subdivided into subunits based on variations in albedo, color (as a proxy for composition), and morphology. Plains make up almost two-thirds of Io's surface, and are composed of old lava flows, pyroclastic deposits, and condensed gases that have been altered by radiation exposure and buried by younger volcanic materials. There are four subunits (white plains, dominated by SO_2; yellow plains, dominated by sulfur-rich materials; red-brown plains, composed of the other two but with more extensive radiation alteration (in high-latitude and polar areas) or otherwise enriched in sulfur polymorphs; and layered plains, composed of the other three units but containing bounding scarps indicative of erosion). Lava flow fields make up almost 30% of Io's surface, and are subdivided into three subunits (dark flows, composed of silicate lavas; bright flows, composed sulfur- or more rarely SO_2 flows; and undivided flows, composed of either of the other two types but with intermediate albedos and showing mantling by other materials suggestive of an older age). Paterae are volcano-tectonic depressions that resemble terrestrial calderas, whose floors are covered in lava flows or lava lakes. 425 paterae have been mapped to date in *Voyager* and *Galileo* images. Like lava flows, they are subdivided into three subunits (dark patera floors, covered in silicate lavas; bright patera floors, covered in sulfur- or more rarely SO_2 flows; and undivided patera floors, covered in either of the other two types but with intermediate albedos and showing mantling by other materials suggestive of an older age). Patera floors are the source of most of Io's detected thermal emission hot spots, even though they cover only a small percentage of Io's surface. Mountains are positive relief topographic edifices ranging from 1 to 18 km in height, and although they cover a small percentage of the surface, they are remarkable expressions of tectonics. Half of resolved mountains are lineated, containing structural features suggestive of tectonic origins. Only 6% have a mottled appearance indicative of mass wasting, and only 4% are classified as tholi (interpreted to be domes or shields of volcanic origin). Diffuse deposits cover about 18% of Io's surface, mantling all other units. These deposits, composed of pyroclastic ash and condensed gases from many explosive eruptions around Io, have five distinct subunits based on their dominant constituents: red (short-chain sulfur), white (SO_2), yellow (sulfur ash), black (silicate ash), and green (sulfur-silicate alteration products). The dominance of gas-derived diffuse deposits (red and white) compared to pyroclast-bearing diffuse deposits (black and yellow) suggests magmatic volatiles escape as gases, re-condense, and bury Io's surface, or are ionized and escape to feed the Io plasma torus. Analysis of topography indicates the generally low-relief plains are consistent with a surface dominated by lavas of low viscosity and mafic to ultramafic compositions that fail to build steep edifices, and that mountains rising from these plains are consistent with a combination of stresses in the lithosphere, compression, transtension and extension.

5.10.2 Future Work

Global geologic mapping and assessment of Io's surface features has raised many questions about the geographical distribution and abundance of various feature types. For example, why are white plains concentrated in the equatorial antijovian region? Why are bright, sulfur-rich flows concentrated in the northern, leading antijovian quadrant relative to the rest of the moon? Does the lack of imaging at consistent resolutions and lighting conditions affect identification and distribution of mapped paterae, mountains, and active hot spots?

To better understand Io's geologic features, what is needed is global imaging coverage at consistent spatial resolutions (~200 m/pixel or better) and lighting conditions (both high- and low-sun), and in both grayscale and color, with stereo data obtained either from imagery or laser altimetry, in a mode equivalent to an orbital mission. In addition, high spatial resolution (few meters to several tens of meters/pixel) images should be obtained over multiple examples of all of Io's various surface features to assess better their origins.

If an Io orbiter is deemed impossible to operate because of radiation hazards, then a multi-flyby mission in which global coverage is obtained by repeated, well-constructed flybys (similar to the planned operations at NASA's Europa Clipper) is desirable. A Discovery-class mission such as the Io Volcano Observer (IVO, McEwen et al. 2021), or a New Frontiers-class Io Observer mission (consisting of a mother craft and multiple CubeSats), both of which would conduct multiple flybys at a range of latitudes and longitudes and image spatial resolutions, would provide much new data that would improve understanding of Io's surface features. Io CubeSats could be used: a) to get very high-resolution images during *Ranger*-like descents at one or more volcanoes; b) to fly through a persistent plume with a mass spectrometer and dust detector to measure plume components; or c) to fly partial orbits with magnetometers to better resolve Io's magnetic induction (Williams et al. 2017).

Published image, topography, geological mapping, hot spot and thermal heat flux data on Io have been assembled into an Io Geographic Information Systems (GIS) Database, to support future research studies and support future Io missions (Williams et al. 2021). It is also accessible in Arizona State University's Java Mission-planning and Analysis for Remote Sensing platform (JMARS: Christensen et al. 2009), and as a Zipped ArcGIS™ file that can be downloaded at: https://rgcps.asu.edu/gis_data/.

Acknowledgments The authors thank Michael Bland for a helpful review. We also would like to recognize all of the engineers, scientists, and managers who worked on NASA's *Voyager*, *Hubble*, *Galileo, Cassini,* and *New Horizons* missions, who enabled collection of the data sets that have allowed us and our colleagues to characterize the geology of Io.

References

Ahern, A.A., Radebaugh, J., Christiansen, E.H., Harris, R.A., Tass, E.S.: Lineations and structural mapping of Io's paterae and mountains: implications for internal stresses. Icarus. **297**, 14–32 (2017)

Bart, G.D., Turtle, E.P., Jaeger, W.L., Keszthelyi, L.P., Greenberg, R.: Ridges and tidal stress on Io. Icarus. **169**, 111–126 (2004)

Becker, T., Geissler, P.: Galileo global color mosaics of Io. In: Lunar and Planetary Science XXXVI, Abstract #1862. Lunar and Planetary Institute, Houston (CD-ROM) (2005)

Bland, M.T., and McKinnon, W.B.: Mountain building on Io driven by deep faulting. Nat. Geosci. **9**, 429–432 (2016)

Bunte, M.K., Williams, D.A., Greeley, R.: Geologic mapping of the Zal region of Io. Icarus. **197**, 354–367 (2008)

Bunte, M.K., Williams, D.A., Greeley, R., Jaeger, W.L.: Geologic mapping of the Hi'iaka and Shamshu regions of Io. Icarus. **207**, 868–886 (2010)

Carlson, R.W., Smythe, W.D., Lopes-Gautier, R.M.C., Davies, A.G., Camp, L.W., Mosher, J.A., Soderblom, L.A., Leader, F.E., Mehlman, R., Clark, R.N., Fanale, F.P.: The distribution of sulfur dioxide and other infrared absorbers on the surface of Io. Geophys. Res. Lett. **24**, 2479–2482 (1997)

Carr, M.H., McEwen, A.S., Howard, K.A., Chuang, F.C., Thomas, P., Schuster, P., Oberst, J., Neukum, G., Schubert, G.: Mountains and calderas on Io: possible implications for lithosphere structure and magma generation. Icarus. **135**, 146–165 (1998)

Chao, K.-H., de Graffenried, R., Lach, M., Nelson, W., Truax, K., Gaidos, E.: Lava worlds: from early Earth to exoplanets. Geochemistry. **81** (2021). https://doi.org/10.1016/j.chemer.2020.12735

Christensen, P.R., Engle, E., Anwar, S., Dickenshied, S., Noss, D., Gorelick, N., Weiss-Malik, M., Piacentine, N.: JMARS – A Planetary GIS. http://adsabs.harvard.edu/abs/2009AGUFMIN22A.06C (2009)

Crown, D.A., Greeley, R., Craddock, R.A., Schaber, G.G.: Geologic map of Io. U.S. Geol. Surv. Misc. Invest. Series Map I-2209, 1:15,000,000. Reston, VA (1992)

Davies, A.G., Keszthelyi, L.P., Williams, D.A., Phillips, C.B., McEwen, A.S., Lopes-Gautier, R.M., Smythe, W.D., Soderblom, L.A., Carlson, R.W.: Thermal signature, eruption style, and eruption evolution at Pele and Pillan on Io. J. Geophys. Res. **106**, 33079–33103 (2001)

Douté, S., Schmitt, B., Lopes-Gautier, R., Carlson, R., Soderblom, L., Shirley, J., the Galileo NIMS Team: Mapping SO_2 frost on Io by the modeling of NIMS hyperspectral images. Icarus. **149**, 107–132 (2001)

Douté, S., Lopes, R., Kamp, L.W., Carlson, R., Schmidt, B., the Galileo NIMS Team: Dynamics and evolution of SO_2 gas condensation around Prometheus-like volcanic plumes on Io as seen by the Near Infrared Mapping Spectrometer. Icarus. **158**, 460–482 (2002)

Douté, S., Lopes, R., Kamp, L.W., Carlson, R., Schmidt, B., the Galileo NIMS Team: Geology and activity around volcanoes on Io from the analysis of NIMS spectral images. Icarus. **169**, 175–196 (2004)

Dundas, C.M.: Effects of lava heating on volatile-rich slopes on Io. J. Geophys. Res. **122**, 546–559 (2017). https://doi.org/10.1002/2016JE005177

Geissler, P.E., McEwen, A.S., Keszthelyi, L., Lopes-Gautier, R., Granahan, J., Simonelli, D.P.: Global color variations on Io. Icarus. **140**, 265–282 (1999)

Geissler, P.E., McEwen, A.S., Phillips, C.B., Keszthelyi, L.P., Spencer, J.S.: Surface changes on Io during the Galileo mission. Icarus. **169**, 29–64 (2004)

Greeley, R., Theilig, E., Christensen, P.: The Mauna Loa sulfur flow as an analog to secondary sulfur flows (?) on Io. Icarus. **60**, 189–199 (1984)

Greeley, R., Spudis, P.D., Guest, J.E.: Geologic map of the Ra Patera area of Io. U.S. Geol. Surv. Misc. Invest. Series Map I-1949, 1:2,000,000. Reston, VA (1988)

Gregg, T.K.P., Lopes, R.M.: Lava lakes on Io: new perspectives from modeling. Icarus. (2008). https://doi.org/10.1016/j.icarus.2007.08.042

Hamilton, C.W., Beggan, C.D., Still, S., Beuthe, M., Lopes, R.M.C., Williams, D.A., Wright, W., Radebaugh, J.: Spatial distribution of volcanoes on Jupiter's moon Io: implications for tidal heating and magma ascent. Earth Planet. Sci. Lett. **361**, 272–286 (2013). https://doi.org/10.1016/j.epsl.2012.10.032

Hapke, B.: The surface of Io: a new model. Icarus. **79**, 56–74 (1989)

Jaeger, W.L., Turtle, E.P., Keszthelyi, L.P., Radebaugh, J., McEwen, A.S., Pappalardo, R.T.: Orogenic tectonism on Io. J. Geophys. Res. **108**(E8), 5093 (2003). https://doi.org/10.1029/2002JE001946

Johnson, R.E.: Polar "caps" on Ganymede and Io revisited. Icarus. **128**, 469–471 (1997)

Johnson, T.V., Cook II, A.F., Sagan, C., Soderblom, L.A.: Volcanic resurfacing rates and implications for volatiles on Io. Nature. **280**, 746–750 (1979)

Keszthelyi, L.P., McEwen, A.S., Phillips, C.B., Milazzo, M., Geissler, P.E., Williams, D.A., Turtle, E., Radebaugh, J., Simonelli, D., the Galileo SSI Team: Imaging of volcanic activity on Jupiter's moon Io by Galileo during GEM and GMM. J. Geophys. Res. **106**, 33025–33052 (2001)

Keszthelyi, L., Jaeger, W.L., Turtle, E.P., Milazzo, M., Radebaugh, J.: A post-Galileo view of Io's interior. Icarus. **169**, 271–286 (2004)

Keszthelyi, L., Jaeger, W., Milazzo, M., Radebaugh, J., Davies, A.G., Mitchell, K.L.: New estimates for Io eruption temperatures: implications for the interior. Icarus. **192**, 491–502 (2007)

Kieffer, S.W.: Dynamics and Thermodynamics of Volcanic Eruptions-Implications for the Plumes on Io. Satellites of Jupiter, pp. 647–723. University of Arizona Press, Tucson, AZ (1982)

Kieffer, S.W., Lopes-Gautier, R., McEwen, A., Smythe, W., Keszthelyi, L., Carlson, R.: Prometheus: Io's wandering plume. Science. **288**, 1204–1208 (2000)

Kirchoff, M.R., McKinnon, W.B., Schenk, P.M.: Global distribution of volcanic centers and mountains on Io: control by asthenospheric heating and implications for mountain formation. Earth Planet. Sci. Lett. **301**, 22–30 (2011)

Lellouch, E., Paubert, G., Moses, J.I., Schneider, N.M., D.F.: Strobel Volcanically-emitted sodium chloride as a source for Io's neutral clouds and plasma torus. Nature. **421**, 45–47 (2003)

Leone, G.A.G., Davies, L., Wilson, D.A., Williams, L.P., Keszthelyi, W.L.J., Turtle, E.P.: Volcanic history, geologic analysis and map of the Prometheus Patera region on Io. J. Volcan. Geotherm. Res. **187**, 93–105 (2009)

Lopes, R.M.C., et al.: Lava lakes on Io? Observations of Io's volcanic activity from Galileo NIMS during the 2001 fly-bys. Icarus. **169**, 140–174 (2004)

McEwen, A.S., Matson, D.L., Johnson, T.V., Soderblom, L.A.: Volcanic hot spots on Io: correlation with low-albedo calderas. J. Geophys. Res. **90**, 12345–12377 (1985)

McEwen, A.S., et al.: Active volcanism on Io as seen by Galileo SSI. Icarus. **135**, 181–219 (1998a)

McEwen, A.S., et al.: High-temperature silicate volcanism on Jupiter's moon Io. Science. **281**, 87–90 (1998b)

McEwen, A.S., et al.: Galileo at Io: results from high-resolution imaging. Science. **288**, 1193–1198 (2000)

McEwen, A.S., Keszthelyi, L.P., Mandt, K.E., the IVO Team: The Io Volcano Observer (IVO). In: 52nd Lunar Planet. Sci. Conf., Abstract #1352. Lunar and Planetary Institute, Houston (2021)

McGovern, P.J., Kirchoff, M.R., White, O.L., Schenk, P.M.: Magma ascent pathways associated with large mountains on Io. Icarus. **272**, 246–257 (2016)

McKinnon, W.B., Schenk, P.M., Dombard, A.J.: Chaos on Io: a model for formation of mountain blocks by crustal heating, melting, and tilting. Geology. **29**, 103–106 (2001)

Milazzo, M.P., Keszthelyi, L.P., McEwen, A.S.: Observations and initial modeling of lava-SO_2 interactions at Prometheus, Io. J. Geophys. Res. **106**, 33121–33128 (2001)

Milazzo, M.P., Keszthelyi, L.P., Radebaugh, J., Davies, A.G., Turtle, E.P., Geissler, P., Klaasen, K.P., Rathbun, J.A., McEwen, A.S.: Volcanic activity at Tvashtar Catena, Io. Icarus. **179**, 235–251 (2005)

Moore, H.J.: Geologic map of the Maasaw Patera area of Io. U.S. Geol. Surv. Misc. Invest. Series Map I-1851, 1:1,003,000. Reston, VA (1987)

Moore, W.B.: The thermal state of Io. Icarus. **154**(2), 548–550 (2001)

Moore, J.M., Mellon, M.T., Zent, A.P.: Mass wasting and ground collapse in terrains of volatile-rich deposits as a solar system-wide geological process: the pre-Galileo view. Icarus. **122**(1), 63–78 (1996)

Moore, J.M., Sullivan, R.J., Chuang, F.C., Head III, J.W., McEwen, A.S., Milazzo, M.P., Nixon, B.E., Pappalardo, R.T., Schenk, P.M., Turtle, E.P.: Landform degradation and slope processes on Io: the Galileo view. J. Geophys. Res. **106**, 33223–33240 (2001)

Moore, W.B., Simon, J.I., Webb, A.A.G.: Heat-pipe planets. Earth Planet. Sci. Lett. **474**, 13–19 (2017)

Morabito, L.A., Synnott, S.P., Kupferman, P.N., Collins, S.A.: Discovery of currently active extraterrestrial volcanism. Science. **204**(4396), 972 (1979)

Peale, S.J., Cassen, P., Reynolds, R.T.: Melting of Io by Tidal dissipation. Science. **203**(4383), 892–894 (1979)

Pearl, J., Hanel, R., Kunde, V., Maguire, W., Fox, K., Gupta, S., Ponnamperuma, C., Raulin, F.: Identification of gaseous SO_2 and new upper limits for other gases. Nature. **280**, 755–758 (1979)

Radebaugh, J., Keszthelyi, L.P., McEwen, A.S., Turtle, E.P., Jaeger, W., Milazzo, M.: Paterae on Io: a new type of volcanic caldera? J. Geophys. Res. **106**, 33005–33020 (2001)

Radebaugh, J., Jaeger, W.L., Keszthelyi, L.P., Turtle, E.P., Milazzo, M.P., Perry, J., McEwen, A.S., Lopes, R., Davies, A.G., Geissler, P.: Relationships between paterae, mountains, and hotspots on Io from a global database. In: 35th Lunar Planet. Sci. Conf., Abstract #2067. Lunar and Planetary Institute, Houston (2004)

Rathbun, J.A., Spencer, J.R.: Loki, Io: new ground-based observations and a model describing the change from periodic overturn. Geophys. Res. Lett. **33**, L17201 (2006). https://doi.org/10.1029/2006GL026844

Ross, M.N., Schubert, G., Spohn, T., Gaskell, R.W.: Internal structure of Io and the global distribution of its topography. Icarus. **85**, 309–325 (1990)

Schaber, G.G.: The surface of Io: geologic units, morphology and tectonics. Icarus. **43**, 302–333 (1980)

Schaber, G.G.: The geology of Io. In: Morrison, D. (ed.) Satellites of Jupiter, pp. 556–597. University of Arizona Press, Tucson, AZ (1982)

Schaber, G.G., Scott, D.H., Greeley, R.: Geologic map of the Ruwa Patera quadrangle (Ji-2) of Io. U.S. Geol. Surv. Geol. Invest. Series Map I-1980, 1:5,000,000. Reston, VA (1989)

Schenk, P.M., Bulmer, M.H.: Origin of mountains on Io by thrust faulting and large-scale mass movements. Science. **279**, 1514–1517 (1998)

Schenk, P.M., Williams, D.A.: A potential thermal erosion lava channel on Io. Geophys. Res. Lett. **31**, L23702 (2004). https://doi.org/10.1029/2004GL021378

Schenk, P., Hargitai, H., Wilson, R., McEwen, A., Thomas, P.: The mountains of Io: global and geological perspectives from Voyager and Galileo. J. Geophys. Res. **106**, 33201–33222 (2001)

Schenk, P.M., Chapman, C.R., Zahnle, K., Moore, J.M.: Ages and interiors: the cratering record of the Galilean satellites. In: Bagenal, F., Dowling, T.E., William, B., McKinnon, W.B. (eds.) Jupiter: The Planet, Satellites and Magnetosphere, pp. 427–456. Cambridge University Press, Cambridge (2004a)

Schenk, P.M., Wilson, R.R., Davies, A.G.: Shield volcano topography and the rheology of lava flows on Io. Icarus. **169**(1), 98–110 (2004b)

Schmitt, B., Rodriguez, S.: Possible identification of local deposits of Cl_2SO_2 on Io from NIMS/Galileo spectra. J. Geophys. Res. **108**(E9), 5104 (2003). https://doi.org/10.1029/2002JE001988

Smith, B.A., et al.: The Jupiter system through the eyes of Voyager 1. Science. **204**, 951–972 (1979a)

Smith, B.A., et al.: The Galilean satellites and Jupiter: Voyager 2 imaging science results. Science. **206**, 927–950 (1979b)

Smythe, W.D., Nelson, R.M., Nash, D.B.: Spectral evidence for the SO_2 frost or adsorbate on Io's surface. Nature. **280**, 766–767 (1979)

Spencer, J.R., Jessup, K.L., McGrath, M.A., Ballester, G.E., Yelle, R.: Discovery of gaseous S_2 in Io's Pele plume. Science. **288**, 1208–1210 (2000)

Spencer, J.R., Stern, S.A., Cheng, A.F., Weaver, H.A., Reuter, D.C., Retherford, K., Lunsford, A., Moore, J.M., Abramov, O., Lopes, R.M.C., Perry, J.E., Kamp, L., Showalter, M., Jessup, K.L., Marchis, F., Schenk, P.M., and Dumas, C. Io volcanism seen by New Horizons: A major eruption of the Tvashtar volcano. Science. **318**, 240–243 (2007)

Tackley, P.J., Schubert, G., Glatzmaier, G.A., Schenk, P., Ratcliff, J.T., Matas, J.-P.: Three-dimensional simulations of mantle convection in Io. Icarus. **149**, 79–93 (2001)

Thomas, P.C., Davies, M.E., Colvin, T.R., Oberst, J., Schuster, P., Neukum, G., Carr, M.H., McEwen, A., Schubert, G., Belton, M.J.S., Galileo Imaging Team: The shape of Io from Galileo limb measurements. Icarus. **135**, 175–180 (1998)

Turtle, E.P., et al.: The mountains of Io: global and geological perspectives from Voyager and Galileo. J. Geophys. Res. **106**, 33175–33200 (2001)

White, O.L., Schenk, P.M., Nimmo, F., Hoogenboom, T.: Topographic mapping of paterae and layered plains on Io using photoclinometry. J. Geophys. Res. **120**, 51–61 (2014). https://doi.org/10.1002/2014JE004672

Whitford-Stark, J.L., Mouginis-Mark, P.J., Head, J.W.: Geologic map of the Lerna region (Ji-4) of Io. U.S. Geol. Surv. Misc. Invest. Series Map I-2055, 1:5,000,000. Reston, VA (1991)

Williams, D.A., Wilson, A.H., Greeley, R.: A komatiite analog to potential ultramafic materials on Io. J. Geophys. Res. **105**, 1671–1684 (2000)

Williams, D.A., Greeley, R., Lopes, R.M.C., Davies, A.G.: Evaluation of sulfur flow emplacement on Io from Galileo Data and numerical modeling. J. Geophys. Res. **106**, 33161–33174 (2001a)

Williams, D.A., Davies, A.G., Keszthelyi, L.P., Greeley, R.: The July 1997 eruption at Pillan Patera on Io: implications for ultrabasic lava flow emplacement. J. Geophys. Res. **106**, 33105–33119 (2001b)

Williams, D.A., Radebaugh, J., Keszthelyi, L.P., McEwen, A.S., Lopes, R.M.C., Douté, S., Greeley, R.: Geologic mapping of the Chaac-Camaxtli region of Io from Galileo imaging data. J. Geophys. Res. **107**(E9), 5068 (2002). https://doi.org/10.1029/2001JE001821

Williams, D.A., Schenk, P.M., Moore, J.M., Keszthelyi, L.P., Turtle, E.P., Jaeger, W.L., Radebaugh, J., Milazzo, M.P., Lopes, R.M.C., Greeley, R.: Mapping of the Culann-Tohil region of Io from Galileo imaging data. Icarus. **169**, 80–97 (2004)

Williams, D.A., Keszthelyi, L.P., Schenk, P.M., Milazzo, M.P., Lopes, R.M.C., Rathbun, J.A., Greeley, R.: The Zamama-Thor region of Io: insights from a synthesis of mapping, topography, and Galileo spacecraft data. Icarus. **177**, 69–88 (2005)

Williams, D.A., Keszthelyi, L.P., Crown, D.A., Jaeger, W.L., Schenk, P.M.: Geologic mapping of the Amirani-Gish Bar region of Io: implications for the global geologic mapping of Io. Icarus. **186**, 204–217 (2007)

Williams, D.A., Keszthelyi, L.P., Crown, D.A., Yff, J.A., Jaeger, W.L., Schenk, P.M., Geissler, P.E., Becker, T.L.: Geologic map of Io. U.S. Geological Survey Scientific Investigations Map 3168, scale 1:15,000,000, 25 p. http://pubs.usgs.gov/sim/3168/ (2011a)

Williams, D.A., Keszthelyi, L.P., Crown, D.A., Yff, J.A., Jaeger, W.L., Schenk, P.M., Geissler, P.E., Becker, T.L.: Volcanism on Io: insights from global geologic mapping. Icarus. **214**, 91–112 (2011b). https://doi.org/10.1016/j.icarus.2011.05.007

Williams, D.A., Lopes, R.M.C., Castillo-Rogez, J., Scowen, P.: CubeSats to support Io exploration. In: 48th Lunar Planet. Sci. Conf., Abstract #1136. Lunar and Planetary Institute, Houston (2017)

Williams, D.A., Nelson, D.M., Milazzo, M.P.: The Io GIS Database, V. 1.0. In: 52nd Lun. Planet. Sci. Conf., Abst. #1504. Lunar and Planetary Institute, Houston (2021)

Chapter 6
Io's Thermal Emission and Heat Flow

Katherine de Kleer and Julie A. Rathbun

Abstract Io's high internal heat flow powers its dramatic silicate volcanism, and leads to a surface and atmosphere that are dominated by sulfurous volcanic products. Io's surface is peppered with constantly-changing thermal hot spots, which are active volcanic features including lava lakes, lava fountains, and lava flow fields; such features are analogous to highly-mafic counterparts on Earth, albeit at much larger scales. Our understanding of Io's hot spots and heat flow has progressed substantially since the end of the *Galileo* mission due to new telescopic datasets, continuing analyses of spacecraft data, and improvements in theoretical models. This chapter reviews advances in our understanding of Io's thermal emission, both volcanic and passive, since the last major review in 2007. The major datasets and observational techniques are reviewed, and the results synthesized and discussed in terms of the volcanology of Io and the mechanisms of tidal heating in its interior.

6.1 Introduction

The intense tides operating in Io's interior generate a tremendous amount of heat that melts Io's mantle materials. This heat is transported by Io's heat pipe volcanism and extruded onto the surface (O'Reilly and Davies 1981). The resultant volcanic features and dramatic eruptions provide the only opportunity to regularly observe active volcanism on an object other than Earth, and hence our only opportunity for generalizing our understanding of volcanism beyond Earth. Comparisons between bodies may constrain how eruption characteristics vary with planetary parameters such as size, composition, atmosphere, and heating mechanism.

K. de Kleer (✉)
California Institute of Technology, Pasadena, CA, USA
e-mail: dekleer@caltech.edu

J. A. Rathbun
Planetary Science Institute, Tucson, AZ, USA
e-mail: rathbun@psi.edu

R. M. C. Lopes et al. (eds.), *Io: A New View of Jupiter's Moon*, Astrophysics and Space Science Library 468, https://doi.org/10.1007/978-3-031-25670-7_6

Erupting magmas, as well as lava lakes and cooling lava flows on Io's surface, produce clear thermal signatures that are detectable from spacecraft and from Earth telescopic observations at a range of thermal wavelengths. Thermally-emitting volcanic features on Io's surface are referred to as "hot spots", a term that is agnostic to the style of volcanism producing the emission; note that this is not the same usage as the term "hot spot" in Earth volcanism. When a hot spot is observed at multiple thermal wavelengths, its average temperature, areal extent, and total power can be measured. The time-evolution of the locations, temperatures, and areal extents of individual volcanic eruptions provides key information on Io's geological processes.

Io's magma temperature is likely in the 1400–1900 K range (McEwen et al. 1997; Davies et al. 2001), and measured temperatures of hot spots, including cooler features such as lava flows and lava lake crusts, typically fall within the 200–1200 K range (McEwen et al. 1997; Veeder et al. 2012; de Kleer and de Pater 2016a). Surfaces at these temperatures produce peak thermal emission at wavelengths of 2–10 μm; Fig. 6.1 illustrates the dominant contributions to Io's disk-integrated spectrum. The 2–5 μm range is particularly well covered by spacecraft and telescopic instrumentation, resulting in an extensive database of observations of Io's volcanic thermal emission (e.g. Veeder et al. 2015; Rathbun and Spencer 2010; de Kleer et al. 2019b) extending back to the discovery of Io's volcanism by *Voyager 1* in 1979 (Morabito et al. 1979; Hanel et al. 1979)—and in fact even prior to the discovery of volcanism (Witteborn et al. 1979), although the thermal signatures were not originally attributed to volcanism (Hansen 1973; Morrison and Cruikshank 1973).

Beyond Io's extrusive volcanism, other potential heat transport mechanisms include intrusive volcanism and conduction through the lithosphere (Spencer et al. 2020). Though not yet directly measured, heat conducted through the crust would contribute to the background thermal emission across Io's surface, raising surface and subsurface temperatures above their solar-heated values. Measurements of

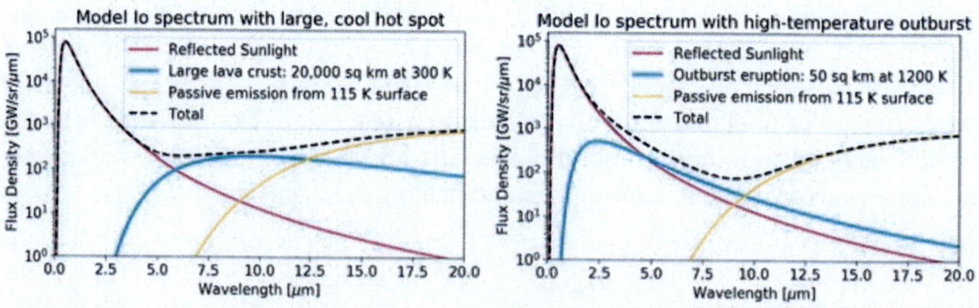

Fig. 6.1 Blackbody spectra illustrating the contributions of different components to Io's disk-integrated infrared spectrum. These include: reflected sunlight off of Io's entire surface; an area of 20,000 km^2 at 300 K to represent a large lava crust such as Loki Patera (left panel); an area of 50 km^2 at 1200 K to represent an outburst eruption (right panel); and passive thermal emission from Io's entire surface at a temperature of 115 K. Higher temperature eruptions lead to greater peak volcanic flux densities and contribute preferentially at shorter wavelengths. Figure after Veeder et al. (1994)

Io's surface temperature, particularly when resolved in time, latitude/longitude, or depth, can therefore place constraints on Io's interior processes and heat generation mechanisms.

6.1.1 The State of Knowledge at the End of the Galileo Mission

The most extensive thermal database that currently exists for Io was provided by the *Galileo* NIMS and PPR instruments in the near- and mid-infrared respectively. By the end of the *Galileo* mission, several properties of Io's hot spots were well established: the dominant magma composition was confirmed to be silicate on the basis of the magma temperature (Carr 1986; Johnson et al. 1988; McEwen et al. 1997); explosive and effusive volcanic styles were identified, including lava flows, lava fountains (Keszthelyi et al. 2001), and intrapatera volcanism such as lava lakes (Lopes et al. 2004); and patterns in the temporal evolution of individual hot spots were identified (Rathbun et al. 2002; Davies et al. 2001). Hints of correlations in volcanic style with latitude were uncovered, though the hot spot database was insufficient for robust statistical studies (McEwen et al. 2000). Io's surface temperature away from volcanic centers was mapped by *Galileo* PPR (Rathbun et al. 2004) and *Voyager* IRIS, and Io's heat flow was measured to be 2.5 W/m^2 (Veeder et al. 1994), of which only about half could be accounted for by hot spots (Veeder et al. 2012). For the remainder of this chapter we will focus on progress made since the last Io review book at the end of the *Galileo* mission (Lopes and Spencer 2007); pre-2007 results are central to our understanding of Io and will be included but without detailed description. For more in-depth coverage of the earlier results, please see Williams and Howell (2007).

6.1.2 Advances Since the Galileo Mission

Data from the *Galileo* mission continue to be analyzed to yield valuable insight into Io's volcanism more than a decade after the end of the mission. Since the *Galileo* mission, the Jupiter system was only visited by spacecraft flying by *en route* to other destinations (Geissler et al. 2004b; Spencer et al. 2007) until the arrival of *Juno* in 2016 (Bolton et al. 2017). At the time of writing, the *Juno* extended mission is just beginning, with dedicated Io flybys still upcoming; the data already obtained during the *Juno* primary mission demonstrate great promise for detecting Io's faint hot spots and mapping polar regions (Mura et al. 2020).

Continuing Io occultation observations have yielded a decades-long cumulative time baseline for individual volcanoes on Io's surface (Rathbun and Spencer 2010). During the past decades, improvements to telescope technologies have also substantially expanded our ability to study Io's volcanism and thermal emission. Near-infrared observations of Io from 8 to 10 m telescope equipped with adaptive optics have provided a database that rivals that of *Galileo* in number of detections and spatial resolution (excepting the closest *Galileo* Io passes; de Kleer et al. 2019b). New interferometric capabilities at both infrared and millimeter wavelengths can now provide an unprecedented level of spatial resolution, resolving emitting components within individual volcanic centers on Io and resolving (sub)surface passive thermal emission (Conrad et al. 2015; de Pater et al. 2020). Current and proposed 3–5 μm imaging spectrometers coupled with adaptive optics systems hold great promise for simultaneously measuring the temperatures of numerous hot spots in a single snapshot observation (Skemer et al. 2015; Deno Stelter et al. 2020). These capabilities have enabled detailed coverage of Io even when a spacecraft is not present in the system, and have provided a multidimensional view into its thermal emission and surface-interior connection.

This chapter reviews new (post-2007) analyses of *Galileo* mission data and observations from more recent spacecraft flybys, as well as results from ground-based observations over the past 10–20 years, and synthesizes this new information with older work into an updated understanding of Io's thermal emission and heat flow. In Sect. 6.2, the major datasets that have been obtained on Io's hot spots over the past two decades will be summarized and the key observational techniques described. Sections 6.3 and 6.4 will discuss what these datasets have revealed about volcanic processes at Io's volcanoes and tidal heating in Io's interior, respectively. Section 6.5 will briefly cover Io's passive emission and global heat flow, and a future outlook including outstanding questions will be presented in Sect. 6.6.

6.2 Hot Spot Observations and Techniques

6.2.1 Earth-Based Observations and Techniques

Since the end of the *Galileo* mission in 2003, ground-based observations have been the primary means of studying Io's hot spot emission and continuing to track the time-evolution of its volcanoes. The bulk of these observations has been obtained using occultation or adaptive optics techniques; these and other techniques are introduced below.

6.2.1.1 Occultations

The technique with the longest time history for measuring the emission from individual volcanic hot spots uses occultation timing to determine the locations of active volcanoes in one spatial dimension (Spencer et al. 1990). Used for over three decades in observations where Io is not (or only minimally) resolved, this technique relies on the total brightness of Io as a function of time while Io is disappearing behind (or reappearing from) Jupiter's limb. The measurements result in an occultation lightcurve (Fig. 6.2) where each step in the curve arises from a

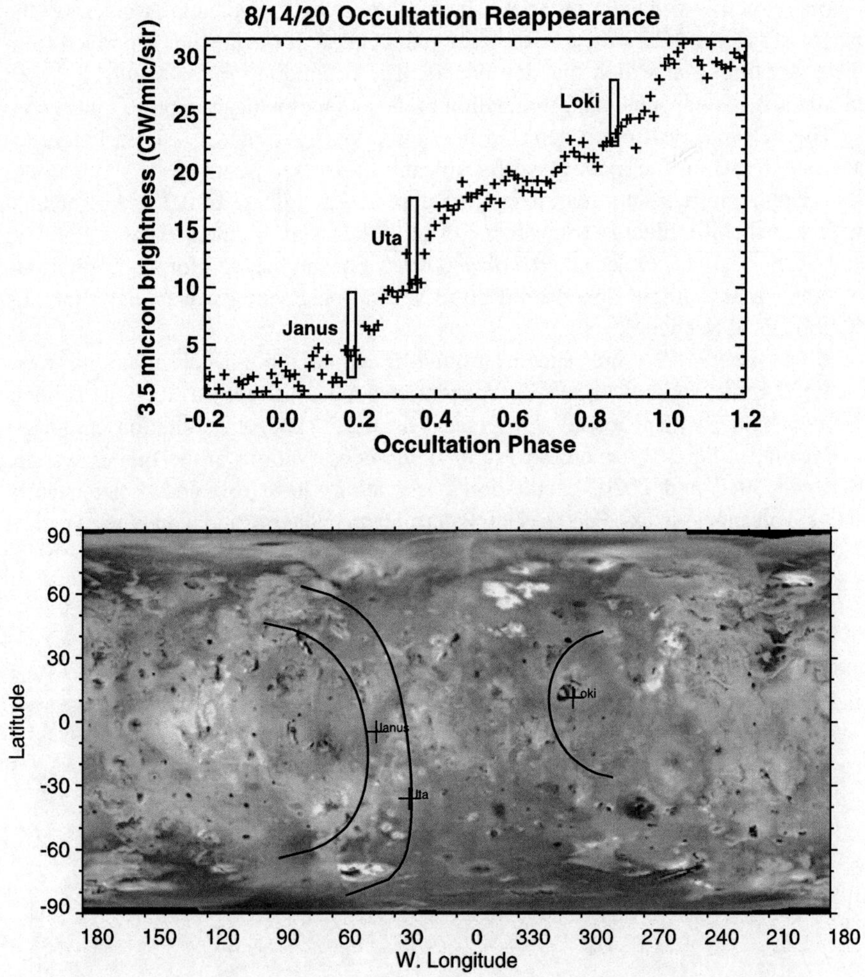

Fig. 6.2 Example occultation light curve and surface tracks. Top: Occultation reappearance light curve obtained on Aug 14, 2020. Each point in the curve is the total brightness of Io observed from a single image. The x-axis indicates the time the image was taken relative to the occultation event. An occultation phase of 0.0 is the time at which the occultation begins and an occultation phase of 1.0 is the time at which the occultation ends. Each step up in the light curve indicates that at least one active volcano reappeared from behind Jupiter at that time. The height of the step is the brightness of that volcano. On this particular night, Janus Patera, Uta Patera, and Loki Patera were active. Bottom: The one-dimensional location of each of the volcanoes identified in the light curve. The location of the curve is given by the projection of Jupiter's limb onto Io at the time the step occurs

thermally-active volcanic hot spot. The brightness of the volcano is given by the height of the step while the one-dimensional location of the hotspot is derived from the exact timing at which the step occurs. The occultation phase at which a step occurs gives a reasonable approximation to the hot spot's location on Io's surface.

The volcano most often detected in occultation light curves is Loki Patera (at a phase of ~0.9), the most powerful volcano on Io (see Sect. 6.3.4.5). A second, less bright, but persistent feature is often observed at a phase of 0.2 and associated with Kanehekilli Fluctus (Stansberry et al. 1997) and/or Janus Patera (Rathbun and Spencer 2010). Other active volcanoes have been observed for shorter periods of time, from a single observation up to several observations over many months (Rathbun and Spencer 2010)

After the *Galileo* era ended, ground-based occultation observations were obtained more sporadically, with concentrated observations in 2007 to support the *New Horizons* spacecraft flyby (see Sect. 6.2.2.2). The occultation campaign was revitalized in 2017 in order to support *Juno* observations of the Jupiter system. Between 2017 and 2020, occultation observations have revealed 4 persistently active volcanoes (Loki Patera, Uta Patera, Janus Patera, and Chalybes Regio), 4 outbursts (see Sect. 6.3.3), and 8 other eruptions lasting less than 5 months in duration (Fig. 6.3).

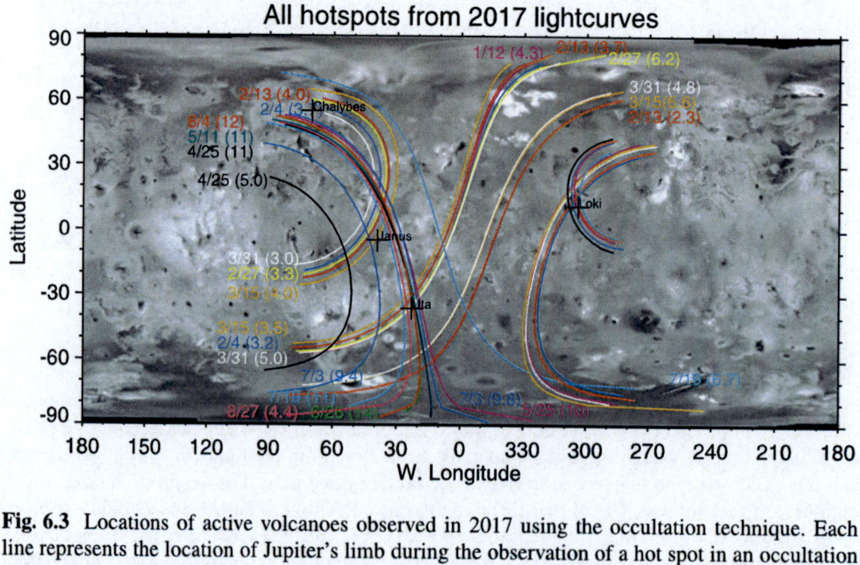

Fig. 6.3 Locations of active volcanoes observed in 2017 using the occultation technique. Each line represents the location of Jupiter's limb during the observation of a hot spot in an occultation light curve. The background is a map of Io from *Galileo* SSI images. The annotations for each curve is the date the observation was obtained, and, in parentheses, the brightness of the hotspot in GW/μm/str. The locations of likely volcanic sources are also plotted. The uncertainty in timing leads to an uncertainty in the location of the line of about 10–15°. So, for example, all of the lines that pass near Loki Patera are interpreted to be Loki Patera and the differences in location due to uncertainties

A recent joint analysis of the entire occultation dataset has demonstrated that if the set of dominant active volcanoes can be assumed to be unchanging, precise locational constraints can be placed on the active hot spots (Bartolić et al. 2021).

6.2.1.2 Adaptive Optics

Io's angular size of 0.8–1.2" as viewed from Earth is only a factor of 1–2× the astronomical seeing in good conditions. Io is therefore only barely resolved by seeing-limited instruments, unless special approaches such as occultation observations or lucky imaging are employed. This limitation is circumvented by adaptive optics (AO) technology, which performs a real-time correction for atmospheric turbulence using a deformable mirror. AO-corrected instruments came online at multiple telescopes beginning around the 1990s, and enabled a significant improvement in the ability to study Io's volcanic hot spots from Earth. Observations are typically made at multiple wavelengths between 1 and 5 microns (see Fig. 6.4), wavelengths sensitive only to the most powerful and highest-temperature eruptions.

The first near-infrared AO observations of Io were obtained from the European Southern Observatory in 1996–1999 (Marchis et al. 2000, 2001) and shortly thereafter from Keck in 2001–2005 (Marchis et al. 2002, 2005; de Pater et al. 2016a). These early observations already demonstrated the power of AO data for directly resolving the emission from multiple hot spots in a single snapshot observation. The rise of "time domain astronomy" has also led to an increase in flexibility in observing modes at telescopes, facilitating simple snapshot observations of bright targets and hence enabling much more frequent observation of Io than was previously achievable. Between 2001 and 2018 Io was observed with AO at 8–10 m telescopes on over 300 occasions (see Figs. 6.4 and 6.5 for examples; de Kleer and de Pater 2016a; Cantrall et al. 2018; de Kleer et al. 2019b). Emission from at least 110 unique hot spots has been detected, between once and over 100 times per hot spot, and at least 25 of these hot spots had not previously been detected by spacecraft.

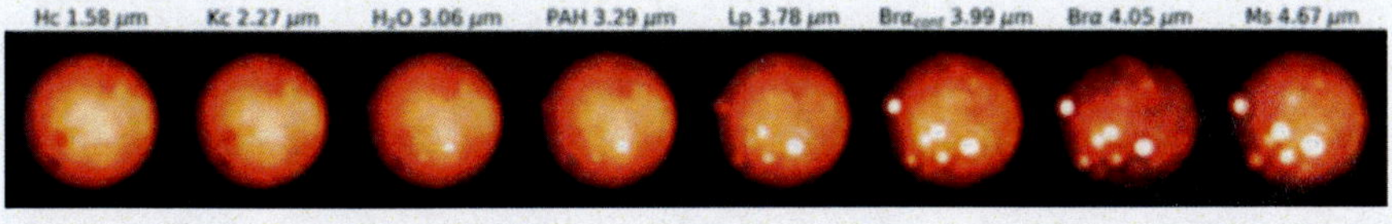

Fig. 6.4 Adaptive optics images of Io in a range of filters/wavelengths from Keck Observatory on 2017 May 28. All images were obtained close enough in time that the same hemisphere is seen. At the shorter wavelengths, reflected sunlight from Io's disk is stronger, and only the high-temperature hot spots are emitting. Figure from de Kleer et al. (2019b)

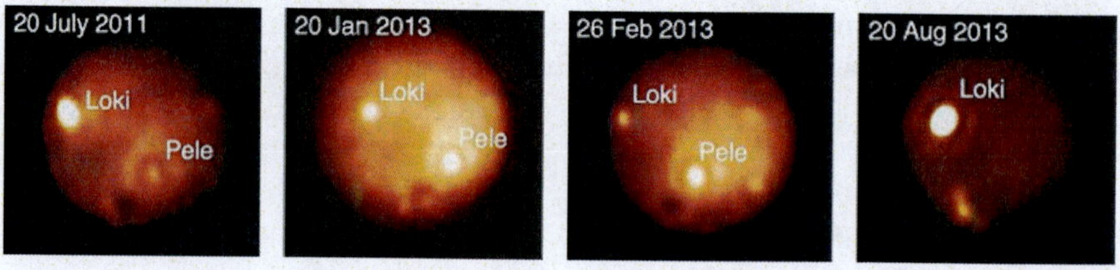

Fig. 6.5 Adaptive optics images of Io from Keck Observatory at a wavelength of 3.8 μm. All four images were taken at different dates with similar viewing geometry, and temporal variability is clearly seen in the two bright volcanoes Loki Patera and Pele. Figure modified from de Pater et al. (2016b)

6.2.1.3 Other Telescopic Techniques

While occultations and adaptive optics imaging have provided the most voluminous datasets of Io's hot spots, several other techniques have pushed to higher spatial resolution and/or greater simultaneous spectral coverage. These techniques fall broadly into the categories of new observational techniques; new instrumentation; and new utilization of unique planetary alignments. Though none have been used extensively, they are worth mentioning here. Howell and McGinn (1985) and Macintosh et al. (2003) used speckle imaging techniques at near-infrared wavelengths to obtain stronger spatial constraints on Io's hot spots. The latter resolved 17 distinct hot spots at 2.2 μm with a positional uncertainty of ~20 km, substantially better than AO or standard occultation observations, and found evidence for a 100 km shift in the peak emission from Loki Patera. Conrad et al. (2015) used near-infrared interferometry with the Large Binocular Telescope Interferometer (LBTI) to directly resolve emission from within Loki Patera into two emitting components, centered in the SW and SE corners of the patera. Skemer et al. (2015) demonstrated the first 3–5 μm AO-corrected integral field unit ALES at Io, showing that simultaneous spectra can be obtained across Io's surface, with the optimal wavelength coverage to make temperature measurements.

Just as the emergence of Io from behind Jupiter can be used to obtain light curves that provide the locations and flux densities of individual hot spots, mutual occultation events between the satellites can also be used. These have the advantage of the fact that the occulting body has a sharp edge, unlike Jupiter, removing the complicating effect of emission passing through Jupiter's atmosphere. In addition, both the ingress and egress provide usable information, in contrast to Jupiter occultation since Io is either entering Jupiter occultation from eclipse or emerging from occultation to eclipse, but not both. However, mutual events have the significant disadvantage of only occurring for a roughly 6 month period every 6 years, when the plane of the galilean satellite orbits is edge-on as viewed from Earth. Descamps et al. (1992) successfully measured emission from Loki Patera and Pele during a mutual occultation event in 1991 using a 1-m telescope. More recently de Kleer et al. (2017, 2021) observed an occultation of Io by Europa with the LBTI using adaptive optics, which allowed the occultation light curve of each hot spot to be extracted independently. Using these data, they derived the temperature distribution within the ~200 km Loki Patera at a spatial resolution of 10 km and inferred the presence of two resurfacing waves traveling around the patera in opposite directions.

While each of the above techniques has only been used a limited number of times, the collective application of novel techniques and instrumentation to Io's hot spots demonstrates the many potential avenues for obtaining higher spatial resolution and greater spectral coverage. As telescope technologies improve and new instrumentation is developed, studies of Io's hot spot emission will be able to achieve an increasing level of detail, enabling more physically and geologically motivated investigations into Io's volcanic processes.

6.2.2 *Space-Based Observations*

6.2.2.1 Recent Analyses of *Voyager* and *Galileo* Data

Thermal emission from Io's hot spots was measured by both the *Voyager* and *Galileo* spacecraft. The most significant space-based dataset of Io's thermal activity is from *Galileo*, whose near-infrared instrument NIMS and mid-infrared instrument PPR measured the emission from numerous hot spots across Io's surface. The state of knowledge at the end of the *Galileo* mission is summarized by Williams and Howell (2007). However, the substantial dataset returned by *Galileo* has been yielding a wealth of results since the end of the mission, with new results still emerging.

Such studies include analyses of individual volcanoes or eruptions (e.g. Howell and Lopes 2011; Davies et al. 2014; Howell et al. 2014; Davies et al. 2018); statistical analyses of hot spots and volcanic centers (e.g. Hamilton et al. 2013; Veeder et al. 2015; Davies et al. 2015; Rathbun et al. 2018; Davies 2021); and the completed geological map of Io, which provides key context for interpretation of thermal emission data (Williams et al. 2011a,b).

Importantly, Veeder et al. (2009, 2011, 2012, 2015) systematically analyzed dark volcanic flow fields, dark paterae, and additional thermal sources observed by *Galileo*. In total, they identified 242 recently active features and quantified their heat flow. Heat flow from most hot spots was determined by fitting 1- or 2-temperature models to NIMS spectra. When the spectra were obtained in sunlight, the emitting area was assumed to be the whole dark area of the volcanic source at that location. The detection limit for each observation was determined, and an upper limit to emitted power placed on all dark surface areas that were not detected by NIMS. This accounts for the fact that dark areas must be at an elevated temperature, even if thermal emission is not detected, because otherwise the SO_2 in Io's atmosphere would condense out and brighten them. Based on this systematic quantification of power output from Io's hot spots, a total volcanic power output of 56.2×10^{12} W is calculated (Veeder et al. 2015), which accounts for about 54% of Io's total heat flow.

6.2.2.2 *Cassini*, *New Horizons*, and *Juno*

The *Cassini* and *New Horizons* spacecraft flew by Jupiter on their way to their respective destinations of Saturn and Pluto. While in the system, each turned their instruments to Io. *Cassini* ISS data were used to understand Tvashtar (Sect. 6.3.4.4; Milazzo et al. 2005), Pele (Radebaugh et al. 2004), Pillan Patera, Wayland Patera, and Loki Patera (Allen et al. 2013). Color temperatures were obtained by using the clear and IR filter with measured temperatures ranging from 700 to 1500 K. Little temporal variation was found at these volcanoes.

Three of *New Horizons*' instruments were particularly useful for understanding Io's active volcanoes: the Long-Range Reconnaissance Imager (LORRI), the Multicolor Visible Imaging Camera (MVIC) and Linear Etalon Infrared Spectral Array (LEISA). All three instruments observed a major eruption of the Tvashtar volcano, including a hotspot and a large plume, as well as a new volcanic hotspot east of the Girru volcano, referred to as 'East Girru' (Spencer et al. 2007). From LEISA (1.25–2.5 μm), 37 active hot spots were detected over nine spectral observations, with 11 hotspots observed more than once (Tsang et al. 2014); fits to the spectra resulted in temperature estimates from 500 to 1240 K, with Tvashtar representing the upper end of the range. From 17 sets of MVIC images and 190 LORRI images, 54 emission sources were detected (Rathbun et al. 2014).

The majority of spots detected by these *New Horizons* instruments were long-lived active volcanoes previously observed as hotspots by either *Galileo* or ground-based instruments. LORRI observations of Io during eclipse confirmed that changes in Ionian volcano brightness, when they occur, generally occur over timescales of hours or longer as no changes were observed over timescales of seconds to minutes (Rathbun et al. 2014; Radebaugh et al. 2004). Ground-based observations obtained in the months leading up to the 2007 *New Horizons* fly-by of Io demonstrated that, with the exception of Tvashtar (see Sect. 6.3.4.4), Io was unusually quiescent during the flyby.

Juno is the most recent spacecraft to enter the Jupiter system, arriving in 2016 with the goal of studying Jupiter's atmosphere and interior (Bolton et al. 2017). While *Juno*'s instruments were not designed to study Io, the Jovian Infrared Auroral Mapper (JIRAM), which can acquire 3–5 μm images and 2–5 μm spectra, has been used to observe Io's active volcanoes (Fig. 6.6; Mura et al. 2020). Given *Juno*'s polar orbit around Jupiter, it is in the unique position to observe Io's polar regions with less foreshortening effects then all past datasets. Given that there is a proposed difference between high- and low-latitude volcanoes (Sect. 6.4.2.1) and that the biggest difference between tidal heating models occurs at the poles (Sect. 6.4.2), future *Juno* data may prove instrumental in resolving current ambiguities in interpreting Io's thermal emission in terms of interior models.

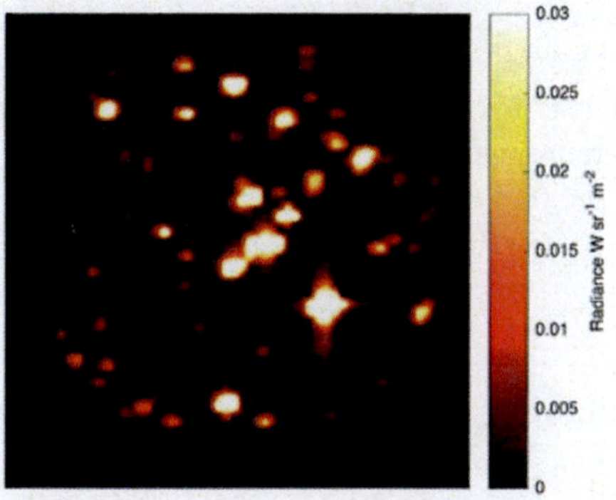

Fig. 6.6 Eclipse image of Io on 2018 Dec 21 from the *Juno* JIRAM instrument at a wavelength of 5 μm. Image credit: NASA/JPL-Caltech/SwRI/INAF

6.3 Volcanic Styles

The eruption characteristics of a given volcanic center depend on numerous factors such as the composition of the magma, including its gas and crystal content; the total volume of the magma; and the pressure of the environment into which it is erupting. Io's lavas appear to be predominantly mafic silicates, broadly basaltic or ultramafic in nature; these compositions produce effusive eruptions with low viscosity lavas that can flow for long distances.

The time-evolution of an eruption, even if the hot spot is unresolved, provides a powerful constraint on the style of volcanism. At the end of the *Galileo* mission, Io's primary eruption styles were identified as flow-dominated, explosion-dominated, and intra-patera (Williams and Howell 2007). Flow-dominated eruptions can be identified by the extensive lava flow fields they generate (e.g. Amirani), which appear to originate in paterae or fissures and expand slowly via repeated small breakouts. These eruptions have been observed to stay active for years. Their associated gas plumes originate at the tip of the flow, likely from vaporization of surface frosts as the hot lava flows over them. In contrast, explosion-dominated volcanism involves brief (hours to weeks in duration), energetic volcanic events that produce explosive plumes and ringed deposits. Intrapatera eruptions are those confined to volcanic-tectonic depressions in the few km to 200 km size range, and include Io's widespread lava lakes. Many of Io's volcanoes exhibit more than one of these styles. For example, the volcano Pele appears to host a lava lake but also produces explosion-dominated eruptions (Lopes et al. 2001; Radebaugh et al. 2004). Effusive SO_2 volcanism has also been suggested, though the evidence is controversial and the volcanic style, if present, is confined to only a few particular regions. This topic is covered by Williams and Howell (2007), and no significant progress has been made since that time.

The volatile content of the magma must play a significant role in eruption characteristics on Io as it does on Earth, and indeed the volatile content of volcanoes on Io varies dramatically between sources. While massive, high-altitude plumes and broad deposits are observed at volcanoes such as Pele and Tvashtar, other hot spots such as Loki Patera have minimal evidence of volatiles despite producing substantial thermal emission. Recent work has been done to interpret volcanic gas composition in terms of mantle oxidation state (Battaglia et al. 2014), and to understand the reincorporation of buried SO_2 into rising magmas (Leone et al. 2011). However, the role of volatiles in setting eruption style, and the causes of variation in volatile content between volcanoes, remains poorly understood.

6.3.1 *Persistent vs. Transient Hot Spots*

It was already recognized by early in the *Galileo* mission that Io's hot spots could be roughly classified into persistent and transient (Lopes-Gautier et al. 1999), and these categories have continued to serve as a useful classification scheme (e.g. de Kleer and de Pater 2016a; Marchis et al. 2005; Rathbun and Spencer 2010). While a classification based on more physically meaningful properties would be preferred, such properties are often not available. However, transient volcanoes also exhibit higher peak emission and are preferentially located at higher latitudes than persistent volcanoes, pointing to true underlying differences between volcanoes in the two categories. Persistent hot spots are those that consistently produce thermal emission over years to decades, typically at a moderate level. In contrast, transient hot spots have short-lived, powerful flare-ups that often last only for days to weeks, and are inactive or active only at a low level between flare-ups. Persistent hot spots are identified by the fact that they are detected during nearly every observation that would have been sensitive to emission (Lopes-Gautier et al. 1999; de Kleer and de Pater 2016a; Rathbun and Spencer 2010). Based on a 5-year study of 75 hot spots, de Kleer et al. (2019b) found that no persistent hot spot on Io consistently exhibits 3.8-μm emission above 20 GW/μm/str, so that this threshold could be used as a proxy to identify transient hot spots. A handful of hot spots exhibit both transient and persistent behavior, emitting consistently at a moderate level but occasionally exhibiting bright eruptions. Examples of this type of behavior include Loki Patera, Pele, and Pillan Patera (see Sects. 6.3.4.5, 6.3.4.2, and 6.3.4.1).

The set of persistent hot spots detected by *Voyager*/*Galileo* in 1979–1999 (Lopes-Gautier et al. 1999), and those detected in 2013–2018 from ground-based AO observations (de Kleer et al. 2019b), is overlapping but not identical. For example, Isum Patera was detected in 9 out of 9 observations during *Galileo* orbits G1-C10, but was only detected 16 times by de Kleer et al. (2019b), compared to 80–113 detections for the most persistent volcanoes. In addition, as will be discussed in more depth in Sect. 6.3.4.2, Pele was one of the most persistently-active volcanoes until the 2010s when its typical thermal emission dropped by a factor of 2–3. Conversely, the hot spot in Chalybes Regio was one of the brightest and consistently-active volcanoes in 2013–2018 (de Kleer et al. 2019b), but was not detected in spacecraft data at all and is not clearly associated with any patera that was present in the *Voyager*/*Galileo* surface imaging data. Thus the 'persistent' label may not hold for longer than ~decadal timescales for the majority of hot spots.

6.3.2 Determining Volcanic Style from Thermal Emission

While observational classification schemes such as the transient vs. persistent categories are useful for pattern identification, the ultimate objective is to understand the underlying volcanic processes generating the observed signatures. The majority of data on Io's hot spots consists of multi-wavelength near/mid-IR photometry or low-resolution spectroscopy. Such measurements, in particular when observations are made at more than one wavelength and multiple times during an eruption, can be used to quantify the total thermal emission, volume of material erupted, and temperature evolution of the eruption. These properties in turn can be used to classify the eruption style.

Figure 6.7 demonstrates one metric for classifying eruption style when the hot spot is unresolved: the 2 vs. 5 μm thermal emission (Davies et al. 2010). This metric captures both the total emission from the hot spot, and the temperature distribution, because higher temperatures produce more emission at shorter wavelengths as shown in Fig. 6.1.

The time evolution of the emission from a hot spot also provides information on the type of volcanism ongoing, particularly when multiple wavelengths are observed and the temperature evolution is constrained. For example, the hot spot Amirani shows thermal emission that is dominated by a low temperature component, with a variable high-temperature component that shifts in location, indicative of an insulated flow field experiencing brief lava breakouts (Davies et al. 2014). Lava lakes, such as Pele, Tupan Patera, or Loki Patera, produce consistent steady thermal

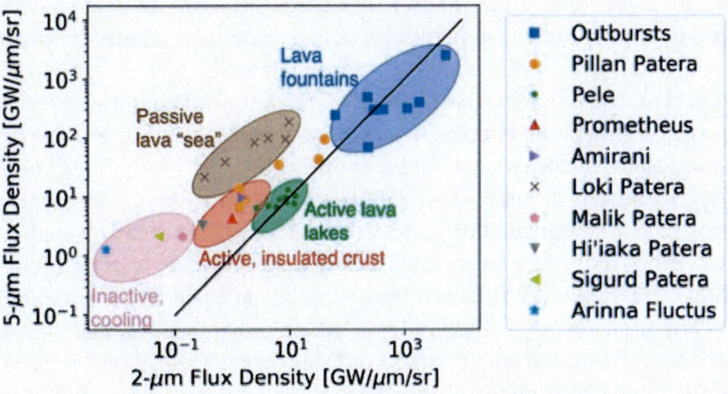

Fig. 6.7 2 and 5 μm flux densities for several Ionian volcanoes thought to host different volcanic styles. The ratio of emission at 2–5 μm is a measure of what fraction of the emitting area is composed of active magma extrusion vs. spreading and cooling lava flow. Where a volcano falls along the line corresponding to a given 2 μm/5 μm emission ratio indicates the total power of the eruption; the very highest eruption powers correspond to vigorous lava fountaining events. Base figure from Davies et al. (2010), with additional data from de Pater et al. (2014) and de Kleer et al. (2014)

emission from a fixed patera location, in some cases interspersed with occasional brightenings that may be due to an overturning crust (Davies et al. 2001; Rathbun et al. 2002; Davies et al. 2012); see Sect. 6.3.4. Io's most dramatic eruptions are thought to be massive lava fountaining events whereby fountains feed open channel flows (Davies 1996; McEwen et al. 2000; Keszthelyi et al. 2007). The time-evolution of these events involves a rapid decrease from peak emitted power and peak temperature, which by analogy to Earth is thought to correspond to a transition from fire fountaining to insulated sheet flows (McEwen et al. 2000; Davies 2007). The flows following such lava fountaining events have been detected for years after event initiation (Davies et al. 2001).

When sufficient observational coverage of an eruption is obtained, the peak volumetric effusion rate—or magma volume erupted per second—can be constrained. This provides one of the clearest comparison points between Ionian and Earth eruptions. Peak effusion rates for outbursts, including the 1990 eruption near Loki, the 1997 Pillan eruption, and the Rarog and Heno outbursts in 2013, were in the 10^4–10^6 m^3/s range (Blaney et al. 1995; Davies 1996; de Pater et al. 2014), similar to effusion rates thought to be present during emplacement of terrestrial flood basalts. Similarly, the volume of magma present in an Ionian lava lake, even excluding the largest lava lake Loki Patera, is roughly an order of magnitude larger than the magma volume in analogous Earth lava lakes (Davies 2007; Lopes et al. 2018).

6.3.3 *Outburst Eruptions*

Io's most powerful eruptions are known as outburst eruptions. In instances where the observations were sufficient to permit modeling, outbursts have been attributed to vigorous lava fountaining (Davies 1996; Stansberry et al. 1997; de Pater et al. 2014; Keszthelyi et al. 2007; McEwen et al. 2000). The canonical definition of an outburst eruption is an eruption that doubles Io's disk-integrated 5-μm flux density including the reflected-sunlight component (Spencer and Schneider 1996). This corresponds to a 5-μm flux density around 300–500 GW/microns/str, based on measurements of Io's quiescent disk-integrated flux density (Veeder et al. 1994; Howell et al. 2001; de Kleer et al. 2014). However, many eruptions labeled as outbursts in the literature fall short of this emission level or lacked measurements at 5 μm. Table 6.1 summarizes 35 bright eruptions reported since 1978, of which only a handful meet the outburst definition above (indicated with asterisks in Table 6.1). The locations of the brightest eruptions over time are shown in Fig. 6.8.

Estimates of the frequency of outburst eruptions on Io have been limited by temporal and wavelength coverage, an inconsistent use of the definition of outburst, and uncertainty regarding whether emission calculations should account for geometric foreshortening, which is appropriate for a lava flow but not a fire fountain. Veeder et al. (1994) found a rate of 4% based on the detection of 2 events over 55 nights of observation, and Spencer and Schneider (1996) compiled 6 outbursts detected prior to 1996 and found a rate of 3.3%. de Pater et al. (2014)

Table 6.1 Outbursts and sub-outbursts

UT date	Location	Identification	Temperature [K]	Power [TW]	Peak 3.8-μm flux density [GW/μm/sr]	Reference
*1978-01-26	293-113 W	–	–	–	–	Spencer and Schneider (1996)
*1978-02-20	341-138 W	–	600	17	500–700	Witteborn et al. (1979)
*1979-06-11	313-125 W	–	550–600	28	–	Sinton (1980)
*1986-08-07	358-140 W	–	1550	58	–	Johnson et al. (1988); Veeder et al. (1994)
*1990-01-09	360-258 W	–	1225–1600	11	–	Veeder et al. (1994)
*1995-03-02	80-110 W 30-60 S	Arusha Patera?	600	3.6	–	Spencer et al. (1995)
1996-10-06	35±15 W 70±15 N	–	1325 ± 45	3.9	150 or 360[a]	Stansberry et al. (1997); Howell et al. (2001)
1997-06-28	244 W 12 S	Pillan Patera	>1600	3.4	83	McEwen et al. (1998a); Davies et al. (2001)
1999-06-22	14±2 W 10±2 S	9906A	–	–	180	Howell et al. (2001)
*1999-08-02	65±5 W 17±7 N	Gish Bar Patera	1247	63	1800	Howell et al. (2001)
*1999-11-13/2000-12-16[b]	120 W 60 N	Tvashtar Catena	>1300	24	–	Marchis et al. (2002); Milazzo et al. (2005)
2000-02-20	80±4 W 64±2 N	–	–	–	200±25 or >800[a]	Howell et al. (2001)
2001-02-19	118 W 27 N	Amirani	990±35	5.7	–	Marchis et al. (2002)
2001-02-22	340 W 41 N	Surt	1240±20	78	–	Marchis et al. (2002)
2004-05-28	1 W 18 S	Tung Yo Patera	>1100	2	72±12	de Pater et al. (2016a)
2004-05-30	3 W 19 S	Sui Jen Patera	1075	3.5	59±9	de Pater et al. (2016a)

(continued)

Table 6.1 (continued)

UT date	Location	Identification	Temperature [K]	Power [TW]	Peak 3.8-μm flux density [GW/μm/sr]	Reference
2004-05-31	272 W 41 S	S. of Babbar	>900	1.2	57±14	de Pater et al. (2016a)
2006-04-17[b]	121 W 61 N	Tvashtar Catena	1200	4.9	–	Laver et al. (2007)
2007-02-27	236 W 21 N	East Girru	1100	0.2	–	Tsang et al. (2014)
2008-07-24	245 W 8 S	Pillan Patera	–	–	157±24	Lellouch et al. (2015)
2013-08-15	306 W 39 S	Rarog Patera	1300±200	10	325±80	de Pater et al. (2014)
2013-08-15	307 W 56 S	Heno Patera	720±100	4.8±2	135±55	de Pater et al. (2014)
*2013-08-29	223.5 W 29.1 N	201308C	>1300	24.5	> 500	de Kleer et al. (2014)
2014-10-22	247 W 65 S	Chors Patera	–	–	57±19	de Kleer and de Pater (2016a)
2015-01-10	266 W 58 S	Mithra Patera	–	–	55±12	de Kleer and de Pater (2016a)
2015-01-26	219 W 49 S	Kurdalagon Patera	1200 ± 150	0.95±0.6	56 ± 9	de Kleer and de Pater (2016a)
2015-02-18	–	Pillan Patera	–	–	80 ± 16	de Pater et al. (2016b)
2015-04-05	224 W 48 S	Kurdalagon Patera	1300 ± 200	1.25 ± 0.03	68 ± 11	de Kleer and de Pater (2016a)
2016-05-17	128 W 10 S	P95	1020 ± 180	–	58 ± 13	de Kleer et al. (2019b)
2016-06-20	151 W 33 S	Shamash Patera	1000 ± 100	–	53 ± 9	de Kleer et al. (2019b)
2016-06-27	180 W 71 S	Illyrikon Regio	1210 ± 690	–	125 ± 69	de Kleer et al. (2019b)
2018-05-10	252 W 37 S	UP 254 W	960 ± 100	–	134 ± 24	de Kleer et al. (2019b)
2018-05-27	205 W 31 N	Isum Patera	1200 ± 220	–	64 ± 16	de Kleer et al. (2019b)

* Meets canonical definition of outburst

[a] With correction for geometric foreshortening

[b] Tvashtar was observed to be active on multiple occasions in 1999/2000 and 2006/2007, which in either time period could represent an extended period of heightened activity, or multiple distinct episodes

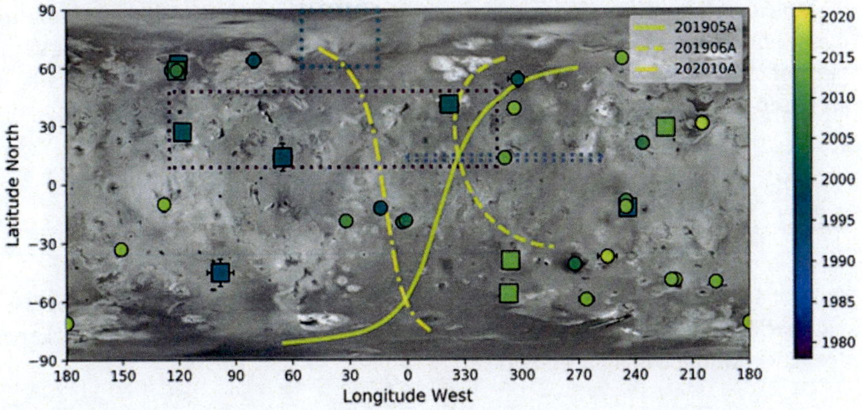

Fig. 6.8 Locations of outbursts on Io over time. Boxes indicate location constraints when observations had no or poor spatial resolution. Curves are the one-dimensional locations of outbursts observed in occultation lightcurves. Colors indicate the year the outburst was detected. Figure courtesy of C. Tate

found a rate of 10% based on the detection of 7 events over 75 nights of observation. However, de Kleer et al. (2019b) subsequently observed Io on 271 nights and detected only 3 outburst eruptions, though 14 volcanoes were detected to host bright, transient activity where bright is defined by an L-band intensity above 30 GW/μm/sr (about an order of magnitude below the threshold for an outburst). Based on the data obtained prior to 2012, Veeder et al. (2012) estimated an average outburst output of 5 TW sustained over a 5-day period, and 25 events per year as a conservative upper bound. Using these numbers, outburst eruptions contribute no more than 2% to Io's total heat flow.

It is clear from Table 6.1 that while outbursts themselves are quite rare, there are many more bright, transient events that reach 3.8-μm flux densities above 100 GW/μm/str but fall short of outbursts, and even events that fall short of this brightness level may reach measured temperatures of 1200 K or higher. These 'mini-outbursts', or 'sub-outbursts', exhibit similar behavior to full outbursts: they exhibit high temperatures, rapid decays in brightness, and occur preferentially at higher latitudes (de Kleer and de Pater 2016a). The emission from outbursts is observed to decay on timescales of hours to days (Stansberry et al. 1997; Howell et al. 2001; de Kleer et al. 2014), and is not yet clear whether sub-outbursts are similar to outbursts but smaller in scale, or whether they are outbursts caught after their emission peak.

An intriguing feature of Io's eruptions is that many of the large eruptions have been first detected within a few days of one another, often at nearby locations. For example, the Amirani and Surt eruptions were detected within 3 days of one another (Marchis et al. 2002); two outbursts were detected near the same longitude in 1996 (Stansberry et al. 1997); and the three outbursts were detected in August of 2013 (de Pater et al. 2014; de Kleer et al. 2014). Several more examples can be seen in Table 6.1. This clustering has not been statistically verified with accounting for observational biases, nor has a physical mechanism been explored, but on Earth the phenomenon of pairs of eruptions occurring within 2 days and 200 km has been proposed (Linde and Sacks 1998; Manga and Brodsky 2006; Williams 1995).

6.3.4 Notable Volcanoes

Each of Io's volcanoes exhibits its own unique combination of volcanic processes and merits detailed geological study. In this section, we highlight a few volcanoes that have exhibited notable activity.

6.3.4.1 Pillan Patera: Source of Repeated Outbursts

Pillan Patera is one of Io's most powerful volcanoes, and exhibits vigorous eruptions that are consistent with the thermal signature of effusive eruptions feeding open channels or sheet flows that result in flow fields. The 1997 eruption at Pillan Patera, observed by *Galileo*, provides one of the strongest pieces of evidence for magma temperatures above 1600 K and hence for the existence of ultramafic magma compositions on Io (McEwen et al. 1998a), although models that treat the fountain as a spray of small particles instead of large clasts find a temperature of 1340 K for the event, within the basaltic range (Keszthelyi et al. 2007). As can be seen in Table 6.1, Pillan Patera is responsible for at least 3 of the 35 bright, transient eruptions that have been detected at Io in the past 40 years, and at least 3 more that weren't bright enough to include in the table. However, while many of Io's most persistent bright volcanoes are lava lakes which remain fixed in location, the precise location of the eruption within the Pillan region has moved by up to 100 km between events, and even between detections within a few-month period such as in February through May of 2015 (de Pater et al. 2016b; de Kleer and de Pater 2016a).

6.3.4.2 Pele: A Persistent Volcano Turns Off

Pele was one of the most notable hot spots on Io during the *Voyager* and *Galileo* missions. It is the source of a massive red plume deposit that is one of the most readily identified features of Io's surface. Pele is thought to be an active lava lake due to its thermal emission spectrum, persistent activity, and fixed location of the emitting center (Davies et al. 2001, 2012), and also exhibits bright episodes thought to be fire fountaining (Howell and Lopes 2011). Prior to 2005, Pele was consistently detected by spacecraft and telescopes whenever the viewing geometry was favorable, frequently with high temperatures (Davies et al. 2005; de Pater et al. 2016b). However, by 2010 it had dropped to half of its former activity level, and has hovered around that level or below since 2013 with no bright activity detected (de

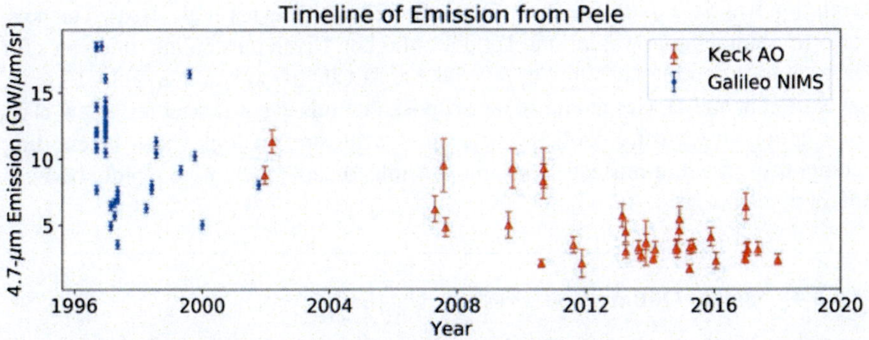

Fig. 6.9 Timeline of thermal emission from Pele, derived from Keck AO and *Galileo* NIMS observations over two decades, demonstrating the systematic decrease in brightness over a 20-year period. Data from Davies et al. (2001, 2012), de Pater et al. (2016b), Cantrall et al. (2018), and de Kleer et al. (2019b)

Pater et al. 2016b; de Kleer and de Pater 2016a; de Kleer et al. 2019b); see Fig. 6.9. This is perhaps the clearest example of a hot spot that was bright and persistently-active for decades that then 'turned off', or dropped drastically in thermal activity and sustained that low activity level. Future observations will reveal whether Pele has in fact turned off, or whether the current activity level is merely a decade(s)-long lull, and will provide at least one clue into the longevity of Io's persistent volcanoes. The red sulfur materials in Pele's plume deposit are thought to have a lifetime of only a few months on Io's surface (Carlson et al. 2007). Future detection (or non-detection) of the plume deposit will tell us whether volatiles are still being released despite the low levels of thermal emission, informing our understanding of the link between volatile release and thermal emission in general, an area that remains poorly understood.

6.3.4.3 Marduk Fluctus: Strombolian-like Activity on Io?

One of the open questions regarding Io's volcanism is the timescales over which eruptions evolve. Aside from the outburst eruptions, which may evolve over hours to days, the timescales for variability are typically on the order of weeks to years. The first truly rapid thermal event was documented at Marduk Fluctus in 1996 (Davies et al. 2018). Marduk Fluctus consists of a patera feeding multiple lava flows, accompanied by a red plume deposit (Williams et al. 2011b), and was detected as a thermal source by *Galileo* (Belton et al. 1996; Lopes-Gautier et al. 1997). It has been one of the most consistently active hot spots over the past decade, detected nearly 100 times from ground-based observation (Cantrall et al. 2018; de Kleer et al. 2019b).

In 1996, *Galileo* NIMS detected a 30-fold increase in 2.5-μm emission over a 2-h period, followed by a threefold decay in emission that took place over just 2 min, and emission was back to the pre-event level after 20 min (Davies et al. 2018). The peak effective temperature was around 1600 K or higher. Given the cooling timescales for lava on Io, a cooling lava flow or lava lake crust cannot explain this rapid decrease in emission, and Davies et al. (2018) propose that this event represents a new class of eruption on Io: a strombolian or vulcanian explosion analog, where the emitting components are mm-m sized clasts that are able to cool much more rapidly because of their small size.

6.3.4.4 Violent Outbursts at Tvashtar Catena

Tvashtar is a chain of large paterae and has exhibited very bright eruptions (classified as an outburst based on ground-based observations; Howell et al. 2001), lava fountaining (Keszthelyi et al. 2001), active surface flows, and a large plume (Porco et al. 2003) and red ring-shaped plume deposit (Geissler et al. 2004a). It was observed through the *Galileo* era, including during the *Cassini* flyby of Jupiter in 2000 and in several ground-based observations (Milazzo et al. 2005). Post-*Galileo*, Tvashtar was the star of the show for the *New Horizons* flyby of the Jupiter system in 2007 (Spencer et al. 2007). The plume was imaged multiple times by *New Horizons* LORRI over 7.8 days and the plume height remained remarkably constant varying only between 320 and 360 km (Spencer et al. 2007). Observations of the Tvashtar hotspot by *New Horizons* LORRI suggested that the activity was due to a surface flow, and MVIC revealed a temperature of 1260 ± 100 K and area of 30–50 km^2 for that flow (Rathbun et al. 2014). LEISA spectra were consistent with a blackbody with a temperature of 1239 ± 19 K and area 37 ± 4 km^2 (Tsang et al. 2014). These temperatures are slightly lower than those obtained for the earlier Tvashtar eruptions, while the area and total power output are substantially higher (Milazzo et al. 2005). Ground-based observations revealed that Tvashtar was erupting at outburst level in the Spring and Summer of 2006 (Laver et al. 2007).

6.3.4.5 Loki Patera: The Trickster Powerhouse

Loki Patera is the most powerful and best-studied volcano on Io, producing roughly 10 TW of thermal emission or 10% of Io's total thermal emission. It has been observed from the ground using the occultation technique (Sect. 6.2.1.1) since the late 1980s and using adaptive optics since the late 1990s, resulting in over 30 years of observations with only a few gaps of more than a few months (Fig. 6.10). Loki Patera has been modeled alternately as a lava flow (Howell 1997; Gregg and Lopes 2008) and as an overturning lava lake (Rathbun et al. 2002; Rathbun and Spencer 2006; Matson et al. 2006; de Kleer and de Pater 2017). The lava lake scenario is able to explain nearly all observed properties of the hot spot, as described below.

Loki Patera was observed at high spatial resolution by the *Galileo* and *Voyager* spacecraft (see Fig. 6.11a,b; Spencer et al. 2000; Davies 2003; Howell and Lopes 2007). Measured temperatures ranged from ~300 K for a low-temperature com-

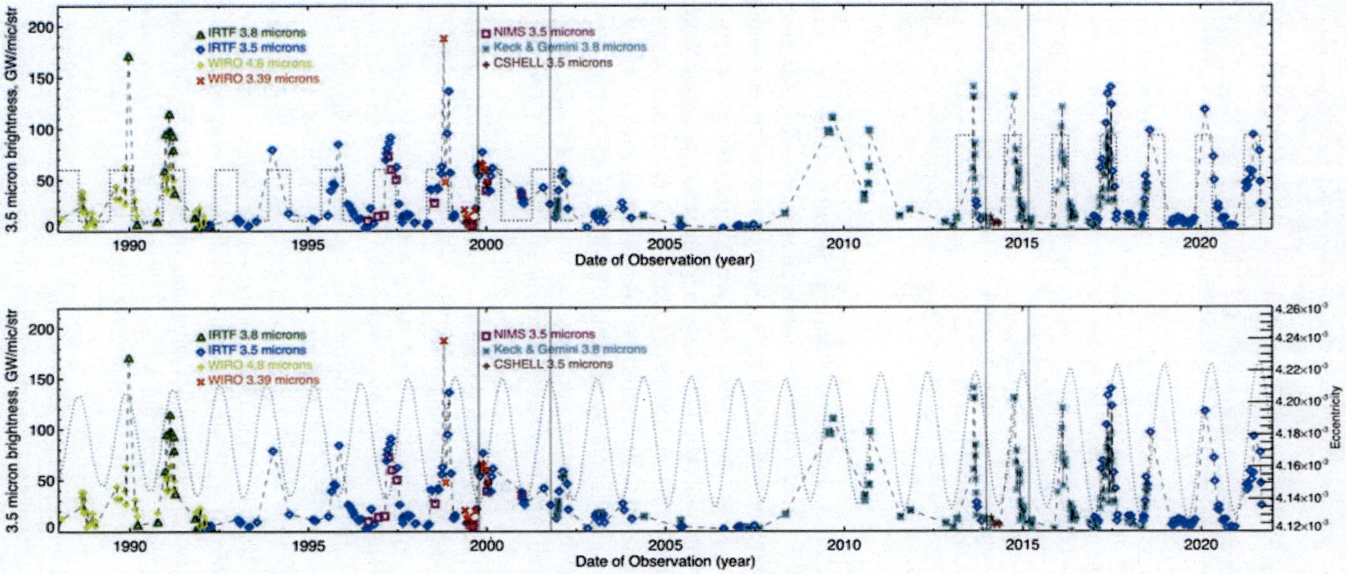

Fig. 6.10 The brightness of Loki Patera as a function of time based on data from Rathbun et al. (2002), Rathbun and Spencer (2006), Rathbun and Spencer (2010), Davies et al. (2012), de Kleer et al. (2019b), de Pater et al. (2017). The vertical lines indicate the times that the observations in Fig. 6.11 were obtained. Top panel: The square wave in the background pre-2002 has a 540-day period while the post-2013 has a 490-day period. Bottom panel: The same data, with Io's eccentricity variations (filtered to isolate the ~470 day period for clarity), indicated by the dashed line (de Kleer et al. 2019a)

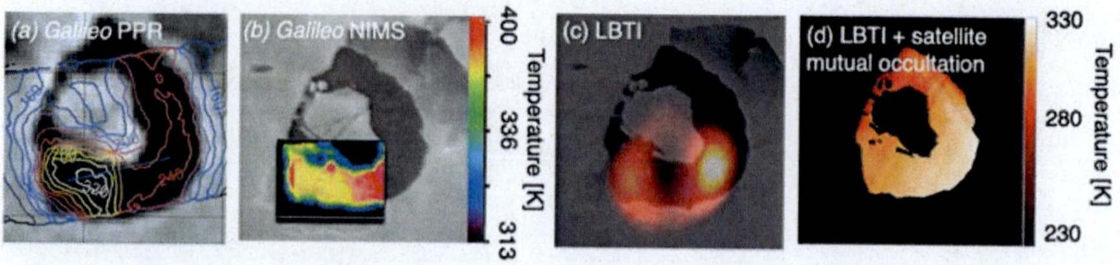

Fig. 6.11 Thermal maps of Loki Patera from observations that spatially resolved the emission. (**a**) Nighttime temperature map of Loki Patera from *Galileo* PPR data obtained near the beginning of a brightening event; figure modified from Spencer et al. (2000); (**b**) Temperature map of the southwest portion of Loki Patera derived from *Galileo* NIMS data obtained near the end of a brightening event; figure modified from Davies (2003); (**c**) Interferometric image at a wavelength of 5 μm from the LBTI obtained in between brightening events and resolving emission from Loki Patera into two distinct emitting regions in the south; figure modified from Conrad et al. (2015); (**d**) Temperature map of Loki Patera derived from LBTI AO data at 5 μm during a mutual satellite occultation with Europa obtained between brightening events; figure modified from de Kleer et al. (2017)

ponent to 600–800 K for a high-temperature component. Shorter-wavelength data obtained by *Cassini* ISS found temperatures as high as ∼1300 K (Allen et al. 2013); the *Cassini* data were obtained at lower emission angles, possibly enabling them to see further into the cracks in Loki Patera's surface than other observations.

The overall low temperatures and the smooth temperature distribution across the patera, combined with the episodic nature of Loki Patera's brightenings, suggest that the brightenings correspond to the overturn of the solidified crust on top of the lava lake when it becomes gravitationally unstable (Rathbun et al. 2002). A quantified version of the crustal foundering model was found to be consistent with ground-based observations from multiple sources (Rathbun and Spencer 2006; Matson et al. 2006; de Kleer and de Pater 2017; Rathbun and Spencer 2010). In the model, apparent changes to eruption cadence can be explained by a change in magma composition or volatile content leading to different crustal overturn velocities.

However, not all aspects of the observations are in full agreement with the overturn model. The smoothness of the temperature profile, combined with large subpixel temperature variations, have also been used to show that the overturn model does not yet encompass all physical processes at play (Howell and Lopes 2007). In addition, the distribution of "bergs" and the persistence of the largest bergs over the 22 years between the *Voyager 1* and *Galileo* missions is difficult to reconcile with the overturn model (Howell et al. 2014). Finally, Gregg and Lopes (2008) present an alternative model of Loki Patera with the same magma production rate and eruption style as Earth's mid-ocean ridges, but with different magma origin and tectonic regime, a type of volcanic regime that may have been present on Earth before plate tectonics.

Adaptive optics observations, which have the spatial resolution to localize the peak emission within the patera, suggested that the overturn wave moves in the opposite direction to that found by Rathbun et al. (2002) (de Kleer and de Pater

2017; de Pater et al. 2017). However, Conrad et al. (2015) and de Kleer et al. (2017) observed Loki Patera at high spatial resolution using the Large Binocular Telescope Interferometer (LBTI) and found that Loki Patera's emission was arising from multiple areas within the patera (Fig. 6.11c,d), so the best-fit locations derived from observations where Loki Patera is not resolved are likely averages of multiple or continuous emitting regions and it is not clear how to interpret them in terms of an overturn event. The LBT observations were obtained in-between Loki Patera's brightening events, giving insight into Loki Patera's behavior between overturn events.

One of the most unique properties of Loki Patera is its temporal behavior. Prior to 2002, Loki Patera appeared to be erupting periodically with a periodicity of 540 days (Rathbun et al. 2002), although the exact interval between subsequent events varied somewhat (Rathbun and Spencer 2006; de Kleer and de Pater 2017; de Pater et al. 2017). Over 2013–2019, the best-fit period was 475 ± 50 days (see Fig. 6.10). The shorter period may arise from a change in magma properties, which is supported by the fact that the more recent eruptions had larger average brightness to go along with their shorter durations (see Fig. 6.10). However, a separate analysis of the full 1987–2018 timeline finds that a single period around 460–480 days matches the entire timeline, although the volcano is found to be only quasi-periodic (de Kleer et al. 2019a). This period matches the timescales for the oscillation of Io's eccentricity and semi-major axis ($\sim$480 and $\sim$460 days; see Fig. 6.10). At the time of writing, both models for Loki Patera's temporal behavior remain consistent with the data, and continued observations of brightening events are needed to discriminate between scenarios.

6.4 Io's Volcanoes and Tidal Heating

Io's volcanism is ultimately driven by tidal heat deposition in its mantle. However, linking specific properties of Io's volcanism back to tidal heating processes has proved challenging, and the specific mechanisms of tidal heat generation have consequently remained elusive. The degree of dissipation, and how deep within the interior the dissipation is primarily occurring, depend strongly on poorly-constrained material properties of Io's interior and cannot be calculated a priori. Instead, observational signatures are used to constrain where and how melting occurs in Io's interior. Two promising signatures for inferring properties of Io's interior from observations outside the Jupiter system (i.e. when gravity and magnetic field observations are not possible) are the temperature of the magma, and the spatial distribution of volcanic heat flow or volcanic styles.

6.4.1 Magma Temperature

Data from the *Galileo* mission confirmed that the dominant magma composition on Io is silicate rather than sulfur (Johnson et al. 1988; Davies et al. 1997; McEwen et al. 1998a). The major outstanding question about Io's magma composition is now whether the silicates are broadly basaltic, as is widespread in Earth volcanism, or of a more ultramafic composition. Which composition dominates Io's eruptions, and whether that composition varies across the surface or between volcanoes, constrains the mantle temperature, degree of mantle melting, and depth of magma generation. Ultramafic magmas have a higher metal content (iron and magnesium) and lower silica content, and are lower viscosity than their less mafic counterparts. Higher mantle temperatures are needed to melt the minerals that form ultramafic magmas, and the temperatures of komatiitic magmas, the magnesium-rich ultramafic magmas that are associated with the highest temperatures and lowest viscosities, are expected to be around 1800 K compared with 1475 K for basalts. This temperature difference provides a potential avenue for distinguishing magma composition observationally: if temperatures above ∼1500 K are observed, the magmas are likely more mafic than basalt. As discussed in Sect. 6.1 of this chapter, near-infrared spectroscopy or spectrophotometry in the 1–5 μm range can provide a temperature measurement (Fig. 6.1). However, that temperature measurement represents an average temperature over a spatial resolution element, which in most ground- or even space-based thermal observations is 100 km or more. Due to the steep cooling curve (Howell 1997; Carr 1986; Davies 1996), magma exposed on Io's surface cools by 400 K in just 2 min (Fig. 6.12; Davies et al. 2005), and the average temperature measured for a hot spot is consequently dominated by the cool component. In addition, constraining the highest temperatures requires optical to very near IR measurements, where

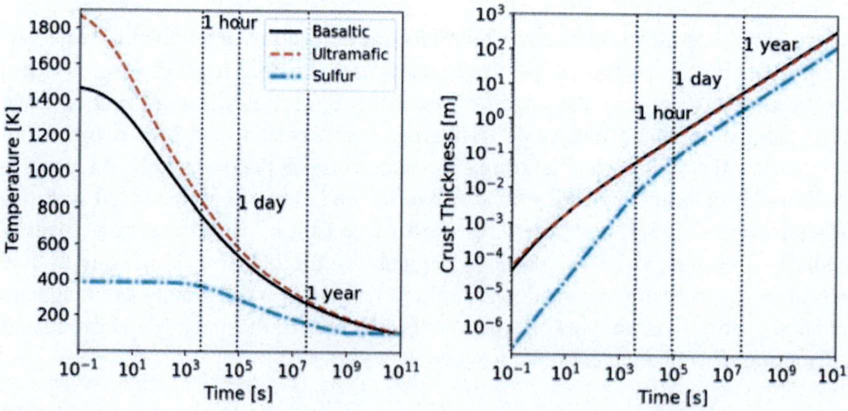

Fig. 6.12 Surface temperature and lava crust thickness as a function of time since the lava was emplaced, for three compositions, using the model of Davies (2007). The basaltic and ultramafic models have a thermal conductivity of 1.0 and the sulfur model has a thermal conductivity of 0.3

reflected sunlight from Io's disk is very bright unless observations are made during eclipse or Io night (see Fig. 6.1). Measuring the actual magma temperature requires either observations at very high spatial resolution to actually isolate the erupting component, or possibly fortuitous observations at the start of a major eruption when active lava fountaining is dominating the thermal emission of the entire volcano.

Indeed, presumed lava fountaining events during outbursts on Io have provided the highest temperature estimates to date. McEwen et al. (1998b) measured temperatures exceeding 1000 K at 15 different hot spots in the *Galileo* SSI eclipse dataset. In a few cases, the best-fit temperatures exceeded 1500 K, although the lower limits on the measured temperatures were never higher than 1500 K. The most powerful eruption ever to be detected at Io, which took place at Surt in 2001 and was detected from Keck Observatory, had a measured temperature of >1400 K using a two-temperature model fit to the emission spectrum, despite the 100 km spatial resolution (Marchis et al. 2002). The eruption labeled 201308C had a best-fit temperature of at least 1900 K, although the lower bound of 1200–1300 K is still consistent with a basaltic composition (de Kleer et al. 2014). Campaigns measuring the temperatures of multiple Ionian hot spots consistently find numerous hot spots above 1000 K, even when measurements are limited to a short period of time (Tsang et al. 2014; de Kleer et al. 2019b). Although none definitively exceeds the 1475 K basalt magma temperature, the high measured temperatures are strongly suggestive of ultramafic compositions, especially when the low spatial resolution and rapid cooling are taken into consideration because these result in measured temperatures that are lower than the actual magma temperature (Keszthelyi et al. 2007). The confirmation of ultramafic magmas actively erupting on Io would be an exciting step forward in terms of understanding the conditions in Io's mantle, and would make Io a promising case study for understanding volcanism on the early terrestrial planets when such volcanism was widespread.

6.4.2 The Spatial Distribution of Io's Volcanism and Heat Flow

Models show that the depth at which the tidal dissipation is occurring within Io's interior determines where on the surface the heat flow should be highest (Segatz et al. 1988). In particular, if dissipation is occurring in the deep mantle, higher heat flow is expected in Io's polar regions, whereas dissipation in the asthenosphere will lead to greater heat flow in the equatorial regions (see Chap. 4). These patterns arise from the fact that the different layers in Io's interior are under different pressures, and have different viscosities and shear moduli. The interior pressure gradient and the fact that Io is tidally locked result in a shear stress and compressional stress that each vary with both depth and latitude/longitude, and the rheology of each modeled interior layer translates these stresses into strains and dissipation rates that vary correspondingly with depth, latitude, and longitude.

The distinctive latitudinal and longitudinal patterns in modeled surface heat flow were searched for during the *Galileo* mission in the distribution of paterae and active hot spots on Io's surface (Lopes-Gautier et al. 1999; Radebaugh et al. 2001; Carr et al. 1998; McEwen et al. 1998b; Hamilton et al. 2013), which are often used as a proxy for volcanic heat flow because heat flow is difficult to quantify. The underlying assumption behind these comparisons is that heat transport is purely radial and that the patterns of heat dissipation in the interior are reflected in the patterns of heat flow at the surface. This assumption breaks down if there is significant convection or lateral heat transport in Io's interior, and some modeling has taken this into consideration (Tackley 2001; Tyler et al. 2015; Steinke et al. 2020, see Chap. 4 for more details on interior modeling). Broadly speaking, the effect of lateral heat transport is to shift the expected patterns in longitude and to blur them out.

There is overall not a close match between any of the simple end-member heat flow models and the distribution of Io's paterae, active hot spots, or volcanic heat flow. However, certain high-level correspondences are present albeit with minor differences between datasets. The population density of paterae shows a clear degree-2 pattern in longitude, peaking 0–30° eastward of the sub- and anti-Jovian points (180° and 360°W longitude; Kirchoff et al. 2011; Hamilton et al. 2013; Radebaugh et al. 2001; see Fig. 6.13). Figure 6.14 shows the spatial distribution of

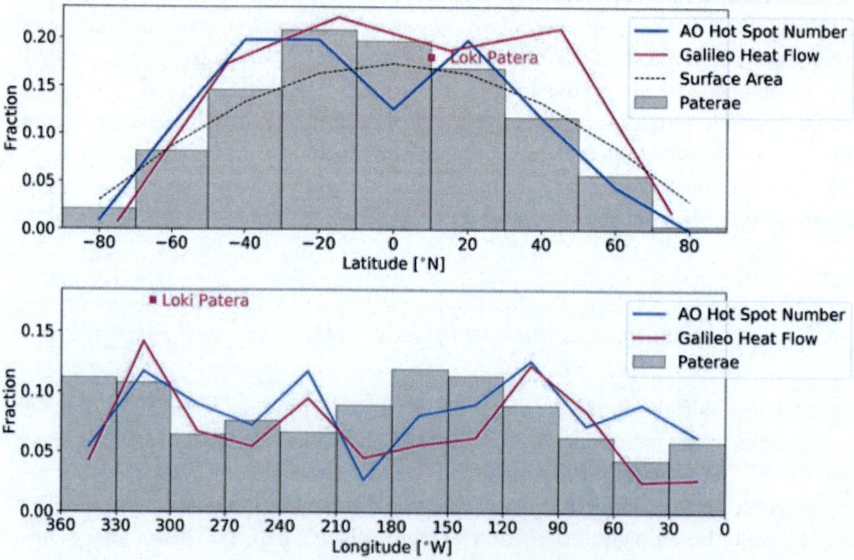

Fig. 6.13 Distribution of paterae, hot spots, and heat flow in latitude and longitude from datasets with large sample sizes and comprehensive longitudinal coverage. The patera histogram is based on 529 patera floor units analyzed by Hamilton et al. (2013). The *Galileo* heat flow measurements are based on analysis of 242 volcanic centers by Veeder et al. (2015). The AO hot spot number is based on detection of 111 hot spots in adaptive optics imaging of Io from 2001 to 2018 tabulated in Cantrall et al. (2018) and de Kleer et al. (2019b). All datasets are plotted by the fraction of the specified quantity falling within that bin. Loki Patera is excluded from the heat flow curve and plotted separately. The latitude plot is not corrected for the surface area within each latitude bin, but the curve of surface area vs. latitude bin is included on the figure for comparison

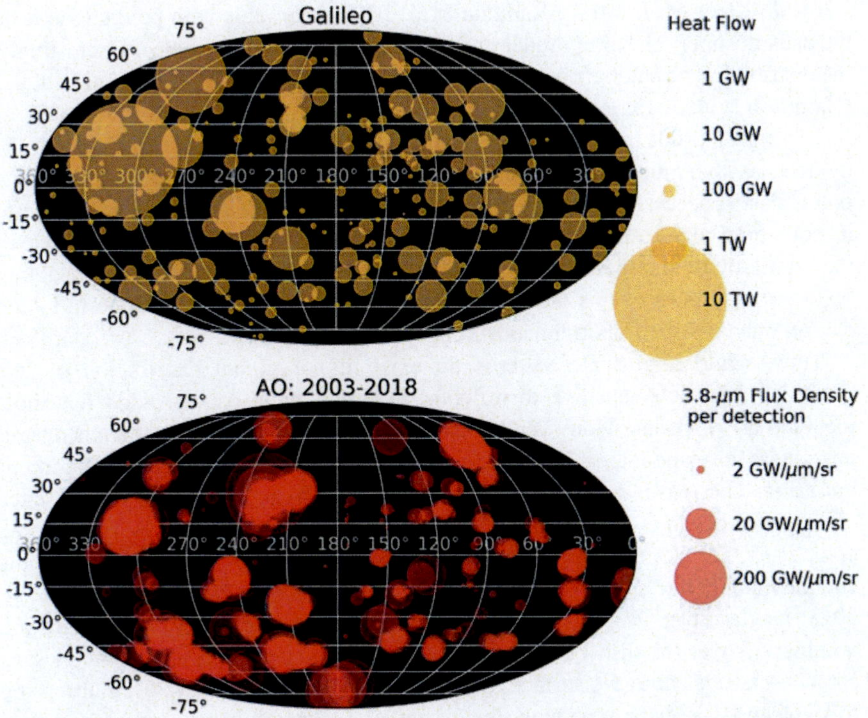

Fig. 6.14 Spatial distribution of volcanic emission from 20 years of observation. Top: Heat flow from 242 paterae and flows in the *Galileo* dataset; one circle is plotted for each volcanic center, with the size corresponding to the power output of that feature. Data from Davies et al. (2015), Veeder et al. (2015). Bottom: 3.8-μm emission measured from adaptive optics images. Roughly 1300 detections of 110 unique hot spots are shown; one circle is plotted for each detection, and the size corresponds to the 3.8-μm intensity. Data as tabulated in Cantrall et al. (2018), de Kleer et al. (2019b)

all 242 volcanic features for which heat flow was calculated by Veeder et al. (2015), alongside the spatial distribution of 110 hot spots measured in over 300 nights of adaptive optics observations (Cantrall et al. 2018; de Kleer et al. 2019b). The longitudinal distribution in both of these datasets show broad peaks roughly 30–90° eastward of the sub- and anti-Jovian points with a weaker secondary peak around 210–240°W (Fig. 6.13; Hamilton et al. 2013; Veeder et al. 2015; Davies et al. 2015; de Kleer and de Pater 2016b). The presence of two dominant peaks and two troughs in the longitudinal distribution of Io's volcanism is more consistent with the deep mantle model, although the clustering of hot spots and paterae near the centers of the leading and trailing hemispheres is most consistent with the asthenospheric heating models. However, models that include bulk in addition to shear dissipation find maximal heat flow centered in the sub- and anti-Jovian hemispheres even for shallow dissipation. The dip in number density (in AO data), and in the near-infrared hot spot emission (in NIMS/PPR data), at latitudes within ±15° of the equator (Cantrall et al.

2018; de Kleer et al. 2019b; Rathbun et al. 2018) is at odds with predictions from the asthenospheric heating model unless bulk dissipation is included. Interestingly, the distribution of paterae does not appear to have such a lack at equatorial latitudes (Hamilton et al. 2013; Radebaugh et al. 2001).

Io's high-latitude heat flow has the potential to discriminate more clearly between models, as the models are most different in their predicted heat flow near the poles. Past spacecraft and Earth-based observations have viewed Io from close to its equatorial plane, and it has not been possible from past data to robustly recover the high-latitude distribution of hot spots. However, *Juno's* polar orbit is enabling high-latitude views of Io's thermal emission (Mura et al. 2020), and the full *Juno* dataset may provide important constraints on polar heat flow.

Thus, while large-scale patterns do exist in the global distribution of Io's volcanism, datasets sensitive to different timescales and processes do not show identical distributions. Moreover, beyond these large-scale patterns, there is overall not a clear correspondence between the tidal heating models and the distribution of hot spots. The reason for this is not yet clear. Perhaps convection or the presence of a magma ocean is erasing the heating patterns at depth (Tackley 2001; Khurana et al. 2011; Steinke et al. 2020). Perhaps heating is occurring at a range of depths and the model effects are averaging out the heat flow across the surface (Tyler et al. 2015; Davies et al. 2015). Perhaps geological processes closer to the surface are dominating over the distribution of tidal heating in determining where the magma actually extrudes onto the surface. Or perhaps the 4 decades we've been observing Io's volcanoes is just not enough time to get a representative view of the heat flow distribution. The paterae trace a longer history, but offer little information on the time-averaged power from each site.

Finally, it is important to remember that the volcanic heat flow appears to be only half of Io's total heat flow, with the other half presumably conducted through the crust below the detection threshold of spatially-resolved data. Future observations may tell us which of the above possibilities is correct, if any. For example: better coverage of hot spots in Io's polar regions may help discriminate between models; thermal measurements with the wavelength coverage and sensitivity needed to map Io's conducted heat flow would answer the question of where the other half of Io's heat is reaching the surface; and in situ measurements of Io's gravity and induced magnitude field can answer the question of Io's interior structure directly, without resorting to these remote signatures at all.

6.4.2.1 Spatial Trends in Volcanic Style

A correlation exists between the locations where volcanoes are found on Io's surface, and the characteristics that they exhibit. Higher latitudes host fewer hot spots, but the hot spots that are there are more powerful. Large eruptions have repeatedly been preferentially seen at high absolute latitudes (Geissler 2003; Lopes-Gautier et al. 1999; McEwen et al. 2000; Howell et al. 2001; Milazzo et al. 2005; Davies et al. 2015; de Kleer and de Pater 2016b). de Kleer et al. (2019b) found that

16 of the 18 bright eruptions they detected took place on the trailing hemisphere, and calculated that the probability of this occurring randomly was <0.001. At the same time, the areal density of hot spots above 60° absolute latitude is half the areal density of hot spots below 60° (Milazzo et al. 2005; Davies et al. 2015). This is consistent with the fact that paterae are also larger and wider spaced at high latitudes (Radebaugh et al. 2001). de Kleer and de Pater (2016b) found the large eruptions in their dataset occurred primarily between 40° and 60° in absolute latitudes, whereas the fainter persistent volcanoes occur primarily between ±30°N. Cantrall et al. (2018) also found a mean absolute latitude of 46.2° for outbursts, significantly higher than the mean latitude assuming a random hot spot distribution, which is $32.7 \pm 9.6°$. The preferential occurrence of large, transient eruptions at high latitudes is suggestive of a deeper magma source for these volcanoes. Conversely, the preferential occurrence of persistent volcanoes towards lower latitudes suggests an asthenospheric source for these volcanoes (Lopes-Gautier et al. 1999).

6.4.3 *Periodicities in Thermal Emission*

The tidal stresses acting on Io are periodic in time. The most prominent variations are on Io's orbital timescale, but the stress amplitudes are modulated by longer-period effects corresponding to the longer-term evolution of Io's orbit. Searches for periodicities at several individual volcanoes, as well as sets of volcanoes in aggregate, have not shown statistically significant periodicities (Davies et al. 2006; de Kleer and de Pater 2016a; de Kleer et al. 2019b). A tentative periodicity is seen in the volcano Loki Patera: the timing of brightenings at Loki Patera between 1987 and 2018 appears to be quasiperiodic, with a period matching the timescales for the evolution of Io's eccentricity and semimajor axis (480 and 460 days; de Kleer et al. 2019a). However, a causal relationship is not possible to determine from current data, and other models exist to explain Loki Patera's behavior (e.g. Rathbun et al. 2002, see Sect. 6.3.4.5).

6.5 Non-volcanic Thermal Emission

6.5.1 *Global Heat Flow*

Io's global heat flow has been estimated many times using different techniques and datasets, all of which find a value in the range of $0.5–1.25 \times 10^{14}$ W (see Moore et al. 2007 for an overview). Early estimates were based on ground-based infrared photometry (Morrison and Telesco 1980; Matson et al. 1981; Sinton 1981, 1982; Johnson et al. 1984), while later calculations were made based on *Voyager 1* IRIS and *Galileo* PPR data, extrapolated from the regions observed to Io's entire surface (Spencer et al. 2000; Rathbun et al. 2004).

The most complete global thermal coverage was obtained by Veeder et al. (1994), who observed Io using near- and mid-infrared photometry over a decade of monitoring. Averaging the observed values to mitigate the impact of Io's time-variable volcanic emission on the total derived power, they found a value of $1.05 \pm 0.12 \times 10^{14}$ W. This total power is derived by modeling Io's thermal emission using a model that includes background thermal emission and 'thermal anomalies', or hot spots. Only 10–20% of Io's emitted power arises from hot spots with temperatures over 200 K, with the remainder arising from sources below 200 K. The derived value is the total power in thermal anomalies required to match the disk-integrated observations, which is a lower limit on Io's total power, as conduction through the crust is not considered.

As discussed in Sect. 6.2.2.1, a careful inventory of hot spots in *Galileo* data finds that only 54% of the total emitted power can be accounted for by identifiable volcanoes. Davies et al. (2015) show that the remaining 48 TW could be produced if the average temperature in Io's high-latitude ($\gtrsim$45°) regions were 90–95 K (rather than the expected 70–80 K) as suggested by *Galileo* PPR data (Rathbun et al. 2004). A detailed comparison is still needed to determine whether the now fully quantified set of near- and mid-IR hot spots (Veeder et al. 2015), combined with warm poles, can match the Veeder et al. (1994) observations, or whether additional emitting components at specific temperatures are still needed.

A thick, cold lithosphere is needed to support the height of Io's mountains, and such a lithosphere limits the amount of global conducted heat flow. However, there must also be enhanced conducted heat near volcanic centers and from intrusive volcanism if present; the heat flow from these sources is difficult to measure and has not been quantified.

6.5.2 Thermal Surface Properties

Spencer and Schneider (1996) summarized the pre-*Galileo* understanding of Io's thermal properties and passive surface temperatures. Because the only available diurnal temperature measurements at that time (from *Voyager*) were difficult to interpret due to contamination by hotspots, most measurements of Io's surface thermal properties were from eclipse cooling observations. These observations suggested that a highly porous surface with low thermal conductivity makes up at least part of Io's surface.

Galileo's PPR was able to obtain temperature measurements at a high enough resolution that the hotspots could be distinguished from the background temperatures (Rathbun et al. 2004). At night, the background temperatures were remarkably constant at 90–95 K regardless of time of night or latitude. The observed diurnal variations near the equator were best fit by a two-component model (a dark component with albedo of 0.34 and thermal inertia of 40 J m^{-2} K^{-1} s$^{-1/2}$, and a bright component with albedo of 0.70 and thermal inertia 1000 J m^{-2} K^{-1} s$^{-1/2}$), though this model was not a perfect match to the flat nighttime temperatures

(Rathbun et al. 2004). A later analysis of the PPR data, particularly those available at higher latitudes, found that while a two-component surface was able to match the observations at lower latitudes it resulted in temperatures too low at high latitudes (Rathbun et al. 2004). This suggests that there is additional heat flow at the poles. Walker et al. (2012) used both the PPR data and Hubble Space Telescope reflectance spectra to fit a three-component (frost, background, and hotspots) model to Io's surface. They found an albedo of 0.55 and thermal inertia of 200 for the frost and an albedo of 0.49 and thermal inertia of 20 for the background surface. A high and low thermal inertia component is thus required by all models, although the albedos and thermal inertias of the components differ.

Io's surface temperature in and out of eclipse also provides a constraint on surface properties. The surface temperature at 19 μm drops ~20 K in eclipse (Tsang et al. 2016), but the temperature observed at millimeter wavelengths (sensitive to the 1–2 cm subsurface) only drops by ~3 K (de Pater et al. 2020). A thermophysical model fit to the eclipse temperature decrease found that a lower emissivity (0.78 in the mm, 0.9 in the IR) and a higher thermal inertia (320 $\mathrm{J\,m^{-2}\,K^{-1}\,s^{-1/2}}$ in the mm; 50 in the IR) were required to match the mm data compared to the IR. This suggests that Io's surface is overlain with a thin low thermal inertia layer as would be expected for dust, fluffy volcanic plume deposits, or surface frost, with the millimeter data sensitive to a denser underlying layer.

6.6 Outlook

The past 20 years of Io observation have yielded a voluminous and multi-dimensional dataset of Io's hot spots and passive thermal emission, enabling more detailed geological studies of individual volcanoes as well as statistical analyses of volcano populations as a whole. However, several outstanding questions remain:

Where is Io's non-volcanic heat flow? Only 54% of Io's heat flow can be accounted for by its volcanoes. If the heat is escaping via conduction through the lithosphere, where is that conduction occurring in latitude and longitude, and from how deep within Io's crust?

How does Io's heat flow vary with time? Processes acting on Io's volcanoes may affect heat flow over a range of timescales, from the 1.77-day orbital period to possible 100 Myr tidal feedback cycles (Ojakangas and Stevenson 1986). While not all timescales are feasible to study, constraints on whether Io's heat flow is steady, episodic, or periodic over a range of timescales provides important information for understanding the mechanisms that power Io's volcanoes.

What is the nature of Io's polar volcanoes? Past datasets have consistently shown that volcanoes are less numerous but larger and more powerful at higher latitudes. However, none of these datasets has had a view of the poles that was not heavily foreshortened. A dataset with statistically meaningful coverage of Io's polar volcanoes would enable a comparison of power output, temperature, and volcanic style across latitudes and may aid in connecting Io's volcanic properties to its internal tidal heating mechanisms.

What is Io's dominant magma composition, and how does magma composition vary between volcanoes? Whether Io's dominant magma composition is closer to basaltic or ultramafic has been one of the major outstanding questions in Io volcanism since the end of the *Galileo* mission. With detailed, high-resolution datasets, it may even be possible to map magma properties across Io's surface and determine whether patterns exist that could point to the underlying mechanisms.

Which, if any, specific properties of Io's volcanoes are directly controlled by tides? Attempts to find spatial or temporal patterns in Io's volcanism that reflect the expected tidal patterns have had mixed success, with some suggestive patterns identified but interpretation challenged by a lack of direct correspondence. Although Io's volcanoes are powered by tidal heating, it has thus remained difficult to directly identify how tidal heating is directly influence the properties of these volcanoes.

What are the characteristics of Io's magma chambers and plumbing systems? While the extruded magma can be studied directly, there is little to no information on the magma plumbing systems powering Io's volcanoes. A longer time baseline of observation may reveal characteristic recharge timescales, or intervals between eruptions at individual sites, that could shed some light on this enigmatic question.

What role does volatile content play in determining eruption style, and what sets variations in volatile content between volcanoes? It is clear that Io hosts both volatile-rich and volatile-poor volcanoes, and that volatile composition varies between sites. However, the origin of these differences and the role they play in eruption style is not yet understood.

With ever-larger telescopes coming online, and the *Juno* extended mission Io flybys still upcoming, there is great potential for advances in our understanding of Io's thermal emission in the coming years. The next generation of ground-based optical/near-infrared telescopes will have the ability to resolve emitting components within a given volcanic center. Passive thermal measurements at mm/cm wavelengths with ALMA or a next-generation cm interferometer could obtain high enough resolution to search for latitudinal and longitudinal trends in surface temperature and perhaps even conducted heat flow. Finally, the 40 years of Io hot spot observation is a drop in the bucket compared to geological timescales, and a continually extending time baseline of observation is sure to yield new discoveries on the temporal behavior of Io's volcanoes.

References

Allen, D.R., Radebaugh, J., Stephens, D.C.: Icarus **226**, 77 (2013). https://doi.org/10.1016/j.icarus.2013.05.026

Bartolić, F., Luger, R., Foreman-Mackey, D., Howell, R.R., Rathbun, J.A.: Planet. Sci. J. **3**, id.67 (2022). https://ui.adsabs.harvard.edu/abs/2022PSJ.....3...67B/abstract

Battaglia, S.M., Stewart, M.A., Kieffer, S.W.: Icarus **235**, 123 (2014). https://doi.org/10.1016/j.icarus.2014.03.019

Belton, M.J.S., Head III, J.W., Ingersoll, A.P., et al.: Science **274**, 377 (1996). https://doi.org/10.1126/science.274.5286.377

Blaney, D.L., Johnson, T.V., Matson, D.L., Veeder, G.J.: Icarus **113**, 220 (1995). https://doi.org/10.1006/icar.1995.1020

Bolton, S.J., Lunine, J., Stevenson, D., et al.: SSRv **213**, 5 (2017). https://doi.org/10.1007/s11214-017-0429-6

Cantrall, C., de Kleer, K., de Pater, I., et al.: Icarus **312**, 267 (2018). https://doi.org/10.1016/j.icarus.2018.04.007

Carlson, R.W., Kargel, J.S., Douté, S., Soderblom, L.A., Dalton, J.B.: Io's surface composition. In: Lopes, R.M.C., Spencer, J.R. (eds.) Io After Galileo, p. 193 (2007). https://doi.org/10.1007/978-3-540-48841-5_9

Carr, M.H.: J. Geophys. Res. **91**, 3521 (1986). https://doi.org/10.1029/JB091iB03p03521

Carr, M.H., McEwen, A.S., Howard, K.A., et al.: Icarus **135**, 146 (1998). https://doi.org/10.1006/icar.1998.5979

Conrad, A., de Kleer, K., Leisenring, J., et al.: AJ **149**, 175 (2015). https://doi.org/10.1088/0004-6256/149/5/175

Davies, A.G.: Icarus **124**, 45 (1996). https://doi.org/10.1006/icar.1996.0189

Davies, A.G.: Geophys. Res. Lett. **30**, 2133 (2003). https://doi.org/10.1029/2003GL018371

Davies, A.G.: Volcanism on Io: A Comparison with Earth, Cambridge, UK: Cambridge University Press (2007). https://ui.adsabs.harvard.edu/abs/2007voio.book.....D/abstract

Davies, A.G.: AJ **163**, 1 (2021). https://doi.org/10.3847/1538-3881/ac3012

Davies, A.G., McEwen, A.S., Lopes-Gautier, R. M.C., et al.: Geophys. Res. Lett. **24**, 2447 (1997). https://doi.org/10.1029/97GL02310

Davies, A.G., Keszthelyi, L.P., Williams, D.A., et al.: J. Geophys. Res. **106**, 33079 (2001). https://doi.org/10.1029/2000JE001357

Davies, A.G., Matson, D.L., Veeder, G.J., Johnson, T.V., Blaney, D.L.: Icarus **176**, 123 (2005). https://doi.org/10.1016/j.icarus.2005.01.015

Davies, A.G., Wilson, L., Matson, D., et al.: Icarus **184**, 460 (2006). https://doi.org/10.1016/j.icarus.2006.05.012

Davies, A.G., Keszthelyi, L.P., Harris, A.J.L.: J. Volcanol. Geothermal Res. **194**, 75 (2010). https://doi.org/10.1016/j.jvolgeores.2010.04.009

Davies, A.G., Veeder, G.J., Matson, D.L., Johnson, T.V.: Icarus **221**, 466 (2012). https://doi.org/10.1016/j.icarus.2012.04.012

Davies, A.G., Veeder, G.J., Hill, S.I., Matson, D.L., Johnson, T.V.: Icarus **241**, 190 (2014). https://doi.org/10.1016/j.icarus.2014.06.034

Davies, A.G., Veeder, G.J., Matson, D.L., Johnson, T.V.: Icarus **262**, 67 (2015). https://doi.org/10.1016/j.icarus.2015.08.003

Davies, A.G., Davies, R.L., Veeder, G.J., et al.: Geophys. Res. Lett. **45**, 2926 (2018). https://doi.org/10.1002/2018GL077477

de Kleer, K., de Pater, I.: Icarus **280**, 378 (2016a). https://doi.org/10.1016/j.icarus.2016.06.019

de Kleer, K., de Pater, I.: Icarus **280**, 405 (2016b). https://doi.org/10.1016/j.icarus.2016.06.018

de Kleer, K., de Pater, I.: Icarus **289**, 181 (2017). https://doi.org/10.1016/j.icarus.2017.01.038

de Kleer, K., de Pater, I., Davies, A.G., Ádámkovics, M.: Icarus **242**, 352 (2014). https://doi.org/10.1016/j.icarus.2014.06.006

de Kleer, K., Skrutskie, M., Leisenring, J., et al.: Nature **545**, 199 (2017). https://doi.org/10.1038/nature22339

de Kleer, K., Nimmo, F., Kite, E.: Geophys. Res. Lett. **46**, 6327 (2019a). https://doi.org/10.1029/2019GL082691

de Kleer, K., de Pater, I., Molter, E.M., et al.: AJ **158**, 29 (2019b). https://doi.org/10.3847/1538-3881/ab2380

de Kleer, K., Skrutskie, M., Leisenring, J., et al.: Planet. Sci. J. **2**, 227 (2021). https://doi.org/10.3847/PSJ/ac28fe

de Pater, I., Davies, A.G., Ádámkovics, M., Ciardi, D.R.: Icarus **242**, 365 (2014). https://doi.org/10.1016/j.icarus.2014.06.016
de Pater, I., Davies, A.G., Marchis, F.: Icarus **274**, 284 (2016a). https://doi.org/10.1016/j.icarus.2015.12.054
de Pater, I., Laver, C., Davies, A.G., et al.: Icarus **264**, 198 (2016b). https://doi.org/10.1016/j.icarus.2015.09.006
de Pater, I., de Kleer, K., Davies, A.G., Ádámkovics, M.: Icarus **297**, 265 (2017). https://doi.org/10.1016/j.icarus.2017.03.016
de Pater, I., Luszcz-Cook, S., Rojo, P., et al.: Planet. Sci. J. **1**, 60 (2020). https://doi.org/10.3847/PSJ/abb93d
Deno Stelter, R., Skemer, A.J., Sallum, S., et al.: Proceedings of the SPIE, 11447, id. 1144764 (2020). https://ui.adsabs.harvard.edu/abs/2020SPIE11447E..64S/abstract
Descamps, P., Arlot, J.E., Thuillot, W., et al.: Icarus **100**, 235 (1992). https://doi.org/10.1016/0019-1035(92)90032-3
Geissler, P.E.: Annu. Rev. Earth Planet. Sci. **31**, 175 (2003). https://doi.org/10.1146/annurev.earth.31.100901.145428
Geissler, P., McEwen, A., Phillips, C., Keszthelyi, L., Spencer, J.: Icarus **169**, 29 (2004a). https://doi.org/10.1016/j.icarus.2003.09.024
Geissler, P., McEwen, A., Porco, C., et al.: Icarus **172**, 127 (2004b). https://doi.org/10.1016/j.icarus.2004.01.008
Gregg, T.K.P., Lopes, R.M.: Icarus **194**, 166 (2008). https://doi.org/10.1016/j.icarus.2007.08.042
Hamilton, C.W., Beggan, C.D., Still, S., et al.: Earth Planet. Sci. Lett. **361**, 272 (2013). https://doi.org/10.1016/j.epsl.2012.10.032
Hanel, R., Conrath, B., Flasar, M., et al.: Science **204**, 972 (1979). https://doi.org/10.1126/science.204.4396.972
Hansen, O.L.: Icarus **18**, 237 (1973). https://doi.org/10.1016/0019-1035(73)90208-X
Howell, R.R.: Icarus **127**, 394 (1997). https://doi.org/10.1006/icar.1997.5686
Howell, R.R., Lopes, R.M.C.: Icarus **186**, 448 (2007). https://doi.org/10.1016/j.icarus.2006.09.022
Howell, R.R., Lopes, R.M.C.: Icarus **213**, 593 (2011). https://doi.org/10.1016/j.icarus.2011.03.008
Howell, R.R., McGinn, M.T.: Science **230**, 63 (1985). https://doi.org/10.1126/science.230.4721.63
Howell, R.R., Spencer, J.R., Goguen, J.D., et al.: J. Geophys. Res. **106**, 33129 (2001). https://doi.org/10.1029/2000JE001382
Howell, R.R., Landis, C.E., Lopes, R.M.C.: Icarus **229**, 328 (2014). https://doi.org/10.1016/j.icarus.2013.11.016
Johnson, T.V., Morrison, D., Matson, D.L., et al.: Science **226**, 134 (1984). https://doi.org/10.1126/science.226.4671.134
Johnson, T.V., Veeder, G.J., Matson, D.L., et al.: Science 242, 1280 (1988). https://doi.org/10.1126/science.242.4883.1280
Keszthelyi, L., Jaeger, W., Milazzo, M., et al.: Icarus **192**, 491 (2007). https://doi.org/10.1016/j.icarus.2007.07.008
Keszthelyi, L., McEwen, A.S., Phillips, C.B., et al.: J. Geophys. Res. **106**, 33025 (2001). https://doi.org/10.1029/2000JE001383
Khurana, K.K., Jia, X., Kivelson, M.G., et al.: Science **332**, 1186 (2011). https://doi.org/10.1126/science.1201425
Kirchoff, M.R., McKinnon, W.B., Schenk, P.M.: Earth Planet. Sci. Lett. **301**, 22 (2011). https://doi.org/10.1016/j.epsl.2010.11.018
Laver, C., de Pater, I., Marchis, F.: Icarus **191**, 749 (2007). https://doi.org/10.1016/j.icarus.2007.06.022
Lellouch, E., Ali-Dib, M., Jessup, K.L., et al.: Icarus **253**, 99 (2015). https://doi.org/10.1016/j.icarus.2015.02.018
Leone, G., Wilson, L., Davies, A.G.: Icarus **211**, 623 (2011). https://doi.org/10.1016/j.icarus.2010.10.016
Linde, A.T., Sacks, I.S.: Nature **395**, 888 (1998). https://doi.org/10.1038/27650

Lopes, R. M.C., Spencer, J.R.: Io After Galileo: A New View of Jupiter's Volcanic Moon (2007). https://doi.org/10.1007/978-3-540-48841-5

Lopes, R. M.C., Kamp, L.W., Douté, S., et al.: J. Geophys. Res. **106**, 33053 (2001). https://doi.org/10.1029/2000JE001463

Lopes, R. M.C., Kamp, L.W., Smythe, W.D., et al.: Icarus **169**, 140 (2004). https://doi.org/10.1016/j.icarus.2003.11.013

Lopes, R. M.C., Gregg, T. K.P., Harris, A., et al.: J. Volcanol. Geothermal Res. **366**, 74 (2018). https://doi.org/10.1016/j.jvolgeores.2018.09.010

Lopes-Gautier, R., Davies, A.G., Carlson, R., et al.: Geophys. Res. Lett. **24**, 2439 (1997). https://doi.org/10.1029/97GL02662

Lopes-Gautier, R., McEwen, A.S., Smythe, W.B., et al.: Icarus **140**, 243 (1999). https://doi.org/10.1006/icar.1999.6129

Macintosh, B.A., Gavel, D., Gibbard, S.G., et al.: Icarus **165**, 137 (2003). https://doi.org/10.1016/S0019-1035(03)00168-4

Manga, M., Brodsky, E.: Annu. Rev. Earth Planet. Sci. **34**, 263 (2006). https://doi.org/10.1146/annurev.earth.34.031405.125125

Marchis, F., Prangé, R., Christou, J.: Icarus **148**, 384 (2000). https://doi.org/10.1006/icar.2000.6506

Marchis, F., Prangé, R., Fusco, T.: J. Geophys. Res. **106**, 33141 (2001). https://doi.org/10.1029/2000JE001376

Marchis, F., de Pater, I., Davies, A.G., et al.: Icarus **160**, 124 (2002). https://doi.org/10.1006/icar.2002.6955

Marchis, F., Le Mignant, D., Chaffee, F.H., et al.: Icarus **176**, 96 (2005). https://doi.org/10.1016/j.icarus.2004.12.014

Matson, D.L., Ransford, G.A., Johnson, T.V.: J. Geophys. Res. **86**, 1664 (1981). https://doi.org/10.1029/JB086iB03p01664

Matson, D.L., Davies, A.G., Veeder, G.J., et al.: J. Geophys. Res. (Planets) **111**, E09002 (2006). https://doi.org/10.1029/2006JE002703

McEwen, A.S., Simonelli, D.P., Senske, D.R., et al.: Geophys. Res. Lett. **24**, 2443 (1997). https://doi.org/10.1029/97GL01956

McEwen, A.S., Keszthelyi, L., Spencer, J.R., et al.: Science **281**, 87 (1998a). https://doi.org/10.1126/science.281.5373.87

McEwen, A.S., Keszthelyi, L., Geissler, P., et al.: Icarus **135**, 181 (1998b). https://doi.org/10.1006/icar.1998.5972

McEwen, A.S., Belton, M.J.S., Breneman, H.H., et al.: Science **288**, 1193 (2000). https://doi.org/10.1126/science.288.5469.1193

Milazzo, M.P., Keszthelyi, L.P., Radebaugh, J., et al.: Icarus **179**, 235 (2005). https://doi.org/10.1016/j.icarus.2005.05.013

Moore, W.B., Schubert, G., Anderson, J.D., Spencer, J.R.: The interior of Io. In: Lopes, R.M.C., Spencer, J.R. (eds.) Io After Galileo, p. 89 (2007). https://doi.org/10.1007/978-3-540-48841-5_5

Morabito, L.A., Synnott, S.P., Kupferman, P.N., Collins, S.A.: Science **204**, 972 (1979). https://doi.org/10.1126/science.204.4396.972

Morrison, D., Cruikshank, D.P.: Icarus **18**, 224 (1973). https://doi.org/10.1016/0019-1035(73)90207-8

Morrison, D., Telesco, C.M.: Icarus **44**, 226 (1980). https://doi.org/10.1016/0019-1035(80)90018-4

Mura, A., Adriani, A., Tosi, F., et al.: Icarus **341**, 113607 (2020). https://doi.org/10.1016/j.icarus.2019.113607

O'Reilly, T.C., Davies, G.F.: Geophys. Res. Lett. **8**, 313 (1981). https://doi.org/10.1029/GL008i004p00313

Ojakangas, G.W., Stevenson, D.J.: Icarus **66**, 341 (1986). https://doi.org/10.1016/0019-1035(86)90163-6

Porco, C.C., West, R.A., McEwen, A., et al.: Science **299**, 1541 (2003). https://doi.org/10.1126/science.1079462
Radebaugh, J., Keszthelyi, L.P., McEwen, A.S., et al.: J. Geophys. Res. **106**, 33005 (2001). https://doi.org/10.1029/2000JE001406
Radebaugh, J., McEwen, A.S., Milazzo, M.P., et al.: Icarus **169**, 65 (2004). https://doi.org/10.1016/j.icarus.2003.10.019
Rathbun, J.A., Spencer, J.R.: Geophys. Res. Lett. **33**, L17201 (2006). https://doi.org/10.1029/2006GL026844
Rathbun, J.A., Spencer, J.R.: Icarus **209**, 625 (2010). https://doi.org/10.1016/j.icarus.2010.05.019
Rathbun, J.A., Spencer, J.R., Davies, A.G., Howell, R.R., Wilson, L.: Geophys. Res. Lett. **29**, 1443 (2002). https://doi.org/10.1029/2002GL014747
Rathbun, J.A., Spencer, J.R., Tamppari, L.K., et al.: Icarus **169**, 127 (2004). https://doi.org/10.1016/j.icarus.2003.12.021
Rathbun, J.A., Spencer, J.R., Lopes, R.M., Howell, R.R.: Icarus **231**, 261 (2014). https://doi.org/10.1016/j.icarus.2013.12.002
Rathbun, J.A., Lopes, R. M.C., Spencer, J.R.: AJ **156**, 207 (2018). https://doi.org/10.3847/1538-3881/aae370
Segatz, M., Spohn, T., Ross, M.N., Schubert, G.: Icarus **75**, 187 (1988). https://doi.org/10.1016/0019-1035(88)90001-2
Sinton, W.M.: ApJL **235**, L49 (1980). https://doi.org/10.1086/183155
Sinton, W.M.: J. Geophys. Res. **86**, 3122 (1981). https://doi.org/10.1029/JB086iB04p03122
Sinton, V.M.: Icarus **51**, 563 (1982). https://doi.org/10.1016/0019-1035(82)90146-4
Skemer, A.J., Hinz, P., Montoya, M., et al.: Society of Photo-Optical Instrumentation Engineers (SPIE) Conference Series, vol. 9605. In: Shaklan, S. (ed.) Techniques and Instrumentation for Detection of Exoplanets VII, 96051D (2015). https://doi.org/10.1117/12.2187284
Spencer, J.R., Schneider, N.M.: Annu. Rev. Earth Planet. Sci. **24**, 125 (1996). https://doi.org/10.1146/annurev.earth.24.1.125
Spencer, J.R., Shure, M.A., Ressler, M.E., et al.: Nature **348**, 618 (1990). https://doi.org/10.1038/348618a0
Spencer, J.R., Spencer, J.E.B., Griep, D.: Science **6149**, 1 (1995)
Spencer, J.R., Rathbun, J.A., Travis, L.D., et al.: Science **288**, 1198 (2000). https://doi.org/10.1126/science.288.5469.1198
Spencer, J.R., Stern, S.A., Cheng, A.F., et al.: Science **318**, 240 (2007). https://doi.org/10.1126/science.1147621
Spencer, D.C., Katz, R.F., Hewitt, I.J.: J. Geophys. Res. (Planets) **125**, e06443 (2020). https://doi.org/10.1029/2020JE006443
Stansberry, J.A., Spencer, J.R., Howell, R.R., Dumas, C., Vakil, D.: Geophys. Res. Lett. **24**, 2455 (1997). https://doi.org/10.1029/97GL02593
Steinke, T., Hu, H., Höning, D., van der Wal, W., Vermeersen, B.: Icarus **335**, 113299 (2020). https://doi.org/10.1016/j.icarus.2019.05.001
Tackley, P.J.: J. Geophys. Res. **106**, 32971 (2001). https://doi.org/10.1029/2000JE001411
Tsang, C.C.C., Rathbun, J.A., Spencer, J.R., Hesman, B.E., Abramov, O.: J. Geophys. Res. (Planets) **119**, 2222 (2014). https://doi.org/10.1002/2014JE004670
Tsang, C.C.C., Spencer, J.R., Lellouch, E., Lopez-Valverde, M.A., Richter, M.J.: J. Geophys. Res. (Planets) **121**, 1400 (2016). https://doi.org/10.1002/2016JE005025
Tyler, R.H., Henning, W.G., Hamilton, C.W.: ApJS **218**, 22 (2015). https://doi.org/10.1088/0067-0049/218/2/22
Veeder, G.J., Matson, D.L., Johnson, T.V., Blaney, D.L., Goguen, J.D.: J. Geophys. Res. **99**, 17095 (1994). https://doi.org/10.1029/94JE00637
Veeder, G.J., Davies, A.G., Matson, D.L., Johnson, T.V.: Icarus **204**, 239 (2009). https://doi.org/10.1016/j.icarus.2009.06.027
Veeder, G.J., Davies, A.G., Williams, D.A., et al.: Icarus **212**, 236 (2011). https://doi.org/10.1016/j.icarus.2010.09.026

Veeder, G.J., Davies, A.G., Matson, D.L., et al.: Icarus **219**, 701 (2012). https://doi.org/10.1016/j.icarus.2012.04.004

Veeder, G.J., Davies, A.G., Matson, D.L., et al.: Icarus **245**, 379 (2015). https://doi.org/10.1016/j.icarus.2014.07.028

Walker, A.C., Moore, C.H., Goldstein, D.B., Varghese, P.L., Trafton, L.M.: Icarus **220**, 225 (2012). https://doi.org/10.1016/j.icarus.2012.05.001

Williams, S.N.: Science **267**, 340 (1995). https://doi.org/10.1126/science.267.5196.340

Williams, D.A., Howell, R.R.: Active volcanism: Effusive eruptions. In: Lopes, R.M.C., Spencer, J.R. (eds.) Io After Galileo, p. 133 (2007). https://doi.org/10.1007/978-3-540-48841-5_7

Williams, D.A., Keszthelyi, L.P., Crown, D.A., et al.: Icarus **214**, 91 (2011a). https://doi.org/10.1016/j.icarus.2011.05.007

Williams, D.A., Keszthelyi, L.P., Crown, D.A., et al.: U.S. Geological Survey Report, 3168 (2011b). https://doi.org/10.3133/sim3168

Witteborn, F.E., Bregman, J.D., Pollack, J.B.: Science **203**, 643 (1979). https://doi.org/10.1126/science.203.4381.643

Chapter 7
The Composition of Io

Laszlo P. Keszthelyi and Terry-Ann Suer

Abstract Io is unlike any other body in the Solar System making questions about its chemical composition especially interesting and challenging. This chapter examines the many different, but frustratingly indirect, constraints we have on the bulk composition of this restless moon. A detailed consideration of Io's lavas is used to illustrate how decades of research have bounded, but not pinned down, the chemistry of Io. A self-consistent model for the core, mantle and crust is constructed based on a conventional chondritic composition but exotic alternatives cannot be ruled out. The study of Io's composition should provide a fertile and exciting realm for future scientists.

7.1 Introduction

The composition of Io is an enigma with many suggestions that it is similar to the other rocky worlds in our solar system but with little hard evidence. This dearth in direct measurements of composition is in stark contrast to many planetary bodies planetary scientists where spectroscopic evidence for specific rock-forming minerals is clear. Unlike the Moon, Mars, and some asteroids, there are no returned samples or recognized meteorites from Io. Unlike worlds from Venus to Titan, there are no *in situ* observations of the surface materials or the atmosphere. Remote sensing observations do not include elemental measurements from gamma ray/neutron spectrometers. UV through infrared spectral observations of the surface are unable to penetrate the practically ubiquitous coating of sulfurous volatiles, meaning that there are no spectroscopic identifications of any rock-forming minerals. In short,

L. P. Keszthelyi (✉)
U.S. Geological Survey, Astrogeology Science Center, Flagstaff, AZ, USA
e-mail: laz@usgs.gov

T.-A. Suer
Department of Earth and Planetary Sciences, Harvard University, Cambridge, MA, USA
e-mail: terrysuer@arizona.edu

R. M. C. Lopes et al. (eds.), *Io: A New View of Jupiter's Moon*, Astrophysics and Space Science Library 468, https://doi.org/10.1007/978-3-031-25670-7_7

attempts to understand the composition of Io work in a realm lacking any direct observations.

Instead, the composition of Io is understood via inferences drawn from a wide array of indirect constraints from geophysics, petrologic modeling, eruption parameters, and the composition of the atmosphere and magnetosphere. These allow Io to have a composition broadly similar to other rocky bodies in the Solar System. However, Io is a confounding world and more exotic compositions that could help explain some unusual phenomena cannot be ruled out. Given this state of affairs, we start by providing an overview of the various data sets that provide the indirect constraints on the composition of Io. We then discuss how a broadly chondritic composition is allowed by these observations. This does not mean that more exotic compositions have been ruled out. Finally, we discuss how different types of future observations could help our understanding of the composition of Io.

7.2 Observations

It is easiest to consider observations relevant to Io's bulk composition by working inward from the magnetosphere and atmosphere to the surface and then continuing downward to the core.

7.2.1 Magnetosphere and Atmosphere

Two chapters in this book discuss the atmosphere and magnetosphere in detail so only the briefest overview is provided here. The fact that Io sits within the intense radiation belts of Jupiter means that molecules in the magnetosphere and atmosphere are often in an excited state and emit photons as they return to their normal, or "ground" state. Quantum mechanics dictates that these photons have discrete energies and thus produce emissions at specific wavelengths characteristic of specific molecules. These emissions can be observed with telescopes on Earth as well as spacecraft. As discussed in Chap. 9, the primary emissions are from various combinations of oxygen and sulfur. Sodium also has a very distinctive and bright emission and is associated with small amounts of potassium and chlorine. Traces of hydrogen as H_2S are also reported.

Other types of observations that do not rely on the emission spectra from the atmosphere and magnetosphere provide additional details but do not identify additional molecular species. One important detail is the estimate of oxygen concentration (usually referred to as oxygen fugacity or f_{O2}). Since oxygen is the single most abundant element in silicate rocks, unusually high or low concentrations of oxygen could create mineral assemblages that would be exotic on Earth. One constraint comes from the S to SO_2 ratio derived from ultraviolet observations of the Loki and Pele plumes using the Hubble Space Telescope (Spencer et al. 2000).

Based on these values, Zolotov and Fegley (1999) suggest that f_{O2} is within a few log units of the Ni-NiO buffer for at least these eruptions. This estimated range falls within the range around quartz-fayalite-magnetite (QFM) and wustite-magnetite (WM) oxygen buffers, which is the typical range for volcanism on the Earth. This would suggest that the silicate minerals in Io are similar to those found in the Earth.

To summarize, the molecules seen in Io's atmosphere and magnetosphere are ones that are commonly seen in gas emissions from volcanoes on Earth. What is also notable is what is not seen. Carbon compounds such as CO and CO_2 have not been reliably detected despite acquiring measurements that should have detected them. While traces of H_2S have been reported, the amount of hydrogen indicated by these observations is much lower than seen elsewhere in the outer solar system. H_2O or CH_4 is found almost everywhere in the outer solar system except Io. Nitrogen also seems to be missing from Io.

7.2.2 *Surface Mantling Volatiles*

Spectroscopic observations, particularly in the ultraviolet and infrared, have identified a number of compounds on the surface of Io. These are similar to those observed in the atmosphere and magnetosphere, that is, they are dominated by sulfur and its oxides. These detections are discussed in great detail in Carlson et al. (2007) and only summarized here (Table 7.1).

The ultraviolet absorptions can be largely explained by SO_2 with possible contributions from elemental sulfur (e.g., Jessup et al. 2002). In the visible, the bright plains generally correspond to regions dominated by SO_2 (e.g., Douté et al. 2001) while various shades of yellow are an indication of significant amounts of sulfur in its most stable form (rings of six or more atoms) (e.g., Kargel et al. 1999). Shorter chains of sulfur molecules can be bright red and are found associated with active volcanic vents. Other compounds that have been suggested by features in the

Table 7.1 Io's spectral features and inferred surface species (Carlson et al. 2007)

Wavelength region	Inferred species
200–300 nm	Primarily SO_2, possible S_8, S_∞
350–500 nm	S_8, S_∞, possible polysulfur oxides
~560 nm	S_4 and/or Na_2, S_2O, Cl_2S
~900 nm	Enstatite
1–1.6 μm	Fe-containing salts, FeS_2, polymeric S with impurities
1.98–5 μm	SO_2
3–5 μm	Possible H_2SO_3
3.15 μm	Possible H_2O in SO_2, bound OH, HCl
3.92 μm	Possible H_2S, Cl_2SO_2
16.7–25 μm	S_8, SO, NaCl, possible KCl

visible to near infrared (0.5–2 micron region) include Na_2S, S_2O, Cl_2S, and FeS_2 as well as some silicate minerals. In the mid-infrared (2–5 micron region), again SO_2 is generally the dominant molecule that is detected but other sulfur compounds are also indicated (e.g., Tosi et al. 2020). Also, various subtle spectral features have sometimes been tentatively interpreted as small amounts of H_2S, H_2SO_3, H_2O as a clathrate within the SO_2 ices, HCl, H_2S, Cl_2SO_2, $ClSO_2$, as well as nitrogen-bearing or organic compounds (Carlson et al. 2007; Tosi et al. 2020).

Interpretation of spectral features is always open to discussion; in the case of Io's surface, there have been over five decades of debate. However, in the broader picture, it is clear that the surface is mantled by frozen volatiles that are similar, if not identical, to those in the atmosphere. Sulfurous compounds dominate with possible minor contributions from halogens and hydrogen. The close linkage between the volatiles on the surface of Io and the plumes and atmosphere is not surprising, but key details remain enigmatic. For example, the relative roles of sublimation of the surface volatiles versus input from volcanic plumes is an area of continuing research. Those interested in the topic are invited to read the following chapter.

7.2.3 *Lava*

Io is widely known as the most volcanically active body in the Solar System. However, the composition of the lavas is poorly constrained by observations. The yellow, orange, and black colors of the lavas seen in the images from the *Voyager* flybys in 1979 (Fig. 7.1) led to the hypothesis that the lavas were composed of sulfur (Sagan 1979). This idea was widespread until the *Galileo* mission returned spectacular images of active silicate lavas. However, the evidence for silicate volcanism had been building starting immediately after the *Voyager* flybys. Based on the fact that near-vertical walls could be seen on the margins of some volcanic depressions (Fig. 7.2), Clow and Carr (1980) concluded that the material had to be stronger than sulfur and was likely to be silicate. However, later studies showed that these walls can also be explained by cold or metastable sulfurous material (Slezak et al. 2014). More direct evidence that at least some lavas were not composed of sulfur came in the form of telescopic observations showing temperatures above the 393 K melting temperature of sulfur (Pearl and Sinton 1982; Johnson et al. 1988; Veeder et al. 1994; Spencer et al. 1997; Stansberry et al. 1997) leading to the explicit hypothesis that silicate volcanism was present and possibly even widespread (Carr 1986; Johnson et al. 1988; Blaney et al. 1995; Davies 1996).

Resolving the role of silicate volcanism on Io was one of the goals of the *Galileo* mission (Carr et al. 1995). However, the *Galileo* instrument suite was selected before the *Voyager* flybys of the Jupiter system and thus were not really designed to take measurement of active volcanism whether it be sulfurous or silicate. The *Galileo* Near Infrared Mapping Spectrometer (NIMS) was nonetheless expected to play the leading role by measuring lava temperatures and determining the composition of the lavas spectrally. By combining 17 detectors and a moving

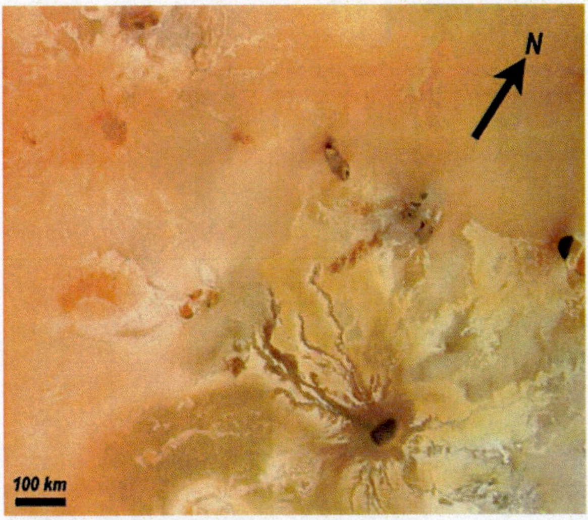

Fig. 7.1 Color image of Io obtained by the *Voyager 1* narrow angle camera on March 5, 1979. The image is approximately 1000 km across. The colors represent the *Voyager* team's best attempt at providing true colors as would be seen by the human eye. However, the *Voyager* camera was blind to red light, so the red tones had to be guessed from data at shorter wavelengths. In reality, Io is not so uniformly orange but instead is even more colorful. This and other limitations of the *Voyager* instruments led to the early interpretation that Io's volcanism was dominated by sulfur (e.g., Sagan 1979) which was not supported by later *Galileo* data (Kargel et al. 1999). (Image credit: NASA/JPL/USGS)

grating, NIMS could acquire data at 408 wavelengths between 0.7 and 5.2 microns (Carlson et al. 1992) which is excellent for both ascertaining lava temperatures and spectral features associated with typical mafic minerals like olivine and pyroxene. Early in the *Galileo* mission, the spacecraft did not obtain data close to Io and interpreting the spectra was hindered by the inability to spatially resolve volcanic features. Furthermore, the short wavelength channels proved too noisy to be usable. By the time *Galileo* made close flybys of Io, NIMS had suffered a series of hardware issues in the harsh radiation environment around Jupiter and was reduced to obtaining useful data at only 13 wavelengths.

As it turned out, nighttime images taken by the *Galileo* Solid-State Imager (SSI) camera provided spectacular direct evidence for widespread high temperature lavas (McEwen et al. 1998). The lavas were hot enough for the incandescence to be visible on the dark side of Io in SSI's infrared filters (0.7–1.0 microns) (Fig. 7.3). When data were obtained in multiple bands, the ratio of the radiance seen at two wavelengths could be converted into a "color" temperature. For typical silicate lavas, this measurement is especially sensitive to the temperature of the hottest lava surfaces. The reason is the shape of the thermal emission spectrum (Fig. 7.4). SSI samples the steep, short-wavelength, side of the spectrum where radiance changes rapidly with both wavelength and temperature. Measurements at about 2–5 microns

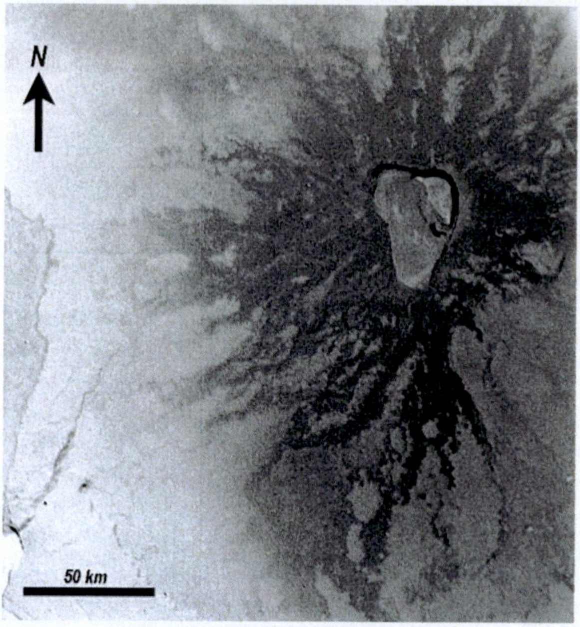

Fig. 7.2 *Voyager 1* image of Maasaw Patera. The term "patera" denotes a broad shallow, bowl-like depression. However, many of Io's paterae actually have steep walls over a kilometer tall. The main patera in this image is 44 km across with a smaller, deeper, depression visible in the upper left part of the main depression. Clow and Carr (1980) argued that these tall steep walls were inconsistent with sulfur which was not strong enough to form such steep slopes if the base was heated. The alternative explanation is that such steep walls are transient features that are collapsing on a geologically rapid timescale. (Image credit: NASA/JPL/USGS)

are more sensitive to the style of the eruption and short-term variations in eruption rate while the 5–20 micron region provided a good view of the long-term heat flux from a volcanic center (e.g., Davies et al. 2010).

Most eruptive centers showed temperatures that could be explained by mafic silicate lavas that are common on Earth, but one spectacular eruption in June 1997 at Pillan Patera indicated even hotter lavas (McEwen et al. 1998). The initial temperature estimate for the Pillan eruption was reported as 1825 K, well above the ~1500 K liquidus temperature of mafic lavas. (Rocks melt over a range of temperatures, starting at a temperature called the *solidus* with complete melting being achieved at a temperature called the *liquidus*.) The eruption of ultramafic lavas, which have even higher melting temperatures, has never been witnessed by humans and the possibility that Io was providing a view into processes last seen on Earth billions of years ago was especially exciting.

However, the key SSI observation (Fig. 7.5) was remarkable not just in what it observed but also in how it was acquired. The eruption was so energetic that it saturated the SSI detector and there was a mistake in the commands sent to the camera. The result was something that would normally have been considered not

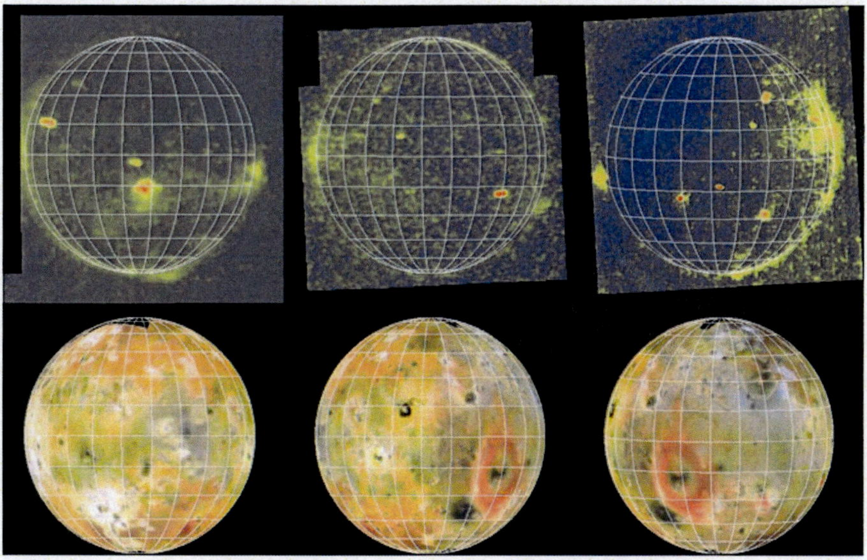

Fig. 7.3 Galileo SSI camera images of Io in eclipse, showing Io's volcanoes glowing red hot in the darkness. The three upper panels are the actual *Galileo* images and the three globes underneath show a map of Io projected into the same geometry. The pixel scales of the Io images are 17.6, 9.1, and 10.5 km, going from left to right. However, the spacecraft was not perfectly stable during these long exposures, so the spots are smeared by several pixels. The data were acquired with the filter wheel in the CLR (or "clear") position, so photons from blue to about 1 micron are all detected. The color in this figure is not the color of the glow but instead is a measure of the intensity of the signal where blue is no signal, green/yellow is low signal, and red is the most intense. These images also show considerable noise from taking long-exposure images in a high radiation environment and auroral emissions from glowing gases. Isolating just the signal from the lava is impossible to do with high precision. (Image credit: NASA/JPL/UA)

worth analyzing. However, since *Galileo* was able to return so little data, each observation, no matter how problematic, was scrutinized in excruciating detail. Given how critical this one observation is for the discussion of Io's composition, it is useful to delve into the details of the observation itself and how it was analyzed. This observation was acquired during the ninth orbit of the *Galileo* spacecraft around Jupiter with a flyby targeting Callisto (making it orbit C9 in the *Galileo* team's jargon). This was a distant observation of Io, acquired at over 1.4 million kilometers from Io, providing a pixel scale of 14.6 km.

The plan was to acquire an image through the 1-micron (1MC) filter followed by one in the clear/panchromatic (CLR) filter in order to measure the color temperature of Io's lavas. By this point in the mission, it had been determined that the CLR/1MC ratio was the best for monitoring typical eruptions on Io. In order to save data volume and acquisition setup time, the two images were to be acquired as a double exposure in the same frame (called an "on-chip mosaic"). This was possible because Io occupied less than ¼ of the 800 × 800 pixel detector and the background was

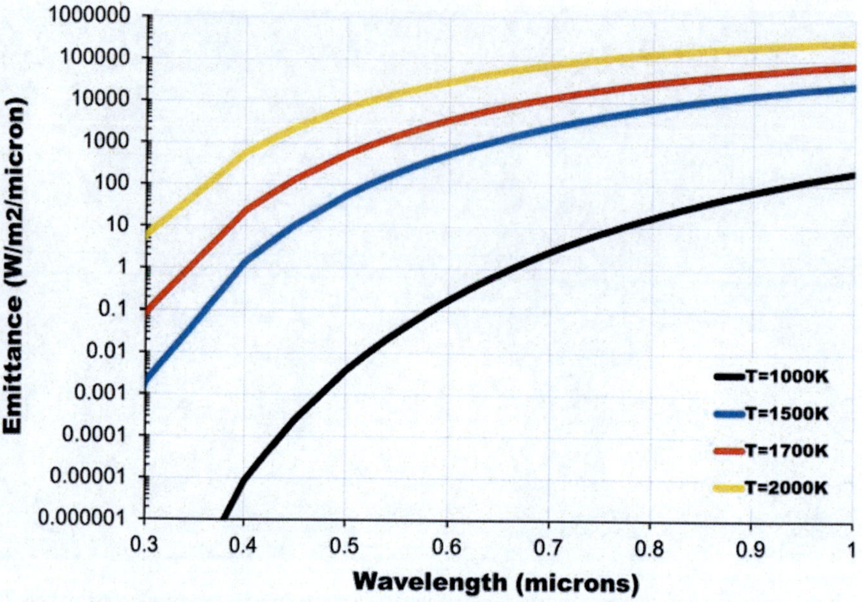

Fig. 7.4 Blackbody thermal spectra for temperatures relevant for incandescent lava on Io. Note the logarithmic vertical axis. The temperature of molten mafic lavas is typically below around 1500 K while ultramafic lavas are expected to be at 1700 K or above. A lava derived from melting a highly refractory mantle (i.e., one dominated by the magnesium-rich olivine called forsterite) would have an eruption temperature around 2000 K. However, after only a few seconds of cooling, the lava surface will drop to around 1000 K. The ratio of thermal emission at visible wavelengths (~0.4–0.7 μm) to near-infrared emission (~1 μm) can be used as a very sensitive thermometer across this temperature range. Note that an experiment set up to measure the hottest lava temperatures in the 1500–2000 K range will have difficulty detecting the much lower signal coming from lava surfaces that are even a few seconds old

dark sky. After acquiring the 1MC image with a 25.6 s exposure in one corner of the frame, the camera was supposed to close the shutter, rotate the filter wheel to the CLR position, slew the camera so Io was in a different corner of the frame, then open the shutter to acquire CLR data for 6.4 s before moving the data off the detector and onto the recorder.

Due to a command error, the slew started too soon—only 13.8 s into the 1MC exposure. The result is that the camera slewed for about 4 s with the shutter open and the filter wheel in the 1MC position. The second image of Io was then a combination of approximately 7.75 s through the 1MC filter and another 6.4 s through the CLR filter. By summing the pixel values along the streak and assuming a 4-s slew, the intensity through the 1MC filter could be estimated. However, there are complications because the slew involved accelerating and decelerating the camera scan platform, so the timing is not known with high accuracy. The intensity of the CLR signal could be estimated because the Pillan hot spot was not only saturated but was also "bleeding." Bleeding is when the detector's pixel becomes so full that

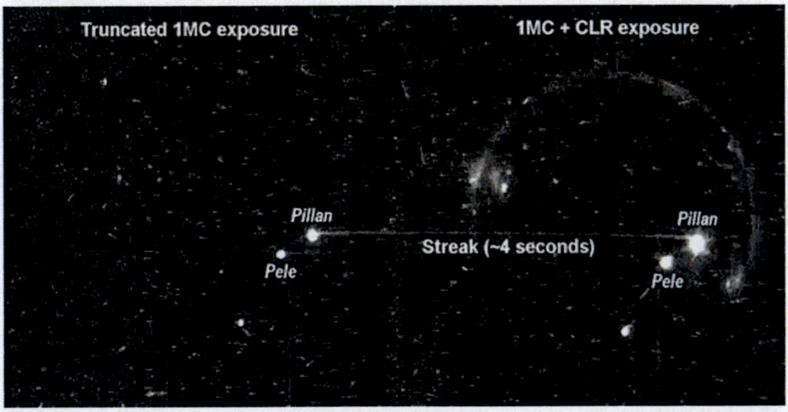

Fig. 7.5 The 1997 Galileo SSI observation that suggested ultramafic eruptions at Pillan. The image has two exposures on the detector, one through the 1-micron filter (1MC), the other should have been through the panchromatic "clear" filter (CLR). As described in the text, the observation did not execute as intended with the camera moving during the 1MC exposure. Some radiation noise has been removed in the process of creating this version of the image, but a careful examination will still show the effects of radiation and data compression done by the camera electronics. The variation in the brightness of the streak was caused by the increase and decrease in the velocity of the scan platform to which the camera was mounted. (Image credit: NASA/JPL/UA/USGS)

electrons spill over into other pixels. Because of how the SSI detector's pixels were wired, the electrons would flow down a column rather than flow in all directions. The signal from 7.75 s through the 1MC filter had to be subtracted from this total to arrive at an estimate of the CLR signal. Obviously, uncertainties are unusually large with such an unconventional observation. The original McEwen et al. (1998) paper reported that temperatures as low as 1500 K were allowed by this observation and that there was no firm upper bound on the temperature. 1500 K is at the upper end of the range that could be considered mafic rather than ultramafic.

The 1825 K temperature estimate in McEwen et al. (1998), which is firmly in the ultramafic range, came from fitting the NIMS spectrum obtained on the same flyby to a three-temperature model. The uncertainty in the fit was reported as only ±25 K which is less than 1.5% of the estimate. However, this was just the uncertainty in the fitting and did not include other, much larger, sources of uncertainty. The NIMS pixel contained not just the Pillan eruption, but also one at Pele. Thus, the thermal emission of Pele needed to be subtracted from the combined spectrum before the Pillan eruption could be analyzed. When NIMS was able to see Pele and Pillan separately, the estimated total thermal output from Pele was remarkably steady but the shapes of the spectra were quite variable (Davies et al. 2001). This means there is no definite way to know what spectrum to subtract from the C9 NIMS observation to isolate the signal from Pillan. This problem was addressed by creating synthetic thermal emission spectra for Pele, based on two-temperature fits to the actual data. This significantly smoothed the spectra, avoiding the problem of doubling the noise by subtracting a noisy spectrum from a noisy spectrum. However, this approach made it impractical to formally estimate uncertainties based on the characteristics of the actual observations. Instead, synthetic thermal spectra derived from observations of Pele at different times were used to obtain a sense of the variability of the thermal emission that could be attributed to Pillan in the C9 NIMS observation. The answer was that the power output of Pillan derived from NIMS had an uncertainty of about a factor of 6 (Davies et al. 2001).

Translating this uncertainty into the temperature at Pillan is even less straightforward. The standard NIMS processing involved fitting two temperatures (and areas) to the data. Subtracting three different plausible spectra for Pele, the Pillan temperatures were estimated as 354–546 K for the lower temperature component and 882–1034 K for the hotter component. When synthetic spectra were generated with these temperatures, they failed to explain the SSI observation. Thus, a three-temperature fit was applied to the C9 NIMS observation of Pillan and Pele. This produced a third, hotter, component at 1825 K for the combined Pillan and Pele spectrum. When the synthetic spectra were generated with this hotter component, the result fit within the error bars of the SSI observation. However, this temperature estimate does not subtract the thermal emission for Pele.

Given this rather unsatisfactory situation, Davies et al. (2001) utilized a fundamentally different approach as well. A thermal model for the distribution of surface temperatures on a lava flow (Davies 1996) was utilized to produce more realistic synthetic spectra. In principle, having a continuous distribution of temperatures, rather than just two temperatures, should have produced significantly superior results. However, the model spectra did not provide a good fit to the measurements. To obtain a good fit, both Pele and Pillan needed to be modeled as two distinct eruptions—one that had been ongoing for hundreds of hours and the other for less than an hour. This is physically plausible because each eruption involved lava fountains as well as lava flows. This technique suggested that Pele was erupting mafic lavas near 1250 K while 1900 K lavas provided a good fit to the Pillan spectra (with the Pele contribution removed). However, later high-resolution SSI observations of Pele suggested much hotter lavas could have been present there (Radebaugh et al. 2004).

There are two points to this extended discussion of the temperature estimates at Pillan. First is to show the lengths to which the *Galileo* team went to try and extract information from even the most challenging observations. The second is to make it clear that the uncertainties are very large, making it very difficult to draw any strong conclusions about whether Io's lavas are mafic or ultramafic. A post-*Galileo* review of the lava temperature estimates (Keszthelyi et al. 2007) identified additional sources of uncertainty in linking lava temperature to composition. There is the issue that lava cools extremely quickly, so the observed surface temperatures will be much lower than the temperature of the rising magma. It was shown that lava flow models, such as Davies (1996) or Keszthelyi and McEwen (1997a) were not appropriate to use for lava fountains where small droplets could cool much faster than a lava flow surface (Fig. 7.6). Furthermore, it was shown that the lavas could become superheated by 50–100 K during rapid ascent, as would be expected for the

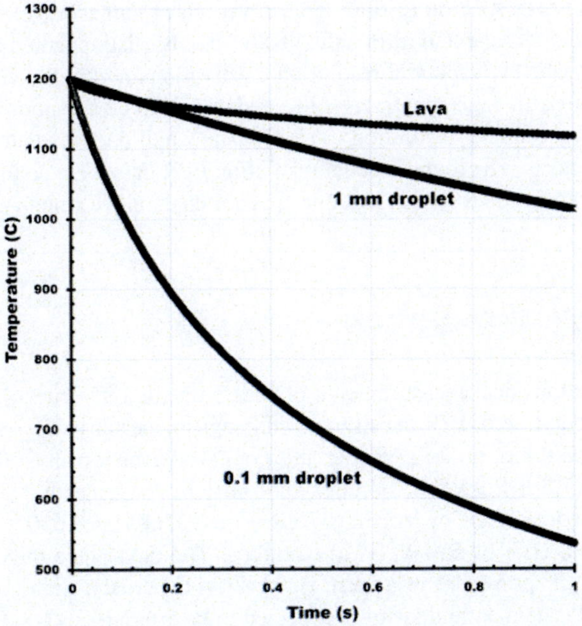

Fig. 7.6 Model cooling of lava surfaces. The "lava flow" curve from a cooling model for an infinite half-space (Keszthelyi and McEwen 1997a). The curves for the 0.1 mm and 1.0 mm droplets are for isolated isothermal spheres cooling by thermal radiation (Keszthelyi et al. 2007). 0.1 mm is similar to the size of particles from ancient lava fountains on the Moon, which are probably good analogs for the lava fountains on Io today. The time evolution of surface temperatures is dramatically different for dispersed lava droplets versus a lava flow. Correctly interpreting the observed temperatures from Io, which must mix thermal emission from surfaces at many different temperatures, depends on correctly identifying the type of volcanic activity and selecting an appropriate thermal model(s)

more energetic eruptions like the June 1997 eruption at Pillan. In the final analysis, Io's lavas are too hot to be anything less than mafic but there are no real upper limits on the eruption temperature. In fact, non-silicate temperatures corresponding to "ceramic" lavas are allowed by the data (Kargel et al. 2003).

Spectral analysis of the lava surfaces proved just as frustrating. NIMS did not clearly detect any silicate minerals (Carlson et al. 2007). SSI showed a dip near 0.9 micron for the dark lavas, and this was interpreted as possible evidence for enstatite, a magnesium-rich pyroxene found in ultramafic lavas (Geissler et al. 1999). However, the contribution from hot lavas to the 1-micron band was not taken into account, making it possible that this feature is related to the elevated temperature of the dark regions rather than their composition. The lack of discernible silicate features from Io's lavas is not entirely surprising since fresh lava surfaces are dominated by glass. The amorphous nature of glass means that distinctive spectral features do not form in the wavelength range that SSI and NIMS were able to observe. *Galileo* also had the PhotoPolarimeter Radiometer (PPR) instrument that was capable of high spectral resolution observations at longer wavelengths (Russell et al. 1992) where mafic and ultramafic glasses can be distinguished (Elachi 1987). Unfortunately, the high spectral resolution observations required considerable time to acquire. Thus, the high spectral resolution data had to be acquired while far from Io and lacked the spatial resolution to resolve the small dark areas that correspond to unmantled lavas. The higher spatial resolution PPR data were acquired with very low spectral resolution (often in just one spectral channel) (Rathbun et al. 2004).

7.2.4 Geophysics

The global geophysical measurements of Io are detailed in Chap. 4 and are only briefly mentioned here. Io's density and moment of inertia place some broad but important constraints on Io's deeper interior (Anderson et al. 2001). Io's bulk density of 3527.8 ± 2.9 kg/m^3 requires an iron-rich core beneath a silicate mantle. The normalized moment of inertia (C/MR^2) of 0.37685 ± 0.00035 is consistent with an Fe-FeS core of 550–900 km in radius. The data place only very limited constraints on the properties of a crust. However, if the crust is mafic/ultramafic and <100 km thick, the mantle density is expected to be around 3300–3400 kg/m^3. For comparison, the density of the lunar mantle is 3360 kg/m^3 (Matsuyama et al. 2016).

7.3 Compositional Models

In the absence of strong direct observational constraints, a wide variety of compositions for Io can be postulated. In this section, we examine two questions: (1) Do any observations require Io to have an exotic composition? and (2) Do any observations rule out an exotic composition for Io? The conventional model starts with a "chondritic" bulk composition whereas the exotic model(s) would remove this assumed constraint.

7.3.1 A Chondritic Model for Io

Since the Sun contains about 99.8% of the mass of the Solar System, it should be no surprise that the planetary bodies that orbit the Sun appear to share a compositional affinity to the Sun and each other with variations largely due to differences in volatiles that are stable at different temperatures. For the rock-forming elements, this composition is referred to as "chondritic" after the name of the most volatile-rich family of meteorites—meteorites with the most Sun-like element ratios (excluding hydrogen and helium). This is the reason why the default assumption is that Io also has a chondritic bulk composition with some volatiles (e.g., H, C, N) missing.

However, as discussed in Chap. 3, different types of chondritic meteorites correspond to material that condensed at different temperatures. The most primitive (and volatile-rich) type of chondritic meteorite is designated CI. However, there is an interesting progression in the amount of volatiles seen on the four large Galilean satellites with less volatiles as one approaches Jupiter. This could correspond to a temperature gradient in the material around Jupiter that accreted to form these moons, or it could be the result of more energetic volatile removal from the more tidally heated worlds (see Chaps. 3 and 4 for more details). The result is that different studies have suggested that less volatile-rich chondritic meteorites could be a better fit to Io's bulk composition. Lewis (1982) argued that "CM" chondrites, which are slightly less primitive than CI chondrites, are the best analog based on the oxidation state of sulfur. However, Anderson et al. (2001) and Kuskov and Konrod (2001) argued that the bulk density and moment of inertia of Io pointed to a lower metal, L or LL chondrite, bulk composition. Others have suggested that the detection of enstatite on Io's surface could point to an enstatite chondrite (EH) composition for Io. This spans the entire range of known chondritic meteorites indicating that there are no definitive constraints on Io's bulk composition.

Models for the composition of Io's core, mantle, and crust critically depend on assumptions about the composition of the core. With most chondritic compositions, Io's sulfur content should be sufficient to form a global surface layer 50 km thick. The fact that such a thick sulfur layer is not seen, and that this is many orders of magnitude more that the amount of sulfur expected to be lost from Io over geologic time, indicates that the bulk of the sulfur must be deeper within Io. Consolmagno (1981) reasoned that >90% of Io's sulfur is in the core. The most common assumption is that the core is at the Fe-FeS eutectic, i.e., the lowest melting temperature iron-sulfur liquid. This would result in a core with a radius of about 950 km, a density of 5150 kg/m^3, and sulfur content of 22.5 wt.% (Anderson et al. 2001). But there is no inherent reason why Io's core needs to have a eutectic composition. Balog et al. (2003) argued that there is no reason to reject a model with Io's core containing only 10 wt.% sulfur.

These *Voyager*- and *Galileo*-era studies can be updated with more recent laboratory studies into the way sulfur partitions between silicates and metals (e.g., Suer et al. 2017). Recent works have also been done to characterize the densities on Fe-S alloys containing low amounts of sulfur (e.g., Morard et al. 2018). Table 7.2 shows the expected properties of the core for different assumed bulk compositions of Io and for two different core sizes. This shows that for Io to have a sulfur-poor core, it would have to be very large (thereby diluting the sulfur in a larger volume of metal). However, such a large dense core is not permitted by the geophysical data, providing strong support for the models where Io has a smaller and more sulfur-rich core. Furthermore, the high-iron enstatite chondrite (EH) bulk composition is also problematic. This composition results in a very iron rich mantle that absorbs the majority of the sulfur, resulting in a core that is again too dense to fit the geophysical constraints. While the available constraints do not precisely pin down the bulk composition of Io, they at least show that a conventional chondritic composition is not inconsistent with the available observations.

Table 7.2 Predictions for Io's core properties based on different initial bulk compositions. Core sulfur contents are based on single stage core formation model using partition coefficients from Suer et al. (2017). The densities of Fe-S alloys are determined using an equation of state based on Morard et al. (2018). Chondrite compositions are from Wasson and Kallemeyn (1988)

Chondrite Model	CI	CM	CO	CV	H	L	LL	EH	EL
Bulk S (wt.%)	5.9	3.3	2.0	2.2	2.0	2.2	2.3	5.8	3.3
Core Mass Fraction = 20%									
Mantle S (wt.%)	0.21	0.094	0.072	0.064	0.053	0.054	0.057	6.6	0.12
Core S (wt.%)	28.6	16.1	9.7	10.7	9.8	10.8	11.3	2.67	16.0
Core Density @ 6 GPa (kg/m^3)	5400	6500	7000	7000	7000	5900	6800	7400	6500
Core radius (km)	930	870	850	850	850	850	832	830	870
Core Mass Fraction = 40%									
Mantle S (wt.%)	0.18	0.078	0.060	0.053	0.044	0.044	0.047	6.5	0.096
Core S (wt.%)	14.6	8.2	5.0	5.5	5.0	5.5	5.7	3.6	8.18
Core Density @ 5 GPa (kg/m^3)	6500	7100	7100	7200	7250	7200	7270	7450	7000
Core radius (km)	1100	1070	1070	1060	1060	1060	1060	1050	1070

Moving upward, we face the mystery of the state of Io's mantle (see Chap. 4 for more discussion of this topic). Multiple lines of evidence suggest that the top of Io's mantle reaches at least 20% partial melting (Keszthelyi et al. 1999; Khurana et al. 2011), but there are also multiple reasons why a completely fluid magma ocean is not favored (Keszthelyi et al. 2007; Moore et al. 2007; de Kleer et al. 2019). Given Io's relatively small size and thus limited increase in pressure with depth, this high degree of partial melting at the top of Io's mantle would suggest that some degree of partial melting will extend at least halfway through the mantle, and the core-mantle boundary temperature is expected to be in the vicinity of the mantle's solidus (Fig. 7.7).

However, this assumes that Io's mantle is homogeneous, but Io's mantle could have become chemically segregated by magmatic differentiation. The estimated current heat flow out of Io (10^{14} W) is the equivalent of over 500 km^3 of lava erupted each year (Blaney et al. 1995). Over 4 billion years, this would be about 100 times the volume of the silicate part of Io. Even if Io has not always been this active, this suggests that all of Io's mantle has undergone many episodes of melt extraction. In each episode, certain elements (especially K and Na, but also Al, Si, and oxidized Fe to a lesser degree) preferentially move into the melt. Keszthelyi and McEwen (1997b) modeled this process and found that, over time, this would leave a residuum that has a very Mg-rich silicate composition dominated by the mineral forsterite, the magnesium-rich version of olivine and enstatite, the magnesium-rich version of pyroxene. This "depleted" part of the mantle would have a very high melting temperature (~2100 K).

A depleted mantle would appear to be an excellent source region for the very high temperature lavas that have been tentatively observed erupting at the surface of Io. The problem is that the Keszthelyi and McEwen (1997b) model makes additional predictions that do not match observations of Io. If there is an extremely magnesium-

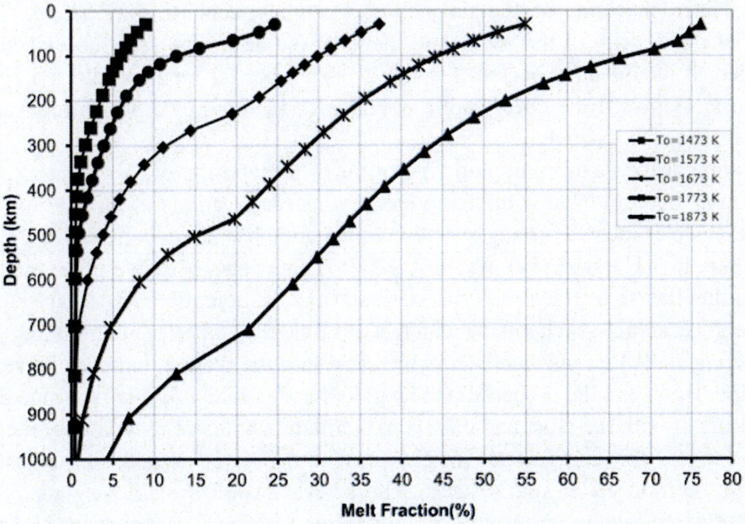

Fig. 7.7 Predicted melt fraction versus depth for various lava eruption temperatures (To). Calculations using the MELTS petrologic modeling software (Ghiorso and Sack 1995), a CM chondrite bulk composition for Io, and assuming an adiabatic temperature gradient that is appropriate if Io is convecting. For Io to efficiently produce tidal heat, much of Io's mantle should have 5–10% partial melting—too much less than that and Io is so rigid that there is little motion in response to tidal stresses but much more than that and Io is too inviscid for the motion to result in much heat. This modeling favors lava eruption temperatures between 1500 and 1600 K (Keszthelyi et al. 2007)

rich part of the mantle, the remainder must be quite iron rich. The prediction is that these rocks would contain >20 wt.% FeO and be relatively dense (~3450 kg/m^3) compared to the forsterite-rich residuum (3200–3300 kg/m^3). The Keszthelyi and McEwen (1997b) model predicts that these denser rocks will descend to form the lower mantle. But, as discussed earlier, putting iron-rich silicates in contact with the core would lead to large amounts of sulfur being extracted from the mantle. The moment of inertia of Io is not consistent with a very dense metallic core overlain by a dense iron-rich mantle topped by a lower-density mantle—at least in the proportions that would be consistent with a chondritic bulk composition. Furthermore, the Keszthelyi and McEwen (1997b) model predicts that the crust would be composed of low-density, low-melting temperature silicates (alkali-rich equivalents of granites). The lavas would be expected to erupt at temperatures significantly lower than the mafic temperatures that are regularly and confidently observed for Io. These departures from the observed Io were considered to be fatal flaws in the Keszthelyi and McEwen (1997b) model (Keszthelyi et al. 1999, 2007).

A more recent set of numerical simulation of magma migration in Io's mantle by Spencer et al. (2020a, b) also comes to the conclusion that the mantle would differentiate and form a zone of high-melting temperature residuum. This work suggested an Io with the depleted mantle at the bottom with melts from the deep mantle ponding at the interface between the upper and lower mantle. While sophisticated in terms of the physics of magma ascent, this modeling did not consider the density of the rocks and melts. If the densities were more accurately modeled, it would be expected that the iron-rich partial melts would not rise but instead sink toward the core-mantle boundary, and would extract sulfur from the core, as mentioned above.

To summarize, petrologic modeling of the differentiation of Io's mantle seem to result in an outcome that is not supported by the observations of the interior or the lavas that are erupted. This suggests that Io's mantle has not experienced significant differentiation (Keszthelyi et al. 1999, 2007). This is only possible if the melts that escape the mantle to form the crust are extremely efficiently cycled back into the mantle. The actual mechanism, which would need to operate without Earth-like subducting slabs that physically mix the crust into the mantle, is not understood at this time. However, the process must be global and is likely to involve some degree of melting of both the crust and mantle to facilitate wholesale recycling of the crust.

Looking at the composition of the crust, the favored hypothesis is that it is primarily composed of mafic lavas. These lavas would be primary melts from the mantle, analogous to basalts on Earth. Compositional variation in the crust is expected to be limited since the compositional variation within the mantle is thought to be limited. Modest differences, especially in alkali and iron contents, would result from local variations in the degree of partial melting and the proportion of crustal versus mantle melts.

Overall, the idea that Io's silicate lavas are minor variations on basalts is supported by the observed volcanic landforms. Rather than the wide variety of types of volcanism seen on Earth, Io's eruptions produce lava lakes when the lavas are confined within paterae and extensive lava flow fields (flucti) when they are not confined. Suggestions of more silica-rich lavas, such as lava domes and stratovolcanoes, are notably absent (see Chap. 5). However, paterae provide a special enigma.

Paterae are often considered to be analogous to terrestrial calderas, which are formed by collapse after the evacuation of a shallow magma chamber (Carr et al. 1979; Radebaugh et al. 2001; Chap. 5). This process is most common when a magma chamber within a volcanic edifice is drained by a flank eruption. But this mechanism cannot work on Io where the paterae are at a lower elevation than the surrounding terrain. Extremely large explosive eruptions provide another means to remove material from a magma chamber and, on Earth, produce calderas similar in scale to Io's paterae. However, this style of eruption is tied to magmas rich in silica and water, and we do not observe other features, like lava domes, that are associated with such lavas (Radebaugh et al. 2001).

This conundrum has led to an alternate model for the formation of paterae as unroofed intrusions into a thick sulfurous volatile layer (Keszthelyi et al. 2007). In this model, mafic magma rising through Io's crust ponds at the base of a near-surface mantling layer of frozen volatiles, forming a horizontal sheet of magma (i.e., a sill). The heat of the intrusion mobilizes the volatiles, forming liquids and gases that move laterally and upward. Where these fluids reach the surface, they would produce a wide variety of sulfurous volcanic features including sulfur dioxide-rich plumes and

Fig. 7.8 Galileo SSI color mosaic of the Chaac-Camaxtli region of Io showing paterae that could correspond to different stages of the "sill excavation" model for patera formation (Keszthelyi et al. 2004). Steropes Patera would be an example of the early phase with some ground collapse and limited mobilization of sulfur dioxide. Balder and Grannos Paterae have deeper collapse and are the first hints of more complex (i.e., colorful) sulfurous fluids. Ababinili Patera has more suggestions of yellowish sulfur flows. Sobo Fluctus shows wide mix of colorful sulfurous flows as well as some dark flows and bright red pyroclastics that are closely associated with silicate lavas and vents. Utu, Chaac, and Ruaumoko Paterae appear to have more silicate than sulfurous flows on their flows. Camaxtli Patera is an example of a silicate-dominated patera floor that could represent a fully exhumed sill. (Image credit: NASA/JPL/Jason Perry (UA)/USGS)

sulfur-rich flows. If the intrusion remains active for sufficient time, the volatiles above the sill would be removed and the mafic silicate lavas exposed. Features seen on Io fit well with this model (Fig. 7.8).

Of course, this model for patera formation posits the existence of thick mantling deposits of volatiles across much of Io's surface. The depth of paterae would require the thickness of this deposit to be locally over a kilometer thick (Clow and Carr 1980; Radebaugh et al. 2001). The existence of such thick volatile layers is supported by observations of thick plateaus that retreat in front of silicate lava flows without leaving any significant lag deposits (Dundas 2017). Similarly, while headscarps for collapses of the walls of paterae are observed, the landslide deposits cannot be found sitting atop the silicate lavas (Fig. 7.9). At the same time, extensive landslide deposits are seen where the material flows onto the cold plains.

In summary, it is possible to construct a self-consistent model for Io starting with a typical chondritic composition. This model has a substantial core with a significant amount of sulfur. The mantle would retain its "primitive" composition by balancing differentiation and remixing. At least the upper part the mantle would contain significant partial melts. The crust would be composed of broadly basaltic rocks topped by a thick layer of sulfurous volatiles.

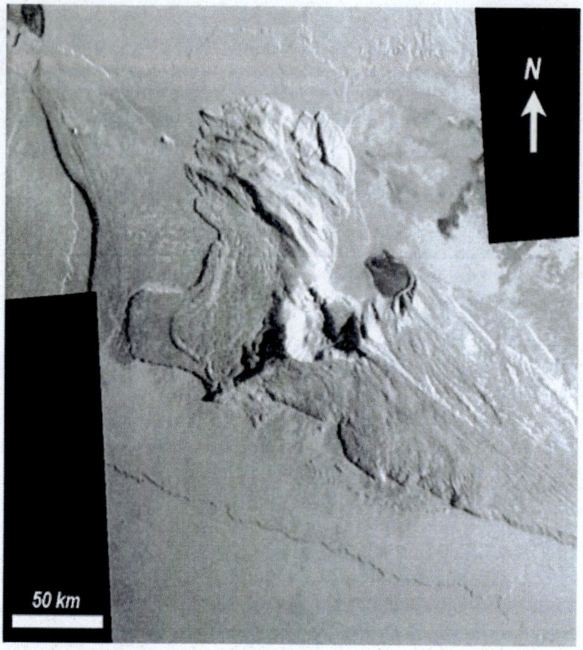

Fig. 7.9 Galileo panchromatic mosaic of the Tohil Mons area. The small dark patera incised into the northeast side of the mountain is Radegast Patera. The sharp tall headscarps formed by large collapses are clearly visible along the contact between Radegast Patera and Tohil Mons. However, the landslide deposit is not visible. This suggests that much of the material that collapsed into Radegast Patera was composed of volatiles that vaporized without leaving a significant lag deposit. (Image credit: NASA/JPL/UA/USGS)

7.3.2 *Exotic Compositional Models for Io*

While a chondritic Io is allowed by the observations, so are other much more exotic compositions. Such exotic compositions could be called upon if the eruption temperatures of the lavas are found to be consistently >2000 K. One possibility is that Io is relatively iron poor and that the iron has been efficiently drawn into the core. The result is a mantle that is magnesium rich and thus dominated by forsterite, which has a melting temperature near 2200 K. However, there are no meaningful limits on how hot Io's lavas may be. As such, it has been speculated that the lavas could be >2200 K, which would be beyond what is possible with conventional silicate rocks. If this were the case, it has been suggested that Io would need to have a ceramic composition, similar to the extremely refractory calcium-alumina inclusions found in chondritic meteorites (Kargel et al. 2003). At this time, there is no need to call upon these exotic compositions, but these innovative ideas can be revisited and developed further if future observations challenge the more conventional chondritic model for Io.

7.4 Future Observations

While there is room for laboratory and theoretical studies to further improve our understanding of Io's composition, the key is to obtain new observations. Landed missions are unrealistic for the foreseeable future (see Chap. 12), so the focus needs to be on data that can be collected from some distance above Io. For the exosphere and atmosphere, the most important new observations would be *in situ* determination of the chemical species. The amounts of volatile species other than sulfur, especially those containing hydrogen and carbon, are of particular interest. Mass spectroscopy remains the best way to accomplish this. It is realistic to expect a mass spectrometer to measure the quantities of not just ionized and neutral molecules but also determine the isotopic ratios for major elements like sulfur and oxygen (de Kleer et al. 2019). This would open a whole new book on the compositional evolution of Io's volatiles and place fundamentally new constraints on how Io formed.

For the silicate portion of Io, the key is to test the currently preferred hypothesis that the lavas erupting on Io are predominantly minor variations on the theme of typical basalts. If Io actually has a diversity of lava compositions, a fundamental rethink of the models for Io's crust and mantle will be necessary. However, measuring the composition of Io's lavas is not easy.

The volatile coatings and glassy nature of the lava surfaces will always be a significant challenge to mineral identification via spectroscopy. However, thermal infrared observations that can resolve clean lava surfaces at modest spectral resolution have the ability to determine the degree of polymerization of silica—and thus distinguish between broad compositional categories (i.e., felsic, intermediate, mafic, and ultramafic lavas) (e.g., Greenhagen et al. 2010). The challenge is that subpixel variations in surface temperature will complicate such spectra.

Ever since the initial estimates of lava temperatures from the *Galileo* SSI observations there has been a desire to obtain color temperatures with properly designed instrumentation to place more robust constraints on eruption temperature. This is technically very challenging for two reasons. First, the surface of lava can change very rapidly, especially for pyroclasts. Therefore, the data from the different colors need to be acquired very close together in time. Even a tenth of a second between the images has been shown to be too long (Davies et al. 2011). The second challenge is that the intensity of the emission is a very strong function of temperature (Fig. 7.4) meaning that the camera system needs a very high dynamic range to acquire useful data in multiple wavelength regions. While difficult, imaging systems under development could obtain robust color temperatures from Ionian eruptions. The Europa Imaging System being developed for the NASA Europa Clipper mission has a detector that has the appropriate capabilities if coupled with a filter set customized for Io (Keszthelyi and McEwen 2021). Such images would need to be acquired at sufficient spatial resolution to also determine the type of volcanic activity being captured. This is essential information in order to select the appropriate thermal model to translate the surface temperatures of the lava to the temperature of the ascending magma.

7.5 Summary

Io's composition is largely conjecture at this point. Occam's Razor suggests a chondritic body with a well-mixed mantle and a crust composed of mafic lavas largely mantled by sulfurous deposits, but new data could alter this. While incremental advances in our understanding of Io's composition can be made with existing data, new data from a mission dedicated to the exploration of Io could allow more fundamental advances in our understanding.

Acknowledgements The study of the composition of Io has relied on the diverse expertise of a wide community of scientists and engineers but we explicitly acknowledge the direct contributions of the teams that operated the *Galileo* spacecraft and continue to operate the *Voyager*, *Hubble Space Telescope*, and *Juno* missions as well as the Infrared Telescope Facility. We are also grateful to Jeff Kargel, Lillian Ostrach, and Mike Bland for their helpful reviews of this chapter.

References

Anderson, J.D., Jacobson, R.A., Lau, E.L., et al.: Io's gravity field and interior structure. J. Geophys. Res. **106**, 32963–32969 (2001). https://doi.org/10.1029/2000JE001367

Balog, P.S., Secco, R.A., Rubie, D.C., Frost, D.J.: Equation of state of liquid Fe-10 wt % S: implications for the metallic cores of planetary bodies. J. Geophys. Res. **108** (2003). https://doi.org/10.1029/2001JB001646

Blaney, D.L., Johnson, T.V., Matson, D.L., Veeder, G.J.: Volcanic eruptions on Io: heat flow, resurfacing, and lava composition. Icarus. **113**, 220–225 (1995)

Carlson, R.W., Kargel, J.S., Doute, S., Soderblom, L.A., Dalton, J.B.: Io's surface composition, in Io after Galileo: a new view of Jupiter's volcanic moon. In: Lopes, R.M.C., Spencer, J.R. (eds.) , pp. 193–229. Springer (2007)

Carlson, R.W., Weissman, P.R., Smythe, W.D., Mahoney, J.C., the NIMS Science and Engineering Teams: Near-infrared spectrometer experiment on Galileo. Space Sci. Rev. **60**, 457–502 (1992)

Carr, M.H.: Silicate volcanism on Io. J. Geophys. Res. **91**, 3521–3532 (1986)

Carr, M.H., Masursky, H., Strom, R.G., Terrile, R.J.: Volcanic features of Io. Nature. **280**, 729–733 (1979)

Carr, M.H., et al.: The Galileo Imaging Team plan for observing the satellites of Jupiter. J. Geophys. Res. **100**(E9), 18935–18955 (1995). https://doi.org/10.1029/95JE00971

Clow, G.D., Carr, M.H.: Stability of sulfur slopes on Io. Icarus. **44**, 729–733 (1980)

Consolmagno, G.J.: Io: thermal models and chemical evolution. Icarus. **47**, 36–45 (1981)

Davies, A.G.: Io's volcanism: thermo-physical models of silicate lava compared with observations of thermal emission. Icarus. **124**, 45–61 (1996)

Davies, A.G., Keszthelyi, L.P., Williams, D.A., et al.: Thermal signature, eruption style and eruption evolution at Pele and Pillan on Io. J. Geophys. Res. **106**, 33079–33104 (2001). https://doi.org/10.1029/2000JE001357

Davies, A.G., Keszthelyi, L.P., Harris, A.J.L.: The thermal signature of volcanic eruptions on Io and Earth. J. Volcanol. Geotherm. Res. **194**, 75–99 (2010). https://doi.org/10.1016/j.jvolgeores.2010.04.009

Davies, A.G., Keszthelyi, L., McEwen, A.S.: Estimating eruption temperature from thermal emission spectra of lava fountain activity in the Erta'Ale (Ethiopia) volcano lava lake: Implications for observing Io's volcanoes. Geophys. Res. Lett. **38**, L21308 (2011). https://doi.org/10.1029/2011GL049418

de Kleer, K., et al.: Tidal Heating: Lessons from Io and the Jovian System. Final Report for the Keck Institute for Space Studies. http://kiss.caltech.edu/programs.html#tidal_heating (2019)

Douté, S., Schmitt, B., Lopes-Gautier, R.M.C., Carlson, R., Soderblom, L., Shirley, J.: Mapping SO2 frost on Io by the modeling of NIMS hyperspectral images. Icarus. **149**, 107–132 (2001)

Dundas, C.M.: Effects of lava heating on volatile-rich slopes on Io. J. Geophys. Res. Planets. **122**, 546–559 (2017). https://doi.org/10.1002/2016JE005177

Elachi, C.: Introduction to the Physics and Techniques of Remote Sensing. Wiley (1987) 413 pp

Geissler, P.E., McEwen, A.S., Keszthelyi, L., Lopes-Gautier, R., Granahan, J., Simonelli, D.P.: Global color variations on Io. Icarus. **140**, 265–282 (1999)

Ghiorso, M.S., Sack, R.O.: Chemical mass transfer in magmatic processes IV: a revised and internally consistent thermodynamic model for the interpolation and extrapolation of liquid-solid equilibria in magmatic systems at elevated temperatures and pressures. Contrib. Mineral. Petrol. **119**, 197–212 (1995)

Greenhagen, B.T., Lucey, P.G., Wyatt, M.B., et al.: Global silicate mineralogy of the Moon from the Diviner lunar radiometer. Science. **329**, 1507 (2010). https://doi.org/10.1126/science.1192196

Jessup, K.L., Spencer, J., Ballester, G.E., Yelle, R., Roessler, F., Howell, R.R.: Spatially resolved UV spectra of Io's Prometheus plume and anti-Jovian hemisphere. Bull. Am. Astron. Soc. **34**, 40.02 (2002)

Johnson, T.V., Veeder, G.J., Matson, D.L., Brown, R.H., Nelson, R.M., Morrison, D.: Io: evidence for silicate volcanism in 1986. Science. **242**, 1280–1283 (1988)

Kargel, J.S., Delmelle, P., Nash, D.B.: Volcanogenic sulfur on Earth and Io: composition and spectroscopy. Icarus. **140**, 249–280 (1999)

Kargel, J.S., et al.: Extreme volcanism on Io: latest insights at the end of the Galileo Era. EOS. **84**, 313–318 (2003)

Keszthelyi, L., McEwen, A.: Thermal models for basaltic volcanism on Io. Geophys. Res. Lett. **24**, 2463–2466 (1997a)

Keszthelyi, L., McEwen, A.: Magmatic Differentiation of Io. Icarus. **130**, 437–448 (1997b)

Keszthelyi, L., McEwen, A.S.: Measuring Ionian lava temperatures via VNIR color imaging. In: 52nd Lun. Planet. Sci. Conf, Abstract #2009 (2021)

Keszthelyi, L., McEwen, A.S., Taylor, G.J.: Revisiting the hypothesis of a mushy global magma ocean in Io. Icarus. **141**, 415–419 (1999)

Keszthelyi, L., Jaeger, W., Milazzo, M., et al.: New estimates for Io eruption temperatures: implications for the interior. Icarus. **192**, 491–502 (2007). https://doi.org/10.1016/j.icarus.2007.07.008

Khurana, K.K., Jia, X., Kivelson, M.G., et al.: Evidence of a global magma ocean in Io's interior. Science. **332**, 1186–1189 (2011). https://doi.org/10.1126/science.1201425

Kuskov, O.L., Kronrod, V.A.: Core sizes and internal structure of the Earth's and Jupiter's satellites. Icarus. **151**, 204–227 (2001)

Lewis, J.S.: Io: geochemistry of sulfur. Icarus. **50**, 103–114 (1982)

Matsuyama, I., Nimmo, F., Keane, J.T., Chan, N.H., Taylor, G.J., Wieczorek, M.A., Kiefer, W.S., Williams, J.G.: GRAIL, LLR, and LOLA constraints on the interior structure of the Moon. Geophys. Res. Lett. **43**, 8365–8375 (2016). https://doi.org/10.1002/2016GL069952

McEwen, A.S., Keszthelyi, L., Spencer, J.R., et al.: High-temperature silicate volcanism on Jupiter's moon Io. Science. **281**, 87–90 (1998). https://doi.org/10.1126/science.281.5373.87

Moore, W.B., et al.: The interior of Io. In: Lopes, J.R., Spencer, J.R. (eds.) Io After Galileo, pp. 89–108. Praxis, Chichester (2007)

Morard, G., Bouchet, J., Rivoldini, A., Antonangeli, D., Roberge, M., Boulard, E., Denoeud, A., Mezouar, M.: Liquid properties in the Fe-FeS system under moderate pressure: tool box to model small planetary cores. Am. Mineralogist J. Earth Planet. Mater. **103**(11), 1770–1779 (2018)

Pearl, J.C., Sinton, W.M.: Hot spots on Io. In: Morrison, D. (ed.) Satellites of Jupiter, pp. 724–755. University of Arizona Press, Tucson, AZ (1982)

Radebaugh, J., Keszthelyi, L.P., McEwen, A.S., Turtle, E.P., Jaeger, W., Milazzo, M.: Paterae on Io: a new type of volcanic caldera? J. Geophys. Res. **106**(E12), 33005–33020 (2001). https://doi.org/10.1029/2000JE001406

Radebaugh, J., McEwen, A.S., Milazzo, M.P., Keszthelyi, L.P., Davies, A.G., Turtle, E.P., Dawson, D.D.: Observations and temperatures of Io's Pele Patera from Cassini and Galileo spacecraft images. Icarus. **169**, 65–79 (2004)

Rathbun, J.A., Spencer, J.R., Tamppari, L.K., Martin, T.Z., Barnard, L., Travis, L.D.: Mapping of Io's thermal radiation by the Galileo photopolarimeter-radiometer (PPR) instrument. Icarus. **169**, 127–139 (2004)

Russell, E.E., Brown, F.G., Chandos, R.A., Fincher, W.C., Kubel, L.F., Lacis, A.A., Travis, L.D.: Galileo photopolarimeter/radiometer experiment. Space Sci. Rev. **60**, 531–563 (1992)

Sagan, C.: Sulfur flows on Io. Nature. **280**, 750–753 (1979)

Slezak, T., Keszthelyi, L.P., Okubo, C., Williams, D.A.: Paterae on Io: compositional constraints from slope stability analysis. In: Lun. Planet. Sci. Conf., 45th, Abstract #1552 (2014)

Spencer, D.C., Katz, R.F., Hewitt, I.J.: Magmatic intrusions control Io's crustal thickness. J. Geophys. Res. Planets. **125**, e2020JE006443 (2020a). https://doi.org/10.1029/2020JE006443

Spencer, D.C., Katz, R.F., Hewitt, I.J., May, D.A., Keszthelyi, L.P.: Compositional layering in Io driven by magmatic segregation and volcanism. J. Geophys. Res. Planets. **125**, e2020JE006604 (2020b). https://doi.org/10.1029/2020JE006604

Spencer, J.R., Stansberry, J., Dumas, C., Vakil, D., Pregler, R., Hicks, M., Hege, K.: A history of high-temperature Io volcanism: February 1995 to May 1997. Geophys. Res. Lett. **24**, 2451–2454 (1997)

Spencer, J.R., Jessup, K.L., McGrath, M.A., Ballester, G.E., Yelle, R.: Discovery of gaseous S2 in Io's Pele plume. Science. **288**, 1208–1210 (2000). https://doi.org/10.1126/science.288.5469.1208

Suer, T.-A., Siebert, J., Remusat, L., Menguy, N., Fiquet, G.: A sulfur-poor terrestrial core inferred from metal-silicate partitioning experiments. Earth Planet. Sci. Lett. **469**, 84–97 (2017)

Stansberry, J.A., Spencer, J.R., Howell, R.R., Dumas, C., Vakily, D.: Violent silicate volcanism on Io in 1996. Geophys. Res. Lett. **24**, 2455–2458 (1997)

Tosi, F., Mura, A., Lopes, R.M.C., Filacchione, G., Ciarniello, M., Zambon, F., et al.: Mapping Io's surface composition with Juno/JIRAM. J. Geophys. Res. Planets. **125**, e2020JE006522 (2020). https://doi.org/10.1029/2020JE006522

Veeder, G.J., Matson, D.L., Johnson, T.V., Blaney, D.L., Goguen, J.D.: Io's heat flow from infrared radiometry: 1983-1993. J. Geophys. Res. **99**, 17095–17162 (1994)

Wasson, J., Kallemeyn, G.: Compositions of chondrites. Philos. Trans. R. Soc. Lond. A Math. Phys. Sci. **325**(1587), 535–544 (1988)

Zolotov, M.Y., Fegley, B.: Oxidation state of volcanic gases and the interior of Io. Icarus. **141**, 40–52 (1999)

Chapter 8
The Plumes and Atmosphere of Io

Imke de Pater, David Goldstein, and Emmanuel Lellouch

Abstract The current state of knowledge of Io's atmosphere and plumes is summarized, with an emphasis on research conducted since the *Galileo* era. While the primary source of Io's atmosphere is sublimation of SO_2 frost, at times and/or locally volcanoes have a substantial effect on the atmosphere (and sometimes dominate), as shown both via observations and model simulations. The effect of sputtering is less clear, but may be important at high latitudes, and perhaps at night and during eclipse. Photodissociation is likely the primary source of SO, Na, K, and Cl but S_2's main source is volcanic, primarily from "Pele-type" volcanism. While the primary source of NaCl and KCl is most likely volcanic, it is not clear how much sputtering may contribute to the emissions. Some of the NaCl and KCl appears to originate from volcanic vents that do not expel much, if any, SO_2 gas.

A comparison between data and models shows that atmospheric collapse upon eclipse ingress and its reformation upon egress happens much faster than models indicate. The observed brightening and expansion of the SO_2 emissions near volcanic plumes upon eclipse egress, however, does agree with models. Data nor models (at the same resolution) show an enhancement in column density above a plume. As shown in this review, although considerable progress has been made towards both a characterization and understanding of Io's atmosphere, there are some fundamental questions that need to get answered before we can truly understand and simulate all physical processes that contribute to both the composition of and dynamics in Io's atmosphere.

I. de Pater (✉)
Department of Astronomy, & Department of Earth and Planetary Science, University of California, Berkeley, CA, USA
e-mail: imke@berkeley.edu

D. Goldstein
Department of Aerospace Engineering, University of Texas at Austin, Austin, TX, USA
e-mail: david@oden.utexas.edu

E. Lellouch
Laboratoire d'Études Spatiales et d'Instrumentation en Astrophysique (LESIA), Observatoire de Paris, Meudon, France
e-mail: Emmanuel.Lellouch@obspm.fr

R. M. C. Lopes et al. (eds.), *Io: A New View of Jupiter's Moon*, Astrophysics and Space Science Library 468, https://doi.org/10.1007/978-3-031-25670-7_8

Keywords Io · Volcanoes · Atmosphere composition · Atmospheric dynamics

8.1 Introduction

The first evidence for the existence of an atmosphere on Io was obtained in 1973, when the *Pioneer 10* spacecraft detected ionospheric layers above Io's surface near its terminator (Kliore et al. 1974). This was soon followed by a detection of sodium (Brown 1974) and potassium (Trafton 1975) in a cloud around Io. This "neutral" cloud extends along a part of Io's orbit, and due to the Keplerian motion of the particles this cloud is shaped like a banana. Upon ionization, the species are swept away by Jupiter's magnetic field, forming the Io plasma torus. The sodium, potassium as well as the sulfur ions detected in the plasma torus (Kupo et al. 1976) were, at the time, hypothesized to originate on Io.

About a decade before the *Pioneer 10* flyby, Binder and Cruikshank (1964) reported that Io was much brighter in reflected sunlight immediately after eclipse egress than 10–20 min later. The authors hypothesized the presence of an atmosphere, condensing into bright frost on the surface while in eclipse, and subliming upon egress, causing the surface to darken again. However, even though we now know that Io has a mostly condensible atmosphere, the reality of the eclipse brightening events remains controversial, as subsequent data have not shown this effect (e.g., Secosky and Potter 1994).

In 1979, when *Voyager 1* flew by, it imaged volcanic plumes above Pele and Loki, the latter a lava flow to the north-east of the horseshoe-shaped caldera Loki Patera (Morabito et al. 1979). Interestingly, just prior to taking these images, Peale et al. (1979) had predicted large amounts of tidal heat dissipation in Io's interior caused by the fact that Io, Europa and Ganymede are locked in a 4:2:1 orbital (Laplace) resonance which creates a significant forced eccentricity in their orbits. Peale et al. (1979) had predicted the *Voyager* spacecraft would detect effects from the resulting high heat flow on Io's surface, a prediction that was confirmed in record time.

In addition to the visible plumes, Pearl et al. (1979) reported the presence of SO_2 gas above Loki based upon an analysis of *Voyager 1*/IRIS spectra, which showed a 7.3 μm absorption feature against a warm thermal background. Since this was the only time such spectra were obtained against a warm background, it was not clear if SO_2 was confined to this volcanically active spot or was more widespread across the satellite. It took another decade before Io's "global" SO_2 atmosphere was observed. This was accomplished from the ground, using the IRAM-30 m telescope at 1.4 and 2.0 mm (222 and 143 GHz) (Lellouch et al. 1990, 1992). In their initial analysis, these data revealed a surface pressure of order 3–40 nbar (column densities of 1.5×10^{17}–2×10^{18} cm^{-2}), covering only a fraction of Io's surface (3–20%) at a high gas temperature (∼500–600 K).

Since that time Io's atmosphere and plumes have been observed across the electromagnetic spectrum, from the UV through mm-wavelengths, from the ground, near-Earth orbit, and the *Galileo*, *Cassini*, and *New Horizons* spacecraft. Excellent early—before *Galileo*—reviews have been published by, e.g., Spencer and Schneider (1996) and Lellouch (1996), while papers by McGrath et al. (2004), Geissler and Goldstein (2007) and Lellouch et al. (2007) present reviews on Io's plumes and atmosphere after completion of the *Galileo* mission. The present review is focused on results obtained since the *Galileo* era, although some early results will be mentioned to present a coherent picture.

We start below with an overview of Io's global atmospheric properties, including a short summary of observational methods, and timescales relevant to atmospheric physics and chemistry. In Sect. 8.3 we present an overview of the spatial distribution of Io's atmospheric gases, including views when Io is in eclipse. Observations and models of plumes are discussed in Sect. 8.4. Finally, in Sect. 8.5, we address the question: "What drives Io's atmosphere", and discuss the arguments in favor of a sublimation-driven and volcanically-driven atmosphere, as well as the conditions under which one might expect sputtering to be important. The chapter ends with conclusions (Sect. 8.6), a list of open questions, and future directions.

8.2 Global Atmospheric Properties

Io's atmosphere has been observed at multiple wavelengths and with a variety of techniques, each with their own advantages and shortcomings. An overview of available data is given in Table 8.1. In order to better assess the interpretation of these data we provide a short summary of the pros and cons for each technique in Sect. 8.2.1. This section is followed by a short review of timescales as relevant both to observations and models of the atmosphere and plumes. We end this section with summaries of the global composition and thermal structure of the atmosphere.

8.2.1 Observational Methods

(1) Millimeter/Submillimeter At mm/submm-wavelengths one detects thermal emission from the atmosphere, typically 20–40 K above the background (sub-)surface brightness temperature (~95 K; Moullet et al. 2008; de Pater et al. 2020b) for disk-integrated data. The emissions occur in local thermodynamic equilibrium (LTE) (Lellouch et al. 1992), i.e. energy levels are populated according to atmospheric physical temperatures. Hence, by observing different transitions and measuring line widths and line contrasts, one should, in principle, be able to determine gas temperature, column density, and fractional coverage of the atmosphere, where the latter refers either to the disk as a whole or to the spatial element provided by the data. However in practice most observed transitions have similar (low) energy levels, which to date has precluded accurate temperature measurements. Moreover, the interpretation of line shapes, indicative of temperature through thermal broadening, is complicated by velocity broadening and/or line

Table 8.1 Molecular observations of Io's atmosphere and plumes

Wavelength	Telescope/Instrument	Sp/Im[a]	Resolution[b]	Constituents	Sensitivity[c]	Sunlight/eclipse	References
Mm (1–3 mm)	IRAM-30m	Sp	Disk	SO_2, SO, NaCl	N, T, v, f	sunlight	17–20
Mm (1 mm)	IRAM PdBI	Sp	Cube	SO_2	N, T, v, f	sunlight	23
Mm (1 mm)	IRAM NOEMA	Sp	Disk	SO_2, NaCl	N, T, v, f	sunlight	29
Submm (0.8 mm)	APEX	Sp	Disk	SO_2, SO, KCl	N, T, v, f	sunlight	25
Submm (0.8 mm)	SMA	Sp	Cube	SO_2, SO, NaCl	N, T, v, f	sunlight	24
Submm (0.8 mm)	ALMA	Sp	Cube	SO_2, SO, NaCl, KCl	N, T, v, f	sunlight/eclipse	7
Mid-IR 19 μm	IRTF/TEXES	Sp	Disk	SO_2 (ν_2 band)	N, T	sunlight/eclipse	34,38,39,40,42
Mid-IR 7.3 μm	Voyager/IRIS	Sp	Res	SO_2 (ν_3 band)	N, T	sunlight	26
Near-IR 4 μm	VLT/CRIRES	Sp	Cube	SO_2 ($\nu_1 + \nu_3$ band)	N, T	sunlight	21
Near-IR 1.7 μm	Keck/NIRSPEC	Sp	Disk, Res	SO	N, T	eclipse	3–5, 16
Near-IR 1.7 μm	Keck/OSIRIS	Sp	Cube	SO	N, T	eclipse	6
Visible	Voyager, Galileo, Cassini, New Horizons	Im	Res	Plumes	Opacity, part. size	Sunlight	10
UV 0.255, 0.336, 0.410 μm	HST/WFC2	Im	Res	Plume, S_2, SO_2	Opacity, part. size	sunlight	14,32,33,
UV 0.232, 0.260 μm	HST/FOC	Im	Res	SO_2	N	sunlight	30,31
UV 0.225–0.330 μm	HST/FOS	Sp	Disk	SO2	N, f	sunlight	2
UV 0.210–0.320 μm	Galileo/UVS	Sp	Res	SO_2	N	sunlight	12
UV, 0.210–0.310 μm	HST/STIS	Sp	Res	SO_2	N, T	sunlight	13,15
UV 0.210–0.234 μm	HST/COS	Sp	Disk	SO_2	N, T, f	post-eclipse	41
UV 0.202-0.233	HST/COS.	Sp.	Disk.	SO_2.	N, T, f.	sunlight.	40
UV 0.197–0.235 μm	HST/FOS and GHRS	Sp	Disk	SO_2	N, T, f	sunlight	1,37
UV 0.180–0.310 μm	HST/STIS	Sp	Res	S_2, SO_2	N, T	transmission	33

UV-Vis 0.175–0.570 μm	HST/STIS	Sp	Res	S, SO, SO_2	N, torus prop.	eclipse	36
UV 0.159–0.231 μm	HST/FOS	Sp	Res	SO_2, S, SO	N, *T*	sunlight	22
UV 0.121 μm (Lyα)	HST/STIS	Im	Res	SO_2	N	sunlight	8,9,11,35,28
UV 0.121 μm (Lyα)	HST/STIS	Im	Res	SO_2	N	transmission	27

References: 1: Ballester et al. 1994; 2: Clarke et al. 1994; 3: de Kleer et al., 2019c; 4: de Pater et al. 2002; 5: de Pater et al. 2007; 6: de Pater et al. 2020a; 7: de Pater et al. 2020b; 8: Feaga et al. 2009; 9: Feldman et al. 2000; 10: Geissler and Goldstein 2007; 11: Giono and Roth 2021; 12: Hendrix et al. 1999; 13: Jessup et al. 2007; 14: Jessup and Spencer 2012; 15: Jessup and Spencer 2015; 16: Laver et al. 2007; 17: Lellouch et al. 1990; 18: Lellouch et al. 1992; 19: Lellouch et al. 1996; 20: Lellouch et al. 2003; 21: Lellouch et al. 2015; 22: McGrath et al. 2000; 23: Moullet et al. 2008; 24: Moullet et al. 2010; 25: Moullet et al. 2013; 26: Pearl et al. 1979; 27: Retherford et al. 2019; 28: Roessler et al. 1999; 29: Roth et al. 2020; 30: Sartoretti et al. 1994; 31: Sartoretti et al. 1996; 32: Spencer et al. 1997; 33: Spencer et al. 2000; 34: Spencer et al. 2005; 35: Strobel and Wolven 2001; 36: Trafton et al. 2012; 37: Trafton et al. 1996; 38: Tsang et al. 2012; 39: Tsang et al. 2013a; 40: Tsang et al. 2013b; 41: Tsang et al. 2015; 42: Tsang et al. 2016

[a] Spectra (Sp) or Images (Im)

[b] Disk-integrated (Disk), spatially resolved (Res), or spectral data cube (Cube)

[c] Sensitivity of observations to the column density (N), velocity (v), fractional coverage (f), and temperature (T), where T refers to direct sensitivity to temperature through the Planck function, while *T* refers to indirect sensitivity through line parameters or cross-sections

shifts due to local or global winds (Sect. 8.4.1). The main drawback of mm/submm observations, namely the lack of spatial resolution for single-dish antennas, is currently being overcome by the use of interferometers (e.g., de Pater et al. 2020b).

(2) Ultraviolet/Visible At UV-visible wavelengths Io is seen in reflected sunlight. Since the solar UV flux drops steeply with decreasing wavelength below ~400 nm, reflected emission from Io's disk decreases sharply as well, with the exception of the Ly-α line (121.6 nm), which shows a spike in solar emission. Both SO_2 gas and surface frost absorb strongly at UV wavelengths. As discussed in Sect. 8.3, the morphology of Io images in the Ly-α line has been used to derive the SO_2 abundance and its spatial extent.

UV data of Io include imaging, disk-averaged and disk-resolved spectroscopy. At low spectral resolution (and even more in imaging), it is difficult to disentangle the surface and atmosphere contributions, mostly because the surface reflectance depends strongly on unknown properties such as grain size and state of mixing of SO_2 frost. Off-limb observations of Io transiting Jupiter are immune to contamination from Io's surface, but require high spatial resolution in the UV/optical (Spencer et al. 2000). Even then, interpretation of plume brightness may be complicated by competing absorption between dust and gas.

At the low brightness of Io's disk at UV wavelengths away from solar lines, Io's aurora are prominent. When the satellite is observed in eclipse, its reflected component disappears completely, and even faint auroral emissions become visible (see Sect. 8.3.3 for details).

Unlike in the thermal range, atmospheric features in the UV are primarily sensitive to the column density of the atmospheric gas and the intensity of the electrical currents that excite auroral emissions (Sect. 8.3.3), but not directly to the gas temperature, although band contrasts, positions and skewness, do show a temperature dependence (Wu et al. 2000).

(3) Mid-infrared SO_2 ro-vibrational lines have been detected in the (thermal) mid-infrared (ν_2 band at 19 μm, ν_3 band at 7.3 μm). Unlike rotational levels, vibrational levels can be populated by direct solar excitation and absorption of surface thermal flux, and are strongly subject to non-LTE effects, with radiative de-excitation timescales shorter than collisional timescales (Lellouch et al. 1992; Spencer et al. 2005). So far, these bands have only been seen in absorption, indicating vibrational temperatures colder than the surface temperature, even though most of the atmosphere is expected to be at physical temperatures higher than the surface (see below Sect. 8.2.4). On the other hand, rotational LTE is expected in the lower atmosphere (see Fig. 8.1), enabling one to determine the translational/rotational temperature of the atmosphere from the relative strengths between the lines in a band, although optical depth effects may occur (see, e.g., Fig. 6 in Tsang et al. (2012)).

(4) Near-infrared SO_2 gas has also been detected in its $\nu_1 + \nu_3$ band at 4.0 μm, with spatial information (Lellouch et al. 2015). Being located in the dominantly solar-reflected part of Io's spectrum, it appears in absorption and interpretation is relatively straightforward in terms of SO_2 columns and temperature, because lines are optically thin (~0.1 opacity at infinite resolution) and numerous energy transitions are observed. Absorptions are rather faint, however, and intermixed with telluric and solar lines, requiring an instrument with both high sensitivity and high resolving power (e.g. VLT/CRIRES).

Observations of Io in eclipse have revealed emissions from SO in the near-infrared, discussed in Sect. 8.3.2.

8.2.2 *Timescales Relevant to Observations and Modeling*

In this section we review the physical time scales relevant to observations and modeling of Io's atmosphere and plumes. Table 8.2 is a breakdown of several such time scales, from shortest at molecular scales up to an Ionian year (#A–O), with comments on associated length scales and physics.

Molecular/Atomic Time Scales Plasma ions traveling at tens of km/s cross spatial density gradients of neutral species in times of O(1E-3 to 1E-2 s) (McDoniel et al. 2019): if it is important to resolve plasma interactions with a density gradient in a simulation then the time step must be less than that (A). Ion or fast neutrals from charge exchange suffer collisions in dense gas regions near the ground with time scales of less than O(1E-2 s) (the mean collision time); however, above the exobase this time scale rises to minutes (Walker et al. 2010) (B). The half-life of SO_2 vibration modes to spontaneous emission varies from milliseconds (i.e., virtually instantaneous on a gas-dynamic time scale) to ~1 s, a time over which hot gas in a plume could move a kilometer or more (Zhang et al. 2003) (C). The cyclotron gyration time of ions in the local magnetic field is roughly half a second—during which time the ion may travel quite a distance, depending on the local gas density and mean free path (McDoniel et al. 2019) (D). The mean collision time for thermal neutrals in a nominal sublimation atmosphere varies from $O(0.1\ \text{s})$ near the surface at equatorial noon to arbitrarily long (escape) above the exobase (E). The residence time of an SO_2 molecule stuck on the surface governs how readily gas condenses on or is released from the surface (Walker et al. 2012); that time scale varies from a few seconds or less (effectively producing a simple diffusely scattering surface) to hours or more on the night side or at high latitudes (those surfaces thus act as local cold traps) (F). Molecules rising from the surface would have a ballistic time (their rise/fall time) determined by gravity and their launch speed and angle. That speed is set by the temperature of the surface. For the cool sources of atmospheric molecules at ~100–150 K, the ballistic time ($\approx 2\sqrt{2RT_{source}/m}/g$) is 3–4 min while in a large Pele-class plume sourced from hot magma, with $T_{source} \approx 1200$ K, the time is $\approx 2\sqrt{2CpT_{source}}/g$. Here, Cp is the specific heat, R the universal gas constant, m the molecular mass, and g the surface gravity, and the resulting ballistic time is about 20 min (G). The photodestruction (dissociation or ionization) time scales for different observable molecules is several hours or more (L).

Table 8.2 Time scales and physics relevant to both observations and models

	Physical process	Timescale	Modeling time step required	Length and velocity scales setting the time scale	Relation to atmospheric or plume dynamics and observations
A	Plasma interactions with atm. and plumes	< 0.1 s	< 0.1 s	Atm/plume density gradients like atm. scale height or plume dimensions and plasma particle speeds	Plasma interactions with atm/plume which are not simply plane-parallel and 1-dimensional
B	Ion/neutral collision rates	$\gtrsim 0.01$ s	$\gtrsim 0.01$ s	Lengths set by neutral density and collision cross sections (ion mean free path) and ion speeds	Determines local chemical rates and ion-impact induced emission
C	Molecular vibration or rotation half-life	millisec-to-sec	Stochastic, based on Einstein-A, but < lifetime of 10^{-3}–1 s	NA—a fundamental molecular property	Determines where thermal emission occurs and whether process is in LTE
D	Cyclotron gyration	~0.5 s for ions	0.1 s	Set by ion mass, charge and $\|\|B\|\|$	Related to Hall effects, emission from atm. and plumes
E	Neutral gas motion	0.1 s +	0.01 s +	Local density through mean free path and local field property gradients	Links between macroscopic gas dynamics (shock waves, scale heights, shear layers, and flows created by surface contrasts) and molecular collision MFP
F	Gas-Surface interaction	Residence time of molecules on solid surfaces	<0.1 s to years	NA—a fundamental material property	Varies with species, surface temperature, surface nature. Governs whether or when surface is a sink or diffuse
G	Molecular ballistic time	1 to >20 minutes	1 s–2 min	Set by surface gravity and vertical launch speed (hence, surface temperature)	Very different in plumes and sublimation atm. Sets effective height of both

H	Plume flow evolution	>20 min for Pele-class plumes	Few sec. since plume details are not determined just by ballistics	Over a cold surface acting as a material sink, set by ballistic time	Plumes initiate (rise and fall) on ballistic times; can be much longer if plume bounces off surface or atmosphere
I	Atmos. vertical flow evolution.	<10 min.	Few seconds, constrained by molecular collision times	Related to atm. ballistic time and hence scale height and speed of sound	Atm. collapse and recovery around eclipse occurs through vertically propagating waves
J	Atmos. horizontal flow evolution.	1 h	Few seconds, constrained by molecular collision times	Related to atm. horizontal pressure gradients, felt at speed of sound and are of planet-scale	Pressure-driven winds form due to weak pressure gradients over long distances
K	Eclipse time	2 h	NA	NA—fundamental property	Affects surface temperatures and photodissociation and thus winds and column
L	SO_2, SO, NaCl, KCl photodestruction (transparent atm)	SO_2: 34 h SO: 17 h NaCl: 3 h KCl: 2.5 h	Stochastic, based on rates at 5 to 5.5 AU	NA - fundamental property	Important for production of daughter species S, O, Na, K, Cl
M	One Io day	42 h	NA	NA—fundamental property	Sublimation components presumably periodic over a day/night
N	Atmos. loss to torus	~10 days	0.1 s for sputtering	NA	Set by atm loss rate and inferred by torus luminosity
O	One Io year	12 Earth years	NA	NA—fundamental property	Sublimation components have seasonal variation

Gas Dynamic Time Scales We next consider gas-dynamic or flow time scales: The flow evolution time in a plume (the time it would take for a plume to develop out to its main deposition ring if the plume were abruptly turned on) is about the same as the ballistic time, ~20 min for Pele or Tvashtar (H). On the other hand, the formation time for portions of the plume which undergo a gasdynamic "bounce" (Zhang et al. 2003) or which spread out atop a lower level sublimation atmosphere (McDoniel et al. 2017) could be appreciably longer. The sublimation component of the atmosphere has at least two different flow evolution time scales, which are associated with two different length scales and the speed of sound, c_s, in SO_2 at the local surface temperature.[1] For vertical motions relevant to atmospheric collapse and reformation around eclipse, the appropriate length scale is the atmospheric scale height H,[2] which is ~10 km near the surface, and the time scale is $\sim H/c_s$, i.e., the atmosphere is expected to collapse on a time scale of ~70 s (Summers and Strobel 1996; de Pater et al. 2002), assuming the surface cools instantaneously (I). The second relevant gasdynamic timescale is the hydrostatic adjustment time c_s/g, which is ~100 s. This few minute time scale is comparable to the thermal response time of the surface to eclipse ingress/egress.

The response of the surface temperature to changes in illumination depends upon the thermal inertia of the surface. Tsang et al. (2016) showed that the surface temperature dropped essentially instantaneously upon eclipse ingress, which has been used to show that the thermal inertia of the surface is quite low (see Sects. 8.3.2 and 8.5.1). The time for the formation of global pressure-driven winds should scale with the speed of sound and a *global* length scale such as Io's radius, R_{Io}, since it is over that length that illumination (and hence pressure) changes appreciably: $R_{Io}/c_s \sim 3.3$ h (J). This scale is comparable to the time scale of a Jovian eclipse itself, ~2 h (K), suggesting that the winds will be changing during eclipse and for a couple of hours thereafter. A second time scale coincidence which may have unrecognized impact on the diurnal evolution of the global scale flow is that the optically thin photo-dissociation time of SO_2, generally into SO and O, is ~34 h (L), comparable to Io's 42 h day/night cycle. The photodestruction time for SO is about half that—17 h, while for NaCl and KCl it is ~3 and 2.5 h, resp. (Moses et al. 2002a,b). Near-surface atmospheric layers will be shielded and dissociate more slowly.

A more complete discussion of the time scales relevant to atmospheric composition is provided in Summers and Strobel (1996).

---

[1] $c_s = \sqrt{\gamma R T_{surf}/m}$, where γ is the ratio of specific heats.

[2] The scale height $H = RT(z)/g/m$, with $T(z)$ the temperature at altitude z. $H \sim 10$ km at temperature $T \simeq 140$ K.

8.2.3 Atmospheric Composition

The primary constituent of Io's atmosphere as observed to date is SO_2 gas, a common terrestrial volcanic species and a species notably able to exist in solid/vapor phases at common surface and near-surface pressure/temperature conditions on Io. Photodissociation of SO_2 produces primarily SO and atomic O (Summers and Strobel 1996; Moses et al. 2002a), both of which have been detected (SO by Lellouch et al. (1996); O by Brown (1981)) at abundance ratios with SO_2 between ~3–10% for SO (Lellouch et al. 2007), and ~10% for O (Roth et al. 2014). It is unclear, though, if all observed SO is produced through photochemistry, since it can also be produced directly in volcanic eruptions (Zolotov and Fegley 1998). SO does not have a condensed form, and its losses include recombination to SO_2 and photolysis to O and S, the latter being detected at ~2% of the SO_2 level (Roth et al. 2014). Although photodissociation of SO_2 also predicts some O_2, the column-integrated production rate of O_2 is about 2 orders of magnitude less than that of SO and O (Moses et al. 2002a). This, combined with its destruction primarily through reactions with atomic S, predicts O_2 to be at the ~1% level of SO_2.

S_2 was directly detected on one occasion via *HST* imaging and spectroscopy of Pele's plume on Io's limb against Jupiter, at an S_2/SO_2 ratio of 0.08–0.3 (Spencer et al. 2000). S_2 has a very short photochemical lifetime, and since it has no solid phase, its presence can only be attributed to volcanic activity (Zolotov and Fegley 1998; Moses et al. 2002a); moreover, it has only been detected above (a few) volcanic vents (Sect. 8.4.1). Based on thermochemical equilibrium models (Zolotov and Fegley 1999, 2000), the measured S_2/SO_2 ratio is consistent with equilibration in silicate magmas at 1400–1800 K.

Early this century, atomic chlorine (Retherford 2002; Feaga et al. 2004) was detected in Io's atmosphere at a typical abundance ratio with SO_2 of $\sim 5 \times 10^{-4}$, followed by the discovery of NaCl (Lellouch et al. 2003) and KCl (Moullet et al. 2013). The latter two constituents are likely the source of the atomic species (Na, K) that were detected in the 1970s in Io's neutral clouds (Sect. 8.1), as well as Cl. The typical relative abundances of NaCl and KCl to SO_2 (assuming co-location, which is now known to not be the case, Sect. 8.5.3) are $\sim 3 \times 10^{-3}$ and $\sim 5 \times 10^{-4}$, respectively. Since the photochemical lifetime for both species is relatively short ($\lesssim$3 h, Table 8.2; Moses et al. 2002b), both species need to be sourced (quasi-) continuously. However, whether NaCl and KCl are purely volcanic in origin, or may also result from sputtering on Io's surface has not yet been resolved (see Sect. 8.5.3 for more discussions).

Finally, a few additional compounds have been unsuccessfully searched for at mm-wavelengths (CO, H_2S, OCS, S_2O, ClO, CS, NaOH, SiO) or in the UV (CS_2). These searches have notably led to an upper limit of 10^{-10} bar for H_2S (Lellouch et al. 1992), another common volcanic species on Earth, ruling out the case for a H_2S global atmosphere on Io.

8.2.4 Thermal Structure

The thermal structure of Io's atmosphere mostly reflects the interplay between heating processes, radiative losses to space, and exchanges (radiative, conductive and convective) between atmospheric layers. As discussed in Sect. 8.5.1, an important fraction of Io's atmosphere is sublimation supported, and initial studies of Io's thermal structure (Lellouch et al. 1992; Strobel et al. 1994) have considered a 1D sublimation atmosphere in hydrostatic equilibrium, with adjustable surface temperature. The surface pressure over SO_2 ice is equal to the saturated vapor pressure of SO_2, which is a steep exponential function of temperature: $P_{vapor} = 1.52 \times 10^8 e^{-4510/T}$ bar (Wagman 1979).

Absorption of sunlight by SO_2 gas occurs at UV and infrared wavelengths, and radiative cooling from ro-vibrational bands and rotational lines at infrared and radio wavelengths. Non-LTE effects, which come into play at altitudes where the radiative de-excitation timescale is shorter than the collisional excitation timescale (Table 8.2), significantly affect the cooling terms, especially vibrational, while rotational LTE is maintained at least in the first few scale heights (see Fig. 8.1). Despite the reduced cooling efficiency associated with non-LTE, Strobel et al. (1994) argued that at large local pressures (>10 nbar), vibrational cooling in the ν_2 19-μm band is sufficient for a ~10 K decrease with altitude in the bottom 10–20 km of the atmosphere. At higher levels, rotational cooling takes over, but also progressively becomes non-LTE, and in the presence of solar heating only, the temperature asymptotically reaches ~275 K. Strobel et al. (1994) demonstrated the importance of Joule and plasma heating at the top of Io's atmosphere, and showed that temperatures could rise up to ~1800 K at several hundreds of kilometers

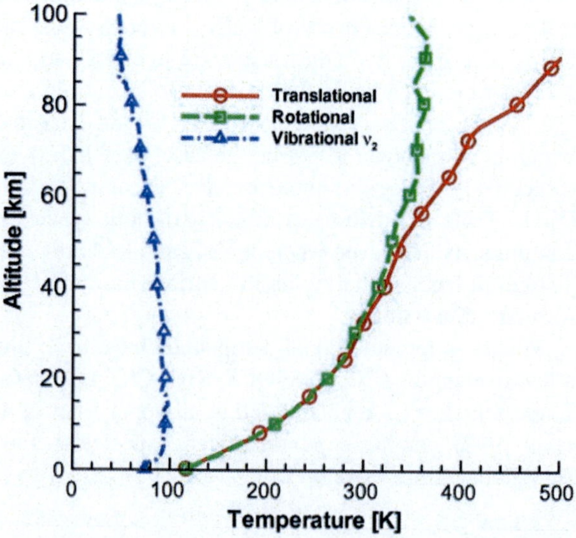

Fig. 8.1 Simulated thermal structure in the lower 100-km of Io's atmosphere above the sub-solar point based on the 3-D simulations by Walker et al. (2010). A 115 K subsolar surface (ice) temperature was adopted (i.e., the surface pressure was ~1.4 nbar), as well as a plasma heating source from above. The vibrational, rotational and translational (i.e. kinetic) temperatures are shown (from Gratiy et al. 2010)

above the surface. These 1-D models were later updated within self-consistent 3-D sublimation models of Io's atmosphere, leading to the prediction of 3-D temperature fields (see Sect. 8.5.1).

While most observations are sensitive to atmospheric temperatures, either explicitly in the thermal range or just from the variation of line-strengths/cross sections in the solar reflected range (see Table 8.1), results remain partial and to some extent contradictory, despite significant progress in the last two decades. Part of the discrepancy is likely related to the fact that so far individual observations have reported a single "temperature" number, which must reflect some weighted means of altitude—and spatially—varying atmospheric temperatures. In principle, multi-transition observations in the thermal range would allow one to reconstruct a vertical temperature profile, as illustrated by the contribution functions for the rotational lines of SO_2 observed by the Atacama Large (sub)Millimeter Array (ALMA) (de Pater et al. 2020b). While the mm-wave data may, in principle, provide the best vertically-resolved constraints on Io's atmospheric profile, it has not yet been possible to determine a temperature profile, or even a representative temperature for the lower atmosphere probed at mm-wavelengths. As mentioned in Sect. 8.2.1, data interpretation is complicated by the additional effect of global winds or plume dynamics on the lineshapes, and the various geometries/lines-of-sight through the atmosphere and plumes with different temperature regimes, as illustrated in Fig. 8.2. As a likely consequence, even though the initial interpretations calling for ~600 K temperatures in the lower few scale heights (Lellouch et al. 1992) are now discarded, temperatures reported from recent mm-observations still range from ~150 K up to 320 K (e.g., Moullet et al. 2010; Roth et al. 2020; de Pater et al. 2020b). Using a high-energy line (580 K lower energy level), Lellouch et al. (2003) inferred $T = 180 \pm 60$ K from disk-average data; it is highly desirable to extend this approach to high angular resolution data.

The range of above temperatures is also generally consistent with UV constraints, that typically favor T = 200–300 K based on the shape of spectrally-resolved UV features (e.g. McGrath et al. 2000). In the infrared, Lellouch et al. (2015) used the structure of the $\nu_1 + \nu_3$ band at 4.0 μm in disk-resolved observations to infer a 170 ± 20 K mean gas temperature (Fig. 8.3). On the other hand, at 19 μm, the shape of the ν_2 band in disk-averaged spectra consistently indicates temperatures below 150 K (Spencer et al. 2005), with best determined temperatures on the anti-Jovian hemisphere (where SO_2 gas abundance is highest) of 108 ± 18 K (Tsang et al. 2012), as illustrated in Figs. 8.4 and 8.5. The contradiction between the two infrared results is puzzling as in a sense both are equivalent to direct rotational temperature measurements. In fact, a gas temperature as low as 108 ± 18 K is not obviously consistent with mm-wave line contrasts of 20–40 K above a ~95 K continuum. We note however that the anti-Jovian hemisphere has not been probed by either mm or 4.0 μm data, and that the association of colder lower atmospheric temperatures with higher columns is, at least qualitatively, predicted by models (Strobel et al. 1994; Walker et al. 2012) in which the lower atmosphere is shielded from plasma heating.

We further expect the temperature profile to vary with latitude, longitude, and time of day. Walker et al. (2012) developed a 3D model for a sublimation

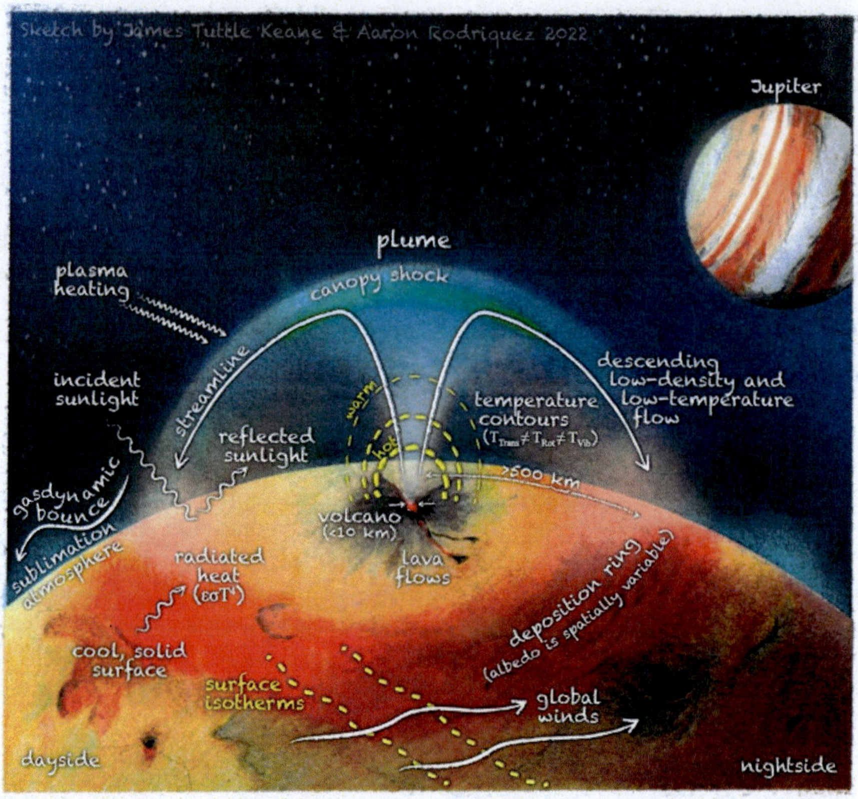

Fig. 8.2 Schematic drawing illustrating some of the complexity involved in interpreting remote observations of volcanic plumes. The drawing shows the plume with the warm high-density canopy at the top, the (white) streamlines, and several (yellow dashed) temperature contours, highest near the magma source. These temperatures include various levels of thermodynamic non-equilibrium (i.e., the translational, rotational and vibrational temperatures are not equal). The volcanic flow will interact with global winds triggered by the temperature distribution at the surface (indicated by the isotherms). The gas and particulate density is expected to be highest in the stem of the plume and the canopy, while, due to the non-equilibrium nature of the system as a whole, the distribution of gas opacity and emission at the various wavelengths is hard to predict. The particulate material falls down to the surface just inside the streamlines and the deposition ring. The temperature of the atmosphere is high where the downward flow "hits" the sublimation atmosphere (the light reddish color). Although spectral resolution may be excellent, the spatial resolution of observations is quite limited. Hence, what one "sees" notably depends on viewing geometry (e.g., view the plumes from any angle from nadir (straight down the plume axis) to large angles if portions of a plume are viewed beyond Io's limb) and whether one can view—or exclude from view—particular features. Illustration by James Tuttle Keane and Aaron Rodriguez

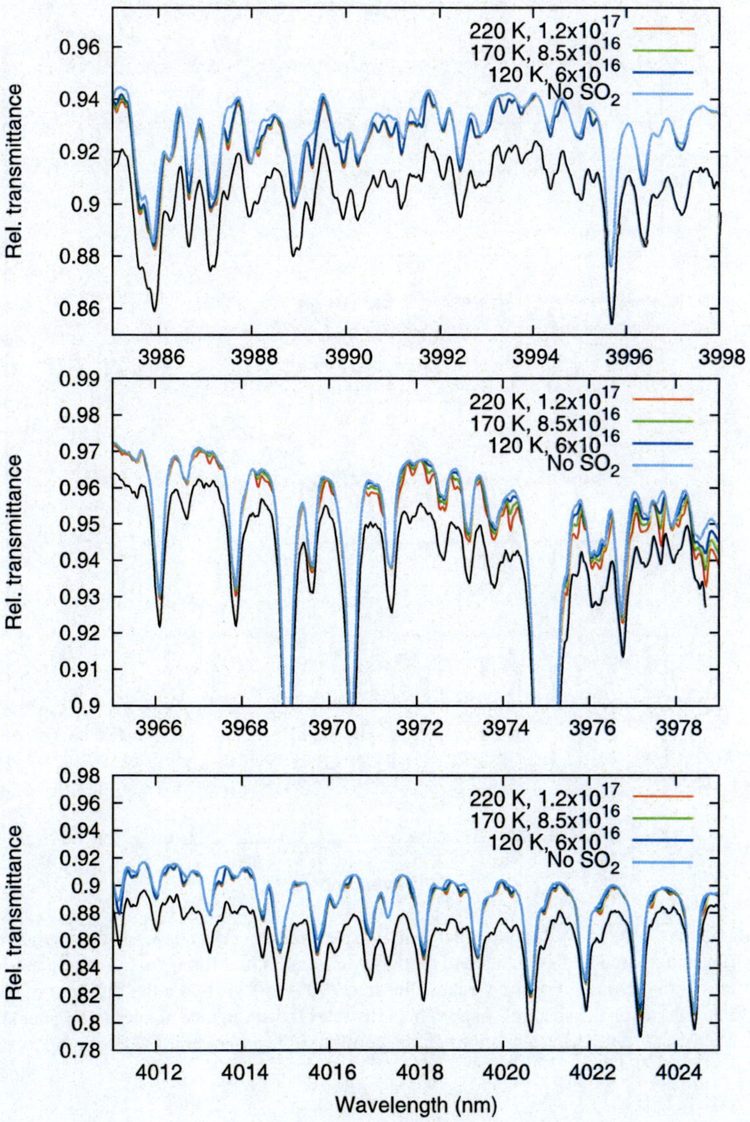

Fig. 8.3 Model fits to Io's reflected intensity at 4 μm. Observations are shown in black. Light blue: model without SO_2 gas. Red, green and dark blue lines show models with gas temperatures of 220, 170 and 120 K, respectively. Top panel: 3985–3998 nm. For each temperature, the SO_2 column density is adjusted to fit the spectrum in this wavelength range. These models are then compared to data at 3966–3978 nm (middle panel) and 4011–4024 nm (bottom panel). The best overall fit is obtained for T = 170 K. All model spectra are offset from the data (from Lellouch et al. 2015)

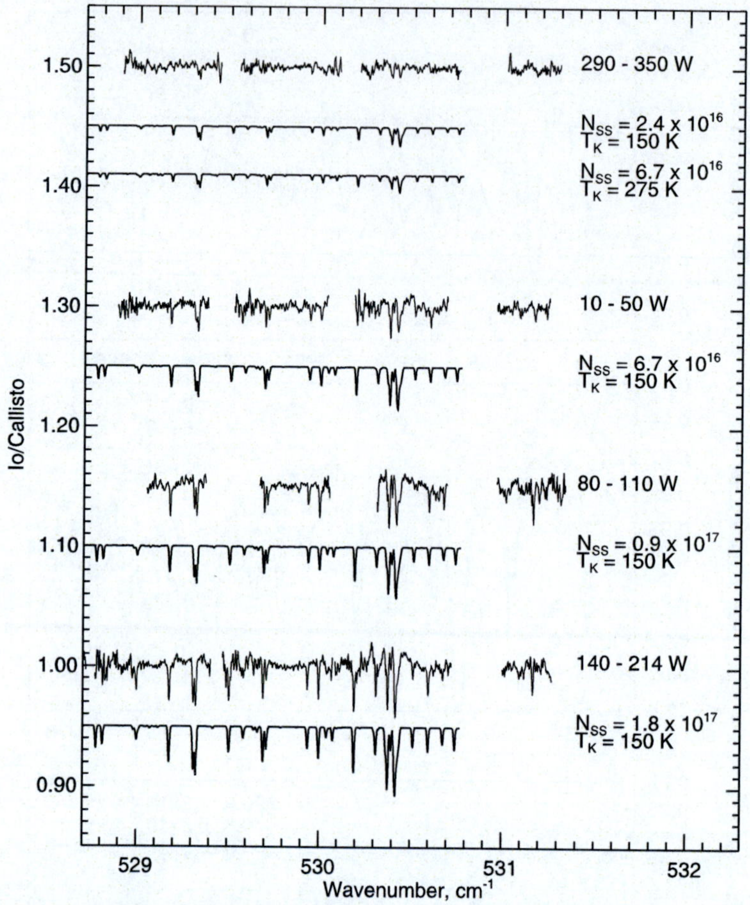

Fig. 8.4 Spectra of the ν_2 vibrational band at 19 μm, combined over several years and averaged by longitude, compared to disk-integrated models with similar line strengths (i.e., these models are not formal fits to the data). For the weakest lines, at 290°–350°W, two quite different models are shown that both match the spectral shape. This illustrates the ambiguity in interpreting some of the spectra. We further note that undulations in the continuum level are not real (from Spencer et al. 2005)

atmosphere to capture these variations and, like Strobel et al. (1994), suggest the presence of a layered thermal structure rather than an isothermal dayside atmosphere. In addition to this sublimation atmosphere with its longitudinal and latitudinal temperature variations, volcanic plumes affect the atmospheric structure (e.g., it is not hydrostatic around plumes) and temperatures dramatically (Sect. 8.4). Hence it is perhaps not surprising that disk-integrated observations at different wavelengths, likely probing different regions in the atmosphere and seeing plumes and the atmosphere as a whole under different viewing geometries (Fig. 8.2),

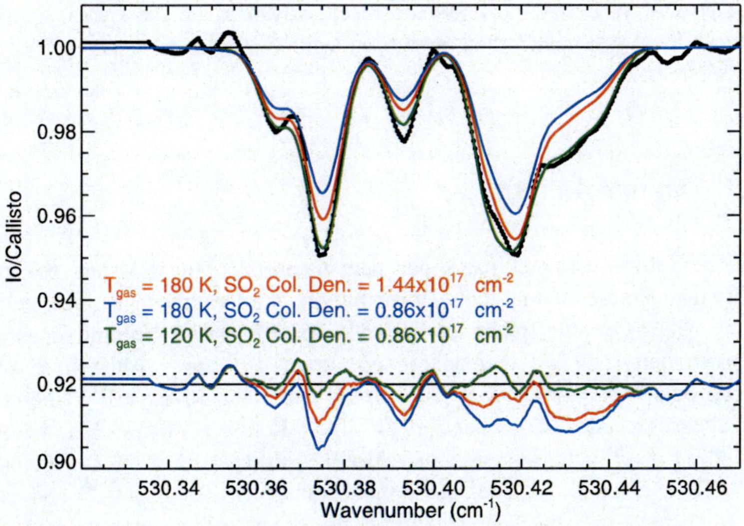

Fig. 8.5 Sensitivity of the 530.43 cm^{-1} wing of the ν_2 19-μm vibrational band to the kinetic temperature of the gas. The observations (black) are compared with synthetic spectra generated at (1) high gas temperatures, low SO_2 abundances (blue), (2) high gas temperatures, high SO_2 abundances (red) and (3) low gas temperatures with low abundances (green). The residuals are plotted at the bottom, offset to 0.92. The dotted lines show the noise level of the data. Although the high gas temperature, high abundance model spectrum fits reasonably well at the shorter wavenumbers, the 530.42 cm^{-1} line is poorly fitted. The lower gas temperature with the low abundance spectrum does match the observed spectrum much better (from Tsang et al. 2012)

give different results from atmospheric models assumed to be isothermal and in hydrostatic equilibrium.

Finally, poorly characterized and assumed isothermal temperature profiles may be the reason for the apparently anomalous $^{34}S/^{32}S$ ratio (2 times higher than terrestrial) derived from $^{32}SO_2$ and $^{34}SO_2$ mm lines (Moullet et al. 2013), since lines of the two isotopic variants have substantially different opacities and therefore probe different atmospheric levels.

8.3 Spatial Distribution and Temporal Variability of Io's Atmosphere

Despite a large observational body of work in the 1990s, it has been difficult to disentangle geographical variations, i.e., "organized" (i.e. diurnal, annual...) from "erratic" (i.e. related to unpredictable volcanic activity) time variations of Io's atmosphere. This dilemma lies at the core of the historic question of the primary source/driver of Io's atmosphere, a question we address in detail in Sect. 8.5.

In the present section we discuss the recent advances on the spatial distribution of the various gases, and their evolution with heliocentric distance and eclipse ingress/egress.

8.3.1 Dayside Atmosphere

Since the 1990s, using in particular new datasets in the UV and thermal IR, a consistent picture of the spatial distribution in Io's atmosphere has started to emerge. The SO_2 atmosphere consistently exhibits larger column densities on the anti-Jovian (180°W) hemisphere—of order $10^{17}\,cm^{-2}$ (though seasonally-variable, Sect. 8.5.1)—versus $\sim 10^{16}\,cm^{-2}$ at the sub-Jovian (0°W) hemisphere (e.g., Spencer et al. 2005; Feaga et al. 2009; Lellouch et al. 2015; Giono and Roth 2021). Millimeter-observations (Moullet et al. 2010, 2013; de Pater et al. 2020b; Roth et al. 2020) indicate typical values of $(0.7–2)\times10^{16}$ and at most $\sim 6 \times 10^{16}\,cm^{-2}$. However, except for the (near)-eclipse measurements of de Pater et al. (2020b) which sample the sub-Jovian hemisphere at 340°W and 20°W, the millimeter observations are typically focused on the leading (90°W) or trailing (270°W) hemisphere, and hence may miss the high anti-Jovian column densities ($\sim$1–$1.5\times10^{17}\,cm^{-2}$) seen at mid-infrared wavelengths. On timescales of a few months, repeated observations of Io's trailing side indicate modest variations ($\lesssim$30%) of the SO_2 columns, seemingly uncorrelated with hot spot activity (Roth et al. 2020).

The most detailed observational picture of the horizontal (latitude/longitude) distribution of SO_2 gas in Io's atmosphere is provided by the *HST*/Ly-α (121.6 nm) images over multiple years (Feaga et al. 2009; Giono and Roth 2021). Such images were first obtained in 1997 (Roessler et al. 1999), and revealed bright polar regions separated by a dark equator. It is now widely accepted that the equatorial region is dark because SO_2 in Io's atmosphere absorbs the surface-reflected solar Ly-α radiation (e.g., Feldman et al. 2000; Strobel and Wolven 2001), and since that time *HST*/Ly-α images have been instrumental in mapping SO_2 gas across Io's surface. The maps reveal that SO_2 gas is mainly confined to latitudes within 30–40° from the equator, with a larger latitudinal extent on the anti-Jovian side, and maximum column densities of $\sim 5 \times 10^{16}\,cm^{-2}$, i.e., intermediate between the mm and 19-μm values, reached at ∼140°W on the equator (Feaga et al. 2009). In a recent paper, Giono and Roth (2021) show that these Ly-α images are only sensitive to SO_2 column densities between $\sim10^{15}$ and $5\times10^{16}\,cm^{-2}$ due to a strong non-linearity in the relationship between SO_2 abundance and Ly-α flux. Based upon a statistical analysis, they show that the maximum SO_2 abundances in the images are closer to $\sim10^{17}\,cm^{-2}$ (Fig. 8.6), i.e. in agreement with mid-infrared results. The drop in column density towards the higher latitudes is interpreted as due to condensation of SO_2 closer to the poles. Attempts have been made to use this observed spatial geographical distribution to constrain the sublimation vs volcanic nature of Io's atmosphere, but with limited success (see Sect. 8.5.1). ALMA maps of the SO_2

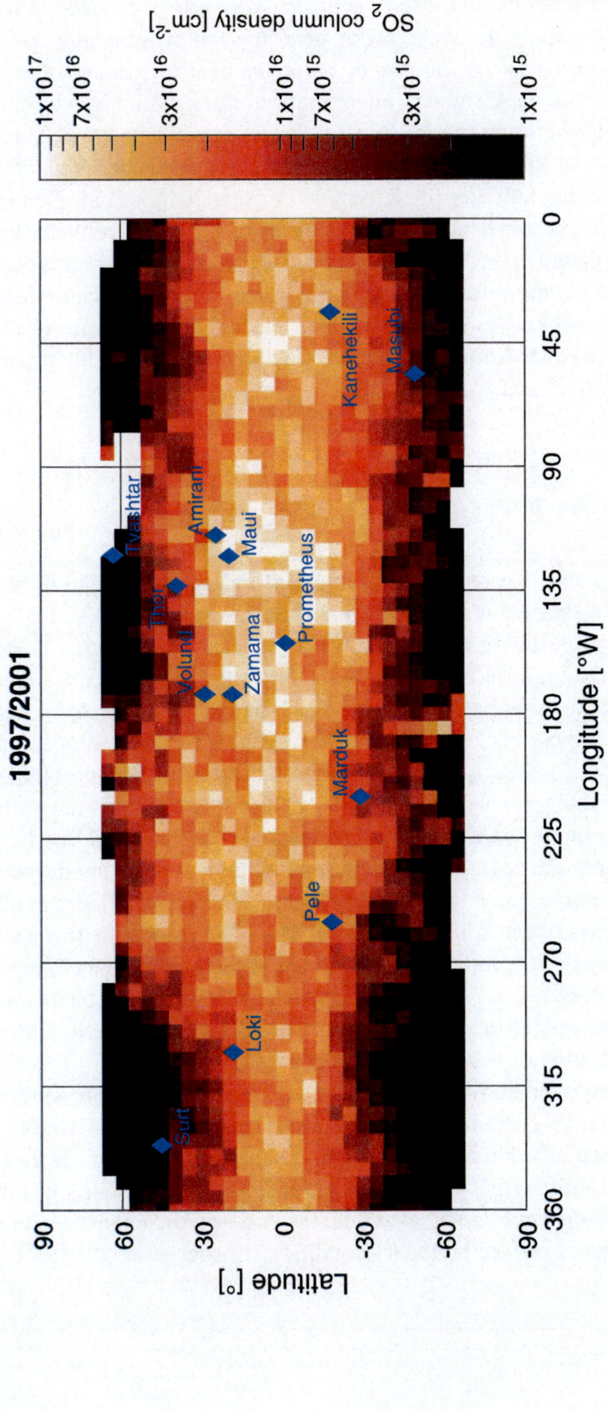

Fig. 8.6 The SO_2 spatial distribution as derived from multiple years of far-UV data obtained with *HST*/STIS. The cylindrical projection shows the obvious difference in abundance between anti- and sub-Jovian hemispheres. Column densities are displayed on a logarithmic color scale (from Giono and Roth 2021)

emission at mm-wavelengths, discussed in Sect. 8.3.2, also show the emissions largely confined to latitudes within 30–40° from the equator (de Pater et al. 2020b).

The spatial distribution of atmospheric gases is also affected by active volcanoes. Most volcanic hot spots as observed at near-infrared wavelengths are at mid-latitudes, with maxima in the number of hot spots near longitudes of ~330° and ~150°, and a relative lack right at equatorial latitudes. Although the number of hot spots is similar between the leading and trailing hemispheres, bright transient eruptions usually take place on the trailing hemisphere (de Kleer et al. 2019a) and at preferentially higher latitudes (de Kleer and de Pater 2016b; Cantrall et al. 2018). At least some of these transient eruptions have been associated with active plumes (e.g., at Tvashtar Patera in the far north, Sect. 8.4.1), but since the plumes themselves cannot be detected at the infrared wavelengths used to map the distribution of hot spots, it is usually not known if a hot spot is connected with a plume. The best way to detect plumes is either from spacecraft or *HST* in the UV (Sect. 8.4.1), or during an eclipse (Sects. 8.3.2 and 8.3.3).

8.3.2 *Eclipse Response*

The evolution of molecular and atomic emissions before, during and right after an eclipse provides a diagnostic tool to investigate the sources and stability of Io's atmosphere. Io's atmosphere is unique in this aspect; no other body is subject to such periodic, huge impulse responses. As soon as the satellite enters an eclipse, the atmospheric and surface temperatures drop, and SO_2 is expected to condense out on a time scale possibly as short as ~70 s (Sect. 8.2.2). This process proceeds from the bottom up: i.e., SO_2 condenses onto the cold surface, destabilizing the atmosphere, so that the gas above it falls down since it is no longer supported from below. Hence, the atmosphere collapses, as illustrated by the solid line in Fig. 8.7. If, however, non-condensible gases are present, as indicated by the dashed line in Fig. 8.7, one may not expect a total collapse of the atmosphere (Moore et al. 2009). In that case, the non-condensible gas rapidly forms a layer along the surface through which the condensing SO_2 must diffuse, perhaps greatly delaying collapse even of the condensible component. During this process the lower atmosphere can warm both due to the recovery of the potential energy of falling gas (compression) as well as the ready penetration of warming plasma nearly to the surface.

The process described above is not symmetric during ingress/egress because the SO_2 subliming off the suddenly warmed surface upon egress can simply lift the non-condensible gas off the ground. In fact, detailed solutions of the Boltzmann equation for this entire ingress/egress scenario indicate the possibility of numerous vertically propagating waves associated with the atmospheric bounce as it responds to an abrupt change in surface boundary condition (Kosuge et al. 2012).

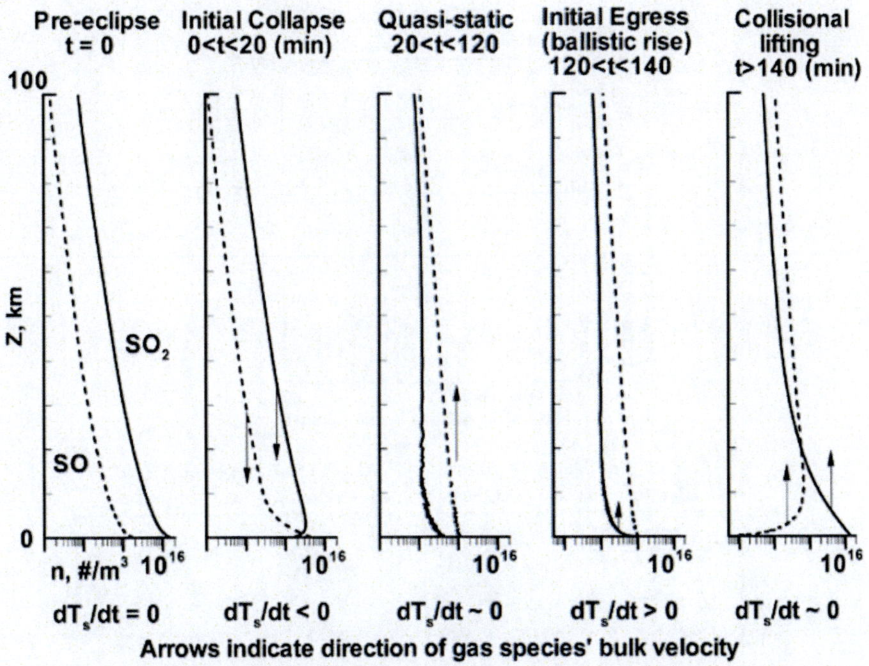

Fig. 8.7 Schematic of the atmospheric dynamics starting before eclipse (left-most panel), through eclipse ingress and until after egress. The altitude is shown on a linear scale on the vertical axis up to 100 km and the number density is shown on a log scale from 10^{13} to $2 \times 10^{16}\,m^{-3}$. The solid lines represent SO_2 and the dashed lines the SO (or any other non-condensible gas) number density. Several distinct periods (labeled at the top) occur during eclipse and early egress and are shown from left to right (From Moore et al. 2009)

8.3.2.1 SO_2 Observations

From an Earth-centered observational point of view, eclipses occur when the sub-Jovian hemisphere is in view, with eclipse ingress (resp. egress) occurring near 340°W (resp. 20°W). The first direct observations of a near-collapse of atmospheric SO_2 when Io entered an eclipse were obtained by Tsang et al. (2016) at a wavelength of 19 μm with the TEXES instrument on the Gemini telescope. They measured a drop in surface temperature from 127 to 105 K within minutes after entering eclipse (Fig. 8.8a). Using different atmospheric cooling models, the analysis of the decrease in band depth from 2.5% down to ~0.2% was interpreted as a decrease in the SO_2 column density by a factor of 5 ± 2 (down from ~2-2.5×$10^{16}\,cm^{-2}$). In contrast, though, Tsang et al. (2015) did not see any change at UV wavelengths when Io went from eclipse into sunlight on two dates in 2011, which was attributed to a larger number of active volcanoes near the egress longitudes compared to ingress (Tsang et al. 2016). However, the UV data were averaged over 10 min time intervals, and on the second day observations did not start until 10 min after egress. As shown

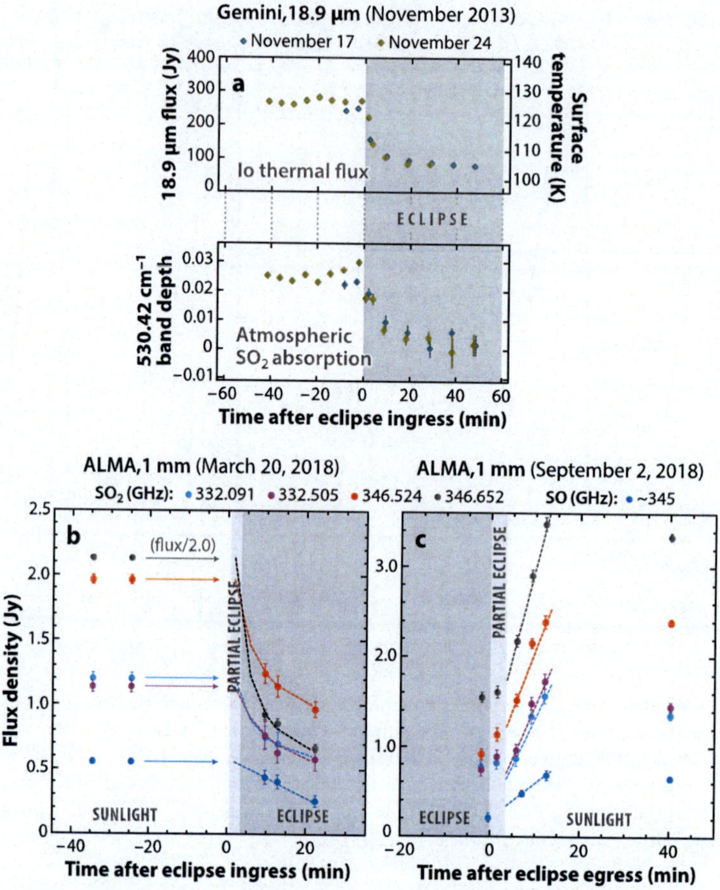

Fig. 8.8 Observations of Io going into and emerging from eclipse. (**a**) Gemini data at 19 μm: (top) The thermal flux of Io and (bottom) the SO_2 absorption depth at 530.42 cm^{-1} in 2013 on Nov. 17 and 24 as a function of time relative to eclipse ingress (Tsang et al. 2016). (**b**, **c**) ALMA data at 1 mm: SO_2 and SO disk-integrated flux densities (from maps integrated over 0.4 km/s, centered on the line) are plotted as a function of time for eclipse ingress (panel **b**) and egress (panel **c**). The dotted lines superposed on the data in panel (**b**) show the exponential decrease in the first few minutes after entering eclipse. In panel (**c**) the dotted lines show the linear increase after emerging from eclipse on September 2. The flux density of the 346.652 GHz data in panel b is divided by a factor of 2. All data are normalized to a geocentric distance of 5.044 AU (de Pater et al. 2020b). Figure reproduced from de Pater et al. (2021)

below, the atmosphere reforms within 10 min after eclipse egress, so we note that the UV observations may have just missed the low in-eclipse abundances right before eclipse egress.

Both eclipse ingress and egress were observed with ALMA in the 1-mm wavelength band (de Pater et al. 2020b). The evolution of the disk-integrated intensities in several transitions of SO_2 together with SO is shown in Fig. 8.8b,c. During eclipse ingress, the SO_2 flux density dropped exponentially, but was re-established in a linear fashion within about 10 min of time after re-emerging in sunlight, with an extra up to ~20% "post-eclipse brightening" after ~10 min. This extra brightening may somehow result from the complex dynamics involved in the interaction of the plumes with the reforming atmosphere (Sects. 8.4.3 and 8.5.2; de Pater et al. 2020b). Disk-integrated in-sunlight flux densities are ~2–3 times higher than in-eclipse, indicative of a roughly 30–50% contribution from volcanic sources, unless the presence of non-condensible gases prevents complete atmospheric collapse as in Fig. 8.7 (Moore et al. 2009), or plasma from the torus, which can now reach parts of the surface, contributes to the atmosphere via surface sputtering.

Maps of Io's emission during eclipse ingress and egress, shown in Fig. 8.9, show an overall collapse of the atmosphere, except for emissions near the known volcanic sites Karei Patera, Daedalus Patera, and North Lerna during ingress, and P207 patera just before egress. The latter is a small visibly dark patera; plumes have never been reported at this site. As soon as sunlight hits the satellite during egress, SO_2 emissions become stronger in particular in the regions where volcanic plumes were present during eclipse, and after ~10 min the SO_2 atmosphere has completely reformed (de Pater et al. 2020b).

Surprisingly, the SO_2 column density ($1.5 \times 10^{16}\,cm^{-2}$) and temperature (~270 K), derived from disk-integrated flux densities under the assumption of an atmosphere in hydrostatic equilibrium, appear to be essentially the same both for the Io-in-sunlight and in-eclipse data; the difference can be explained entirely by a factor of 2–3 decrease in fractional coverage over the disk when in-eclipse (de Pater et al. 2020b). These findings may agree with the factor-of-5 drop in column density at mid-infrared data as reported by Tsang et al. (2016), since they cannot distinguish between a high column density with low fractional coverage and a low column density with a high fractional coverage. Similar results were typically obtained for individual plumes, where the fractional coverage within a beam centered on the plume decreased by a factor of 2–3 when going into eclipse, while the column density and temperature stayed more or less the same (de Pater et al. 2020b). The authors stressed, however, that the models used to fit the data were hydrostatic models, and during an eclipse and in plumes the applicability of such models is very limited.

The ALMA maps can be compared with the DSMC simulations for a purely sublimation-sourced atmosphere before/during/after eclipse (Walker et al. 2012), shown in Fig. 8.10. These simulations are based on a parametric study of Io's thermophysical surface properties, using three thermal units: (1) frosts/ices with surface areas as in Douté et al. (2001), with a best-fit albedo A = 0.55 and thermal inertia $\Gamma = 200\,J\,m^{-2}\,K^{-1}\,s^{-1/2}$ (hereafter referred to as MKS units), (2) non-frosts with A = 0.49, $\Gamma = 20$ MKS, and (3) hot spots. The thermophysical properties were derived by fitting the model to observations at mid- to near-UV wavelengths, and assuming that the column density must be in vapor pressure equilibrium with the surface temperature. The modeled images are centered at 10°N, 350°W, and show the predicted changes over time from ~2 h before local noon to ~4.5 (Earth) h later, with an eclipse in between. The subsolar point is indicated by the white

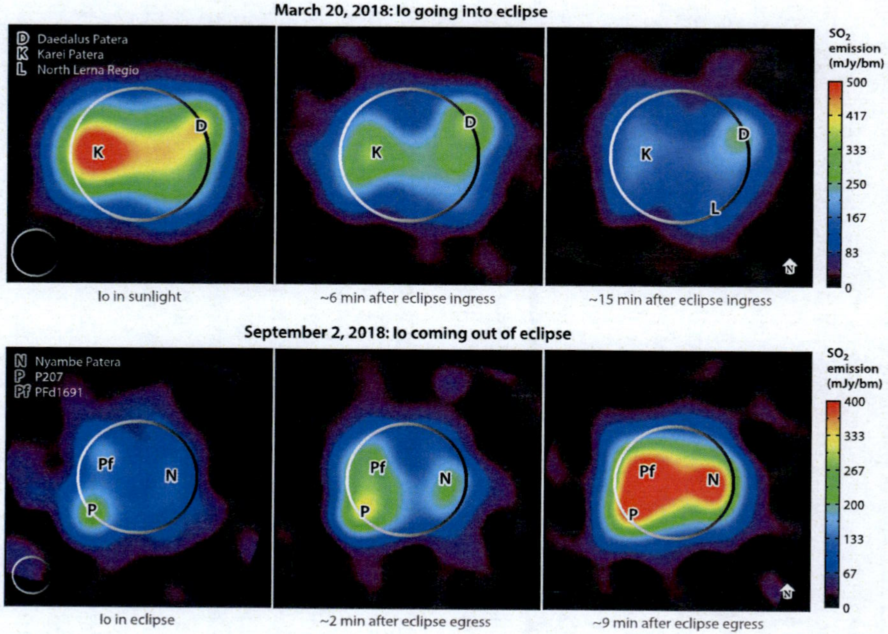

Fig. 8.9 Individual frames of a series of ALMA SO_2 maps constructed from data at 346.652 GHz when Io went into eclipse (20 March 2018) and emerged from eclipse (2 September 2018). The data are averaged over 0.4 km/s (∼0.45 MHz). The large circle shows the limb of Io, and the small circle in the lower left shows the resolution of the data (0.35″ or 1205 km in March; 0.30″ or 1235 km in September). The effect of volcanoes on Io's SO_2 emission is clearly seen (de Pater et al. 2020b). Figure reproduced from de Pater et al. (2021)

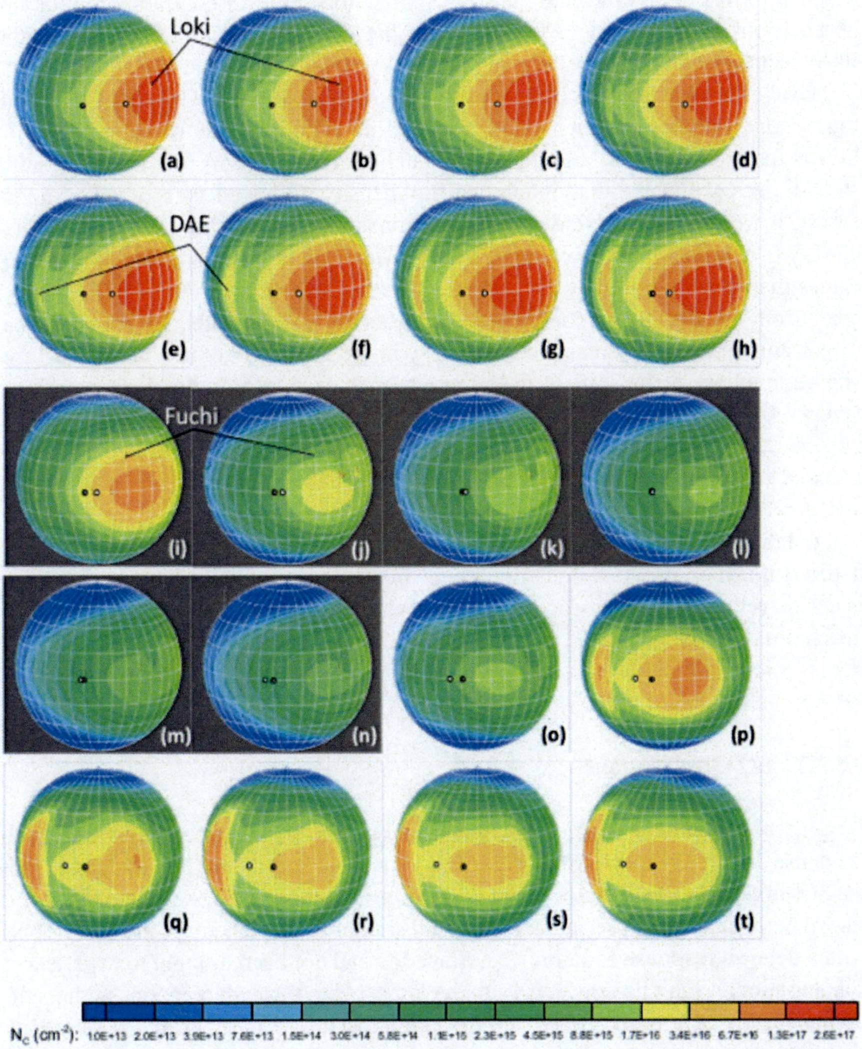

Fig. 8.10 Column density contours with a view centered at 10°N, 350°W as a function of time. The 20 snapshots at intervals of 1250 s starting approximately 2 h and 40 min prior to eclipse and ending ~ 2 h after egress from eclipse. Hence the first in-eclipse panel is 6.7 min after eclipse ingress, and the first panel upon egress is taken after Io was 11 min in sunlight. So they correspond to the top middle and lower right panels in Fig. 8.9. Two hot spots (Loki Patera and Fuchi Patera) are highlighted in (a), (b), (i) and (j). The white dot denotes the location of the subsolar point while the black dot denotes the sub-Jovian point. DAE refers to the dawn atmospheric enhancement (From Walker et al. 2012)

dot, and moves across the satellite from sunrise to sunset. In contrast, Earth-based observations show Io rotate over time, while the subearth and subsolar point remain fixed near the center of Io.

Despite these differences in geometries, the ALMA and DSMC results both trend toward an equatorial confinement of SO_2 outside of eclipse due to condensation of SO_2 at the higher colder latitudes. During eclipse, both ALMA and DSMC results show a decrease in the disk-integrated flux density, followed by a recovery upon egress, if we use the modeled column density as a proxy for flux density. The flux density in the ALMA data changes much faster, however, both upon ingress and egress than the models show. Moreover, the structure shown in the ALMA maps, apart from the equatorial confinement, appears to be dominated by the presence of volcanic plumes, whereas the bimodality in the modeled column density before and after eclipse arises largely from the presence of an atmospheric enhancement at dawn (DAE) coupled to the adopted distribution of frost coverage on the surface, i.e., the DAE is located over a low thermal inertia region ($\Gamma = 20$); the SO_2 gas, once released from the surface shortly after sunrise, will move towards lower pressures and condense if/when meeting lower surface temperatures, such as at night, at higher latitudes, and above SO_2 still-cool frost along the equator. Hence this feature depends much on the spatial distribution of frost and bare rock in the models; there is, as of yet, no clear evidence of DAE in the data. Clearly, there are a number differences between data and models that need to be reconciled in future work (see Sect. 8.5 for more discussion).

8.3.2.2 SO Observations

Despite the fact that SO, as a non-condensible gas, is not expected to significantly condense during an eclipse (Sect. 8.2.2), Fig. 8.8 shows a gradual (linear) decrease in the SO flux density by a factor of ~2 upon eclipse ingress. ALMA maps of SO (not shown here) during eclipse-ingress look very similar to the SO_2 maps in Fig. 8.9, but with a delayed response as in Fig. 8.8. Since SO had not been thought to experience condensation, it may be removed from the atmosphere through reactions with itself on the surface at a much faster rate than anticipated (de Pater et al. 2020b). The chemical reaction rate may be increased due to an increase in the SO partial pressure at the surface, because SO is forced into a thin layer by the collapsing SO_2 column of gas, which increases the collision rate of SO molecules in the atmosphere and with the surface. Some SO may also get trapped in porous surface layers through this process. Upon eclipse egress, SO is restored about three times more slowly than SO_2, as expected if SO is formed primarily through photolysis (de Pater et al. 2020b), perhaps augmented by a slow release from the surface.

SO has also been observed and mapped while in eclipse at near-infrared wavelengths. Such SO emissions were first detected in 1999 at 1.707 μm (de Pater et al. 2002). They were attributed to the SO forbidden electronic $a^1\Delta \rightarrow X^3\Sigma^-$ transition, and these first disk-integrated measurements were indicative of a rotational temperature of ~1000 K. The authors hypothesized the emissions to

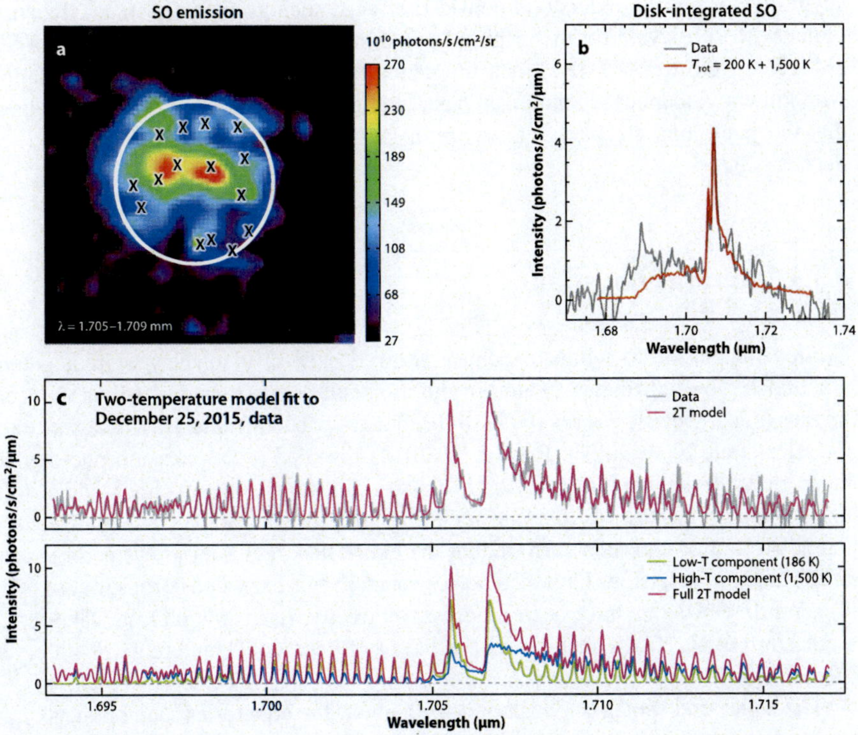

Fig. 8.11 (**a**) Keck image of the forbidden 1.707 μm emission band of SO obtained with the field-integral spectrometer OSIRIS on the Keck 2 telescope on 25 Dec. 2015. The image is obtained by integrating over the center channels of the emission band (see panel (**b**)). Superposed are the location of a number of volcanic centers (note the absence of a clear one-on-one correlation), and the limb of Io's disk. (**b**) Disk-integrated OSIRIS spectrum of the SO data in panel a, with a model consisting of two temperatures (200 and 1500 K, in approximately equal proportions) superposed. Note that the 1.69 μm feature cannot be matched. (**c**) Disk-integrated spectrum at a high spectral resolving power (R ~ 25,000) taken simultaneously with the data in panels (**a**) and (**b**). A very similar 2-temperature model is superposed. The individual components of the model are shown in the bottom panel (panels a,b from de Pater et al. 2020a) (panel c from de Kleer et al. 2019b). Figure reproduced from de Pater et al. (2021)

originate at Loki Patera, which was exceptionally bright in the near-infrared at the time. They discussed many potential explanations, including the electron impact mechanism which causes the auroral glows on Io (Sect. 8.3.3), and concluded that the SO emission must result from excited SO molecules directly ejected from the vent at a thermodynamic quenching temperature of ~1500 K.

More recent observations at a higher spectral resolution (Fig. 8.11c) indicate the presence of gas at both a low (~200 K) and high (~1500 K) temperature (de Kleer et al. 2019b). This combination is required to fit the detailed band shape over 1.695–1.715 μm (Fig. 8.11c), but the interpretation of these two temperatures is uncertain. Furthermore a secondary emission at 1.69 μm remains unexplained (Fig. 8.11b), suggestive of poorly understood non-LTE effects, such as expected in gas dynamic plumes. Most relevant for the origin of this emission, the spatial distribution of SO as derived from Keck/OSIRIS measurements (Fig. 8.11a) shows that the correlation with known volcanoes is tenuous at best, leading de Pater et al. (2020a) to suggest that the emissions are likely caused by a large number of "stealth" plumes (See Sect. 8.4.2).

8.3.3 Auroral Emissions

Galileo images of Io while in eclipse showed a colorful display of red, green, and bluish glows attributed to atomic and molecular emissions excited via electron impact (e.g., Geissler et al. 1999). Subsequent spacecraft and *HST* images (e.g., Geissler et al. 2001, 2004a; Roth et al. 2014) revealed a complex morphology of these glows, as shown in Fig. 8.12: (1) equatorial "spots", one on either side of Io's disk, usually referred to as Io's "aurora"; (2) bluish glows from volcanic plumes; (3) a reddish ring of emission surrounding the entire disk; (4) faint glows across parts of Io's disk; *v*) emissions have also been seen from Io's extended corona, out to ~10 R_{Io}. While emissions have been reported, indirectly, from Io's plasma wake (e.g., Retherford et al. 2007), such emissions have not been confirmed (e.g., Roth et al. 2014).

The equatorial spots rock back-and-forth about the equator as seen on the sky in response to the changing orientation of Jupiter's magnetic field. The spots track the tangent points of the Jovian magnetic field lines with Io, and are produced by electrons impacting the various atmospheric gases. Most of these emissions originate within 100 km from the surface, and the variations can be explained by a combination of the local plasma environment and the changing viewing geometry of Io in Jupiter's magnetosphere (e.g., Roth et al. 2014).

Spectra of the emissions, obtained primarily from *HST*/STIS observations, yield information on the composition and abundance of these glowing gases, and the intensity of the electrical currents that excite the emissions (e.g., Geissler et al. 2004a; Trafton et al. 2012; Roth et al. 2014). The bluish glows from aurora and volcanic plumes are dominated by emissions from molecular SO_2. Some of the atomic species, e.g., O, Na, and K, produce line emissions at longer visible and near-infrared wavelengths, resulting in more reddish glows. These glows, which are brighter on the side of Io closest to the center of the plasma torus, surround the entire disk (the limb glows), and hence indicate that these species (O, Na, K) are spread across Io's surface and are not only confined to the equatorial regions, in contrast to the near-equatorial distribution of SO_2 gas, which condenses at the colder higher latitudes, as discussed above.

Since the auroral emissions depend on the column density of the emitting species as well as the impinging electron flux and temperature, the latter of which is controlled by the penetration depth into the atmosphere of the impacting electrons (the electrons, originating in the plasma torus, cool after entering the atmosphere),

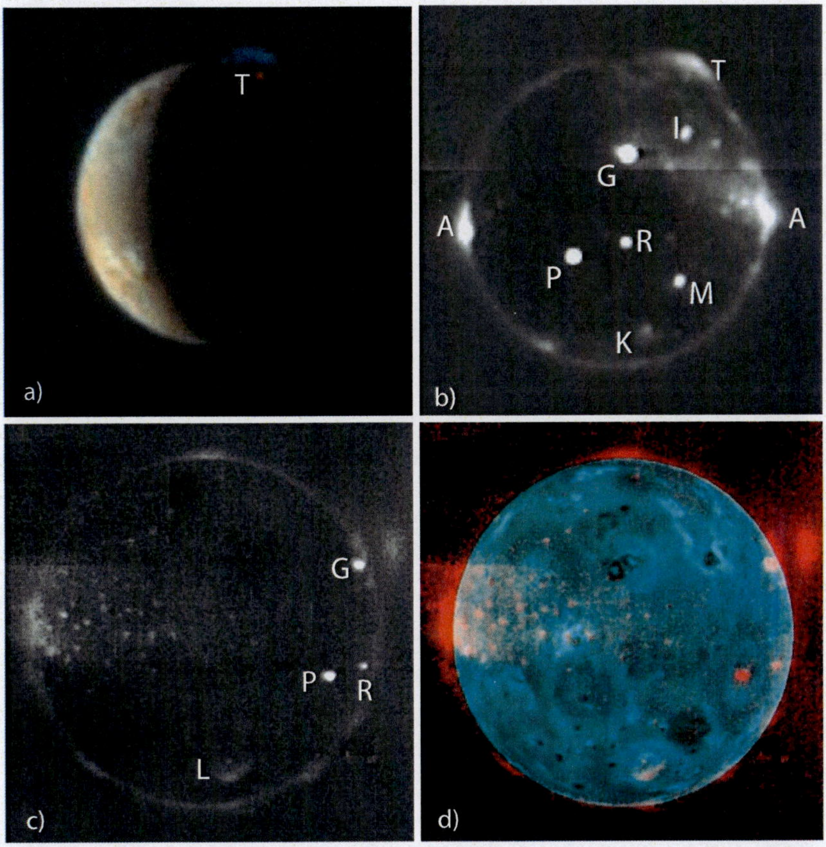

Fig. 8.12 (**a**) Nighttime glow of the north-polar Tvashtar volcano (T) and its plume rising 330 km above Io's surface. This image was taken with the blue and methane filters of the Multispectral Visible Imaging Camera (MVIC) of the imaging instrument Ralph on the *New Horizons* spacecraft on March 1, 2007. The image shows an intense red color (methane-band image) of the glowing lava at the plume source, and the contrasting blue (blue-filter) of the fine dust particles in the plume. The lower part of the plume is in Io's shadow, and hardly visible in this image. (**b, c**) *New Horizons* images of Io-in-eclipse. The brightest spots on the disk are "hot spots", thermal emissions from hot lava at active volcanoes. The brightest spots are indicated: P: Pele, R: Reiden Patera, M: Marduk Fluctus, G: East Girru Patera, I: Isum Patera. A plume is seen over a hot spot at N. Lerna (L) in panel (**c**), and over Kurdalagon Patera (K) in panel (**b**). The plume above Tvashtar (T) rises out above the limb in panel (**b**) (Tvashtar itself is not visible; it is just over the limb). Diffuse glows and faint spots are from gas in the plumes and atmosphere. On either side of the satellite, along the equator, are auroral spots (A), where the eastern spot might be enhanced by the Prometheus plume, and the western one by Ra Patera, which are both right on the limb. The edge of Io's disk is outlined by a faint glow. (**d**) The eclipse image from panel (**c**) (in red) overlain on a sunlit image (cyan). The numerous point-like sources near the equator in both (**b**), (**c**) might be manifestations of stealth volcanism (PIA09254, PIA09354, PIA10100) (NASA/JHU/APL/SwRI)

the change in emissions during an eclipse provide information on the sources and losses of the emitting gases, as well as changes in the atmospheric density. Disk-averaged observations of Io have shown a factor-of-3 decrease in the far-UV atomic S and O emissions ~20 min. after eclipse ingress (Clarke et al. 1994), and a factor-of-2 increase after egress (Wolven et al. 2001). Sodium emissions decreased by a factor-of-4 during eclipse ingress, and recovered after egress (Grava et al. 2014). Most of the changes in auroral glows happened in the equatorial spots, while the limb glow and extended corona did not seem to change much (Retherford 2002). Hence one might attribute a decrease in these aurora to a (temporary) "break" in the production rate. Indeed, the decrease in S, O, and Na glows have been attributed to a lack of photodissociation (from SO_2 and NaCl) when the satellite is in Jupiter's shadow.

Figure 8.13 shows two spectra, one taken during the first 14 min after eclipse ingress, and a second one averaged over the subsequent (almost) equal time period. The decrease in all emissions during the eclipse is clearly visible, indicative of ongoing atmospheric collapse due to freeze-out. In addition to these identified species (SI multiplets, SO, SO_2), there are unidentified emissions between 0.33 and 0.57 μm, seemingly caused by a tri-atomic molecule like SO_2, S_2O, or perhaps

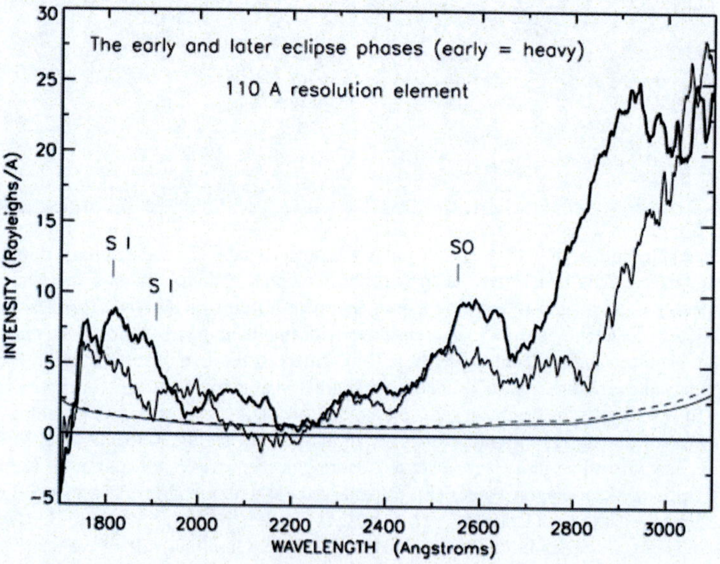

Fig. 8.13 Change in spectra during an eclipse, as observed with the MAMA UV (0.175–0.320 μm) detector of *HST*/STIS on August 18, 1999. The heavy line shows the spectrum as averaged over the first 14 min upon eclipse ingress; the thinner line shows a spectrum averaged over a time from 17 to 29 min after eclipse ingress. The curves near the bottom of the plot represent the 1-σ errors for the early (solid line) and later (dashed line) observations. The SI lines and SO are indicated; the broad SO_2 band rises to the right above 2200 Å across the plot (with SO superposed). The sharper rise on the right likely includes Io's attenuated continuum, which becomes weaker deeper into eclipse (and with declining UV wavelength) (Adapted from Trafton et al. 2012)

caused by positive or negative ions of SO_2 and its daughter species (Trafton et al. 2012). For these spectra, Trafton et al. (2012) showed that dissociative excitation of SO_2 by electrons in the plasma torus is a significant source of emission by its daughter products S and SO.

Saur and Strobel (2004) modeled the response of auroral emissions upon entering and exiting eclipse, assuming the emissions are caused by electrons from the (upstream, i.e., trailing hemisphere) plasma torus impacting the atmospheric gases. They assumed a column density of 1.5×10^{16} cm^{-2} before eclipse, and calculated the response in auroral emissions throughout atmospheric collapse. They showed that the auroral glows can only decrease in intensity, as observed for the equatorial spots, if the atmosphere collapses down to column densities <3–5×10^{14} cm^{-2}. At such low densities, the impacting electrons have kept their high plasma temperature (~5 eV), and emissions vary linearly with atmospheric column density. At atmospheric densities over ~5×10^{14} cm^{-2}, the auroral emissions will brighten upon eclipse ingress. A delay of the plasma interaction upon eclipse egress, when sublimation of surface frost increases the atmospheric density, may therefore result in a post-eclipse brightening in the UV. We note that whether the emissions dim or brighten is a very non-intuitive process, since the electron temperature affects the emissions in an extremely non-linear fashion, so that small changes in temperature can have large effects in the emissions. Also, the intensities of the emissions depend on the fraction of upstream Io torus flux tubes that intercept and feed energy into the atmosphere. This fraction is controlled by the strength of Io's electrodynamic interaction that depends on the ratio of the Alfvén conductance to the ionospheric conductances, adding further non-linearity to the auroral emissions' response. When modeling the aurora as observed with *New Horizons* when Io was in eclipse, Roth et al. (2011) derived an order of magnitude decrease in the atmospheric density compared to in-sunlight, in agreement with the above theory; their derived densities, however, were about two times higher than the ~ 5×10^{14} cm^{-2} maximum value mentioned above for the equatorial spots, perhaps indicative of the complexity of the interaction.

In contrast to the aurora, plumes have been seen to brighten in eclipse (Geissler et al. 1999), which is caused by the same process discussed above: the background atmosphere is collapsing, but the plume column density is high. So any change may brighten the plume emissions, but certainly not dim it (Saur and Strobel 2004).

Several authors (e.g., Sauer et al. 2002; Roth et al. 2011; Dols et al. 2012; Blöcker et al. 2018) have modeled the magnetic field and plasma perturbations near Io to derive diagnostics on Io's atmosphere. In particular, they find a longitudinal asymmetry very similar to that derived from the UV and mid-IR data (Sect. 8.3.1). These simulations further suggest that the atmosphere's radial extension is limited upstream (scaleheight ~60 km) and at least several times larger on the anti-Jovian downstream side, where simulations support a very extended corona ($\gtrsim 6\ R_{Io}$) of SO_2 and SO.

8.3.4 Atmospheric Escape

Although the source of Io's atmosphere can ultimately be attributed to volcanism, it must be continuously replenished since Io loses ~1 ton/s (~3×10^{28} atoms/s) of material to its neutral clouds and the magnetosphere, primarily through sputtering by ions in the plasma torus (e.g., Spencer and Schneider 1996). Most sputtered products, however, will have velocities much less than Io's escape speed of 2.6 km/s, and populate Io's corona or exosphere, out to the boundary of the satellite's Hill sphere (~6 R_{Io}). Those that do have higher velocities form Io's neutral clouds. Other important processes that lead to a loss from Io's atmosphere (and its corona and neutral clouds) are electron impact ionization of an atmospheric atom by an electron from the plasma torus (electron impact on a molecule often leads to dissociation), and charge exchange between an atmospheric atom or molecule with an ion in the torus; upon ionization the new ions are accelerated and supply the plasma torus with fresh material, while the newly formed neutral will keep its high velocity and populate extended neutral clouds (e.g., Mendillo et al. 1990; Schneider and Bagenal 2007; See also Chap. 9 in this book). Given the inferred supply rates to the torus for O and S, the atmospheric lifetime is of order 10 days for a 1 nbar atmosphere covering 25% of the surface (Lellouch 1996).

Mendillo et al. (2004) had reported a positive correlation between Io's infrared brightness and the brightness of the extended sodium cloud, but an increase in Io's infrared brightness does not necessarily imply plume activity. Moreover, direct ejection of material from volcanoes should not be important, since the ejection speeds (at most ~1 km/s) are well below Io's escape speed. McDoniel et al. (2019) show that the interaction of plasma from Io's plasma torus with volcanic plumes depends much on the location of the plume due to the direction of the impinging plasma. They show that, although plasma does inflate plume canopies, the rising plume itself is not much affected and the canopy height barely changes. A large, diffuse neutral cloud may form above the canopy, and some SO_2 and its dissociated daughter products may escape the plume and add material to Io's corona and exosphere. Upon ionization, these may escape Io's direct environment, and hence form a potential source of material for the plasma torus.

The Japan Aerospace Exploration Agency (JAXA) *Hisaki* satellite has been studying UV emissions from ions and neutrals in the Jovian system from Earth's orbit since 2013 (Yoshikawa et al. 2014). In January–March 2015, using a combination of groundbased telescopes and *Hisaki*, a brightening of Io's extended sodium cloud and plasma torus was observed (Tsuchiya at al. 2015; Yoneda et al. 2015), while Io's extended neutral oxygen cloud spread outward from Jupiter, with a more than doubling of its number density (Koga et al. 2019). During this time a sudden brightening at near-infrared wavelengths was observed at Kurdalagon Patera. Although plumes could not be detected directly in these observations, plumes have been detected here before (e.g., by *New Horizons*, Spencer et al. 2007). de Kleer and de Pater (2016a) therefore suggested that the changes observed in the Jovian system may have been caused by an influx of neutral material from a plume at Kurdalagon Patera, perhaps through a process related to that modeled by McDoniel et al. (2019). In addition, the process by which dust streams in Jupiter's magnetosphere, which are primarily composed of salt (NaCl, Postberg et al. 2006), are expelled from Io's volcanoes is also unknown (e.g., Krüger et al. 2004).

8.4 Plumes: Characteristics, Deposits, and Models

An excellent review of plumes and their deposits is provided by Geissler and Goldstein (2007). Since then more research has been conducted. For example, Geissler and McMillan (2008) summarized *Galileo* observations of Io's plumes, Jessup and Spencer (2012) analyzed *HST*/WFPC2 data of the plumes above Pele, Tvashtar, and Pillan as observed between 1995 and 2007, de Pater et al. (2020b) observed plumes with ALMA during an eclipse, and there have been several developments in the modeling of volcanic plumes. In the next subsection we discuss observations of plumes and their deposits, followed by sections on the thermodynamic properties and on hydrodynamic models of plumes.

8.4.1 Observations of Plumes and Their Deposits

Plumes are easiest to see in sunlight through light scattered off dust particles and condensates in the plume; the plumes typically have a bluish color (e.g., Fig. 8.12a) indicative of light scattered off small ($\lesssim$sub-μm-sized) particles. The plume material coats the surface, resulting in a colorful display, including bright red rings surrounding the vent for Pele-type plumes (Fig. 8.14). The variety of colors is attributed to SO_2-frost, a variety of sulfur allotropes (S_2–S_{20}), and metastable polymorphs of elemental sulfur mixed in other species (Moses and Nash 1991; Carlson et al. 2007). The colors and coverage change on time scales of months–years due to burial by new eruptions, thermal metamorphism (such as annealing of fine-grained frost into coarse-grained ice at the equator), and slow chemical alterations on the surface, such as the change from red short-chain sulfur allotropes to the more stable yellow S_8 which cause a fading of the red rings around plumes once the volcano is no longer active (e.g., Geissler et al. 2004b).

Historically, plumes have been divided into two classes: "Pele-type" plumes reach altitudes over ~400 km and are surrounded by red rings of deposits; in contrast, "Prometheus-type" plumes do not extend much higher in altitude than ~100 km. A plume's radial extent is typically two times larger than its altitude. In Sect. 8.4.2 we expand more on plume classes.

HST observations of large plumes (Pele, Tvashtar, Pillan) show a higher reflectivity (I/F) at 0.33 μm than at 0.26 and 0.41 μm. Based upon Mie calculations, Jessup and Spencer (2012) suggest particle radii of order 0.05–0.1 μm for the particulates in these plumes. Geissler and McMillan (2008) suggested somewhat larger particle radii (~0.1 μm) for dust in Prometheus-type plumes, as derived from the linear decrease in I/F between 0.4 and 0.76 μm seen in *Galileo* data. These particles are referred to as "coarse-grained ash" and make up the central columns of Prometheus-type plumes; this ash is entrained in the gas flow when it leaves the surface. The dust mass is typically of order 10^6 to a few $\times 10^7$ kg, or ~1–10% of the gas (SO_2) mass, with the low end for Pele-type, and high end for Prometheus-type plumes (Geissler and McMillan 2008; Jessup and Spencer 2012). This implies a dust production rate of order 10^3–10^4 kg/s assuming a dynamical (in-flight) lifetime of $\sim 10^3$ s. Some (and perhaps all) Prometheus-type plumes have a halo of much smaller-sized (radii $\lesssim$10 nm) particles, with a mass similar or larger than the mass in the gas ($\gtrsim 10^8$ kg); these may be sulfurous snowflakes or droplets condensed from the gas during flight, while the gas is cooling through adiabatic expansion and radiation (Geissler and McMillan 2008).

When observed during an eclipse or at night a plume glows due to bluish gas emissions, likely dominated by SO_2 emissions as seen in the aurora discussed in Sect. 8.3.3. Gas in Prometheus-type plumes reaches altitudes up to ~200–400 km above the surface, i.e., 2–4 times higher than the dust in these plumes, although the halo of tiny snowflakes or droplets covers a similar extent in altitude and radius as the gas (Geissler and McMillan 2008). In Pele-type plumes the dust and gas reach similar (~400 km) altitudes, indicative of the somewhat smaller sized dust grains mentioned above. The smaller-sized particles and $\gtrsim$10 times less dust mass explains why these Pele-type plumes are more difficult to detect.

Eruptions may last for decades, such as for Pele and Prometheus, which were active during both the *Voyager* and *Galileo* era's. Tvashtar has been erupting intermittently on decade-timescales, being active for months once erupting; a plume and red ring were seen during the *Galileo/Cassini* era (Fig. 8.14). A "re-awakening" was observed in April 2006 with the Keck telescope through a brightening at 1.5–2.4 μm, indicative of a hot spot with a temperature at ~1240 K (Laver et al. 2007); about 10 months later (February/March 2007) a plume and red ring were detected by the *New Horizons* spacecraft (Spencer et al. 2007).

Although large outburst-style eruptions on timescales of hours–days have been reported from data at near-infrared wavelengths, which are sensitive to the temperature of the lava (e.g., Chap. 6), not much is known about potentially short-lived plumes. The presence of plume activity missed by spacecraft has occasionally been inferred through observations of new deposits (see, e.g., the review by Geissler and Goldstein 2007), but this does not provide information on the duration of such plumes. However, although plumes may be active over periods of months, *New Horizons* provided a 5-frame "movie" of Tvashtar's plume showing unsteady dynamics in the particulate canopy with large fluctuations on time scales of minutes suggesting dynamics of the source processes on similar time scales.

The gaseous content of the plumes has been measured from imaging and/or spectroscopy on a few occasions, but the quantitative interpretation of imaging data is complicated by the competing effects of gas and dust, or of different gases, in producing the opacity. Observations include the direct detection of SO_2, S_2, S, and SO over Pele's plume (McGrath et al. 2000; Spencer et al. 2000; Jessup et al. 2007),

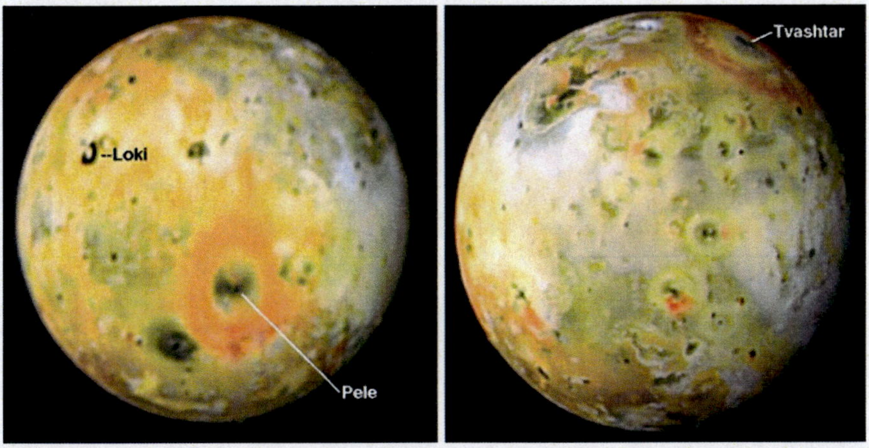

Fig. 8.14 *Galileo* images showing the abundance of colors on Io. The volcanic sites of Loki Patera, Tvashtar Patera and Pele are indicated. The latter two are surrounded by rings of red material, deposits from gigantic plumes that were imaged simultaneously by the *Galileo* and *Cassini* spacecraft. The images were taken in late December 2000 and early January 2001 (PIA02588; NASA/JPL/University of Arizona)

SO_2 at Loki and Pillan (Pearl et al. 1979; Jessup et al. 2007), and more indirect (imaging) evidence of SO_2 and S_2 in Tvashtar's plume (Jessup and Spencer 2012). Although McGrath et al. (2000) reported a SO_2 column density (3.25×10^{16} cm^{-2}) over a region encompassing Pele to be several times larger than in two other regions, this is not direct evidence for volcanically-emitted gas, but may simply reflect variations with longitude and latitude in the overall distribution of SO_2 gas. Roth et al. (2011) modeled the (auroral) emission from Tvashtar's plume while Io was in eclipse to derive a column density in the plume of $\sim 5 \times 10^{15}$ cm^{-2}.

Gaseous plumes can also be discerned in the ALMA spectral maps discussed previously, both in sunlight and in eclipse, taking advantage of the spectral resolution. Figure 8.15 (de Pater et al. 2020b) shows a series of ALMA images at different velocities for the in-sunlight SO_2 data. This series of images reveals that volcanic plumes (in this case above the P207 patera) dominate the emission at large velocities from the line center, ~ -0.8 km/s in frame 1 and $\sim +0.4$ km/s in frame 3, implying that volcanic plumes shape the high-velocity wings of the disk-integrated line profile (fourth frame). Since the high-density "core" or "stem" of the plume only covers a small area compared to the beamsize of the telescope, we most likely see the front-side of the large umbrella-shaped plume in frame 1, and the far-side of the canopy moving away from us in frame 3. These high speeds match the expected gas velocities associated with large plumes when simulating its shape using ballistic trajectories. Occasionally, an entire disk-integrated line profile had been observed to be red-shifted by several tens m/s (Lellouch 1996; de Pater et al. 2020b), attributed to the downward flow of an umbrella-shaped canopy of a plume on the disk.

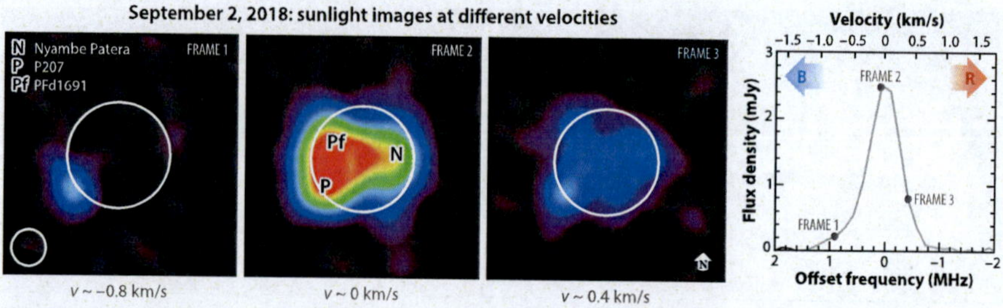

Fig. 8.15 Individual frames at different offset frequencies (velocities) obtained with ALMA on 2 September 2018, when Io was mapped while in-sunlight. Data at two transitions (346.652 and 346.524 GHz) were combined to increase the signal-to-noise. Each frame is averaged over 0.142 km/s (∼0.16 MHz), and the line is centered on Io's frame of reference. As in Fig. 8.9, the large circle shows the limb of Io, and the small circle the resolution of the data. The fourth panel shows the disk-integrated line profile, and the grey dots indicate the offset frequency (velocity) of each image in frame 1–3. The symbols B (blueshift) and R (redshift) show the velocities of gas moving towards (B) or away from us (R). The approximate positions of several volcanoes are indicated on frame 2 (de Pater et al. 2020b). Figure reproduced from de Pater et al. (2021)

Eclipse response in the line profiles of regions associated with plumes in ALMA data (Fig. 8.16) show a similar behaviour as the eclipse response of disk-integrated line profiles. In eclipse, the atmospheric columns and temperatures obtained from best fit isothermal hydrostatic models to the spectra remain roughly constant, but the fractional coverage of the atmosphere in the beam decreases. Note that even in plume regions, the fractional coverage of the atmosphere is not unity in sunlight, suggesting that the plume emitting region is not resolved in the observations. The emission shoulders related to plume emissions are clearly visible in the associated spectra. It is clear from the line profiles of the plumes, but also for the disk-integrated profiles in-eclipse, that simple hydrostatic models are not sufficient (de Pater et al. 2020b).

Io's active volcanism must lead to a constant resurfacing of its crust, whether caused by plume deposits, or lava pouring out of vents. A lack of impact craters on Io's surface suggests an upper limit of 10^6–10^7 years on Io's surface age, which implies a global resurfacing rate of 0.1–1 cm/yr (e.g., Carr 1986). Based upon the above mentioned dust production rates in plumes, Geissler and McMillan (2008) conclude that the high resurfacing rate based on the obliteration of all impact craters is likely caused by the emplacement of lava flows rather than deposition of dust from plumes, unless many plumes were missed in *Galileo* observations, or that other material in addition to dust fall-out might be important, such as SO_2 snowfall or direct condensation from the gas phase onto the surface.

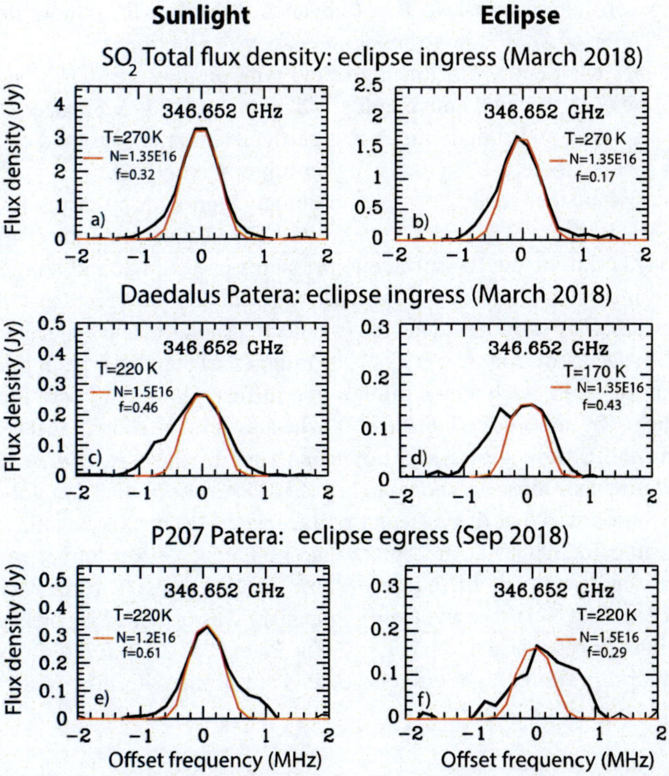

Fig. 8.16 SO_2 line profiles (in black) with superposed the best-fit hydrostatic models (in red). The column density N, temperature T, and fractional coverage f used for fitting are indicated. Both data and models are at a frequency of 346.652 GHz. (**a**) Disk-integrated flux density for Io in sunlight. (**b**) Disk-integrated flux density for Io in eclipse, after eclipse-ingress. (**c**) In-sunlight data for Daedalus Patera, integrated over 1 beam diameter. (**d**) In-eclipse data for Daedalus Patera, integrated over 1 beam diameter, after eclipse-ingress. (**e**) In-sunlight data for P207, integrated over 1 beam diameter. (**d**) In-eclipse data for P207, integrated over 1 beam diameter, before eclipse-egress (Adapted from de Pater et al. 2020b)

8.4.2 Thermodynamic Properties of Plume Classes

Kieffer (1982) investigated potential reservoirs and thermodynamic properties of Io's diverse volcanic plumes. She composed a temperature–entropy diagram, and suggested 5 potential entropy ranges or reservoirs for Io's plumes, varying from low-entropy (reservoir I) to extremely high entropy (reservoir V) eruptions. In connecting these models with observations of plumes, one can distinguish three types of plumes. The majority of plumes fall in the category of the dust-rich Prometheus-type plumes. These appear to "wander" in location (the Prometheus plume migrated over 80 km over a 20-year time interval, Kieffer 2000), and may originate when hot silicate lava flows advance through a SO_2 snow field. These plumes are referred to as "low-to-moderate entropy" eruptions.

The highly energetic $\gtrsim$400 km high Pele-type plumes are rich in sulfur gases (S_2 and S are both detected above Pele; McGrath et al. 2000; Spencer et al. 2000; Jessup et al. 2007), and contain much less particulate matter (dust and condensates) (Sect. 8.4.1); they therefore are likely higher-entropy eruptions.

A third type are the "stealth" plumes, extremely high-entropy eruptions (Kieffer's reservoir V), from a reservoir of superheated SO_2 vapor in contact with silicate melts about 1.5 km below the surface at pressures of ~40 bar and temperatures of ~1400 K. Since such plumes would consist of essentially pure gas, i.e., without dust or condensates, they cannot be detected in reflected sunlight, and hence were usually not seen by spacecraft. Johnson et al. (1995) suggested that this type of plume might be widespread on Io, such as the plumes and diffuse glows that were imaged over Acala Fluctus by the *Galileo* spacecraft when Io was in eclipse (McEwen et al. 1998), and the diffuse glows and point-like sources the *New Horizons* mission captured during an eclipse, as shown in Fig. 8.12 (Spencer et al. 2007). Johnson et al. (1995) also proposed that these stealth plumes were responsible for the millimeter SO_2 emission, for which an interpretation (Lellouch 1996; Moullet et al. 2008) called for a large number of un-seen plumes in comparison to the visible ones. More evidence for wide-spread stealth volcanism was provided by observations of SO emissions, as discussed in Sect. 8.3.2.2 (de Pater et al. 2020a). This phenomenon could, perhaps, prevent a total collapse of Io's atmosphere during eclipse (de Pater et al. 2020b).

8.4.3 Models of Plumes

In order to learn more about the underlying sources of volcanic explosions, we need to model the plumes and hot spots, the two resulting phenomena that can be observed from afar. In this section we discuss models of plumes. The large umbrella-shaped plumes seen from afar (Fig. 8.17a,b) arise from a vastly smaller, geometrically complex source region through a sequence of non-LTE processes. The overall plume size reflects the source energy (see previous section) in that the thermal energy at the source, which depends on the SO_2 stagnation temperature, T_{stag}, is converted to directed kinetic energy (velocity) during the gasdynamic expansion into the near-vacuum just above the surface. The gas subsequently rises and falls, exchanging directed kinetic energy for potential energy and returning again to kinetic energy before it strikes the surface or shocks and expands further. The peak velocity/altitude is determined by T_{stag} and by whether the gas mass flow rate is sufficient for the gas to be collisional; if it is dense enough to be collisional at high altitudes, the falling gas encounters rising gas and an umbrella-shaped canopy shock wave forms at a height determined by conservation of mass, momentum and energy (App. B in McDoniel 2015), keeping a lid on the canopy size. Such a shock-bound canopy is thus about a factor of two lower in altitude than a simple ballistic

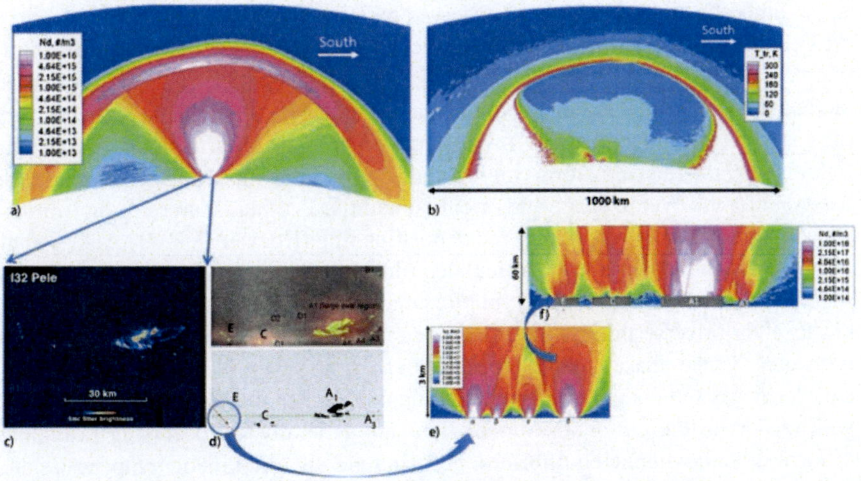

Fig. 8.17 Composite figure from McDoniel et al. (2015) illustrating the development of the Pele plume from multiple source regions into the grand dome canopy seen from a distance. At the top are (**a**) density and (**b**) temperature fields in the umbrella-shaped plume over Io's limb. (**c**) Galileo SSI image of the Pele caldera and (**d**) archipelago of simulated sources. Many little gas/particle jets merge through a sequence of oblique shock and rarefaction waves which, at a set of ever increasing scales, produce a variety of strong density and temperature gradients which appear differently depending on the view direction and cutting plane. (**e**) The circled four tiny sources $\alpha, \beta, \gamma, \delta$ that make up region E in panel (**d**) lead to four interacting jets. A sideways view of the lower 3 km of the plumes is shown along the red dashed line in panel (**d**). These plumes interact with more distant jets higher up, shown in panel (**f**). (**f**) A sideways view of the plumes seen along the green dashed line in panel (**d**), up to an altitude of 60 km. Even at 60 km the region constitutes only a small portion of the "stem" of the grand Pele plume draped over the limb of Io

calculation would indicate. The canopy width depends on whether the gas shocks and also on the initial jet spreading angle near the surface which is determined by geometric details of the vent (McDoniel 2015; Hornung 2016).

Simulations conducted since the review by Geissler and Goldstein (2007) have incorporated most of the physics that was missing before. These simulations include fully three dimensional simulations (McDoniel et al. 2015; Ackley et al. 2021), unsteady plumes interacting with a changing sublimation atmosphere (McDoniel et al. 2017) or undergoing 3D dynamic pulses (Hoey et al. 2021), and plumes at different locations on Io interacting with impinging streams of Jovian plasma and sunlight (Blöcker et al. 2018; McDoniel et al. 2019).

In Fig. 8.17 the large umbrella-shaped plume is sourced from a large number of compact vents. McDoniel et al. (2015) examined how the geometrically complex near-surface sources of Pele's plume of gas and particles merge together through a sequence of oblique gas-dynamic shock and rarefaction waves which turn the direction of the rising and falling gas and particulate streams away from the midline. Sheets of concentrated gas and particles are evident in simulations at low altitudes due to multiple jet-jet interactions, but at higher altitudes they combine into the broad umbrella shaped canopy as observed by *Voyager, Galileo* and *New Horizons*. At Pele, those detailed near-surface jet-jet interactions leave their imprint on the visible canopy structure, the shape of the reddish deposition ring and the darker silicate deposits which fall inside of the main ring. While the smallest particles track the gas well ($\lesssim$10 nm radii; Sect. 8.4.1), larger micron-scale particles do not and tend to fall shy of the red deposition rings. Similarly, the same modeling approach can be used to infer the overall shape and orientation of the Tvashtar source region from the shape of the surrounding primary red deposition ring (Hoey et al. 2021; Ackley et al. 2021). Sunlight scattered off particulates (dust, condensates, droplets) is typically seen in images (Sect. 8.4.1). The volatile condensates may change size or sublimate during their traverse through the plume, depending on the details of the local gas dynamics. Those phase changes are not obvious, however, in that they depend on, e.g., local gas temperatures, densities, radiation fields, and the particulate motion with respect to the gas. Those interactions are generally not occurring in regions of local thermodynamic equilibrium. For example, the gas kinetic temperature and internal grain/droplet temperature are not likely to be the same at a given point in the plume. Similarly for the gas velocity and particulate velocity. This disequilibrium occurs because the gas density is not large enough outside of the central plume core to sustain enough molecule-particulate collisions for equilibrium to occur.

At lower altitudes within plumes, radiative exchange is important to subsequent plume evolution. The hot SO_2 gas leaving the surface is at a sufficiently high density that it is opaque to radiative emission from the ν_1, ν_2 and ν_3 bands. That is, most of the volume of hot rising gas cannot cool by spontaneous emission from those bands because the respective photon mean free path for re-absorption is insufficient for it to escape the plume core. The gas thus initially expands and cools adiabatically. As the expansion proceeds, however, the gas density and temperature drop and once the density drops enough, the plume core becomes successively more transparent to cooling from the three vibrational bands. For Tvashtar or Pele, the resulting emission surfaces occur around 20–30 km above the surface (Zhang et al. 2003; Hoey et al. 2021). This energy loss ultimately diminishes the ability of the plume to rise to the same height as it would if the flow were everywhere adiabatic. The radiation process depends on the details of the total mass flux, the sizes and shapes of the volcanic sources, the collision cross sections for translation/rotation/vibration energy exchange, the Einstein A coefficients for spontaneous emission, and the photon absorption cross sections.

This dependence of plume gas dynamics on radiative loss is related to what is ultimately observed of the plume from afar. By observing the Doppler shifts of SO_2 lines, one can determine localized flow velocities along the observational line of sight. Moullet et al. (2008) used this method to infer planetary-scale winds (which turned out to mimick a prograde zonal flow), and de Pater et al. (2020b) to determine gas velocities of/in (averaged over the plume area) plumes (e.g., Figs. 8.15 and 8.16). However, as illustrated in Fig. 8.2, because of the huge variations in gas density associated with temperature and velocity changes in the plumes, extremely high spatial resolutions are required to observationally separate velocities in a plume. For comparison, the collisional portion of the sublimation atmosphere only spans a few scale heights at most and gas temperatures vary by factors of 2-to-5, while the sublimation atmosphere along lines of sight remains nearly transparent to most radiation. Therefore, when looking remotely at a Pele-class plume that is not well resolved spatially, it is not obvious what is observed: hot gas at 1500 K above the vent is expected to expand from a few hundred meters per second to over 1 km/s while its density drops by orders of magnitude. It is not clear whether the gas has reached nearly its ultimate speed ($\sim \sqrt{2 * Cp * T_{stag}}$ for adiabatic flow) before its density drops enough for it to become transparent or whether it becomes transparent much higher up. That is, it is not clear to what depth into the plume core one probes when observing remotely. Similarly, when examining the plume canopy in vibrational/rotational lines it is not clear how those are related to gas velocity or translational (kinetic) temperatures since molecular collision rates are quite low and spontaneous emission losses are important. Thus, interpreting existing observations remains tricky. But simulations remain tricky as well, not so much due to the shortcomings of included physical models, but due to the uncertainty in the many necessary physical boundary conditions, specifically the real "gas-surface interaction" in the broadest sense. While the launch-to-landing gas dynamics can be simulated including the full non-equilibrium physics, the reality of those simulations remains in doubt due to the uncertainties in the many boundary conditions—i.e., mostly the details of the plume source region.

8.5 What Drives Io's Atmosphere?

In the following subsections we address the historic question of the primary "driver" of Io's atmosphere (sublimation, volcanism, sputtering). This question can in fact be interpreted in different ways: (1) what is the "immediate source" of the atmosphere? Or (2) what drives Io's atmospheric dynamics? After reviewing the three possible sources, we will return to this question in Sect. 8.6.

8.5.1 Sublimation-Supported Atmosphere

The fundamental aspect of a sublimation-driven atmosphere that is amenable to observational characterization is that it reacts to changes in insolation on timescales from minutes (eclipse ingress/egress) to hours/days (Io day = 42 h) and years (Io's year = 11.9 years). Io's low surface pressure, typically a few nanobars at most, means that latent heat exchanges are inefficient at redistributing energy, so that individual areas of SO_2 frost assume largely independent, solar-insolation driven, temperatures (Ingersoll et al. 1985; Ingersoll 1989). In a sublimation-driven atmosphere we therefore expect collisionally thick conditions on the day-side in low-latitude regions collapsing to a thin atmosphere on the nightside, in eclipse and at high latitudes.

To determine if the primary immediate source of Io's atmosphere is sublimation, the response of Io's atmosphere to time scales on minutes–hours–years is even more diagnostic than the use of atmospheric bulk properties or spatial distribution, although these bring useful additional constraints. Considerable progress has been achieved in this area since the 2007 Io book, thanks to a large body of thermal-infrared 19-μm observations (Spencer et al. 2005; Tsang et al. 2012; Tsang et al. 2013a,b), as well as 19-μm and sub-millimeter observations of Io going into and coming out of eclipse (Tsang et al. 2016; de Pater et al. 2020b). The latter data were summarized in Sect. 8.3.2.

From a broad coverage of the 19-μm SO_2 observations along Io's orbit around Jupiter, Spencer et al. (2005) determined factor-of-10 larger SO_2 columns on the anti-Jovian hemisphere (180°W) compared to 300°W, in general agreement with the Ly-α maps shown in Fig. 8.6 (Feaga et al. 2009; Giono and Roth 2021). Extending the coverage over a full ionian year, from 2001 to 2013, Tsang et al. (2012, 2013b) established that the SO_2 columns on the anti-Jovian atmosphere peaked at the time of the March 2011 perihelion at 4.955 AU, and reached a 3 times smaller minimum at the March 2005 aphelion (5.459 AU), yielding direct evidence for a largely sublimation-driven atmosphere on the anti-Jovian hemisphere (Fig. 8.18). These annual variations of the SO_2 columns were fit by a combination of a volcanic component ($6.5\times10^{16}\,\mathrm{cm}^{-2}$, making up most of the aphelion atmosphere) and a sublimation component calculated for a frost albedo $A = 0.535$ and thermal inertia $\Gamma = 350$ MKS.

Observational results regarding the diurnal variability of the SO_2 atmosphere are not as definite. "Instantaneous" maps of the SO_2 columns do feature spatial

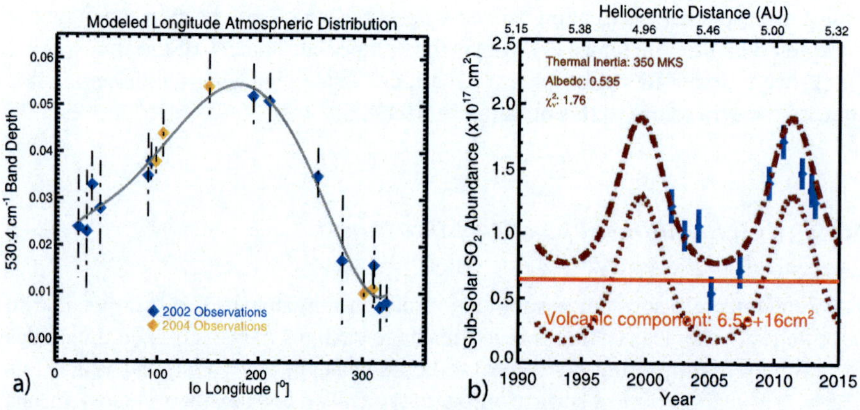

Fig. 8.18 (**a**) Disk-averaged band depth at 530.4 cm^{-1} observed in 2002 (blue) and 2004 (yellow) from Spencer et al. (2005), with a 5th order least-squares polynomial (gray). (**b**) A comparison between retrieved SO2 column densities, after scaling to an Io central longitude of 180° (blue) and a modeled atmosphere in vapor pressure equilibrium with surface frost (dotted), after adding a constant volcanic component (dot-dash), as a function of solar insolation (from Tsang et al. 2013b)

variability, but whether they are purely geographical or also affected by diurnal variations is ambiguous. For example, the individual Ly-α maps of Feaga et al. (2009) do not reveal any obvious decrease of the SO_2 columns towards the terminator. From their 4-μm data, Lellouch et al. (2015) inferred moderate diurnal variations of the equatorial atmosphere, with factor-of-2 lower densities 2.7 h before and 5.3 h after noon, compared to local noon. The most convincing study is from Jessup and Spencer (2015) who used spatially-resolved *HST*/STIS observations to observe regions at 200–250°W longitudes at 2 distinct times of day, differing by 50° of rotation (3.3 h in local time) of Io, disentangling in this manner geographical and diurnal variations, while carefully avoiding volcanic areas. Negligible variation of the SO_2 column with local time was found, consistent with the dawn-to-dusk presence of the atmosphere as derived from Ly-α images. This, however, should not be taken as an argument against a sublimation-driven atmosphere as it can still be consistent with sublimation support by high thermal inertia frost (Spencer et al. 2005).

On the theoretical side, Moore et al. (2009) and Walker et al. (2012), amongst others, show that the surface temperature of the sub-Jovian hemisphere is colder than that of the anti-Jovian side due to the 2-h-long eclipses every orbit (42.5 h), as demonstrated in Fig. 8.19. Since the SO_2 surface pressure virtually follows the saturated vapor curve, based on this figure one would therefore naturally expect a $\sim 4\times$ higher column of gas above the anti-Jovian hemisphere, assuming a more or less uniform ice coverage. The ice coverage, however, varies across the surface. SO_2 fine-grained frost, as indicated by the strong 4.07 μm band, is generally widespread, but most abundant on the anti-Jovian hemisphere and at mid-latitudes (Douté et al. 2001); thick SO_2 frost, traced by the weak 2.12 μm band, is enhanced at equatorial

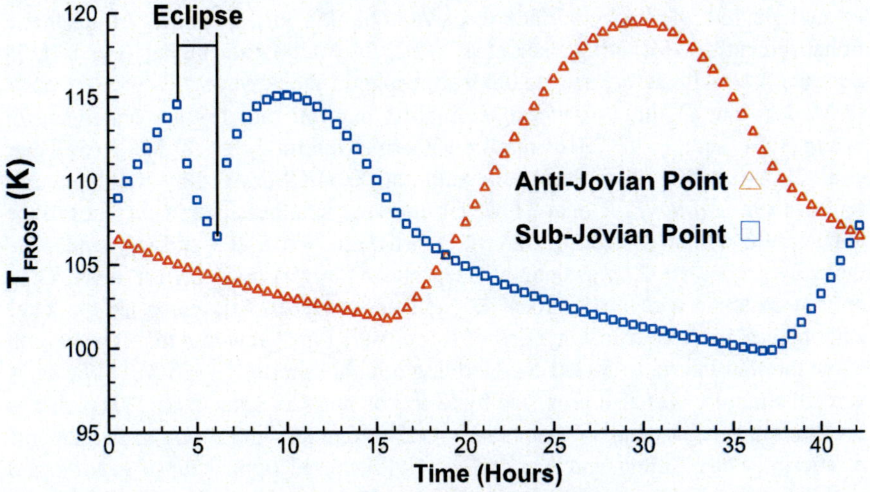

Fig. 8.19 A comparison between the simulated surface frost temperature at the anti-Jovian and sub-Jovian points as a function of time of day (from Walker et al. 2012)

latitudes near 120–180°W (Laver and de Pater 2009; de Pater et al. 2020a). These asymmetries in ice coverage should also contribute to the longitudinal variations in column density. The lower ice coverage on the sub-Jovian hemisphere might be caused by an increased loss due to sputtering during eclipse (Sect. 8.5.3).

The observations of Io's atmosphere in response to eclipse ingress/egress discussed in Sects. 8.3.2 and 8.3.3 provide information on the sublimation aspect of Io's atmosphere on the shortest timescales. As discussed in Sect. 8.3.3, Saur and Strobel (2004) modeled the response of auroral emissions when Io enters an eclipse. They concluded, based on the factor-of-3 decrease in Io's atomic emissions upon eclipse ingress observed by Clarke et al. (1994), that Io's global (background) atmosphere must strongly decrease during eclipse and must therefore be mostly driven by sublimation rather than volcanoes. Similarly, observations of eclipse-ingress both in the mid-IR and sub-mm (ALMA) showed a substantial collapse of the SO_2 atmosphere, and (in the sub-mm) reformation upon egress within about 10 min. A comparison of the ALMA maps with Walker et al. (2012)'s sublimation models (Sect. 8.3.2.1) shows that both collapse and reformation of the atmosphere occur much faster than modeled (∼10 min vs ∼2 h).

Walker et al. (2012)'s model is based upon thermophysical properties of Io's surface (Sect. 8.3.2.1), with horizontal variations in the thermal inertia Γ. Since Γ depends both on composition and compactness (density) of the material, one might expect variations both across the surface (e.g., areas covered by frost and non-frost or rock) and with depth (e.g., deeper layers will be more compacted, and might have a different composition). The thermal inertia for rock and dust is relatively low, while areas covered with SO_2 ice will have a higher Γ. Hence rocky/dusty areas will respond much faster to changes in illumination than areas covered by frost. Hence one expects nearly instantaneous SO_2 condensation on the "bare" rocky/dusty surface when it cools, and sublimation (or desorption) as soon as sunlight hits the rocky surface at dawn. This led to the "Dawn Atmospheric Enhancement" (DAE) in Walker et al. (2012)'s sublimation model (Fig. 8.10 in Sect. 8.3.2.1), with associated standing atmospheric shock waves.

Models that fit the 19-μm eclipse ingress data on minute-time scales shown in Fig. 8.8 (Tsang et al. 2016) require a thermal inertia $\Gamma \simeq 50$ MKS (de Pater et al. 2020b), which is similar to the value of 70 MKS derived by Rathbun et al. (2004) from *Galileo*/PPR data. However, the very small change in the subsurface temperature obtained with ALMA (i.e., ∼1–2 cm below the surface, since one typically probes 10–20 wavelengths deep into the crust) upon eclipse ingress can only be matched with $\Gamma = 320$ MKS (de Pater et al. 2020b), which led the latter authors to suggest a vertical layering of Io's crust: a top ($\lesssim$few mm thick) layer with a low thermal inertia (∼50 MKS) overlying a higher inertia (∼320 MKS) layer. A vertical structure was also proposed by Morrison and Cruikshank (1973) (with Γ's ∼4 times lower) based upon an analysis of eclipse ingress and egress measurements at 20 μm, while Sinton and Kaminsky, (1988), based upon eclipse heating and cooling curves at wavelengths between 3.5 and 30 μm, suggested surface areas with different albedos and Γ's (A = 0.1, Γ = 5.6 MKS; A = 0.47, Γ = 50 MKS), with the brighter surface overlying a higher Γ layer.

Based upon the above mentioned discrepancy between observations and sublimation models of eclipse ingress and egress, we suggest that Io's surface is covered everywhere by a low thermal inertia layer, perhaps only a few mm thick, even over areas covered by frost (perhaps due to dust or fluffy plume deposits). We propose that a multi-layer thermophysical model based upon proper surface albedo maps, with a low few-mm thick thermal inertia layer on top (with potentially horizontal variations, but still at a low Γ), overlying higher Γ layers will match the timescales in the data (i.e., a model as, e.g., in de Kleer et al. 2021).

8.5.2 Volcanically-Supported Atmosphere

As discussed in Sect. 8.4.1, numerous volcanic plumes have been detected at seemingly random latitudes. All plumes contain gases, primarily SO_2 as in Io's global atmosphere, with smaller fractions of SO, S, and S_2, depending on plume type. Some volcanoes may expel NaCl and KCl. The dust-to-gas relative content varies from ~10% in mass for Prometheus-type plumes, to ~1% in Pele-type plumes, and no dust in stealth volcanoes. The gases in both Prometheus- and Pele-type plumes typically reach altitudes of up to ~100–400 km, which is well above the exobase (which is at several tens of km; McDoniel et al. 2017). In the following we investigate the effects of volcanic plumes on Io's atmosphere by comparing plume models (Sect. 8.4.3) with data.

McDoniel et al. (2017) examined the plume/atmosphere interaction over a full ionian day for plumes at different latitudes, and showed that one cannot simply add the volcanic and sublimation components to understand an observation. While the total atmospheric mass of SO_2 observable on the (Earth-facing) day side of Io is well controlled by a vapor pressure equilibrium atmosphere over an ice surface in radiative equilibrium with sunshine and in simple hydrostatic equilibrium, the actual material above the surface may be dominated by gas which was just exhausted from a volcano. If a plume is strong enough to launch gas above the nominal exobase, the falling canopy gas mostly settles atop the sublimated component. This raises the surface pressure and drives some of the sublimated gas back into the icy surface until vapor pressure equilibrium at the surface is restored, but leaves a broad layer of volcanic gas atop a sublimated component. If both the sublimated and volcanic components are pure SO_2, the material layers would be distinguishable only based on their temperatures; if the plume material included other gases (e.g., SO, S_n, NaCl, KCl) or nanoparticles, or if the atmospheric component had photodissociation daughter species, the layered components may be chemically distinguishable. These theoretical findings may explain why observed SO_2 column densities above a plume are usually hardly different from their surrounding areas (Lellouch et al. 2015; de Pater et al. 2020b), unless the spatial resolution in the observations would be high enough to distinguish the core of rising gas in a plume.

When the falling canopy of gas from a large plume meets the atmosphere, the plume gas may "bounce" off the atmosphere and thereby heat and increase its areal extent by a factor of 2–3 (Zhang et al. 2003; McDoniel et al. 2017). This might explain the concentrations of SO_2 emissions near the plumes in ALMA maps (Fig. 8.9).

There can be further modifications on local scales to the plume/sublimation atmosphere as well. As shown by Zhang et al. (2003) in 2D simulations and McDoniel et al. (2017) in 3D simulations, a large Pele-class plume can produce a ring of *reduced* vertical column density above where the main plume deposition ring (i.e., Pele's red ring) is seen. This occurs due to the dynamics of the down-pouring material passing through a curved "re-entry" shock wave (the "bounce"), heating the gas up to levels similar to those seen above the vent. The high pressure will push material away, which results in a depression in the column density at the intersection of the canopy with the sublimation atmosphere, and is also associated with local scouring of surface ice at some times of day.

Finally, there may be a number of active plumes in close proximity to each other (e.g., Fig. 8.17, in which individual jets of Pele are close enough that they interact while still rising) with different strengths, which would complicate the overall interaction between a plume and the sublimation atmosphere. Needless to say that if we consider all these different modeled aspects, together with the variety of viewing angles through a plume (Fig. 8.2) and the spatial resolution of observations, it would be a heroic task to identify parameters in the model to tweak to "match" observations.

8.5.3 *Sputtering*

Sputtering off a body's surface has been identified as a source of atmosphere for e.g., Mercury and the Moon (where the impinging particles are from the solar wind and micrometeorites), and Giant Planet icy satellites, where the mechanism can coexist with a sublimation source, see e.g. Milillo, et al. (2011) for the Moon and Mercury; Marconi, (2007), Leblanc et al. (2017), Roth et al. (2021) for Ganymede. Sputtering collectively refers to the processes (e.g. direct knocking of atoms from the surface, electronic excitation in the ice) that can lead to the ejection of neutral molecules or atoms into an atmosphere, and is usually accompanied by chemical alteration of the surface (radiolysis).

Sputtering as a source of planetary atmospheres has been reviewed by Cheng and Johnson (1989). Early sputtering models demonstrated that the impact of energetic magnetospheric particles from the Io plasma torus (ions and electrons) can generate an extended, weakly bound, rarefied atmosphere ("corona") around Io. The process is self-limited to column densities of $\sim 10^{15-16}\,cm^{-2}$ (Lanzerotti et al. 1982) as large gas production halts further penetration of energetic particles, at which point the atmosphere is itself sputtered, with consequences for the upper atmosphere thermal structure (Sect. 8.2.4) and escape (Sect. 8.3.4).

Significant contributions to Io's atmosphere through sputtering can only be expected at times and locations where sublimation and direct volcanic venting are inefficient in producing an atmosphere. This may occur at high latitudes away from volcanic centers/plumes, or during eclipse or at night. Since surface sputtering also leads to a chemical alteration of the surface, through, e.g., preferential loss of volatile material, sputtering at Io's high latitudes might explain the much darker surface there and the apparent absence of SO_2 ice. Moreover, it may also have led to a preferential loss of volatile ice from Io's sub-Jovian side, since every Io day this side experiences 2-h long eclipses during which time most of the atmosphere is collapsed (Sect. 8.3.2) and hence ions and electrons have more time to impact the sub- versus anti-Jovian hemisphere.

The origin of the sodium and potassium atoms in Io's atmosphere has long been an outstanding puzzle. Although we now know these atoms are derived from NaCl and KCl (Sect. 8.2.3), the question still remains whether the dominant source is sputtering from Io's surface, or volcanically produced. The spatial distribution of NaCl and KCl as mapped with ALMA in 2015 and shown in Fig. 8.20 (Moullet et al. 2015), shows the strongest emission for both species near the volcano Isum Patera (~206°W longitude, ~30° latitude). Interestingly, no SO_2 emissions were detected at this location. The alkalis have been mapped with ALMA at several other occasions, and are often localized, do not coincide with SO_2 or SO, and vary drastically over Io's globe (Redwing et al. 2022), yet overall disk-averaged column abundances do not vary much over time (Roth et al. 2020).

The maps in Fig. 8.20 suggest a ratio of ~5–6 in NaCl/KCl column densities, (Redwing et al. 2022) which is a factor of 2-3 lower than the Na/K ratio in chondrites, but consistent with that measured in Io's extended atmosphere (Brown 2001; Redwing et al. 2022). If sourced volcanically, the difference in the observed

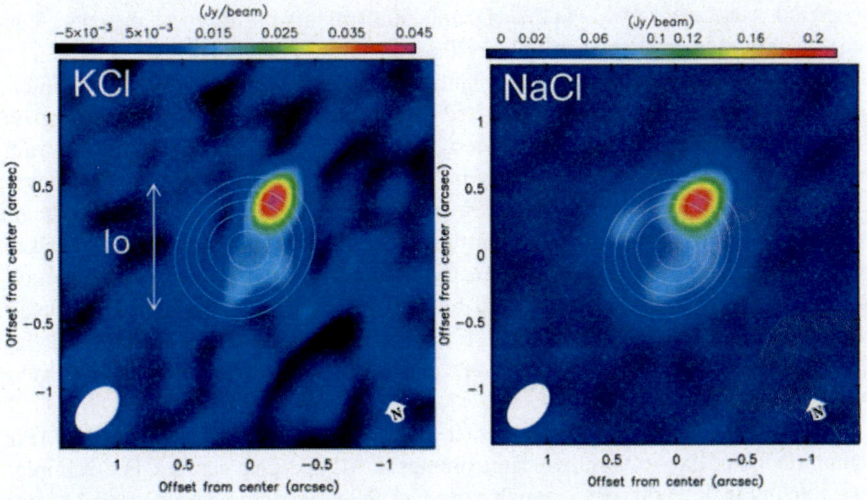

Fig. 8.20 ALMA maps of NaCl and KCl in Io's atmosphere. The data were taken in June 2015, at a spatial resolution of $0.25'' \times 0.4''$. The beam is indicated in the lower left of each figure, and Io's north pole by the arrow in the lower right (Courtesy A. Moullet)

vs chondritic ratio may be caused by the fact that KCl has a lower condensation temperature than NaCl (1173 vs 1363 K), in which case the data may suggest a magma temperature of ~1300 K (Fegley and Zolotov 2000). Moreover, if indeed volcanically sourced, the magma in the chambers that power volcanoes must have different melt compositions, and/or the magma has access to different surface/subsurface volatile reservoirs, since SO_2 gas is usually not detected at the same locations as NaCl and KCl (de Pater et al. 2020b).

However, once released by volcanoes, NaCl and KCl may also condense/freeze out on Io's surface, in which case plasma impact from the torus could sputter these species back into Io's atmosphere in areas where the atmospheric column density is low ($\lesssim 10^{15-16}$ cm^{-2} Lanzerotti et al. 1982).We note that, regardless of their origin, the short lifetimes require a continuous source to explain the almost-continuous presence (Roth et al. 2020) in Io's atmosphere (Sect. 8.2.3).

8.5.4 *Dynamics/winds*

Large horizontal variations in SO_2 densities and temperature give rise to horizontal pressure gradients, and hence winds. As suggested by Ingersoll et al. (1985) and Ingersoll (1989), for a sublimation-driven atmosphere this would result in winds away from the subsolar point towards the night side (day-to-night winds) and towards high latitudes. Even if the atmospheric density does not show any clear dusk–dawn variations (Sect. 8.5.1), winds near the terminator might still blow from the day-to-night side, as the atmosphere is still expected to collapse on the nightside due to condensation.

Winds can be observed via their Doppler shift in line profiles, i.e., the entire line can be shifted in frequency, or line profiles can show "shoulders" (e.g., Sect. 8.4.1). Such shifts only provide a line-of-sight component to the wind direction. Interestingly, Moullet et al. (2008) reported strong horizontal winds in the prograde direction from IRAM Plateau de Bure maps, with a beam-integrated limb-to-limb difference of 330 ± 100 m/s. Such winds are hard to reconcile with the above-mentioned models. An enhancement in atmospheric SO_2 near the dawn terminator due to molecules desorbed from a warming rock surface (DAE in Fig. 8.10; Walker et al. 2010) may produce a pressure wind mimicking a prograde zonal wind, although synthetic wind maps based on optical ray tracing through a simulated atmosphere generated by Gratiy et al. (2010) do not closely match those of Moullet et al. (2008). Millimeter-wavelength data at a higher spatial resolution are needed to observe wind profiles in more detail.

It should be understood, however, that winds are not as simple as just some linear combination of day-to-night, volcanic plumes, DAE disturbances, etc. For example, we note in Fig. 8.21 that even though a massive Pele-class plume produces relatively localized column density disturbances (positive in the center, negative in the ring), its influence on the winds may extend over half of the hemisphere visible from Earth.

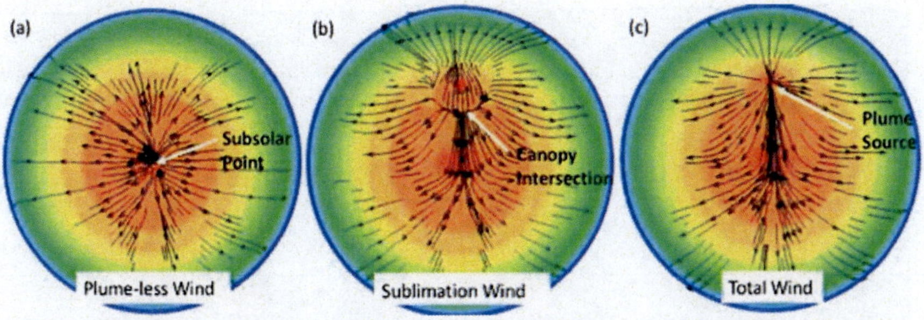

Fig. 8.21 Simulation of a hypothetical Pele-class plume at 30° N on the daylit side of Io. (**a**) Column density color contours and wind vectors in the absence of a plume. (**b**) Color contours show column density with a Pele-class plume included at 30° N but the wind vectors only show the sublimation component of the atmosphere; note how the plume drags or deforms the sublimation component. (**c**) Again, contours represent column density of the composite atmosphere. Now, the wind vectors represent the column-mass-averaged composite value showing broad influence of the plume over northern hemisphere (McDoniel et al. 2017)

Even without the presence of plumes, full planet-scale simulations of an SO_2 and SO atmosphere suggest possibly grossly different wind patterns approaching the terminator and on the night side for the two species. This highlights the fact that a wind only can be represented by a single vector value at each point in space if the gas flow is dense enough to be fully continuum—But Io's plumes and atmosphere are sufficiently rarefied that the wind speed or its direction may not be a unique value.

Figure 8.22 is from Walker's 2012 dissertation, the most physically comprehensive atmosphere simulation to date, notably including photochemistry and plasma-impact chemistry for several species. It also includes plasma pressure due to electric and magnetic fields around Io which can drive winds. The simulated details are remarkable and include a prograde equatorial jet of gas extending much of the way from the day side across the leading hemisphere towards the cold nightside. On the leading side the jets (one prograde, one retrograde) collapse into the solid surface, being viscously dissipated as the atmosphere thins out. Associated with the jet are cyclonic and anticyclonic winds at higher nighttime latitudes. The wind fields are similar for SO_2, S, O, and O_2 where the atmosphere is dense enough to be collisional. These complexities, especially for species other than SO_2, are driven largely by the specifics of the species' chemical source/sink mechanisms and the (residence time) interaction of the species with the solid surface. Perhaps future ALMA observations at high spatial resolution may observe these predicted species' wind and concentration patterns.

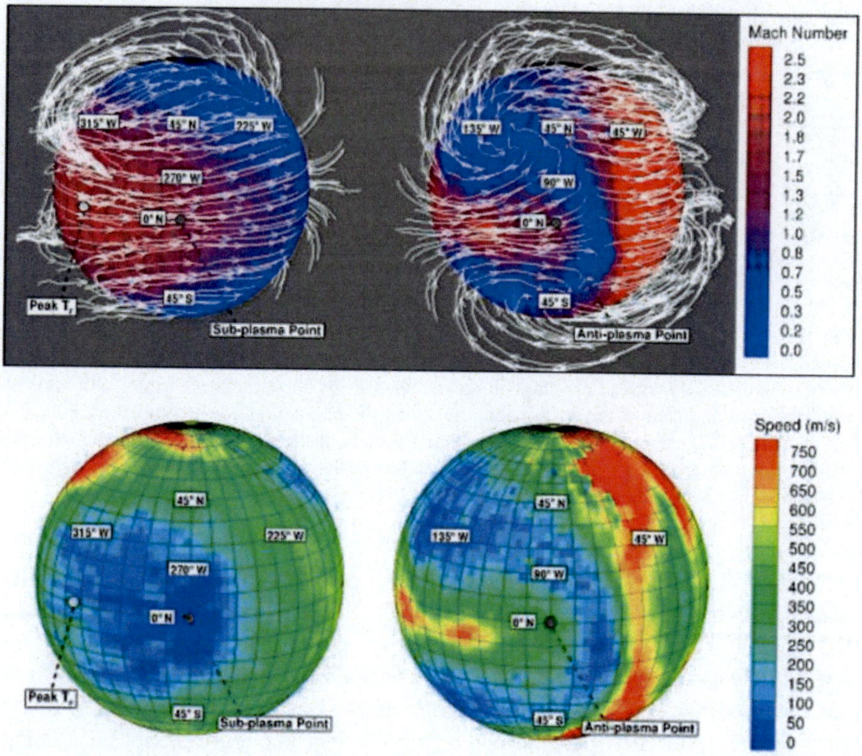

Fig. 8.22 Wind simulations from Walker (2012). In the top two subfigures are wind stream traces overlaid on contours of SO_2 flow Mach number at an altitude of 10 km for the (left) trailing and (right) leading hemispheres. The sub-plasma and anti-plasma points are shown. The subsolar longitude is 342°, a time ~9° prior to eclipse. Hence, the trailing hemisphere shows mostly the day side, and the leading hemisphere the night side. The lower two subfigures illustrate actual wind speeds at 10 km altitude

8.6 Conclusions and Next Steps

As shown by the data and models presented in this review, we have learned a great deal about Io's atmosphere since the *Galileo* era. However, many questions still remain, and new questions have surfaced, as summarized below.

In Sect. 8.5 we addressed the historic question of the primary "driver" of Io's atmosphere (sublimation, volcanism, sputtering), and distinguished between the immediate source of gas and dynamics in Io's atmosphere. Based upon the large body of data combined with their analyses since the 2007 Io book, it is clear that the immediate source of Io's atmosphere is sublimation/condensation of SO_2. Yet, volcanoes do have a substantial impact on its atmosphere and surface, and may at times and certainly locally be dominant.

While the main source of the observed SO is most likely photodissociation of SO_2, some fraction must be sourced from volcanoes, including stealth volcanism. The main source of Na, K and Cl is photodissociation from NaCl and KCl. Some small fraction of sulfur may also be produced via photochemistry, but the main source of S_2 is volcanic; without it there would not be red deposition rings around Pele-type volcanoes. NaCl and KCl are also most likely volcanic in origin, though some of what we see at higher latitudes, where the column density of SO_2 gas is low enough for ions/electrons from the plasma torus to reach the surface, could have been produced via surface sputtering on salts condensed on the surface. Interestingly, if indeed NaCl and KCl originate from volcanic vents, these do not expel much, if any, SO_2 gas, indicative of variations in magma composition between different volcanoes.

Observations during eclipse ingress and egress show a near-instantaneous decrease in surface temperature ($\sim$2 min) and collapse of a large fraction of the SO_2 atmosphere ($\lesssim$10 min). This is much faster than current models predict. The not-complete atmospheric collapse has been predicted by models, though it is not clear what prevents the SO_2 atmosphere from complete collapse: a layer of non-condensibles as predicted by models, and/or volcanic activity. In addition, although SO itself is not condensible, its significant decrease during eclipse shows that self-reactions on the surface occur at a much faster rate than anticipated.

The near-collapse of Io's SO_2 atmosphere during eclipse and on the night side must trigger winds in the atmosphere: condensation onto the surface must cause a down-pour of gas towards the surface when it is no longer supported from below, while high gas densities during the day and low densities at night and at high latitudes result in winds from high-to-low pressure regions. However, none of the predicted winds have been measured directly. If anything, at millimeter wavelengths a zonal wind has been measured (Moullet et al. 2008), rather than a day-to-night flow.

Volcanic gas plumes are easiest to discern and compare to models at night or during an eclipse; as soon as the SO_2 atmosphere re-forms during the day or after eclipse egress, the falling plume "meets" the atmosphere, and models show a complex interaction, with lateral and vertical variations in density and temperature, as well as the formation of winds. With the relatively low spatial resolution of the present data from Earth it is difficult to compare the models and data in detail, in particular with the realization that plumes are seen under different viewing geometries (e.g., Fig. 8.2), where the line of sight "slices" through different regimes of density and temperature. Perhaps more progress on the observational side (short of sending a spacecraft) can be made by mapping Io at mm-wavelengths at a higher spatial resolution and sensitivity with ALMA using transitions over a broad range of energies. Ideally, observations at high spatial resolution will be obtained during eclipse and ingress/egress at a minute-time resolution.

Yet, it is comforting that models and data agree on some aspects: when the models are convolved down to the same resolution as the data, neither the model nor data show an enhancement in column density above a plume (Lellouch et al. 2015; de Pater et al. 2020b); the models show an enhancement directly over the vent, which requires a much higher spatial resolution than currently observationally available. Also, the observed brightening and expansion of the SO_2 emissions near volcanic plumes upon eclipse egress agree between models and data, when comparing a modeled plume during an Io-day (McDoniel et al. 2017) with the ALMA data. Note that the "plume bounces" expand the area of volcanic-sourced gas around a plume considerably compared to the case of a nightside plume, which may explain the morphology of ALMA maps in sunlight.

Despite considerable advances in our understanding of Io's atmosphere and plumes, there still are a large number of unsolved problems and questions:

1. What is the impact of Io's volcanic eruptions on its global atmosphere?
2. How do volcanic plumes and sublimated components interact, and how can we best compare the models with data?
3. Io's SO_2 atmosphere does not completely collapse during eclipse; what is the reason?
4. Io's atmosphere collapses during eclipse ingress and reforms upon egress much faster than models predict; how to update models to match the data?
5. How much does surface sputtering contribute to the atmosphere? Are there areas on Io where the atmosphere is collisionally thin, so that ions and electrons from the plasma torus can penetrate down to the surface?
6. What is the origin of NaCl and KCl, and why are they not co-located with SO_2?
7. What is the atmospheric temperature profile (in particular close to the surface), and how does it vary across Io, and how is it affected by volcanic plumes?
8. What are the atmospheric wind patterns? Are there large-scale zonal winds, day-to-night winds, rapid winds away from plumes, and can one discern the effect of the plasma flow?
9. What are the details of atmospheric loss processes (gas and dust), including atmosphere—magnetosphere connections during quiescent periods and volcanic eruptions?

Acknowledgments We like to thank Joachim Saur and Darrell Strobel for helping us understand the variations in auroral and plume emissions upon eclipse ingress, Laurence Trafton for providing Fig. 8.13, and Lorenz Roth for providing a detailed and helpful review of our manuscript. DG acknowledges support in part from NASA award number 80NSSC21K0830

References

Ackley, P., Hoey, W., Trafton, L., Goldstein, D.B., Varghese, P.L.: Hybrid dust-tracking method for modeling Io's Tvashtar volcano plume. Icarus **359**, 114274 (2021)

Ballester, G.E., McGrath, M.A., Strobel, D.F., Zhu, X., Feldman, P.D., Moos, H.W.: Detection of the SO_2 Atmosphere on Io with the Hubble Space Telescope. Icarus, **111**, 2–17 (1994)

Binder, A.P., Cruikshank, D.P.: Evidence for an atmosphere on Io. Icarus **3**, 299–305 (1964)

Blöcker, A., et al.: MHD modeling of the plasma interaction with Io's asymmetric atmosphere. J. Geophys. Res. Space Phys. **123**(11), 9286–9311 (2018)

Brown, R.A.: Optical line emission from Io. In: Woszczyk, A., Iwaniszewska, C. (eds.) Exploration of the Planetary System (IAU Symposium 65), pp. 527–531. Reidel, Dordrecht (1974)

Brown, R.A.: The Jupiter hot plasma torus: Observed electron temperature and energy flows. Astrophys. J. **244**, 1072–1080 (1981)

Brown, M.E.: Potassium in Europa's atmosphere. Icarus **151**, 190–195 (2001)

Cantrall, C., de Kleer, K., de Pater, I., Williams, D.A., Davies, A.G., Nelson, D.: Variability and geologic associations of volcanic activity on Io in 2001–2016. Icarus **312**, 267–294 (2018)

Carlson, R.W., Kargel, J.S., Douté, S., Soderblom, L.A., Dalton, B.: Io's surface composition. In: Lopes, R.M., Spencer, J.R. (eds.) Io after Galileo: A New View of Jupiter's Volcanic Moon, pp. 193–229. Springer. ISBN 3-540-34681-3 (2007)

Carr, M.H.: Silicate volcanism on Io. J. Geophys. Res. **91**, 3521–3532 (1986)

Cheng, A.F., Johnson, R.E.: Effects of magnetospheric interactions on origin and evolution of atmospheres. In: Atreya, S.K., Pollack, J.B., Matthews, M.S. (eds.) Origin and Evolution of Planetary and Satellites Atmospheres, pp. 683–722. University of Arizona Press, Tucson (1989)

Clarke, J.T., Ajello, J., Luhmann, J., Schneider, N., Kanik, I.: Hubble Space Telescope UV spectral observations of Io passing into eclipse. J. Geophys. Res. **99**, 8387–8402 (1994)

de Kleer, K., de Pater, I.: Time variability of Io's volcanic activity from near-IR adaptive optics observations on 100 Nights in 2013–2015. Icarus **280**, 378–404 (2016a)

de Kleer, K., de Pater, I.: Spatial distribution of Io's volcanic activity from near-IR adaptive optics observations on 100 nights in 2013–2015. Icarus **280**, 405–414 (2016b)

de Kleer, K., de Pater, I., Molter, E., Banks, E., Davies, A.G., et al.: Io's volcanic activity from time-domain adaptive optics observations: 2013–2018. Astron. J. **158**, 129 (14 pp.) (2019a).

de Kleer, K., de Pater, I., Ádámkovics, M.: Emission from volcanic SO gas on Io at high spectral resolution. Icarus **317**, 104–120 (2019b)

de Kleer, K., Butler, B., de Pater, I., Gurwell, M., Moullet, A., Trumbo, S., Spencer, J.: Thermal properties of Ganymede's surface from millimeter and infrared emission. PSJ **2**, 5 (2021). https://doi.org/10.3847/PSJ/abcbf4

de Pater, I., Roe, H.G., Graham, J.R., Strobel, D.F., Bernath, P.: Detection of the forbidden SO $a^1\Delta \rightarrow X^3\Sigma^-$ Rovibronic transition on Io at 1.7 μm. Icarus Note **156**, 296–301 (2002)

de Pater, I., Laver, C., Marchis, F., Roe, H.G., Macintosh, B.A.: Spatially resolved observations of the forbidden SO $a^1\Delta \rightarrow X^3\Sigma^-$ Rovibronic transition on Io during an eclipse. Icarus **191**, 172–182 (2007)

de Pater, I., de Kleer, K., Ádámkovics, M.: High spatial and spectral resolution observations of the forbidden 1.707 μm Rovibronic SO emissions on Io: Evidence for widespread stealth volcanism. Planet. Sci. J. **1** (2020a). https://doi.org/10.3847/PSJ/ab9eb1

de Pater, I., Luszcz-Cook, S., Rojo, P., Redwing, E., de Kleer, K., Moullet, A.: ALMA observations of Io going into and coming out of Eclipse. Planet. Sci. J. **1**, 60 (25 pp.) (2020b)

de Pater, I., Keane, J.T., de Kleer, K., Davies, A.G.: A 2020 observational perspective of Io. Annu. Rev. Earth Planet. Sci. **49**, 633–668 (2021). https://doi.org/10.1146/annurev-earth-082420-095244

Dols, V.J., Delamere, P.A., Bagenal, F., Kurth, W.S., Paterson, W.R.: Asymmetry of Io's outer atmosphere: Constraints from five Galileo flybys. J. Geophys. Res. **117**, E10010 (2012). https://doi.org/10.1029/2012JE004076

Douté, S. Schmitt, B., Lopes-Gautier, R., Carlson, R., Soderblom, L., Shirley, J., Galileo NIMS Team: Mapping SO_2 frost on Io by the modeling of NIMS hyperspectral images. Icarus **149**, 107 (2001)

Feaga, L.M., McGrath, M., Feldman, P.D., Strobel, D.F.: Detection of atomic chlorine in Io's atmosphere with the Hubble Space Telescope GHRS. Astrophys. J. **610**, 1191–1198 (2004)

Feaga, L.M., McGrath, M., Feldman, P.D.: Io's dayside SO_2 atmosphere. Icarus **201**, 570–584 (2009)

Fegley, B., Zolotov, M.Yu.: Chemistry of sodium, potassium, and chlorine in volcanic gases on Io. Icarus **148**, 193–210 (2000)

Feldman, P.D., Strobel, D.F., Moos, H.W., Retherford, K.D., Wolven, B.C., McGrath, M.A., Roesler, F.L., Woodward, R.C., Oliversen, R.J., Ballester, G.E.: Lyman-α imaging of the SO_2 distribution on Io. Geophys. Res. Lett. **27**, 1787–1790 (2000)

Geissler, P.E., Goldstein, B.D.: Plumes and their deposits. In: Lopes, R.M., Spencer, J.R. (eds.) Io after Galileo: A New View of Jupiter's Volcanic Moon, pp. 163–192. Springer. ISBN 3-540-34681-3 (2007)

Geissler, P., McMillan, M.T.: Galileo observations of volcanic plumes on Io. Icarus **197**, 505–518 (2008)

Geissler, P.E., McEwen, A.S., Ip, W., Belton, M.J.S., Johnson, T.V., Smyth, W.H., Ingersoll, A.P.: Galileo imaging of atmospheric emissions from Io. Science **285**, 870–874 (1999)

Geissler, P.E., Smyth, W.H., McEwen, A.S., Ip, W., Belton, M.J.S., Johnson, T.V., Ingersoll, A.P., Rages, K., Hubbard, W., Dessler, A.J.: Morphology and time variability of Io's visible aurora. J. Geophys. Res. **106**, 26,137–26,146 (2001)

Geissler, P., McEwen, A., Porco, C., Strobel, D.F., Saur, J., et al.: Cassini observations of Io's visible aurorae. Icarus **172**, 127–140 (2004a)

Geissler, P., McEwen, A.S., Phillips, C.B., Keszthelyi, L.P., Spencer, J.: Surface changes on Io during the Galileo mission. Icarus **169**, 29–64 (2004b)

Giono, G., Roth, L.: Io's SO_2 atmosphere from HST Lyman-α images: 1997 to 2018. Icarus **359**, 114212 (2021)

Gratiy, S.L., Walker, A.C., Levin, D.A., Goldstein, D.B., Varghese, P.L., et al.: Multi-wavelength simulations of atmospheric radiation from Io with a 3-D spherical-shell backward Monte Carlo radiative transfer model. Icarus **207**, 394–408 (2010)

Grava, C., Schneider, N.M., Leblanc, F., Morgenthaler, J.P., Mangano, V., Barbieri, C.: Solar control of sodium escape from Io. J. Geophys. Res. **119**, 404–415 (2014)

Hendrix, A.R., Barth, C.A., Hord, C.W.: Io's patchy SO_2 atmosphere as measured by the Galileo ultraviolet spectrometer. JGR, **104**, 11817–11826 (1999)

Hoey, W., Trafton, L., Ackley, P., Goldstein, D.B., Varghese, P.L.: Variations in the canopy shock structures of massive extraterrestrial plumes: Parametric DSMC simulation of the 2007 Tvashtar observations. Icarus **363**, 114431 (2021)

Hornung, H.G.: A model problem for a supersonic gas jet from a moon. J. Fluid Mech. **795**, 950–971 (2016)

Ingersoll, A.P.: Io meteorology: How atmospheric pressure is controlled locally by volcanos and surface frosts. Icarus **81**, 298–313 (1989)

Ingersoll, A.P., Summers, M.E. Schlipf, S.G.: Supersonic meteorology of Io: Sublimation-driven flow of SO_2. Icarus **64**, 375–390 (1985)

Jessup, K.L., Spencer, J.R.: Characterizing Io's Pele, Tvashtar and Pillan plumes: Lessons learned from Hubble. Icarus **218**, 378–405 (2012)

Jessup, K.L., Spencer, J.R.: Spatially resolved HST/STIS observations of Io's dayside equatorial atmosphere. Icarus **248**, 165–189 (2015)

Jessup, K.L., Spencer, J.R., Yelle, R.: Sulfur volcanism on Io. Icarus **192**, 24–40 (2007)

Johnson, T.V., Matson, D.L., Blaney, D.L., Veeder, G.J., Davies, A.: Stealth plumes on Io. Geophys. Res. Lett. **22**, 3293–3296 (1995)

Kieffer, S.W.: Ionian volcanism. In: Morrison, D. (ed.) Satellites of Jupiter, pp. 647–723. Univ. of Ariz. Press (1982)

Kieffer, S.W., Lopes-Gautier, R., McEwen, A., Smythe, W., Keszthelyi, L., Carlson, R.: Prometheus: Io's wandering plume. Science **288**, 1204–1208 (2000)

Kliore, A., Cain, D.L., Fjeldbo, G., Seidel, B.L., Rasool, S.I.: Preliminary results on the atmospheres of Io and Jupiter from the *Pioneer 10* S-Band occultation experiment. Science **183**, 323–324 (1974)

Koga, R., Tsuchiya, F., Kagitani, M., Sakanoi, T., Yoshioka, K., Yoshikawa, I., et al.: Volcanic change of Io's neutral oxygen cloud and plasma torus observed by Hisaki. J. Geophys. Res. Space Phys. **124**, 10318–10331 (2019)

Kosuge, S., Aoki, K., Inoue, T., Goldstein, D.B., Varghese, P.L.: Unsteady flows in Io's atmosphere caused by condensation and sublimation during and after eclipse: Numerical study based on a model Boltzmann equation. Icarus **221**, 658–669 (2012)

Krüger, H., Horányi, M., Krivov, A.V., Graps, A.L.: Jovian dust: streams, clouds and rings. In: Bagenal, F., Dowling, T.E., McKinnon, W. (eds.) Jupiter: Planet, Satellites & Magnetosphere. Cambridge University Press, Cambridge, UK (2004)

Kupo, I., Mekler, Yu., Eviatar, A.: Detection of ionized sulfur in the Jovian magnetosphere. Astrophys. J. Lett. **205**, L51–L53 (1976)

Lanzerotti, L.J., Brown, W.L., Augustyniak, W.M., Johnson, R.E., Armstrong, T.P.: Laboratory studies of charged particle erosion of SO_2 ice and applications to the frosts of Io. Astrophys. J. **259**, 920–929 (1982)

Laver, C., de Pater, I., Marchis, F.: Tvashtar awakening detected in April 2006 with OSIRIS at the W.M. Keck Observatory. Icarus **191**, 749–754 (2007)

Laver, C., de Pater, I.: The global distribution of Sulfur Dioxide ice on Io, observed with OSIRIS on the W. M. Keck telescope. Icarus **201**, 172–181 (2009)

Leblanc, F., Oza, A.V., Leclercq, L., Schmidt, C., Cassidy, T., Modolo, R., Chaufray, J.Y., Johnson, R.E.: On the orbital variability of Ganymede's atmosphere. Icarus **293**, 185–198 (2017)

Lellouch, E.: Urey Prize Lecture. Io's atmosphere: not yet understood. Icarus **124**, 1–21 (1996)

Lellouch, E., Belton, M.J.S., de Pater, I., Gulkis, S., Encrenaz, T.: Io's atmosphere from microwave detection of SO_2. Nature **346**, 639–641 (1990)

Lellouch, E., Belton, M., de Pater, I., Paubert, G., Gulkis, S., Encrenaz, T.: The structure, stability, and global distribution of Io's atmosphere. Icarus **98**, 271–295 (1992)

Lellouch, E., Strobel, D., Belton, M.J.S., Summers, M.E., Paubert, G., Moreno, R.: Detection of sulfur monoxide in Io's atmosphere. Astrophys. J. Lett. **459**, L107–L110 (1996)

Lellouch, E., Paubert, G., Moses, J.I., Schneider, N.M., Strobel, D.F.: Volcanically-emitted sodium chloride as a source for Io's neutral clouds and plasma torus. Nature **421**, 45–47 (2003)

Lellouch, E., McGrath, M.A., Jessup, K.L.: Io's atmosphere. In: Lopes, R.M., Spencer, J.R. (eds.) Io after Galileo: A New View of Jupiter's Volcanic Moon, pp. 231–264. Springer. ISBN 3-540-34681-3 (2007)

Lellouch, E., Ali-Dib, M., Jessup, K.-L., Smette, A., Käufl, H.-U., Marchis, F.: Detection and characterization of Io's atmosphere from high-resolution 4-μm spectroscopy. Icarus **253**, 99–114 (2015)

Marconi, M.L.: A kinetic model of Ganymede's atmosphere. Icarus **190**, 155–174 (2007)

McDoniel, W.J.: Realistic simulation of Io's Pele Plume and its effect on Io's atmosphere. University of Texas PhD Dissertation (2015)

McDoniel, W.J., Goldstein, D.B., Varghese, P.L., Trafton, L.M.: Three-dimensional simulation of gas and dust in Io's Pele Plume. Icarus **257**, 251–274 (2015)

McDoniel, W.J., Goldstein, D.B., Varghese, P.L., Trafton, L.M.: The interaction of Io's plumes and sublimation atmosphere. Icarus **294**, 81–97 (2017)

McDoniel, W.J., Goldstein, D.B., Varghese, P.L., Trafton, L.M.: Simulation of Io's plumes and Jupiter's plasma torus. Phys. Fluids **31**, 077103 (2019)

McEwen, A.S., Keszthelyi, L., Spencer, J.R., Schubert, G., Matson, D.L., et al.: High-temperature silicate volcanism on Jupiter's Moon Io. Science **281**, 87–90 (1998)

McGrath, M.A., Belton, M.J.S., Spencer, J.R., Sartoretti, P.: Spatially resolved spectroscopy of Io's Pele plume and SO_2 atmosphere. Icarus **146**, 476–493 (2000)

McGrath, M.A., Lellouch, E., Strobel, D.F., Feldman, P.D., Johnson, R.E.: Satellite atmospheres. In: Bagenal, F., Dowling, T.E., McKinnon, W. (eds.) Jupiter: Planet, Satellites & Magnetosphere, pp. 457–483. Cambridge University Press, Cambridge, UK (2004)

Mendillo, M., Baumgardner, J., Flynn, B., Hughes, J.W.: The extended sodium nebula of Jupiter. Nature **348**, 31–314 (1990)

Mendillo, M., Wilson, J., Spencer, J., Stansberry, J.: Io's volcanic control of Jupiter's extended neutral clouds. Icarus **170**, 430–442 (2004)

Milillo, A., Orsini, S., Hsieh, K.C., Baragiola, R., Fama, M., Johnson, R., Mura, A., Plainaki, C., Sarantos, M., Cassidy, T.A., De Angelis, E., Desay, M., Goldstein, R., Ip, W.-H., Killem, R.,

Livi, S.: Observing planets and small bodies in sputtered high-energyatom fluxes. JGR **116**, AO7229 (2011)

Moore, C., Goldstein, D.B., Varghese, P., Trafton, L., Stewart, B.: 1-D DSMC simulation of Io's atmospheric collapse in eclipse. Icarus **201**, 585–597 (2009)

Morabito, L.A., Synnott, D.P., Kupferman, P.N., et al.: Discovery of currently active extraterrestrial volcanism. Science **204**, 972 (1979)

Morrison, D., Cruikshank, D.P.: Thermal properties of the Galilean satellites. Icarus **18**, 224–236 (1973)

Moses, J.I., Nash, D.B.: Phase transformations and the spectral reflectance of solid sulfur - Can metastable sulfur allotropes exist on Io? Icarus **89**, 277–304 (1991)

Moses, J.I., Zolotov, M.Y., Fegley, B.: Photochemistry of a volcanically driven atmosphere on Io: Sulfur and oxygen species from a Pele-type eruption. Icarus **156**, 76–106 (2002a)

Moses, J.I., Zolotov, M.Y., Fegley, B.: Alkali and chlorine photochemistry in a volcanically driven atmosphere on Io. Icarus **156**, 107–135 (2002b)

Moullet, A., Lellouch, E., Moreno, R., Gurwell, MA., Moore, C.: First disk- resolved millimeter observations of Io's surface and SO_2 atmosphere. Astrophys. Astron. **482**, 279–292 (2008)

Moullet, A., Gurwell, M.A., Lellouch, E., Moreno, R.: Simultaneous mapping of SO_2, SO, NaCl in Io's atmosphere with the submillimeter array. Icarus **208**, 353–365 (2010)

Moullet, A., Lellouch, E., Moreno, R., Gurwell, M., Black, J.H., Butler, B.: Exploring Io's atmospheric composition with APEX: First measurement of $^{34}SO_2$ and tentative detection of KCl. Astrophys. J. **776**, 32, 9 pp. (2013). https://doi.org/10.1088/0004-637X/776/1/32

Moullet, A., Lellouch, E., Gurwell, M., Moreno, R., Black, J., Butler, B.: Distribution of alkali gases in Io's atmosphere. In: AAS DPS Meeting #47, Abstract 311.31 (2015)

Peale, S.J., Cassen, P., Reymolds, R.T.: Melting of Io by tidal dissipation. Science **203**, 892–894 (1979)

Pearl, J.C., Hanel, R., Kunde, V., Maguire, W., Fox, K., et al.: Identification of gaseous SO_2 and new upper limits for other gases on Io. Nature **288**, 757–758 (1979)

Postberg, F., Kempf, S., Srama, R., Green, S.F., Hillier, J.K., McBride, N., Grün, E.: Composition of jovian dust stream particles. Icarus **183**, 12–134 (2006)

Rathbun, J.A., Spencer, J.R., Tamppari, L.K., Martin, T.Z., Barnard, L., Travis, L.D.: Mapping of Io's thermal radiation by the Galileo photopolarimeter-radiometer (PPR) instrument. Icarus **169**, 127–139 (2004)

Redwing, E., de Pater, I., Luszcz-Cook, S., de Kleer, K., Moullet, A., Rojo, P.: NaCl and KCl in Io's atmosphere. Planet. Sci. J., **3**:238, 11 (2022)

Retherford, K.D.: Io's aurora: HST/STIS observations. PhD thesis, Johns Hopkins University, Baltimore, MD (2002)

Retherford, K.D., Spencer, J.R., Stern, S.A., Saur, J., Strobel, D.F., et al.: Io's atmospheric response to eclipse: UV aurorae observations. Science **318**, 237–240 (2007)

Retherford, K.D., Roth, L., Becker, T.M., Feaga, L.M., Tsang, C.C.C., Jessup, K.L., Grava, C.: Io's atmosphere silhouetted by Jupiter Ly-α. Astron. J. **158**, 154 (2019)

Roesler, F.L., Moos, H.W., Oliversen, R.J., Woodward, Jr, R.C., Retherford, K.D., Scherb, F., McGrath, M.A., Smyth, W.H., Feldman, P.D., Strobel, D.F.: Far-ultraviolet imaging spectroscopy of Io's atmosphere with HST/STIS. Science **283**, 353–356 (1999)

Roth, L., Saur, J., Retherford, K.D., Strobel, D.F., Spencer, J.R.: Simulation of Io's auroral emission: Constraints on the atmosphere in eclipse. Icarus **214**, 495–509 (2011)

Roth, L., Saur, J., Retherford, K.D., Feldman, P.D., Strobel, D.F.: A phenomenological model of Io's UV aurora based on HST/STIS observations. Icarus **228**, 386–406 (2014)

Roth, L., Boissier, B., Moullet, A., Sánchez-Monge, Á, de Kleer, K. et al.: An attempt to detect transient changes in Io's SO_2 and NaCl atmosphere. Icarus **350**, article id. 113925 (2020)

Roth, L., Ivchenko, N., Gladstone, G.R., Saur, J., Grodent, D., Bonfond, B., Molyneux, P.M., Retherford, K.D.: Evidence for a sublimated water atmosphere on Ganymede from Hubble Space Telescope observations. arXiv:2106.03570 (2021)

Sartoretti, P., McGrath, M.A., Paresce, F.: Disk-resolved imaging of Io with the Hubble Space Telescope. Icarus, **108**, 272–284 (1994)

Sartoretti, P., Belton, M.J.S., McGrath, M.A.: SO_2 distributions on Io. Icarus, **122**, 273–287 (1996).

Saur, J., Neubauer, F.M., Strobel, D.F., Summers, M.E.: Interpretation of Galileo's Io plasma and field observations: I0, I24, and I27 flybys and close polar passes. J. Geophys. Res. **107**(A8), 1422 (2002). https://doi.org/10.1029/2001JA005067

Saur, J., Strobel, D.F.: Relative contributions of sublimation and volcanoes to Io's atmosphere inferred from its plasma interaction during solar eclipse. Icarus **171**, 411–420 (2004)

Schneider, N.M., Bagenal, F.: Io's neutral clouds, plasma torus, and magnetospheric interaction. In: Lopes, R.M., Spencer, J.R. (eds.) Io after Galileo: A New View of Jupiter's Volcanic Moon, pp. 265–286. Springer. ISBN 3-540-34681-3 (2007)

Secosky, J.J., Potter, M.: A Hubble Space Telescope study of posteclipse brightening and albedo changes on Io. Icarus **111**, 73–78 (1994)

Sinton, W.M., Kaminsky, C.: Infrared observations of eclipses of Io, its thermophysical parameters, and the thermal radiation of the Loki volcano and environs. Icarus **75**, 207–232 (1988)

Spencer, J.R., Schneider, N.M.: Io on the eve of the Galileo mission. Annu. Rev. Earth Planet. Sci. **24**, 125–190 (1996)

Spencer, J.R., Sartoretti, P., Ballester, G.E., McEwen, A.S., Clarke, J.T., McGrath, M.A.: The Pele plume (Io): observations with the hubble space telescope. Geophys. Res. Lett. **24**, 2471–2474 (1997)

Spencer, J.R., Jessup, K.L., McGrath, M.A., Ballester, G.E., Yelle, R.: Discovery of gaseous S_2 in Io's Pele Plume. Science **288**, 1208–1210 (2000)

Spencer, J.R., Lellouch, E., Richter, M.J., López-Valverde, M.A., Lea Jessup, K., et al.: Mid-infrared detection of large longitudinal asymmetries in Io's SO_2 atmosphere. Icarus **176**, 283–304 (2005)

Spencer, J.R., Stern, S.A., Cheng, A.F., et al.: Io volcanism seen by *New Horizons*: A major eruption of the Tvashtar volcano. Science **318**, 240–243 (2007)

Strobel, D.F., Wolven, B.C.: The atmosphere of Io: abundances and sources of sulfur dioxide and atomic hydrogen. Astrophys. Space Sci. **277**, 271–287 (2001)

Strobel, D.F., Zhu, X., Summers, M.E.: On the vertical structure of Io's atmosphere. Icarus **111**, 18–30 (1994)

Summers, M.E., Strobel, D.F.: Photochemistry and vertical transport in Io's atmosphere and ionosphere. Icarus **120**, 290–316 (1996)

Trafton, L.M.: Detection of a potassium cloud near Io. Nature **258**, 690–692 (1975)

Trafton, L.M., Caldwell, J.J., Barnet, C., Cunningham, C.C.: The gaseous sulfur dioxide abundance over Io's leading and trailing hemispheres: HST spectra of Io's C^1B_2–X^1A_1 band of SO_2 near 2100 Angstrom. Astrophys.J., **456**, 384–392 (1996)

Trafton, L.M., Moore, C.H., Goldstein, D.B., Varghese, P.L., McGrath, M.A.: HST/STIS observations and simulation of Io's emission spectrum in Jupiter shadow: Probing Io's Jupiter-facing eclipse atmosphere. Icarus **220**, 1121–1140 (2012)

Tsang, C.C.C., Spencer, J.R., Lellouch, E., Lopez-Valverde, M.A., Richter, M.J., Greathouse, T.K.: Io's atmosphere: Constraints on sublimation support from density variations on seasonal timescales using NASA IRTF/TEXES observations from 2001 to 2010. Icarus **217**, 277–296 (2012)

Tsang, C.C.C., Spencer, J.R., Jessup, K.L.: Synergistic observations of Io's atmosphere in 2010 from HST-COS in the mid-ultraviolet and IRTF-TEXES in the mid-infrared. Icarus, **226**, 604–616 (2013a)

Tsang, C.C.C., Spencer, J.R., Lellouch, E., Lopez-Valverde, M.A., Richter, M.J., Greathouse, T.K., Roe, H.: Io's contracting atmosphere post 2011 perihelion: Further evidence for partial sublimation support on the anti-Jupiter hemisphere. Icarus **226**, 1177–1181 (2013b)

Tsang, C.C.C., Spencer, J.R., Jessup, K.L.: Non-detection of post-eclipse changes in Io's Jupiter-facing atmosphere: Evidence for volcanic support? Icarus **248**, 243–253 (2015)

Tsang, C.C.C., Spencer, J.R., Lellouch, E., Lopes-Valverde, M.A., Richter, J.J.: The collapse of Io's primary atmosphere in Jupiter eclipse. JGR **121**, 1400–1410 (2016)

Tsuchiya, F., Kagitani, M., Yoshioka, K., Kimura, T., Murakami, G., Yamazaki, A., et al.: Local electron heating in the Io plasma torus associated with Io from the Hisaki satellite observation. J. Geophys. Res. Space Phys. **120**, 10,317–10,333 (2015)

Wagman, D.D.: Sublimation Pressure and Enthalpy of SO_2. Chem. Thermodynamics Data Center, Nat. Bureau of Standards, Washington, DC (1979)

Walker, A.C., Gratiy, S.L., Goldstein, D.B., Moore, C.H., Varghese, P.L., et al.: A comprehensive numerical simulation of Io's sublimation-driven atmosphere. Icarus **207**, 409–432 (2010)

Walker, A.C., Moore, C.H., Goldstein, D.B., Varghese, P.L., Trafton, L.M.: A parametric study of Io's thermophysical surface properties and subsequent numerical atmospheric simulations based on the best fit parameters. Icarus **220**, 225–253 (2012)

Wolven, B.C., Moos, H.W., Retherford, K.D., Feldman, P.D., Strobel, D.F., Smyth, W.H., Roesler, F.L.: Emission profiles of neutral oxygen and sulfur in Io's exospheric corona. J. Geophys. Res. **106**, 26155–26182 (2001)

Wu, R.C.Y., Yang, B.W., Chen, F.Z., Judge, J., Caldwell, J., Trafton, L.M.: Measurements of high-, room-, and low-temperature photoabsorption cross sections of SO_2 in the 2080- to 2950-A region, with applications to Io. Icarus **145**, 289–296 (2000)

Yoneda, M., Kagitani, M., Tsuchiya, F., Sakanoi, T., Okano, S.: Brightening event seen in observations of Jupiter's extended sodium nebula. Icarus **261**, 31–33 (2015)

Yoshikawa, I., Yoshioka, K., Murakami, F., aYmazaki, A., Tsuchiya, F., et al.: Extreme ultraviolet radiation measurement for planetary atmospheres, magnetospheres from Earth-orbiting spacecraft (EXCEED). Space Sci. Rev. **184**, 237–258 (2014)

Zhang, J., Goldstein, D.B., Varghese, P.L., Gimelshein, N.E., Gimelshein, S.F., Levin, D.A.: Simulation of gas dynamics and radiation in volcanic plumes of Io. Icarus **163**, 182–187 (2003)

Zolotov, M.Y., Fegley Jr, B.: Volcanic production of sulfur monoxide (SO) on Io. Icarus Note **132**, 431–434 (1998)

Zolotov, M.Y., Fegley Jr., B.: Oxidation state of volcanic gases and the interior of Io. Icarus **141**, 40–52 (1999)

Zolotov, MY., Fegley Jr., B.: Eruption conditions of Pele Volcano on Io inferred from chemistry of its volcanic plume. GRL **27**, 2789–2792 (2000)

Chapter 9
Space Environment of Io

Fran Bagenal and Vincent Dols

Abstract The moon Io plays major role in the giant magnetosphere of Jupiter. At the same time, the magnetospheric particles and fields affect the moon. The impact of magnetospheric ions on Io's atmosphere supplies clouds of escaping neutral atoms that populate a substantial fraction of its orbit. At the same time, ionization of atoms in the neutral cloud is the primary source of magnetospheric plasma. The stability of this feedback loop depends on the plasma-atmosphere interaction. The purpose of this review is to describe the physical processes that shape the space environment around Io and to show the impact from the planet Jupiter out into interplanetary space.

9.1 Introduction

In the 60s and 70s ground-based observations suggested Io was peculiar: the moon triggered radio emissions and optical emissions indicated clouds of sodium atoms and sulfur ions around Io (see Chap. 2 detailed history). Further hints of Io's peculiarity (specifically, its ionosphere) were indicated by *Pioneers* in 1973–1974. Such strange behavior became more understandable when *Voyager* 1 and 2 flybys of Jupiter in 1979 revealed Io's remarkable volcanism. The *Voyagers* also detected strong UV emissions from a torus of sulfur and oxygen ions surrounding Jupiter at Io's orbit, and measured *in situ* the torus plasma. When *Voyager* 1 passed close to Io, perturbations in the plasma and magnetic field showed Io generating Alfven waves propagating away from the moon, carrying million-Ampere electrical currents along the magnetic field lines towards Jupiter. The *Voyager*s also characterized a plethora of radio and plasma waves, several associated with Io.

The *Galileo* spacecraft went into orbit around Jupiter in 1995, making five successful close flybys of Io between 1995 and 2001, measuring magnetic field

F. Bagenal (✉) · V. Dols
Laboratory for Atmospheric & Space Physics, University of Colorado, Boulder, CO, USA
e-mail: bagenal@colorado.edu; vincent.dols@colorado.edu

R. M. C. Lopes et al. (eds.), *Io: A New View of Jupiter's Moon*, Astrophysics and Space Science Library 468, https://doi.org/10.1007/978-3-031-25670-7_9

and particle perturbations. In late 2000, on its journey to Saturn, the *Cassini* UVIS instrument provided months of high-quality observations of the torus UV emissions, revealing the plasma composition and how the torus changed after a volcanic eruption on Io. The aim of the 2004 monograph *Jupiter: The Planet, Satellites and Magnetosphere* (Bagenal et al. 2004) was to summarize the post-*Galileo* view of the jovian system. Chapters by Kivelson et al. (2004) and Saur et al. (2004) reviewed the current understanding of the electrodynamics of the plasma-moon interactions, and Thomas et al. (2004) reviewed Io's neutral clouds and plasma torus, describing how ~1 ton/s of sulfur and oxygen escapes Io, becomes ionized and trapped in the magnetic field, and moves out to fill Jupiter's vast magnetosphere.

In the meantime, while the *Hubble Space Telescope* (HST) and ground-based telescopes continue to observe the system, new measurements and models have emerged. In spring 2007 the *New Horizons* spacecraft got a gravity assist from Jupiter to speed its journey to Pluto, observing Io's volcanism including an eruption of Tvashtar (Spencer et al. 2007; Rathbun et al. 2014; Tsang et al. 2014) and aurora on Io's flanks (Retherford et al. 2007).

The Japanese space agency (JAXA) put the *Hisaki* satellite into orbit around Earth in September 2013 (Yoshikawa et al. 2014). *Hisaki*'s UV spectrometer has been observing emissions from the Io plasma torus, determining composition, mapping the oxygen neutral cloud, and measuring temporal variations (Yoshioka et al. 2017, 2018; Hikida et al. 2018, 2020; Tsuchiya et al. 2015, 2018, 2019; Koga et al. 2018a, b, 2019; Yoshikawa et al. 2017).

The *Juno* spacecraft went into orbit around Jupiter in July 2016. While the prime *Juno* mission did not bring the spacecraft to the equator inside Europa's orbit, the particles and fields instruments are measuring the consequences of Io's activity both in the middle magnetosphere and close-in over the poles (Bagenal et al. 2017; Bolton et al. 2017a, b). For Juno's extended mission seven flybys of Io are planned. Future missions to Europa further emphasize the need to consider the possible impacts that the changing environment near Io might have on the Europa system.

In the following sections we summarize the primary processes that influence Io's space environment and list the outstanding questions. For a review of the full Io system see de Pater et al. (2021). For a more extensive review of Io's space environment, including comparison with Europa, see Bagenal and Dols (2020).

9.2 Plasma Interactions with Io

Figure 9.1 illustrates the main components of plasma interaction with Io. Figure 9.1a presents the basic geometry and Fig. 9.1b indicates some of the plasma-neutral reactions as the plasma flows around the moon and impacts the atmosphere. Figure 9.1c shows the UV emissions excited in Jupiter's atmosphere by beams of electrons that are generated by the interaction and stream along the magnetic field to the

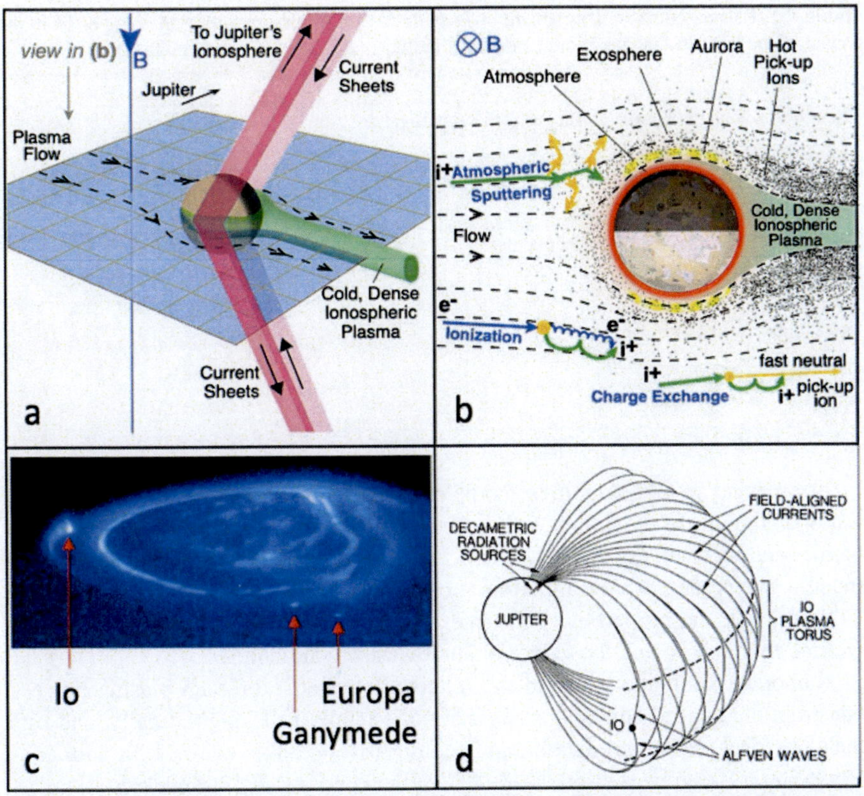

Fig. 9.1 Plasma-Io Electrodynamic Interaction. (**a**, **b**) Schematic of interaction between the surrounding plasma and Io (from Schneider and Bagenal 2007). (**c**) Hubble Space Telescope image of Jupiter's UV aurora (Clarke et al. 2002) showing auroral footprints of Io, Europa and Ganymede. (**d**) Pattern of Alfven waves generated by Io, bouncing between hemispheres of Jupiter where bursts of radio emission are generated above the ionosphere (Gurnett and Goertz 1981). (From Bagenal and Dols 2020)

planet. Figure 9.1d shows the train of waves generated by Io propagating around Jupiter.

Table 9.1 summarizes the range of plasma conditions upstream of Io. During a jovian synodic period (13 h at Io), the moon experiences variable upstream plasma densities and thus, a variable strength of the interaction. When looking at the plasma conditions listed in Table 9.1, we note that the atmosphere/plasma interaction is intrinsically time-variable for two main reasons. The first is the effect of the ~10° tilt of Jupiter's magnetic field which produces an inclination of the torus centrifugal equator by 7° relative to Io's orbital plane. Secondly, the moon's atmosphere is also variable depending on the atmospheric sources. At Io, the SO_2 atmosphere is primarily sublimation-supported, varying with illumination and when there are particularly strong volcanic eruptions (see Chap. 8).

Table 9.1 Plasma conditions upstream of Io with the range corresponding to changes from the center of the torus to the maximum centrifugal latitude. (Based on Kivelson et al. 2004)

Property upstream of Io	
Range from magnetic equatorRange from centrifugal equator	$\pm 1\ R_J \pm 0.65\ R_J$
Local jovian magnetic field	1720–2080 nT
Plasma-moon relative velocity	53–57 km/s
Electron density (Ne)	1200–3800 cm^{-3}
Alfven Mach numberPlasma sheet center	0.3
Thermal electron temperature	5 eV
Hot electrons	0.2% at 40 eV
Major ions (Ni/Ne)	S^{++} (20%) O^{+} (25%)
Average Ion temperature	20–90 eV
Average ion gyroradius	~3 km

Approaches to modeling these complex plasma-moon interactions tend to either focus on the electrodynamics with limited chemistry or assume simple electrodynamics and focus on the physical chemistry. Beyond matching the observations, the modeling approach allows an exploration of the main features of the interaction. Table 9.2 lists the properties of the interaction as they have been observed on five *Galileo* flybys of Io and the values of some measured quantities.

Numerical simulations reveal the relative importance of each plasma process, constrain the neutral atmosphere distribution radially and longitudinally, and estimate the local plasma production and neutral losses (Linker et al. 1998; Saur et al. 1999, 2002, 2003; Dols et al. 2008, 2012; Blocker et al. 2018; Šebek et al. 2019). Such models are also used to explore the presence of an induced magnetic field either in Io's asthenosphere (Khurana et al. 2011) and/or core (Roth et al. 2017), as discussed further in Sect. 9.2.5.

9.2.1 *Electrodynamics*

Io's apparent control of radio emissions provoked early ideas of the moon acting as a electrically-conducting object moving through a magnetic field, with currents flowing from Io to Jupiter and back to Io. The first models of the interaction assumed a stationary current loop between Io and the Jovian ionosphere—described as a "unipolar inductor" by Goldreich and Lyden-Bell (1969). When the dense plasma torus was discovered by Voyager, this model of the interaction was replaced by the Alfven wing model illustrated in Fig. 9.1a, d. Flying south of Io, Voyager detected significant magnetic and flow perturbations propagating south from Io (Acuna et al. 1981; Belcher et al. 1981) consistent with a large (~3 MAmp) current flowing from Io to Jupiter.

The plasma-atmosphere interaction produces these large magnetic and velocity perturbations at Io, which propagate along field lines as Alfven waves towards the

Table 9.2 Plasma-atmosphere interactions. (1) Saur et al. 1999; (2) Kivelson et al. 1996; (3) Dols et al. 2008; (4) Saur et al. 2002; (5) Bagenal 1997; (6) Frank and Paterson 1999; (7) Retherford et al. 2003, 2007; (8) Wolven et al. 2001; Ballester et al. 1987; (9) Bouchez et al. 2000; Geissler et al. 2004; (10) Roth et al. 2011; (11) Williams et al. 1996, Williams and Thorne 2003; Frank and Paterson 1999; (12) Mauk et al. 2001; (13) Bonfond et al. 2008

Plasma-Io interaction	
Galileo flybys	J0, I24, I27, I31, I32
Flow diversion	95% (1)
Magnetic perturbation	700 nT (2)
Main ion-neutral process	Collision and chex (1, 3)
Ionization	• Thermal electron impact (2, 3) • Photoionization ~15% (3) • Electron beams (2, 4)
Main ion produced	SO_2^+ ~200 kg/s (1, 3, 5)
Wake	N_e ~ 30,000 cm^{-3} Slow flow (~1 km/s) Cold ions (~1 eV) (5, 6)
Auroral emissions	Mainly O (7) Also S (8), Na, K (9)
Aurora location	• Equatorial flanks (10) • Polar emission stronger towards plasmasheet (7) • Corona (7)
Electron beams location	In wake J0, above poles (I31, I32) (11)
Electron beams energy	>150 eV Power law distribution (12)
Footprint emissions	Multi-spot structure + long tail (13)

ionosphere of Jupiter and forms a stationary structure in the reference frame of Io called an Alfven wing (Goertz 1980; Neubauer 1980; Belcher 1987). The plasma flow is diverted not only around the solid body of the moon but also around the entire Alfven wing extending from Io to the ionosphere of Jupiter (see Fig. 21.2 in Kivelson et al. 2004). Thus, instead of a single closed current loop between Io and Jupiter, the long timeframe for Alfven waves to propagate through the plasma torus results in the standing structure illustrated in Fig. 9.1d.

Saur et al. (2013) propose an estimate of the relative strength of the interaction at the moons as the ratio of the ionospheric electric field to the corotation electric field (called $\overline{\alpha}$). This ratio is based on the calculation of the Alfven, Pedersen and Hall conductances of the moon ionosphere (see also Southwood et al. 1980). This $\overline{\alpha}$ parameter is a quantitative estimate of the divergence and slowing of the flow around the moon's ionosphere: $\overline{\alpha} = 0$ corresponds to the case of no electrodynamical interaction (plasma interaction with an insulating moon) where the plasma impinges, undiverted, onto the moon's surface, while $\overline{\alpha} = 1$ corresponds to the strongest interaction (the moon as a perfect conducting body) where the plasma is fully diverted around the moon by the electrodynamic interaction and does not reach the surface. The interaction at Io has $\overline{\alpha}$ ~ 0.96 when it is located in the center of the torus (Saur et al. 2013).

9.2.2 *Physical Chemistry*

In Fig. 9.1b, the impact of thermal electrons on the neutral atmosphere is the main local source of new ions (called "mass-loading"), which are mostly molecular SO_2^+ at Io. Photoionization represents only ~10–15% of the total ionization rate (Saur et al. 1999). A molecular atmosphere is very efficient at cooling the impinging thermal electrons via ionization and dissociation of molecular neutrals and also by exciting molecular electronic, vibrational and rotational levels (Saur et al. 1999; Dols et al. 2008, 2012). These cooling processes are so efficient that the primary electrons would rapidly become too cold to provide any further ionization if they were not replenished by the content of the flux tube above and below the moon.

Ion charge exchange processes do not provide new charge density but they can change the ion composition and constitute an important sink of momentum for the upstream plasma (called "momentum-loading"). After an ionization or a charge exchange reaction, the new ion is initially at rest in the frame of the moon. It is then "picked up" by the background bulk plasma flow and also starts a gyro-motion at the local flow velocity. The pickup process also affects the average ion temperature of the plasma. Depending on the location of this pickup, the local flow velocity is larger than the upstream velocity (on the flanks) or smaller (in the deep atmosphere) and the pickup process is either a gain or a loss of energy that affects the average ion temperature. For instance, an SO_2^+ ion picked up by a 60 km/s flow at Io will gain 1080 eV, which represents a net heating of the upstream plasma (with a typical temperature of ~100 eV). In Fig. 9.1b, the ion/neutral process called "atmospheric sputtering" refers to ion/neutral elastic collisions without exchange of charge and after multiple collisions. In this process, a neutral is ejected from the atmosphere (McGrath and Johnson 1987), feeding an extended corona and a neutral cloud. The resulting neutral clouds are discussed in Sect. 9.3 of this chapter.

Figure 9.2 illustrates an example approach to modeling the physical chemistry of the plasma-atmosphere interaction at Io. Numerical simulations of this interaction are constrained by in-situ observations of the plasma properties close to the moon and by remote observations of the auroral emissions in Io's atmosphere.

The *Galileo* spacecraft made five flybys of Io between 1995 and 2001 revealing very strong plasma and field perturbations. The relative velocity of the torus plasma in Io's frame is ~60 km/s. The plasma is mainly composed of S^{++} and O^+ ions at temperatures of ~100 eV. The gyroradius of the thermal ions is ~3 km and the gyrofrequency is ~1.5 Hz (see Table 9.1). When Io is in the centrifugal plane of the torus, the flow is quasi-stagnated at the point where the plasma impinges on the atmosphere, with 95% of the plasma diverted around the flanks. The electron flow close to Io is strongly twisted towards Jupiter because of the Hall conductivity of the ionosphere (Saur et al. 1999). The magnetic perturbation reaches ~700 nT in a background field of ~1800 nT (Kivelson et al. 1996).

Downstream of Io, *Galileo* detected a wake of plasma that was very dense (~30,000 cm^{-3}), very slow (flow speed <1 km s^{-1}) and very cold (T_i ~ few eV) (Frank et al. 1996; Bagenal 1997). Bi-directional parallel electron beams above the

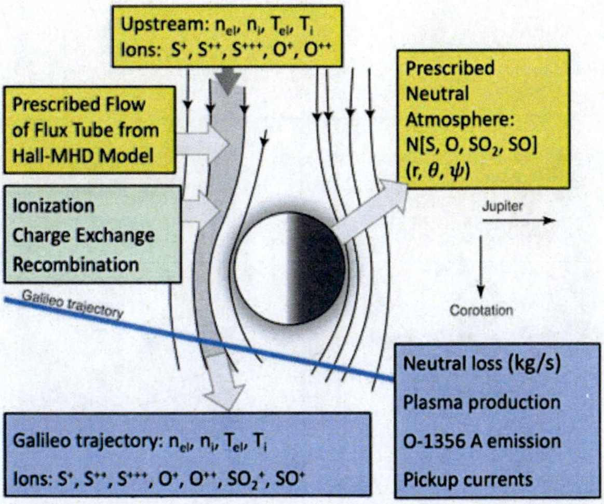

Fig. 9.2 Sketch of the concept of the multi-species chemistry approach of the local interaction at Io. A parcel of plasma of prescribed composition and energy is carried by the prescribed plasma flow into the atmosphere of Io. The ion-electron-neutral processes change the composition and energy of the plasma in the parcel, which are then collected along a specific Galileo flyby of Io (here the J0 flyby in red) and compared to the observations. Such simulations aim at constraining the atmospheric distribution and composition of Io's atmosphere. Based on Dols et al. (2012)

poles and in Io's wake were also detected with an energy ranging from ~140 eV to several 10s keV (Frank and Paterson 1999; Williams et al. 1996; Williams and Thorne 2003; Mauk et al. 2001). Finally, Electro-Magnetic Ion Cyclotron (EMIC) waves were observed downstream, far from Io at the $SO_2{}^+$ and SO^+ ion gyro-frequencies, suggesting a pickup process far (>10 R_{Io}) from Io (Warnecke et al. 1997; Russell and Kivelson 2000, 2001).

In Figs. 9.3, 9.4 and 9.5, we illustrate the plasma-neutral interaction at Io using the multi-species chemistry approach sketched in Fig. 9.2. Although the model results vary with the assumptions of the simulation, they provide reasonable estimates of the contribution of each process and the resulting plasma properties. The simulations presented here are 2D simulations in the equatorial plane of Io similar to Dols et al. (2008) and first published in Bagenal and Dols (2020). The plasma flow is prescribed as an incompressible flow around a conducting obstacle. The simulation shown in Fig. 9.3 does not include the effects of parallel electron beams detected by *Galileo*. These parallel electron beams probably provide much of the ionization in the wake, particularly when a dense atmosphere is present (Saur et al. 2002; Dols et al. 2008).

Figure 9.3 shows the plasma properties in the equatorial plane of Io. The SO_2 distribution is assumed cylindrically symmetrical and thus does not include any day-night asymmetry resulting from a collapse of the SO_2 atmosphere at night, nor other asymmetries from plumes, etc. The radial distribution is based on Saur et al. (1999). These authors propose a radial hydrostatic atmosphere with a surface scale height of

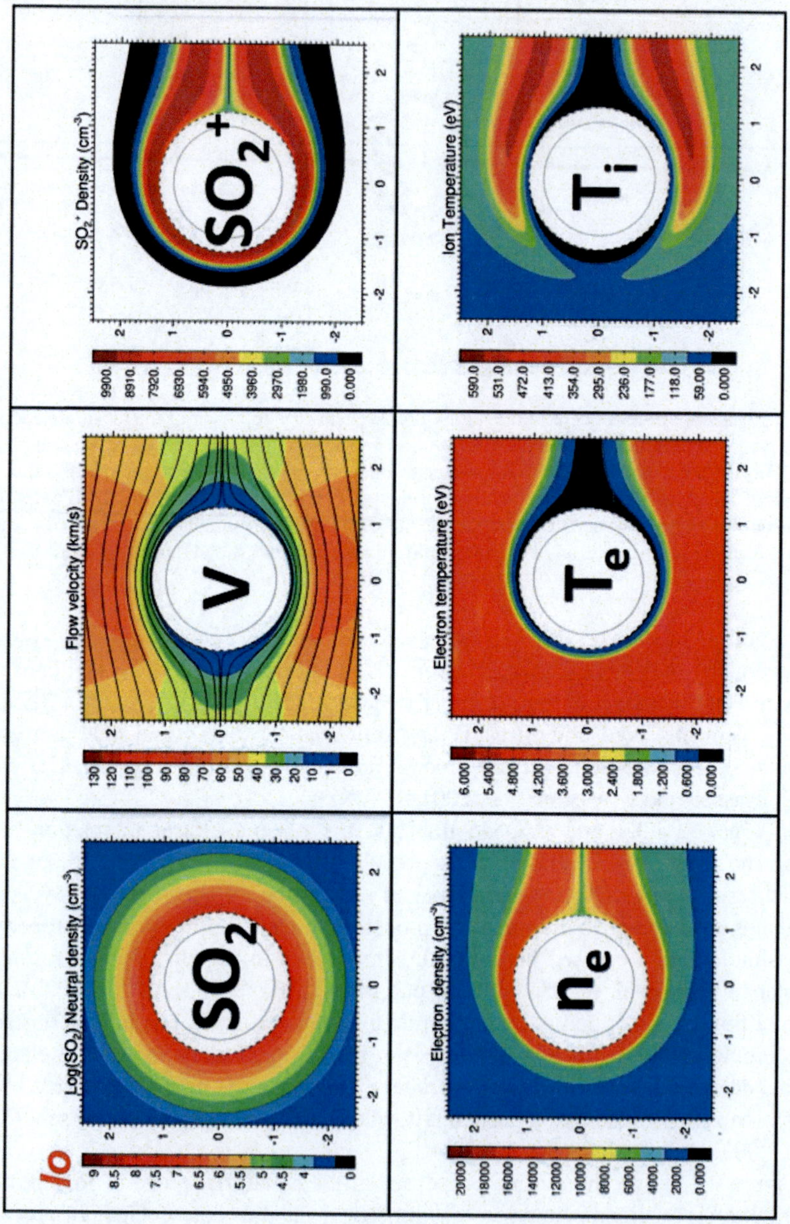
Io
Log(SO2) Neutral density (cm-3)
SO2
Flow velocity (km/s)
V
SO2+ Density (cm-3)
SO2+
Electron density (cm-3)
ne
Electron temperature (eV)
Te
Ion Temperature (eV)
Ti

Fig. 9.3 Plasma-atmosphere interaction at Io. Plasma properties at Io from a multi-species chemistry approach of the atmosphere-plasma interaction sketched in Fig. 9.1. The plasma flow is prescribed as an incompressible flow around a conducting ionosphere, which is assumed to extend to 1.26 R_{Io} (see text). From upper left in clockwise direction: the prescribed SO_2 neutral density of the corona; the plasma flow slowed upstream and downstream and accelerated on the flanks; the resulting average ion temperature, which increases where SO_2^+ ions are picked up by a fast flow; the electron density resulting from the torus thermal electron ionization; the electron temperature, which decreases in the molecular atmosphere because of the efficient molecular cooling processes; and the resulting SO_2^+ density. These simulations do not include the ionization caused by the parallel electron beams detected in the wake of Io and nor do they include a day-night asymmetry of the SO_2 atmosphere. (From Bagenal and Dols 2020)

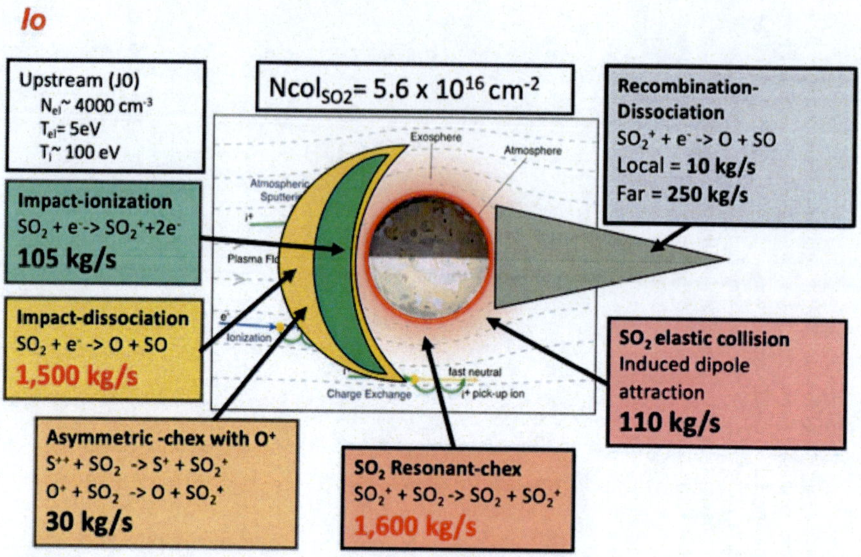

Fig. 9.4 Net Production at Io. Estimate of the contribution of each plasma-neutral process above an exobase (at 150 km altitude) using the multi-species modeling approach described in Fig. 9.1. Although these rates depend on the flow and the atmosphere prescribed, they provide a reasonable estimate of the relative contribution of each process. Photoionization likely contributes 10–15% of the total ionization (Saur et al. 1999). The most important loss processes are the SO_2 electron-impact dissociation, which provide slow atomic neutrals (which are actually prescribed in our chemistry simulations) and a cascade of SO_2 resonant charge-exchange reactions, which provide faster neutrals depending where they occur. (Based on Bagenal and Dols 2020)

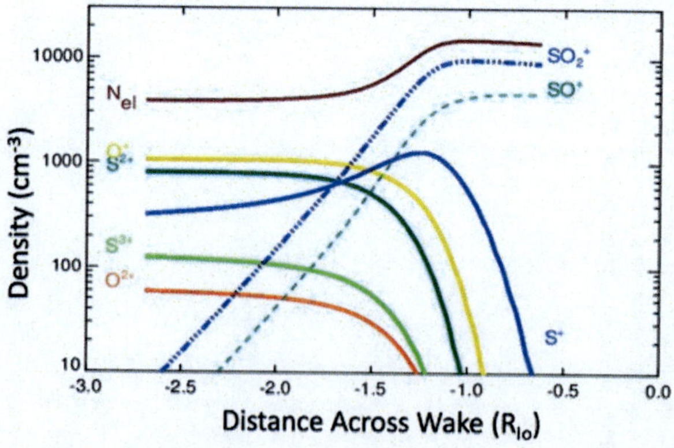

Fig. 9.5 Densities of the plasma downstream of the interactions with Io based on Dols et al. (2008, 2012). We plot the simulated composition of the plasma along the inbound J0 Galileo flyby with a closest approach at ~0.5 Rj. Note that chemistry in the dense wake is not included in these simulations. From Bagenal and Dols (2020)

100 km, vanishing at a distance of 3.5 R_{Io} above the surface. The resulting vertical column at the equator (6×10^{16} cm^{-2}) is consistent with dayside observations. To account for the limited latitudinal distribution of the SO_2 atmosphere (Strobel and Wolven 2001) in this 2D approach, we assume that the SO_2 atmosphere extends 1 R_{Io} perpendicular to the equatorial plane. The S and O atmosphere beyond 1 R_{Io} is assumed to be spherically symmetric based on the UV emission radial profiles of Wolven et al. (2001).

The plasma flow is slowed upstream and downstream and accelerated on the flanks. Following Dols et al. (2008), we prescribe a radial slowing of the flow consistent with the ion temperature observed along the *Galileo* J0 (first) flyby. The ion temperature increases on the flanks because of the pickup of SO_2^+ ions in the local fast flow. The increased electron and SO_2^+ densities are carried along the flow while the electron temperature decreases because of the efficient cooling process provided by the molecular atmosphere. The SO_2 electron-impact ionization and dissociation are located mainly on the upstream hemisphere because of the efficient electron cooling processes. The SO_2 resonant charge-exchange rate ($SO_2 + SO_2^+$ => $SO_2^+ + SO_2$) is greater on the flanks.

9.2.3 *Atmospheric Loss*

Early theoretical studies suggested that various neutral species could be removed from Io's atmosphere by many different processes, with variable efficiency: Jeans escape, charged particle sputtering of its surface, atmospheric sputtering, charge-exchange, direct single collisional ejection (Matson et al. 1974; Haff et al. 1981; Kumar 1982; Ip 1982; Johnson and Strobel 1982; Sieveka and Johnson 1984; McGrath and Johnson 1987; most completely compiled by Summers et al. 1989). More recently, electron impact molecular dissociation reactions are added to the list (Dols et al. 2008).

In Fig. 9.4 we show the volume-integrated rates of each plasma-SO_2 process. These results are computed using a 3D MHD simulation of the plasma flow around Io and a SO_2 atmosphere description presented in Dols et al. (2012) with an equatorial vertical column of 5.6×10^{16} cm^{-2}. The MHD flow allows for a better description of the slowing of the flow in Io's atmosphere, which affects the reaction rates and the results are thus somewhat different in detail to the 2D simulations shown in Fig. 9.3. The dominant SO_2 loss process is the electron-impact dissociation, which ultimately provides slow atomic neutrals that feed an extended corona of S and O. Another significant process is a cascade of resonant charge exchange reactions, which provides slow to fast SO_2 molecules, depending on the local flow speed where the charge exchange reaction occurs. Table 9.3 lists the sources of neutrals and plasma at Io derived via the physical chemistry model described above.

In the wake of Io, along the *Galileo* J0 flyby, the flux tubes are emptied of the upstream S and O ions because of charge-exchange reactions with SO_2, with

Table 9.3 Neutral and plasma sources at Io (1) see Figs. 9.3 and 9.4; (2) Thomas et al. 2004; (3) Smyth and Marconi 2005; Smith et al. 2019; (4) Thomas et al. 2004; Wilson et al. 2002; (5) Bagenal 1997; Saur et al. 2003; Dols et al. 2008; (6) Delamere et al. 2004; Nerney et al. 2017, Nerney and Bagenal 2020

Sources and losses at Io	
Local Loss processes (1)	• Exospheric collisions, chex • Atmospheric sputtering • SO_2 electron impact dissociation • Direct volcanic injection negligible • Surface sputtering negligible • Ionization by torus electrons • Photoionization
Net Neutral loss (1)	~3000 kg/s
Neutral clouds—detected (2, 3)	S, O, Na, K
Neutral clouds—expected	SO_2, SO Other?
Sodium sources (4)	10–30 kg/s
Local plasma source (5)	~200 kg/s
Extended plasma source (6)	700–2000 kg/s

$SO_2{}^+$ becoming the dominant ion (Fig. 9.5). This cascade of charge exchange and dissociation reactions potentially feeds neutral clouds of S, O atoms and SO_2 molecules similar to the observed Na structures (discussed in Sect. 9.3), which extend along Io's orbit and possibly through the whole magnetosphere.

Such modeling efforts require a consistent and robust set of observational constraints, currently provided mainly by *Galileo* instruments. The plasma measurements are essential to infer the plasma flow perturbation, the ion temperature, density and composition. Unfortunately, the Plasma Science (PLS) sensitivity evolved during the *Galileo* mission and PLS plasma densities are often not consistent with those inferred from the Plasma Wave Sensor (PWS) measurements (Dols et al. 2012). This lack of reliable plasma observations limits the reach of the numerical modeling of the local interaction at Io. The *Galileo* flybys were also relatively far from the denser part of the atmosphere (~200–900 km) and thus mostly constrain the tenuous corona. Moreover, no *Galileo* flybys were achieved in eclipse and on the nightside. Thus, the spatial and temporal variabilities of the plasma-atmosphere interaction remain poorly constrained.

Important extensions of the interaction modeling described above will need to include (a) the electron beams in the wake; (b) full molecular chemistry of SO_2, and products thereof; (c) more sophisticated treatment of the dependence of charge exchange reactions on the relative speed of the ions and neutrals (updating McGrath and Johnson 1989); (d) self-consistent production of the S and O corona from the plasma interaction with Io's atmosphere (currently prescribed in the models); and eventually, (e) further development of asymmetric atmosphere models and cases for different upstream magnetic field and plasma conditions. We aim to take a more consistent approach in future studies.

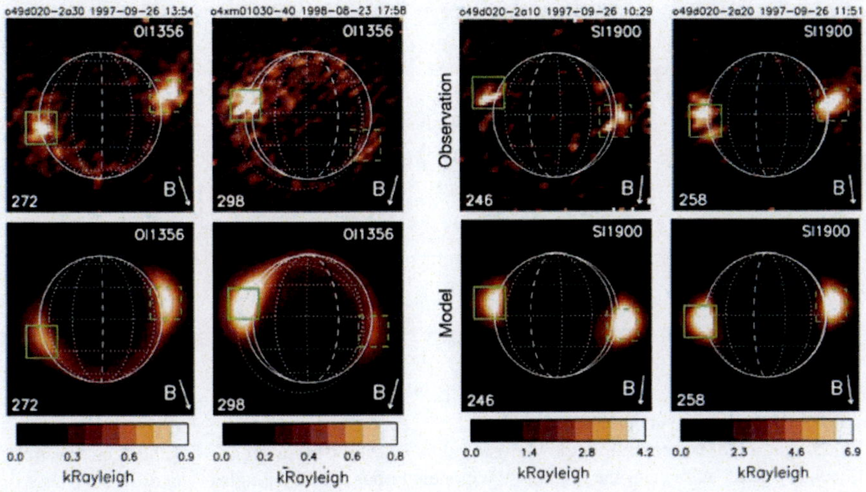

Fig. 9.6 Examples of extracted $3R_{Io} \times 3R_{Io}$ (above) HST-STIS observations of (left) 1356 Å emissions from atomic O and (right) 1900 Å emissions from atomic S with (below) corresponding model prediction. The arrow in the lower right corner shows the direction of the background magnetic field. The green boxes indicate the sub-jovian (solid) and anti-jovian (dashed) equatorial spots. The disk of the brightest multiplet line is shown in white with dotted longitudes and latitudes. The dashed longitude indicates the upstream/downstream meridian (270°/90° west longitude). The sub-jovian/anti-jovian meridian (0°/180° west longitude) is shown as solid line. From Roth et al. (2014)

9.2.4 Aurora at Io

Galileo images of Io in eclipse show a complex morphology of auroral emissions, both local and global, resulting from electron impact on the diverse components of Io's atmosphere. Figure 9.6 shows example images of FUV emissions by atomic O and S from Roth et al. (2014). These and several other HST-STIS images (Roesler et al. 1999; Geissler et al. 2001, 2004; Retherford et al. 2000, 2003, 2007; Roth et al. 2014, 2017) reveal the global morphology of S and O neutral emissions: equatorial spots on the flanks of Io that rock with the changing direction of the jovian background field at Io during Jupiter's synodic period, a limb emission that is stronger on the polar hemisphere facing the center of the torus, faint diffuse emissions that extend several 10s of Io radii, and possibly emissions in the wake downstream of Io (Oliversen et al. 2001; Retherford et al. 2007).

With *Galileo* particle detectors measuring beams of electrons streaming along the magnetic field in the proximity of Io, it was tempting to infer such electron beams were responsible for the auroral emissions. Michael and Bhadwaj (2000) simulated these emissions with an 1D energy degradation model of the field-aligned electron beams (0.1–5 keV) detected by the *Galileo* plasma instrument (Frank and Paterson 1999). They argue that while the detected flux of high-energy (15–100 keV) beams (Williams et al. 1996, 1999) is not sufficient to excite the observed emission, the

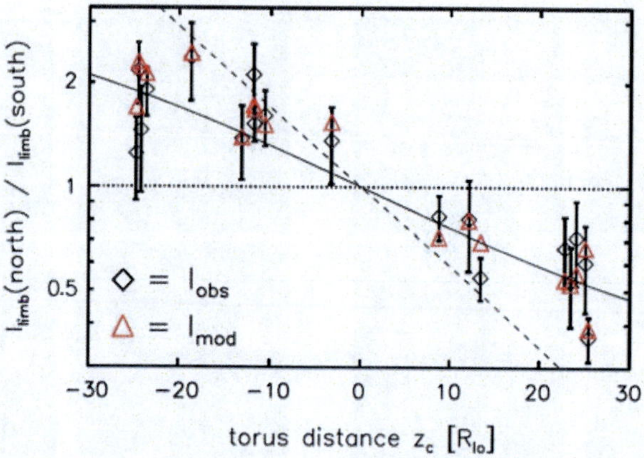

Fig. 9.7 Ratio of the polar limb glow brightness as measured in the atomic O 1356 Å observations (black diamonds with error bars) and in the modeled images (red triangles) versus Io's distance to the torus equator. With an appropriate scale length of Hc = 1.1 R_J for the background plasma, the theoretical profile for the ratio of the global emission only is steeper than the observed ratios. When the coronal emission is taken into account (solid) the theoretical profile is in reasonable agreement with the observations. From Roth et al. (2014)

energy supplied by the low-energy (0.1–5 keV) beams (Frank and Paterson 1999) in Io's atmosphere possibly contributes to the low altitude emissions.

Several observed geometric aurora properties are not in agreement, however, with the expected beam structures. The beam intensity is expected to be lower in altitude and shifted towards the wake on the hemisphere facing the torus equator (Bonfond et al. 2008; Jacobsen et al. 2010). This would lead to a north-south asymmetry exactly opposite to the one observed. The electron beams observed by Frank and Paterson (1999) peaked directly in the wake region, but were not detected along the elongated flank regions. Alternatively, auroral enhancements detected in the wake of Io by Retherford et al. (2007) can be explained by elongated flank emissions (Roth et al. 2011). The beams are also not expected to peak around the flanks of Io, but rather towards the center of the interaction region (Jacobsen et al. 2010).

To determine the emission around the northern and southern limb, Retherford et al. (2003) developed a method to compute the average brightness over angular ranges of 90° centered at Io's north and south poles, and with inner and outer radii of 0.75 and 1.25 R_{Io}. Figure 9.7 shows the modulation of these limb emissions with the displacement of Io from the center of the torus at the centrifugal equator. Figure 9.7 illustrates how modeling by Roth et al. (2014) can match the modulation of the emissions due to the vertical (centrifugal latitude) displacement of Io, revealing a close correspondence with the electron energy stored in the flux tubes above and below Io, when the near-surface emission is analyzed separated from the background.

Overall, the auroral emissions provide additional constraints on the atmospheric distribution, composition and plasma interaction. Saur et al. (2002) and Roth et al. (2011) simulate these emissions with a two-fluid model of the plasma/atmosphere interaction and conclude that these emissions result from thermal electrons of the torus impacting the atomic S and O components of Io's atmosphere. The flank emission coincides exactly with the ionospheric high-density regions formed by the corotating thermal plasma (Saur et al. 1999, 2002; Roth et al. 2011). Roth et al. (2014) claim *"the main morphological features of the aurora and their variation can be solely explained by the changes of the plasma environment and by the viewing perspective."*

9.2.5 *Induction*

When combined with MHD simulations of the plasma-atmosphere interaction, *Galileo* magnetic field observations along the close flybys of Io revealed a surprising component of the magnetic perturbations. Khurana et al. (2011) conclude that part of the observed magnetic field perturbations could be explained by a magnetic field induced in the subsurface magma of Io by the periodic variation of the background Jovian magnetic field during its 13-h synodic period. Roth et al. (2017) claim that the latitudinal amplitude of the rocking of the equatorial auroral spots is not consistent with an induction in a subsurface magma but is better matched with an induction deeper in the core. Bloeker et al. (2018) further claim that the magnetic perturbation observed by *Galileo* could also be explained by longitudinal asymmetries of Io's atmosphere. This debate from limited data suggests that extensive observations with multiple close flybys will be needed to resolve the induction issue at Io.

9.3 Neutral Clouds

While radio astronomers were still scratching their heads about Io triggering bursts of radio emission, in 1973 optical astronomers detected puzzling emission that were eventually interpreted as evidence of atmospheric escape from Io. The first observational evidence of neutral atoms escaping Io was the detection of optical sodium D-line emission from a cloud in the vicinity of Io by Brown (1974). The efficient resonant scattering of sunlight by sodium produced the bright emission (Trafton et al. 1974; Bergstrahl et al. 1975; Brown and Yung 1976). Many images of the sodium cloud have been acquired (see Fig. 9.8 and reviews by Thomas et al. 2004; Schneider and Bagenal 2007). *Voyager* 1 observations revealed Io's SO_2 atmosphere (Pearl et al. 1979) and a sulfur-dominated surface (Sagan 1979), indicating that Na was just a trace element in a neutral cloud dominated by sulfur and oxygen atoms and their compounds.

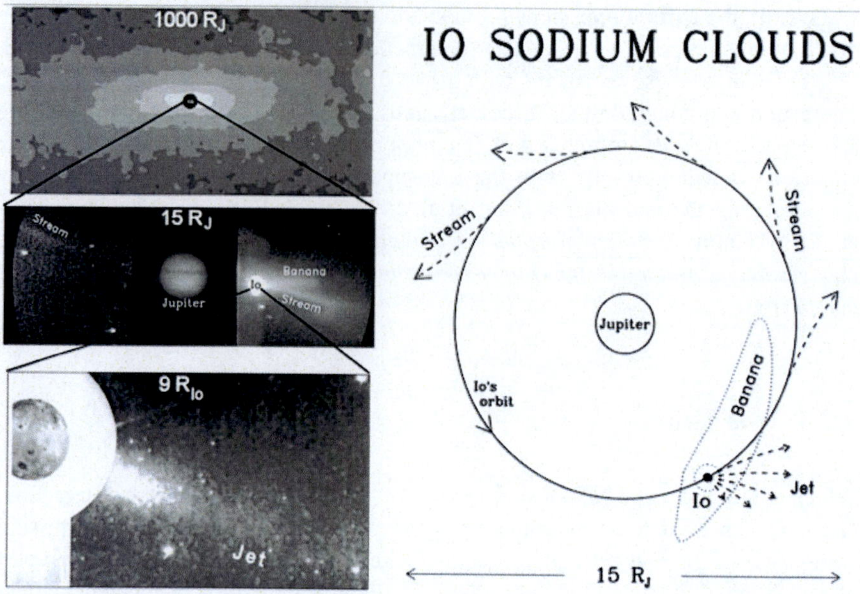

Fig. 9.8 Neutral Clouds of Io (based on Wilson et al. 2002). Left—Io's sodium cloud on three spatial scales, as imaged by ground-based observations of sodium D-line emission. The features observed on the left are explained by the three atmospheric escape processes shown schematically on the right

9.3.1 Observations

While sodium is a trace component of the atmosphere (~few %), its large cross section for scattering visible sunlight makes it the most easily observed species in the neutral clouds (30 times brighter than potassium and orders of magnitude brighter than more abundant species' emissions). Figure 9.8 illustrates how sodium's extreme efficiency at scattering visible sunlight reveals a rich sodium morphology: from jets close to Io, to the banana-shaped cloud extending around Io's orbit, and from streams along Io's orbit, to the large Mendillodisk nebula (defined in the last section) that extends for many hundreds of R_J. Other neutral species are much harder to detect, particularly molecular species, so that our picture of the iogenic neutral clouds are largely based on the behavior of sodium and on models.

After the initial detection by Brown (1981) of atomic oxygen emissions from Io's neutral cloud at the wavelength of 630.0 nm leading to an estimate of a ~30 cm^{-3} oxygen atoms, further observations of these optical emissions from the Io cloud by Thomas (1992) and close to Io by Oliversen et al. (2001) showed substantial variations with longitude, local time and Io phase. Meanwhile, atomic emissions were also detected in the UV. Atomic sulfur emissions at 142.9 nm and atomic oxygen emissions at 130.4 nm were observed with a rocket-borne telescope by Durrance et al. (1983). The first detections of neutral O and S were around 180°

in Io phase away from Io itself and analyzed by Skinner and Durrance (1986) to infer densities of neutral O and S of 29 ± 16 cm^{-3} and 6 ± 3 cm^{-3} respectively.

Lagg et al. (1998) noted the presence of dense neutral clouds along the orbit of Io is also consistent with losses of energetic sulfur and oxygen ions (~10s keV/nucleon) with a 90° pitch angle measured by the *Galileo* Energetic Particle Detector (EPD) instrument. These ions are thought to be removed by charge exchange with extended neutral clouds with a density ~35 cm^{-3}, consistent with previous estimates of Brown (1981) and Skinner and Durrance (1986).

The International Ultraviolet Explorer (IUE) satellite detected UV emissions from O and S near Io (Ballester et al. 1987). Using 30 observations by the HST-STIS, Wolven et al. (2001) mapped out O emission at 135.6 nm in the UV to 10 R_{Io}. They also demonstrated that these neutral emissions downstream of Io were brighter than those upstream, that there seemed to be a weak, erratic System III longitude effect, and that the local time (dawn versus dusk) asymmetry depends on the Io phase angle.

The *Hisaki* UV observatory in Earth orbit has provided new oxygen observations, mapping out Io's neutral oxygen cloud to show it comprises a leading cloud inside Io's orbit and an azimuthally uniform region extending to 7.6 R_J (Koga et al. 2018a, b, 2019). The peak number density of oxygen atoms is estimated at 80 cm^{-3}, spreading ~1.2 R_J vertically. The *Hiskaki* team estimate the source rate of oxygen ions at 410 kg/s, roughly consistent with previous studies (Smyth and Marconi 2003; Delamere and Bagenal 2003; Yoshioka et al. 2018). When Io exhibited a volcanic eruption in 2015, lasting ~90 days, Koga et al. (2019) used *Hisaki* observations of the time variations of O and O^+ to show that volcanism shortens the lifetime of O^+ and that Io's neutral oxygen cloud spreads outward from Jupiter during the 2015 volcanic event. The number density of O in the neutral cloud at least doubles during the active period.

9.3.2 Models

The neutral cloud morphologies are addressed by numerical simulations that take typical escape fluxes of different species at Io's exobase and tracks particles under Jupiter's gravity. A background plasma model is also used to predict when along its trajectory a neutral particle of a particular species will be changed or removed (via dissociation, ionization or charge exchange). From 1977 to 2011 William H. Smyth and colleagues published ~25 papers applying a neutral cloud model to study satellite atmospheres and neutral clouds (Smyth and MacElroy 1977; Smyth et al. 2011). The Smyth neutral cloud model involved solving the kinetic equations for a population of neutral particles embedded in prescribed torus plasma. For each small time-step, the state (location, velocity, ionization state) of the different particle species is updated following the consequences of the effects of gravity and collisions (elastic, inelastic, and chemically reactive), which are calculated for the average conditions in each model bin. They applied their model, including various refinements, to study Io's Na, K, O, S, SO_2 and SO corona, plus the plasma torus properties (Smyth and Marconi 2003, 2005; Smyth et al. 2011).

Detailed measurements of Io's sodium cloud require models to explain the features of the cloud (Fig. 9.8). The main sodium cloud is consistent with a roughly uniform corona of sodium atoms (a little above escape speed) around Io producing a uniform source of an extended cloud that is shaped by interaction with the surrounding plasma (Schneider et al. 1991a; Smyth and Combi 1997; Burger et al. 1999, 2001). Jets of fast sodium neutrals require a process that includes acceleration, probably as an atomic or molecular ion, followed by a neutralizing charge exchange (or dissociative recombination) reaction that sends out a fast (~100 km/s) neutral or 1.2 keV ENA (Schneider et al. 1991b; Wilson and Schneider 1994, 1999). Recently, Grava et al. (2021) discovered fast sodium jets emitted toward Jupiter. These jets are attributed to negatively charged Na-bearing dust particle accelerated toward Jupiter. We return to the extended sodium nebula in Sect. 9.6.

The recent Smith et al. (2019) model shown in Fig. 9.9 is conceptually similar to the Smyth models, with updated reaction cross-sections, updated neutral fluxes from the exobase at Io and Europa and incorporating current data on the variability of conditions at the orbit of Io. The main goal of Smith et al. (2019) was to show that Io's neutral material, whatever its composition, would not significantly reach Europa's orbit. The Smith et al. (2019) model illustrated in Fig. 9.9 assumes that only SO_2 escapes directly from Io with a canonical rate of 1 ton/s and a prescribed low velocity distribution. They produce iogenic neutral clouds that are consistent with observations of O and S emissions (Skinner and Durrance 1986; Koga et al. 2018a). But there are no observations of the neutral SO_2 cloud around Io. Future

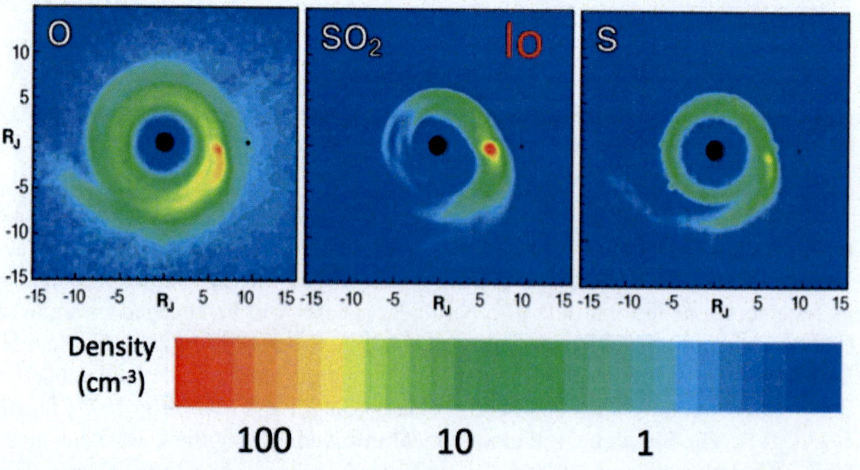

Fig. 9.9 Io neutral clouds as modeled by Smith et al. (2019) assuming a single SO_2 source of 1 ton/s from Io's atmosphere. The color contours show local densities (cm^{-3}). The black dot to the right is Europa, showing that little of the Io neutral cloud reaches Europa. Based on Smith et al. (2019)

models of the neutral clouds need to include sources of SO_2, S and O based on quantitative results from models of the plasma-atmosphere interaction.

Smyth and Marconi (2003) and Smith et al. (2019) study the formation of extended neutral structures but to this day, a self-consistent approach of their formation, from the processes in the atmosphere of Io, to the neutral cloud and finally to the torus ion supply, has not yet been carried out. Such extended neutral structures of S, O and SO_2 are notably difficult to observe beyond the minor Na species and we hope that in a near future, new creative observational methods will help to constrain quantitatively these extended neutral structures and improve our understanding of their atmospheric sources.

9.3.3 The Mendillodisk

By the late 80s it was realized that sodium atoms escaping from the Io environment could span many degrees across the sky, and could be imaged by wide-field cameras designed to study the Earth's atmospheric emissions. Mendillo et al. (1990) reported emission extending >400 R_J from Jupiter in a disk (Fig. 9.8). They proposed that sodium ions in the Io plasma torus charge exchange with neutrals around Io's orbit. This means that when neutralized via charge-exchange the atoms spray mostly outwards in a disk of neutral atoms with energies of ~400 eV. To honor the discoverer, we call this nebula the "Mendillodisk". Flynn et al. (1994) modeled 2-years of observations of the extended sodium nebula and found that the flaring angle of the disk (20–27°) anti-correlated with the source production rate (as indicated by Io's surface IR output or brightness of near-Io emissions).

Mendillo et al. (2004, 2007) show a variation in brightness and shape with Io's volcanic activity—the disk having angular edges (like an anulus) when Io is particularly active compared with an oblate spheroid during quiet times. The suggested explanation is that under quiet times the sodium ENAs are produced (from neutralized Na^+ in the torus) as a stream (Fig. 9.8). When Io is more active, production as jets local to Io increases, e.g. via dissociative recombination of molecular sodium ions such as $NaCl^+$ (Mendillo et al. 2007). A recent extensive survey of various data compiled by Roth et al. (2020), however, questions the relationship between Io's volcanic activity and any magnetospheric response.

Similar extended disks of escaping S and O atoms, as well as SO_2 and SO molecules also probably exist. Physical chemistry models of the torus (e.g. Delamere et al. 2005) predict charge exchange processes will produce a flux of escaping neutralized atoms of 1–5 $\times$ 10^{28} atoms per second, mostly oxygen. While this is ~100 times greater than the sodium flux, the radiation efficiency of other species is so much weaker that any neutral disk would be very hard to detect. Direct *in situ* detection of these escaping neutrals would require an instrument that measures ENAs with energies less than ~300 eV (e.g., <20 eV/nucleon for oxygen atoms moving at 70 km/s). Instruments on JUICE will provide such measurements.

While corotating torus ions that are neutralized via charge exchange with Io's neutral clouds will come off as an outward stream, charge exchange or molecular ion recombination close to Io could come off as a jet or spray (Fig. 9.8). For example, a recently-picked-up ion produced in the plasma-atmosphere interaction that charge exchange could be directed towards Jupiter where re-ionization on impacting the planet's atmosphere would be a source of cold heavy ions in the planet's ionosphere (Valek et al. 2019).

9.4 Plasma Torus

The Io plasma torus comprises three main regions: (1) the outer region that has a roughly circular cross-section—sometimes called the "doughnut"—that contains 90% of the mass and emits most of the UV emission; (2) just inside Io's orbit there is narrow but vertically-extended region—sometimes called the "ribbon"—that is bright in UV and particularly visible wavelengths; and (3) extending inwards from the ribbon is a thin disk or "washer" that emits visible (but no UV) light. Figure 9.10 illustrates these three regions and their properties are listed in Table 9.4. The warm torus extends past the orbit of Europa and merges into the plasma sheet that extends through Jupiter's vast magnetosphere.

Optical emissions from a toroidal cloud of S^+ ions surrounding the orbit of Io were first detected in ground-based observations by Kupo et al. (1976), which Brown (1976) recognized as coming from a cold, dense plasma. The *Voyager* 1 flyby of Jupiter in 1979 revealed Io spewing out SO_2 from active volcanoes and provided detailed measurements of the Io plasma torus both from the strong emissions in the EUV, observed remotely by the *Voyager* Ultraviolet Spectrometer (UVS) (Broadfoot et al. 1979), as well as *in situ* measurements made by the Plasma Science (PLS) instrument (Bridge et al. 1979) and the Planetary Radio Astronomy (PRA) instrument (Warwick et al. 1979).

Since *Voyager*, the Io plasma torus has been observed remotely via visible (Morgan, 1985a, b; Brown 1994; Schneider and Trauger 1995; Thomas et al. 2001; Schmidt et al. 2018) and UV emissions (Thomas 1992; Hall et al. 1994; Gladstone and Hall 1998; Feldman et al. 2001, 2004; Steffl et al. 2004a, b, 2006) as well as *in situ* by the *Galileo* spacecraft (Gurnett et al. 1996, 2001; Frank and Paterson 2000). A summary of observations obtained by the end of *Galileo* mission is provided by Thomas et al. (2004) and Nerney et al. (2017) revisit Voyager, Gallileo, and Cassini datasets to compare the three epochs. Each of the techniques for measuring the plasma properties in the torus has its pros and cons. The remote sensing techniques provide good temporal and spatial coverage but suffer from being integral measurements along the line of sight as well as being dependent on calibration of the instrument and accurate atomic data for interpretation of the spectra. Moreover, a very broad wavelength range is needed to cover emissions from all the main ion species. The *in situ* plasma measurements provide detailed velocity distributions but suffer from limited spatial and temporal coverage as well as poor

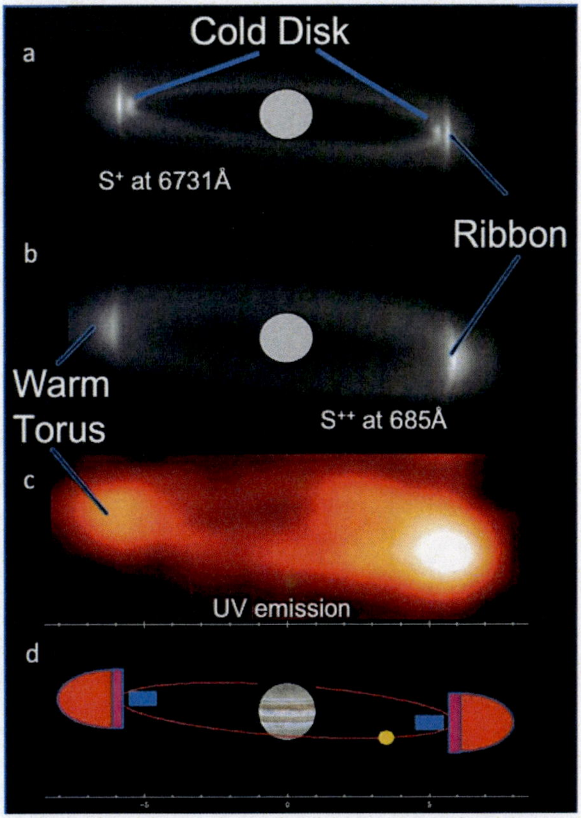

Fig. 9.10 Three regions of the Io plasma torus: Cold disk, ribbon, warm torus. (**a**) This simulated image shows optical S^+ emission (**b**) EUV S^{++} emission. Note that S^+ dominates the cold torus and S^{++} dominates the warm torus. The ribbon is a tall, narrow ring which appears bright at the torus ansa because of projection effects. The ribbon is typically the most prominent of the three regions for S^+, while in S^{++} emission the ribbon is a slight brightening at the inner edge of the warm torus. (**c**) Net emission across the EUV spectrum as observed by Cassini UVIS. (**d**) Cartoon showing the three regions. The structure of the torus can exhibit strong longitudinal variations, and the relative brightnesses of different regions can vary with time. From Bagenal and Dols (2020)

determination of parameters for individual ionic species in the warm region of the torus where the spectral peaks for different species overlap. Since these two data sets are complementary, they can be combined to construct a description of the plasma conditions in the torus, i.e. an empirical model, that can be compared to theoretical models based on the physical chemistry of the torus plasma. Gathered under the term "physical chemistry" are key physical processes that control the plasma in Io's space environment: ion pick-up, centrifugal confinement to the equatorial region, radial transport and collisional processes such as dissociation, ionization, charge-exchange, radiation, etc., that drive the radial distribution of different ion species as well as the energy of the plasma. See review by Bagenal and Dols (2020) for full discussion of these processes and physical chemistry models of the Io plasma torus.

The dominance of sulfur and oxygen composition of heavy ions throughout the magnetosphere tell us that these products of volcanic gases fill the vast volume of Jupiter's magnetosphere, extending 60–100 R_J on the dayside and stretching in a magnetotail away from the Sun, at times reaching as far a Saturn's orbit. At the same

Table 9.4 Io Plasma Torus properties (from Bagenal and Dols 2020). The conditions at Europa's orbit are shown for comparison. At Europa there is not a distict torus but continuation of the plasma sheet so the mass is for a disk around Europa's orbit of 1 R_J width

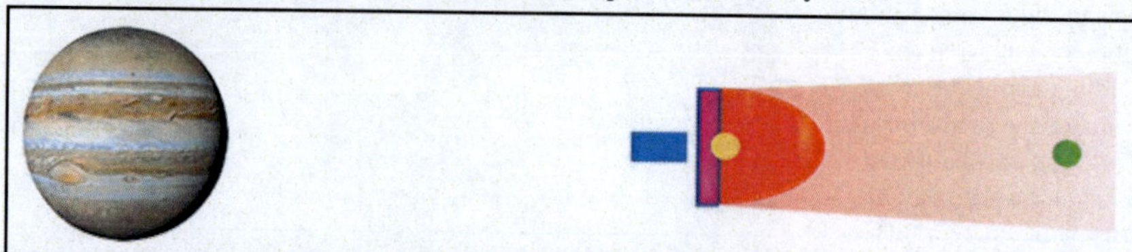

	Cold Disk	Ribbon	Warm Torus	Europa Orbit
Location (R_J)	4.7 - 5.7 Height 0.2	5.6 - 5.9 ? Width 0.2	6 - 8	9.4
Mass (% of net torus)	35 kton 1%	200 kton 10%	2 Mton 90%	~72 kton/R_J
Radial Transport (/R_J)	Inward: months	??	Outward: 20 - 60 days	~20 hrs
N_e (cm^{-3})	~1000	~3000	~2000	~160
T_e (eV)	<2	~5	~5	~10 - 30
Hot Electrons	~0	<0.1%	0.2% @ 40eV	2-10% @ 250eV
T_i (eV)	<2	20 - 40	20 - 90	50 - 500
Ions	S^+, O^+	O^+, S^+, S^{++}	O^+, S^{++}, S^+ S^{+++}, O^{++}	O^+, S^{++}, O^{++} S^{+++}, S^+

time, energetic particles from the outer magnetosphere are transported inwards, bringing in supra-thermal ions that charge-exchange with the neutral clouds, plus electrons that ionize and excite UV emissions.

9.5 Conclusions

The peculiar role of Io in the magnetosphere of Jupiter was first noticed in 1964. A half century and a thousand or so papers later it is a good time to consider our understanding of the system and what are the key open questions.

9.5.1 Outstanding Questions

Plasma-Atmosphere Interactions The aurora produced via electron bombardment of the atmosphere can provide useful information about the interaction of the surrounding plasma with a moon's atmosphere. But to explore the full physics we need a combination of in situ measurements along close flybys and detailed modelling. Specific open questions are:

- How much of the plasma interaction goes into heating the atmosphere? Where? What is the impact of the heating sources on the atmospheric distribution, in altitude, latitude and longitude?
- How do seasonal changes in distance from the Sun (that drives sublimation and photoionization rates) modify the plasma-atmosphere interaction?
- What are the roles of electron impact ionization, charge exchange, collisions, and electron beams in the plasma-atmosphere interaction? What are their effects on the radial composition of Io's atmosphere? What are the net ionized products of these reactions that escape into the space beyond?
- How do the plasma interactions vary around Io with the asymmetries of the moon's atmosphere (e.g. upstream vs. downstream, sub/anti-Jupiter, day/night), with magnetic latitude and longitude and with local time/phase of the moon along its orbit?
- How do the plasma interactions vary (qualitatively and quantitatively) with volcanic activity of Io? What types and/or locations of volcanos affect the interaction?
- What is the strength of the internal induction signal compared to perturbations in the fields and flow caused by the plasma/atmosphere interaction? How much from the conducting asthenosphere and/or the core?
- What are the composition, velocity distribution, and fluxes of the neutrals that escape the moon's gravity and into the space beyond?
- How much of the iogenic material reaches the surface of Europa?

Neutral Clouds While alkali elements are clearly observed around Io, the major species are poorly measured, except Io's oxygen cloud that is now being mapped out by the *Hisaki* mission. Specific open questions are:

- What are the amounts and trajectories of different neutral species that escape Io? In particular, are there molecular clouds of SO_2, SO, O_2? What roles do they play? Is there a way to detect them? Are there any features in other atomic and molecular neutral clouds similar to the Na neutral structures such as jets, streamers and the extended Mendillo-disk?
- How are these neutral clouds shaped by the plasma that flows through them? At Io, how much of the neutral clouds reach inwards of Io's orbit (as molecules and/or atoms) to provide an extended source for the ribbon and inner torus?

- What role do the neutral clouds play in controlling the influx of energetic particles (electrons, protons and heavy ions) from the middle magnetosphere through to the radiation belts within a few R_J of Jupiter?
- How are dust streams ejected from Io? What is their composition and size distribution? Does pick-up of smaller aerosol sized statically-charged particles influence the near Io plasma environment?

Plasma Torus The general structure and systematic modulations (System III, IV, local time) of the three components of the torus—cold inner torus, ribbon, warm outer torus and plasma sheet—are fairly well described but underlying physical processes remain unclear. Specific open questions are:

- What is the spatial distribution of the production of plasma in the three regions of the torus? Where is the separation between outward and inward transport? What is the nature and role of the torus ribbon region?
- What processes drive and control the radial transport rate? How do these processes vary with radial distance and (if at all) with plasma production rate?
- What is (are) the source(s) of hot electrons? Do waves within the torus accelerate local electrons? Are hot electrons injected from outside?
- What are the physical processes whereby changes in sources affect the System III/IV modulations, radial transport rate, dawn/asymmetries?
- Can a direct causal connection be made between the specific nature of Io volcanic eruptions that might cause enhanced sources of plasma?
- How is plasma heated as it moves out from the torus to the plasma sheet? How do the supra-thermal populations (both ions and electrons) evolve?

Extended Neutral Nebulae The only Mendillodisk nebula that has been detected directly is that of sodium from Io. But it is clear that charge-exchange reactions must be generating clouds of Energetic Neutral Atoms—and perhaps even molecules. Specific open questions are:

- What are the fluxes of different neutral species (composition, energy, direction) out of the jovian system? Are these neutrals re-ionized? Where? And what is their fate?
- What are the fluxes of different neutral species (composition, energy, direction) inwards towards Jupiter and impacting the planet's atmosphere? Are these neutrals re-ionized? Where? And what is their fate?

The components of the Io space environment are clearly highly coupled. While there are modulations and temporal variations, such changes seem to be limited to factors of a few (rather than wild swings of orders of magnitude). Thus, there must be both negative and positive feedback systems. Full understanding of this complex system will require both comprehensive observations as well as systematic modeling—of each component separately as well as carefully coupled—to explore what factors control (qualitatively and quantitatively) the different components. The beauty of the Io system, however, is that the timescales for the different processes make the components separable: plasma takes a minute or so to pass the moon, neutrals survive ~hours orbiting Jupiter, and plasma moves through the magnetosphere over many days. This separation of time scales allows us to study each component independently and then couple them together.

9.5.2 Future Observations

The *Juno* mission has come to the end of its primary mission, mapping out the polar regions as well as outer to middle magnetosphere (for review of magnetospheric science see Bagenal et al. 2017; for updated orbits see Bolton et al. 2017a). As the line of apsides of the orbit precesses southward, the spacecraft crosses the jovigraphic equator closer to the planet. By the end of the primary mission (34 orbits, spring 2021) *Juno* reached between the orbits of Ganymede and Europa. Moreover, the tilted magnetic field allows the spacecraft to cross magnetic fluxshells intersecting Europa's and Io's orbits and sample the torus. NASA has extended the *Juno* mission through mid-2025. As the orbit further evolves, the spacecraft will directly pass through the Io plasma torus many times, as well as likely intersect the Alfven wings that stretches downstream of Io (Fig. 9.1).

While *Juno in situ* measurements of particles and fields are very valuable, it is key that the torus emissions are also monitored from the ground (Schmidt et al. 2018; Morgenthaler et al. 2019) and from the *Hisaki* satellite (Yoshikawa et al. 2014). Ground-based telescopes seem to be ever expanding in wavelength and aperture. Io's volcanism is monitored in the IR and the extension into the millimeter range of telescope systems such as Atacama Large Millimeter Array are allowing detection of the molecular species such as SO_2, SO, O_2, NaCl, KCl, S_2, etc. for Io (de Pater et al. 2020). The James Webb Space Telescope will also provide valuable information in the IR with high resolution and sensitivity that may provide the temperature and behavior of erupting lavas, plus changes in the SO_2 atmosphere at Io (Keszthelyi et al. 2016). In the meantime, we urge the community to support semi-continuous monitoring of the more observable species (Na, K, S, S^+, O, O^+ ...) with modest-sized telescopes.

Looking farther to the future, the development of ESA's *JUICE* and NASA's *Europa Clipper* missions promise close exploration of the Europa system with perhaps some forays closer to Io. But to properly address the key scientific questions of Io's peculiar role in the jovian system we need a mission that makes multiple close flybys of Io (de Pater et al. 2021; Thomas 2021).

Acknowledgements The authors thank Kurt Retherford for his review of the paper and acknowledges fruitful discussions with Nick Thomas, Nick Schneider, Frank Crary, Peter Delamere, Tim Cassidy and Edward Nerney. We thank Crusher Bartlett for graphics. This work was supported at the University of Colorado as a part NASA's Juno mission supported by NASA through contract 699050X with the Southwest Research Institute.

References

Acuna, M.H., Neubauer, F.M., Ness, N.F.: Standing Alfven Wave current system at Io: Voyager 1 observations. J. Geophys. Res. **86**, A10 (1981)

Bagenal, F.: Ionization source near Io from Galileo wake data. Geophys. Res. Lett. **24**, 2111–2114 (1997)

Bagenal, F., McKinnon, W.B., Dowling, T.E. (eds.): Jupiter: Planet, Satellites, Magnetosphere. Cambridge University Press (2004)

Bagenal, F., Adriani, A., Allegrini, F., Bolton, S.J., Bonfond, B., Bunce, E.J., et al.: Magnetospheric science objectives of the Juno mission. Space Sci. Rev. **213**, 219–287 (2017)

Bagenal, F., Dols, V.: The space environment of Io and Europa. J. Geophys. Res. **125**, e2019JA027485 (2020)

Ballester, G.E., Moos, H.W., Feldman, P.D., Strobel, D.F., Summers, M.E., Bertaux, J.L., Skinner, T.E., Festou, M.C., Lieske, J.H.: Detection of neutral oxygen and sulfur emissions near lo using IUE. Astrophys. J. **319**, L33–L38 (1987)

Belcher, J.W., Goertz, C.K., Sullivan, J.D., Acuna, M.H.: Plasma observations of the Alfven wave generated by Io. J. Geophys. Res. **86**, 8508 (1981)

Belcher, J.W.: The Jupiter-Jo connection: an Alfven engine in space. Science. **238**, 170–176 (1987)

Bergstralh, J.T., Matson, D.L., Johnson, T.V.: Sodium D-line emission from Io: synoptic observations from Table Mountain Observatory. Astrophys. J. **195**, L131–L135 (1975)

Blöcker, A., Saur, J., Roth, L., Strobel, D.F.: MHD modeling of the plasma interaction with Io's asymmetric atmosphere. J. Geophys. Res. **123**, 9286–9311 (2018)

Bolton, S.J., Lunine, J., Stevenson, D., Connerney, J.E.P., Levin, S., Owen, T.C., Bagenal, F., Gautier, D., Ingersoll, A.P., Orton, G.S., Guillot, T., Hubbard, W., Bloxham, J., Coradini, A., Stephens, S.K., Mokashi, P., Thorne, R., Thorpe, R.: The Juno Mission. Space Sci. Rev. **213**, 5–37 (2017a)

Bolton, S.J., et al.: Jupiter's interior and deep atmosphere: the initial pole-to-pole passes with the Juno spacecraft. Science. **356**, 821–825 (2017b)

Bonfond, B., Grodent, D., Gerard, J.-C., Radioti, A., Saur, J., Jacobsen, S.: UV Io footprint leading spot: a key feature for understanding the UV Io footprint multiplicity? Geophys. Res. Lett. **35**, L05107 (2008)

Bouchez, A.H., Brown, M.E., Scheider, N.M.: Eclipse spectroscopy of Io's atmosphere. Icarus. **148**, 316–319 (2000)

Bridge, H.S., Belcher, J.W., Lazarus, A.J., Sullivan, J.D., McNutt, R.L., Bagenal, F., Scudder, J.D., Sittler, E.C., Siscoe, G.L., Vasyliunas, V.M., Goertz, C.K., Yeates, C.M.: Plasma observations near Jupiter: initial results from Voyager 1. Science. **204**, 987–991 (1979)

Broadfoot, A.L., et al.: Extreme ultraviolet observations from Voyager I encounter with Jupiter. Science. **204**, 979–982 (1979)

Brown, R.A.: Optical line emission from lo. In: Brown, R.A. (ed.) Exploration of the Planetary System, pp. 527–531. Reidel, Dordrecht (1974)

Brown, R.A.: A model of Jupiter's sulfur nebula. Ap. J. **206**, LI79–L183 (1976)

Brown, R.A., Yung, Y.L.: Io, its atmosphere and optical emissions. In: Jupiter, T.G. (ed.) , pp. 1102–1145. University of Arizona Press (1976)

Brown, R.A.: The Jupiter hot plasma torus: observed electron temperature and energy flows. Ap. J. **244**, 1072–1080 (1981)

Brown, M.E.: Observation of mass loading in the Io plasma torus. Geophys. Res. Lett. **21**, 847–850 (1994)

Burger, M.H., Schneider, N.M., Wilson, J.K.: Galileo's closeup view of the Io sodium jet. Geophys. Res. Lett. **26**, 3333–3336 (1999)

Burger, M.H., Schneider, N.M., De Pater, I., Brown, M.E., Bouchez, A.H., Trafton, L.M., Sheffer, Y., Barker, E.S., Mallama, A.: Mutual event observations of Io's sodium corona. Ap. J. **563**, 1063–1074 (2001)

Clarke, J.T., Ajello, J., Ballester, G., Ben Jaffel, L., Connerney, J., Gerard, J.-C., Gladstone, G.R., Grodent, D., Pryor, W., Trauger, J., Waite, J.H.: Ultraviolet emissions from the magnetic footprints of Io, Ganymede and Europa on Jupiter. Nature. **415**, 997–1000 (2002)

Delamere, P.A., Bagenal, F.: Modeling variability of plasma conditions in the Io torus. J. Geophys. Res. **108**, 1276 (2003)

Delamere, P.A., Bagenal, F., Steffl, A.: Radial variations in the Io plasma torus during the Cassini era. J. Geophys. Res. **110**, A12223 (2005)

Delamere, P. A., Steffl, A., & Bagenal, F.: Modeling temporal variability of plasma conditions in the Io torus during the Cassini era. J. Geophys. Res. **109**, A10216 (2004)

de Pater, I., Luszcz-Cook, S., Rojo, P., Redwing, E., de Kleer, K., Moullet, A.: ALMA observations of Io going into and coming out of Eclipse. Planet. Sci. J. **1**, 60 (2020)

de Pater, I., Keane, J.T., de Kleer, K., Davies, A.G.: A 2020 observational perspective of Io. Annu. Rev. Earth Planet. Sci. **49**, 633–668 (2021)

Dols, V.J., Delamere, P.A., Bagenal, F.: A multi-species chemistry model of Io's local interaction with the plasma torus. J. Geophys. Res. **113**, 9208 (2008)

Dols, V.J., Delamere, P.A., Bagenal, F., Kurth, W.S., Paterson, W.R.: Asymmetry of Io's outer atmosphere: constraints from five Galileo flybys. J. Geophys. Res. **117**(E16), 10010 (2012)

Durrance, S.T., Feldman, P.D., Weaver, H.A.: Rocket detection of UV emission from neutral oxygen & sulfur in the Io torus. Ap. J. **267**, L125–L129 (1983)

Feldman, P.D., et al.: Detection of chlorine ions in the FUSE spectrum of the Io plasma torus. Ap. J. **554**, L123 (2001)

Feldman, P.D., et al.: The far-ultraviolet spectrum of the Io plasma torus. Ap. J. **601**, 583–591 (2004)

Flynn, B., Mendillo, M., Baumgardner, J.: The Jovian sodium nebula: two years of ground-based observations. J. Geophys. Res. **99**, 8403–8409 (1994)

Frank, L.A., Paterson, W.R., Ackerson, K.L., Vasyliunas, V.M., Coroniti, F.V., Bolton, S.J.: Plasma observations at Io with the Galileo spacecraft. Science. **274**, 394–395 (1996)

Frank, L.A., Paterson, W.R.: Intense electron beams observed at Io with the Galileo spacecraft. J. Geophys. Res. **104**, 28657–28669 (1999)

Frank, L.A., Paterson, W.R.: Observations of plasmas in the Io Torus with the Galileo spacecraft. J. Geophys. Res. **105**, 16017–16034 (2000)

Geissler, P.E., Smyth, W.H., McEwen, A.S., Ip, W., Belton, M.J.S., Johnson, T.V., Ingersoll, A.P., Rages, K., Hubbard, W., Dessler, A.J.: Morphology and time variability of Io's visible aurora. J. Geophys. Res. **106**, 26137–26146 (2001)

Geissler, P., McEwen, A., Porco, C., Strobel, D.F., Saur, J., Ajello, J., West, R.: Cassini observations of Io's visible aurorae. Icarus. **172**, 127–140 (2004)

Gladstone, G.R., Hall, D.T.: Recent results from EUVE observations of the Io plasma torus and Jupiter. J. Geophys. Res. **103**, 19927–19933 (1998)

Goertz, C.K.: Io's interaction with the Plasma Torus. J. Geophys. Res. **85**, 2949–2956 (1980)

Goldreich, P., Lynden-Bell, D.: Io, a Jovian unipolar inductor. Astrophys. J. **156**, 59 (1969)

Grava, C., Cassidy, T.A., Schneider, N.M., Hsu, H.-W., Morgenthaler, J.P., Leblanc, F., Mangano, V., Retherford, K.D., Burger, M.H., Barbieri, C.: A possible dust origin for an unusual feature in Io's sodium neutral clouds. Astron. J. **162**, 190 (2021)

Gurnett, D.A., Goertz, C.K.: Multiple Alfvén wave reflections excited by Io: origin of the Jovian decametric arcs. J. Geophys. Res. **86**, 717–722 (1981)

Gurnett, D.A., Kurth, W.S., Roux, A., Bolton, S.J., Kennel, C.F.: Galileo plasma wave observations in the Io plasma torus and near Io. Science. **274**, 391–392 (1996)

Gurnett, D.A., Persoon, A., Kurth, W.: Electron densities near Io from Galileo plasma wave observations. J. Geophys. Res. **106**, 26225–26232 (2001)

Haff, P.K., Watson, C.C., Yung, Y.L.: Sputter ejection of matter from Io. J. Geophys. Res. **86**, 6933–6983 (1981)

Hall, D.T., Gladstone, G.R., Moos, H.W., Bagenal, F., Clarke, J.T., Feldman, P.D., McGrath, M.A., Schneider, N.M., Shemansky, D.E., Strobel, D.F., Waite, J.H.: Extreme Ultraviolet Explorer satellite observation of Jupiter's Io plasma torus. Ap. J. **426**, L51–L54 (1994)

Hikida, R., Yoshioka, K., Murakami, G., Kimura, T., Tsuchiya, F., Yamazaki, A., et al.: Identification of extreme ultraviolet emission lines of the Io plasma torus observed by Hisaki/EXCEED. J. Geophys. Res. **123**, 1723–1731 (2018)

Hikida, R., Yoshioka, K., Tsuchiya, F., Kagitani, M., Kimura, T., Bagenal, F., et al.: Spatially asymmetric increase in hot electron fraction in the Io plasma torus during volcanically active period revealed by observations by Hisaki/EXCEED from November 2014 to May 2015. J. Geophys. Res. **123** (2020). https://doi.org/10.1029/2019JA027100

Ip, W.-H.: On charge-exchange and knock-on processes in the exosphere of Io. Astrophys. J. **262**, 780–785 (1982)

Jacobsen, S., Saur, J., Neubauer, F.M., Bonfond, B., Gérard, J., Grodent, D.: Location and spatial shape of electron beams in Io's wake. J. Geophys. Res. **115**, 4205–4214 (2010)

Johnson, R.E., Strobel, D.F.: Charge exchange in the Io torus and exosphere. J. Geophys. Res. **87**, 10385–10393 (1982)

Keszthelyi, L., Grundy, W., Stansberry, J., Sivaramakrishnan, A., Thatte, D., Gudipati, M., Tsang, C., Greenbaum, A., McGruder, C.: Observing Outer Planet Satellites (except Titan) with the James Webb Space Telescope: science justification and observational requirements. Astro. Soc. Pacific. **128**, 018006 (2016). https://doi.org/10.1088/1538-3873/128/959/018006

Khurana, K.K., Jia, X., Kivelson, M.G., Nimmo, F., Schubert, G., Russell, C.T.: Evidence of a Global Magma Ocean in Io's interior. Science. **332**, 1186 (2011)

Kivelson, M.G., Khurana, K.K., Walker, R.J., Linker, J.A., Russell, C.R., Southwood, D.J., Polanskey, C.: A magnetic signature at Io: initial report from the Galileo magnetometer. Science. **273**, 337–340 (1996)

Kivelson, M.G., Bagenal, F., Kurth, W.S., Neubauer, F.M., Paranicas, C., Saur, J.: Magnetospheric interactions with satellites. In: Bagenal, F., Dowling, T.E., McKinnon, W.B. (eds.) Jupiter: Planet, Satellites. Magnetosphere. Cambridge University Press (2004)

Koga, R., Tsuchiya, F., Kagitani, M., Sakanoi, T., Yoneda, M., Yoshioka, K., Kimura, T., Murakami, G., Yamazaki, A., Yoshikawa, I., Smith, H.T.: The time variation of atomic oxygen emission around Io during a volcanic event observed with Hisaki/EXCEED. Icarus. **299**, 300–307 (2018a)

Koga, R., Tsuchiya, F., Kagitani, M., Sakanoi, T., Yoneda, M., Yoshioka, K., Yoshikawa, I., Kimura, T., Murakami, G., Yamazaki, A., Smith, H.T., Bagenal, F.: Spatial distribution of Jovian Moon Io's oxygen neutral cloud observed by Hisaki. J. Geophys. Res. **123**, 3764–3776 (2018b)

Koga, R., Tsuchiya, F., Kagitani, M., Sakanoi, T., Yoshioka, K., Yoshikawa, I., Kimura, T., Murakami, G., Yamazaki, A., Smith, H.T., Bagenal, F.: Volcanic change of Io's neutral oxygen cloud and plasma torus observed by Hisaki. J. Geophys. Res. **124** (2019)

Kumar, S.: Photochemistry of SO_2 in the atmosphere of Io and implications on atmospheric escape. J. Geophys. Res. **87**, 1677–1684 (1982)

Kupo, I., Mekler, Y., Eviatar, A.: Detection of ionized sulphur in the jovian magnetosphere. Ap. J. **205**, L51–L54 (1976)

Lagg, A., Krupp, N., Woch, J., Livi, S., Wilken, B., Williams, D.J.: Determination of the neutral number density in the Io torus from Galileo-EPD measurements. Geophys. Res. Lett. **25**, 4039–4042 (1998)

Linker, J.A., Khurana, K.K., Kivelson, M.G., Walker, R.J.: MHD simulations of Io's interaction with the plasma torus. J. Geophys. Res. **103**, 19867–19877 (1998)

Matson, D.L., Johnson, T.V., Fanale, F.P.: Sodium D-line emission from Io: sputtering and resonant scattering hypothesis. Ap. J. **192**, L43–L46 (1974)

Mauk, B.H., Williams, D.J., Eviatar, A.: Understanding Io's space environment interaction: recent energetic electron measurements from Galileo. J. Geophys. Res. **106**, 26195–26208 (2001)

McGrath, M.A., Johnson, R.E.: Magnetospheric plasma sputtering of Io's atmosphere. Icarus. **69**, 519–531 (1987)

McGrath, M.A., Johnson, R.E.: Charge exchange cross sections for the Io plasma torus. J. Geophys. Res. **94**, 2677–2683 (1989)

Mendillo, M., Baumgardner, J., Flynn, B., Hughes, W.J.: The extended sodium nebula of Jupiter. Nature. **348**, 312–314 (1990)
Mendillo, M., Wilson, J., Spencer, J., Stansberry, J.: Io's volcanic control of Jupiter's extended neutral clouds. Icarus. **170**, 430–442 (2004)
Mendillo, M., Laurent, S., Wilson, J., Baumgardner, J., Konrad, J., Karl, C.: The sources of sodium escaping from Io revealed by spectral high definition imaging. Nature. **448**, 330–332 (2007)
Michael, M., Bhardwaj, A.: FUV emissions on Io: role of Galileo-observed field aligned energetic electrons. Geophys. Res. Lett. **27**, 3137–3140 (2000)
Morgan, J.S.: Temporal and spatial variations in the Io torus. Icarus. **62**, 389–414 (1985a)
Morgan, J.S.: Models of the Io torus. Icarus. **63**, 243–265 (1985b)
Morgenthaler, J.P., Rathbun, J.A., Schmidt, C.A., Baumgardner, J., Schneider, N.M.: Volcanic event on Io inferred from Jovian sodium nebula brightening. Ap. J. **871**, L23 (2019)
Nerney, E., Bagenal, F., Steffl, A.: Io plasma torus ion composition: Voyager, Galileo, Cassini. J. Geophys. Res. **122** (2017)
Nerney, E., Bagenal, F.: Combining UV spectra and physical chemistry to constrain the hot electron fraction in the Io Plasma Torus. J. Geophys. Res. **125** (2020)
Neubauer, F.M.: Nonlinear standing Alfven wave current system at Io: theory. J. Geophys. Res. **85**, 1171 (1980)
Oliversen, R.J., Scherb, F., Smyth, W.H., Freed, M.E., Woodward, R.C., Marconi, M.L., Retherford, K.D., Lupie, O.L., Morgenthaler, J.P.: Sunlit Io atmospheric [O I] 6300 A emission and the plasma torus. J. Geophys. Res. **106**, 26183–26194 (2001)
Pearl, J., Hanel, R., Kunde, V., Maguire, W., Fox, K., Gupta, S., Ponnamperuma, C., Raulin, F.: Identification of gaseous SO_2 and new upper limits for other gases on Io. Nature. **280**, 755–758 (1979)
Rathbun, J.A., Spencer, J.R., Lopes, R.M., Howell, R.R.: Io's active volcanoes during the New Horizons era: insights from New Horizons imaging. Icarus. **231**, 261–272 (2014)
Retherford, K.D., Moos, H.W., Strobel, D.E., Wolven, B.C.: Io's equatorial spots: morphology of neutral UV emissions. J. Geophys. Res. **105**, 27157–27165 (2000)
Retherford, K.D., Moos, H.D., Strobel, D.F.: Io's auroral limb glow: Hubble Space Telescope FUV observations. J. Geophys. Res. **108**, 1333 (2003)
Retherford, K.D., Spencer, J.R., Stern, S.A., Saur, J., Strobel, D.F., Steffl, A.J., Gladstone, G.R., Weaver, H.A., Cheng, A.F., Parker, J.W., Slater, D.C., Versteeg, M.H., Davis, M.W., Bagenal, F., Throop, H.B., Lopes, R.M.C., Reuter, D.C., Lunsford, A., Conard, S.J., Young, L.A., Moore, J.M.: Io's atmospheric response to eclipse: UV aurorae observations. Science. **318**, 237 (2007)
Roesler, F.L., Moos, H.W., Oliversen, R.J., Woodward Jr., R.C., Retherford, D.K., Scherb, F., McGrath, M.A., Smyth, W.H., Feldman, P.D., Stobel, D.F.: Far UV imaging spectroscopy of Io's atmosphere, with HST/STIS. Science. **283**, 353–357 (1999)
Roth, L., Saur, J., Retherford, K.D., Strobel, D.F., Spencer, J.R.: Simulation of Io's auroral emission: constraints on the atmosphere in eclipse. Icarus. **214**, 495–509 (2011)
Roth, L., Saur, J., Retherford, K.D., Feldman, P.D., Strobel, D.F.: A phenomenological model of Io's UV aurora based on HST/STIS observations. Icarus. **228**, 386–406 (2014)
Roth, L., Saur, J., Retherford, K.D., Blöcker, A., Strobel, D.F., Feldman, P.D.: Constraints on Io's interior from auroral spot oscillations. J. Geophys. Res. **122**, 1903–1927 (2017)
Roth, L., Boissier, J., Moullet, A., Sánchez-Monge, Á., de Kleer, K., Yoneda, M., Hikida, R., Kita, H., Tsuchiya, F., Blöcker, A., Gladstone, G.R., Grodent, D., Ivchenko, N., Lellouch, E., Retherford, K.D., Saur, J., Schilke, P., Strobel, D., Thorwirth, S.: An attempt to detect transient changes in Io's SO2 and NaCl atmosphere. Icarus. **350** (2020)
Russell, C.T., Kivelson, M.G.: Detection of SO in Io's exosphere. Science. **287**, 1998–1999 (2000)
Russell, C.T., Kivelson, M.G.: Evidence for sulfur dioxide, sulfur monoxide, and hydrogen sulfide in the Io exosphere. J. Geophys. Res. **106**, 267–273 (2001)
Sagan, C.: Sulphur flows on Io. Nature. **280**, 750–753 (1979)
Saur, J., Neubauer, F.M., Strobel, D.F., Summers, M.E.: Three dimensional plasma simulation of Io's interaction with the Io plasma torus: asymmetric plasma flow. J. Geophys. Res. **104**, 25105–25126 (1999)

Saur, J., Neubauer, F.M., Strobel, D.F., Summers, M.E.: Interpretation of Galileo's Io plasma and field observations: I0, I24, and I27 flybys and close polar passes. J. Geophys. Res. **107**, 1422 (2002)

Saur, J., Strobel, D.F., Neubauer, F.M., Summers, M.E.: The ion mass loading rate at Io. Icarus. **163**, 456–468 (2003)

Saur, J., Neubauer, F.M., Connerney, J.E.P., Zarka, P., Kivelson, M.G.: Plasma interaction of Io with its Plasma Torus. In: Bagenal, F., McKinnon, W.B., Dowling, T.E. (eds.) Jupiter: Planet, Satellites. Magnetosphere. Cambridge University Press (2004)

Saur, J., Grambusch, T., Duling, S., Neubauer, F.M., Simon, S.: Magnetic energy fluxes in sub-Alfvenic planet star and moon planet interactions. Astron. Astrophys. **552**, A119 (2013)

Schmidt, C., Schneider, N., Leblanc, F., Gray, C., Morgenthaler, J., Turner, J., Grava, C.: A survey of visible S^+ emission in Io's plasma torus during the Hisaki Epoch. J. Geophys. Res. **123**, 5610 (2018)

Schneider, N.M., Hunten, D.M., Wells, W.K., Schultz, A.B., Fink, U.: The structure of Io's corona. Ap. J. **368**, 298–315 (1991a). https://doi.org/10.1086/169694

Schneider, N.M., Trauger, J.T., Wilson, J.K., Brown, D.I., Evans, R.W., Shemansky, D.E.: Molecular origin of Io's fast sodium. Science. **253**, 1394–1397 (1991b)

Schneider, N.M., Trauger, J.T.: The structure of the Io torus. Ap. J. **450**, 450–462 (1995)

Schneider, N.M., Bagenal, F.: Io's neutral clouds, plasma torus, and magnetospheric interaction. In: Lopes, R.C., Spencer, J.R. (eds.) Io After Galileo, pp. 265–286. Springer, New York (2007)

Šebek, O., Trávníček, P.M., Walker, R.J., Hellinger, P.: Dynamic plasma interaction at Io: multispecies hybrid simulations. J. Geophys. Res. **124**, 313–341 (2019). https://doi.org/10.1029/2018JA026153

Sieveka, E.M., Johnson, R.E.: Non-isotropic coronal atmosphere of Io. J. Geophys. Res. **90**, 5327–5331 (1984)

Skinner, T.E., Durrance, S.T.: Neutral oxygen and sulfur densities in the Io torus. Ap. J. **310**, 966–971 (1986)

Smith, H.T., Mitchell, D.G., Johnson, R.E., Mauk, B.H., Smith, J.E.: Europa neutral torus confirmation and characterization based on observations and modeling. Ap. J. **871**, 69 (2019)

Smyth, W.H., McElroy, M.B.: The sodium and hydrogen gas clouds of Io. Planet. Space Sci. **25**, 415–431 (1977)

Smyth, W.H., Combi, M.R.: Io's sodium corona and spatially extended cloud: a consistent flux speed distribution. Icarus. **126**, 58–77 (1997)

Smyth, W.H., Marconi, M.L.: Nature of the iogenic plasma source in Jupiter's magnetosphere. I. Circumplanetary distribution. Icarus. **166**, 85–106 (2003)

Smyth, W.H., Marconi, M.L.: Nature of the iogenic plasma source in Jupiter's magnetosphere. II. Near-Io distribution. Icarus. **176**, 138–154 (2005)

Smyth, W.H., Peterson, C.A., Marconi, M.L.: A consistent understanding of the ribbon structure for the Io plasma torus at the Voyager 1, 1991 ground-based, and Galileo J0 epochs. J. Geophys. Res. **116**, A07205 (2011)

Southwood, D.J., Kivelson, M.G., Walker, R.J., J.A.: Io and its plasma environment. J. Geophys. Res. **85**, 5959–5968 (1980)

Spencer, J.R., Stern, S.A., Cheng, A.F., Weaver, H.A., Reuter, D.C., Retherford, K., Lunsford, A., Moore, J.M., Abramov, O., Lopes, R.M.C., et al.: Io volcanism seen by New Horizons: a major eruption of the Tvashtar volcano. Science. **318**, 240 (2007)

Steffl, A.J., Bagenal, F., Stewart, A.I.F.: Cassini UVIS observations of the Io plasma torus: I. Initial results. Icarus. **172**, 78–90 (2004a)

Steffl, A.J., Bagenal, F., Stewart, A.I.F.: Cassini UVIS observations of the Io plasma torus: II. Initial results. Icarus. **172**, 91–103 (2004b)

Steffl, A.J., Delamere, P.A., Bagenal, F.: Cassini UVIS observations of the Io plasma torus: III. Observations of temporal and azimuthal variability. Icarus. **180**, 124–140 (2006)

Strobel, D.F., Wolven, B.C.: The atmosphere of Io: abundances and sources of sulfur dioxide and atomic hydrogen. Astrophys. Space Sci. **277**, 271–287 (2001)

Summers, M.E., Strobel, D.F., Yung, Y.L., Trauger, J.T., Mills, F.: The structure of Io's thermal corona and implications for atmospheric escape. Ap. J. **342**, 468–480 (1989)

Thomas, N.: Optical observations of Io's neutral clouds and plasma torus. Surveys Geophys. **13**, 91–164 (1992)

Thomas, N., Lichtenberg, G., Scotto, M.: High-resolution spectroscopy of the Io plasma torus during the Galileo mission. J. Geophys. Res. **106**, 26277–26291 (2001)

Thomas, N., Bagenal, F., Hill, T.W., Wilson, J.K.: The Io neutral clouds and plasma torus. In: Jupiter: The Planet, Satellites, and Magnetosphere. Cambridge University Press, New York (2004)

Thomas, N.: A comprehensive investigation of the Galilean moon, Io, by trasing mass and energy flows. Exp. Astron. (2021). https://doi.org/10.1007/s10686-021-09768-y

Trafton, L., Parkinson, T., Macy, W.: The spatial extent of sodium emission around lo. Ap. J. **190**, L85 (1974)

Tsang, C.C.C., Rathbun, J.A., Spencer, J.R., Hesman, B.E., Abramov, O.: Io's hot spots in the near-infrared detected by LEISA during the New Horizons flyby. J. Geophys. Res. **119**, 2222–2238 (2014)

Tsuchiya, F., Kagitani, M., Yoshioka, K., Kimura, T., Murakami, G., Yamazaki, A., Nozawa, H., Kasaba, Y., Sakanoi, T., Uemizu, K., Yoshikawa, I.: Local electron heating in the Io plasma torus associated with Io from the Hisaki satellite observation. J. Geophys. Res. **120**, 10317–10333 (2015)

Tsuchiya, F., Yoshioka, K., Kimura, T., Koga, R., Murakami, G., Yamazaki, A., et al.: Enhancement of the Jovian magnetospheric plasma circulation caused by the change in plasma supply from the satellite Io. J. Geophys. Res. **123**, 6514–6532 (2018)

Tsuchiya, F., Arakawa, R., Misawa, H., Kagitani, M., Koga, R., Suzuki, F., Hikida, R., Yoshioka, K., Steffl, A.J., Bagenal, F., Delamere, P., Kimura, T., Kasaba, Y., Murakami, G., Yoshikawa, I., Yamazaki, A.: Azimuthal variation in the Io plasma torus observed by the Hisaki satellite from 2013 to 2016. J. Geophys. Res. **124**, 3236–3254 (2019)

Valek, P.W., Allegrini, F., Bagenal, F., Bolton, S.J., Connerney, J.E.P., Ebert, R.W., Kim, T.K., Levin, S.M., Louarn, P., Mccomas, D.J., Szalay, J.R., Thomsen, M.F., Wilson, R.J.: Jovian high-latitude ionospheric ions: Juno in situ observations. Geophys. Res. Lett. **46** (2019)

Warnecke, J., Kivelson, M.G., Khurana, K.K., Huddleston, D.E., Russell, C.T.: Ion cyclotron waves observed at Galileo's Io encounter: implications for neutral cloud distribution and plasma composition. Geophys. Res. Lett. **24**, 2139 (1997)

Warwick, J.W., Pearce, J.B., Riddle, A.C., Alexander, J.K., Desch, M.D., Kaiser, M.L., Thieman, J.R., Carr, T.D., Gulkis, S., Boischot, A., Harvey, C.C., Pedersen, B.M.: Voyager 1 planetary radio astronomy observations near Jupiter. Science. **204**, 995–998 (1979)

Williams, D.J., Mauk, B.H., McEntire, R.E., Roelof, E.C., Armstrong, T.P., Wilken, B., Roederer, J.G., Krimigis, S.M., Fritz, T.A., Lanzerotti, L.J.: Electron beams and ion composition measured at Io and in its torus. Science. **274**, 401–403 (1996)

Williams, D.J., Thorne, R. M., Mauk, B.H.: Energetic electron beams and trapped electrons at Io, J. Geophys. Res. **104**(14), 739–753 (1999)

Williams, D.J., Thorne, R.M.: Energetic particles over Io's polar caps. J. Geophys. Res. **108**, 1397 (2003)

Wilson, J.K., Schneider, N.M.: Io's fast sodium: implications for molecular and atomic atmospheric escape. Icarus. **111**, 31–44 (1994)

Wilson, J.K., Schneider, N.M.: Io's sodium directional feature: evidence for ionospheric escape. J. Geophys. Res. **104**, 16567–16583 (1999)

Wilson, J.K., Mendillo, M., Baumgardner, J., Schneider, N.M., Trauger, J.T., Flynn, B.: The dual sources of Io's sodium clouds. Icarus. **157**, 476–489 (2002)

Wolven, B.C., Moos, H.W., Retherford, K.D., Feldman, P.D., Strobel, D.F., Smyth, W.H., Roesler, F.L.: Emission profiles of neutral oxygen and sulfur in Io's exospheric corona. J. Geophys. Res. **106**, 26155 (2001)

Yoshikawa, I., Yoshioka, K., Murakami, G., Yamazaki, A., Tsuchiya, F., Kagitani, M., Sakanoi, T., Terada, N., Kimura, T., Kuwabara, M., Fujiwara, K., Hamaguchi, T., Tadokoro, H.: Extreme ultraviolet radiation measurement for planetary atmospheres, magnetospheres from Earth-orbiting spacecraft (EXCEED). Space Sci. Rev. **184**, 237–258 (2014)

Yoshikawa, I., Suzuki, F., Hikida, R., Yoshioka, K., Murakami, G., Tsuchiya, F., Tao, C., Yamazaki, A., Kimura, T., Kita, H., Nozawa, H., Fujimoto, M.: Volcanic activity on Io and its influence on the dynamics of the Jovian magnetosphere observed by EXCEED/Hisaki in 2015. Earth, Planets Space. **69**, 110 (2017)

Yoshioka, K., Tsuchiya, F., Kimura, T., Kagitani, M., Murakami, G., Yamazaki, A., Kuwabara, M., Suzuki, F., Hikida, R., Yoshikawa, I., Bagenal, F., Fujimoto, M.: Radial variation of sulfur and oxygen ions in the Io plasma torus as deduced from remote observations by Hisaki. J. Geophys. Res. **122**, 2999–3012 (2017)

Yoshioka, K., Tsuchiya, F., Kagitani, M., Kimura, T., Murakami, G., Fukuyama, D., Yamazaki, A., Yoshikawa, I., Fujimoto, M.: The influence of Io's 2015 volcanic activity on Jupiter's magnetospheric dynamics. Geophys. Res. Lett. **45**, 10193–10199 (2018)

Chapter 10
Io as an Analog for Tidally Heated Exoplanets

Amy C. Barr, Ramon Brasser, Vera Dobos, and Lynnae C. Quick

Abstract Jupiter's moon Io is unique in our Solar System. A rocky body orbiting close to its parent planet, in resonant orbits with its sibling satellites, Io experiences intense tidal heating on an ongoing basis. These conditions also occur in at least one system of a star and its planets and perhaps others. The TRAPPIST-1 exoplanetary system harbors seven planets with compositions that could include rock, ice, and metal, which orbit close enough to their parent star to experience tidal heating at levels close to that observed at Io. Here, we use simple geophysical models, which reproduce observed behaviors of Io, to show that the TRAPPIST-1 bodies may be in a similar geophysical regime as Io. Several of the bodies could have magma oceans, but any volcanic activity could be hidden beneath layers of H2O or rock at their surfaces. However, if the planets are water-free, next-generation space telescopes could detect infrared emission from exoplanets experiencing tidal heating, as well as the spectroscopic signatures of sulfur compounds due to surface eruptions. Observations of Io that constrain where the tidal dissipation occurs in its interior would help make more accurate predictions about the behavior of exoplanets experiencing tidal heating. Hopefully, more groups of resonant, close-

A. C. Barr (✉)
Planetary Science Institute, Tucson, AZ, USA
e-mail: amlinar@psi.edu

R. Brasser
Origins Research Institute, Research Centre for Astronomy and Earth Sciences, Budapest, Hungary
e-mail: ramon.brasser@csfk.org

V. Dobos
Kapteyn Astronomical Institute, University of Groningen, Groningen, The Netherlands

MTA-ELTE Exoplanet Research Group
e-mail: vera.dobos@rug.nl, Szombathely, Hungary

L. C. Quick
NASA Goddard Space Flight Center and The Nexus for Exoplanet System Science (NExSS), Greenbelt, MD, USA
e-mail: lynnae.c.quick@nasa.gov

R. M. C. Lopes et al. (eds.), *Io: A New View of Jupiter's Moon*, Astrophysics and Space Science Library 468, https://doi.org/10.1007/978-3-031-25670-7_10

orbiting exoplanets will be discovered in the future, and improved knowledge of the interior geophysics of Io can help guide interpretations of observations.

10.1 Introduction

Among the outer planets of our Solar System, it is common to find groups of satellites orbiting close to their parent planet, in resonant, eccentric orbits. As the satellite-to-planet distance changes during each orbit, a tidal bulge is raised and lowered on each satellite's surface. The height of the bulge depends on the satellite-to-primary mass ratio, the distance to the primary, and the internal structure of the satellite. In the absence of perturbations from other moons in the system, this process circularizes the orbits of the satellites. The friction associated with the raising and lowering of the bulge results in heat, and can drive volcanic activity. If multiple satellites are in resonant orbits, gravitational perturbations from the system's other large moons can replenish orbital eccentricities, allowing tidal heating to persist over geologically long time scales.

Heating of planetary satellites by tides, and the associated tectonics and volcanism has occurred on other Solar System bodies. However, among Solar System bodies, Jupiter's innermost large satellite Io is in a class of its own. Most of the bodies that have experienced tidal heating and the associated activity are icy—for example, Jupiter's moon Europa, Saturn's moon Enceladus, and Uranus' moon Miranda. Io is the only rocky planetary body experiencing extensive volcanism driven by tidal dissipation. This process may occur in other solar systems, as well, and not just within satellite systems of giant planets. Groups of exoplanets that orbit close to their parent star in resonant orbits can also experience ongoing tidal dissipation. Thus, the intense volcanism and tidal heating experienced by Io may be occurring on some exoplanets, as well, suggesting that Io may be a nearby, observable analog for a rocky, tidally heated exoplanet.

The similarities between Io's unique geophysical environment and exoplanets became apparent only recently, after the discovery of seven roughly Earth-sized planets around the ultracool M-dwarf star TRAPPIST-1 (Gillon et al. 2017). The planets have roughly earthlike masses, and mean densities that allow for interior compositions including metal and perhaps even water ice, provided the planets' surface temperatures due to stellar insulation and greenhouse effects are low (Unterborn et al. 2017; Barr et al. 2018; Turbet et al. 2020; Agol et al. 2021). Owing to the low luminosity of TRAPPIST-1, the effective surface temperatures on each planet are quite low, suggesting that the planets could be solid, rather than completely molten. Some of the planets orbit within the star's habitable zone, raising the possibility of solid surfaces, possible liquid water surface layers, and atmospheres.

Table 10.1 compares the latest planetary parameters for the TRAPPIST-1 bodies to the properties of the Galilean satellites. Each TRAPPIST-1 planet has a mean density less than that of Earth, indicating a composition that is iron-poor, water-rich,

Table 10.1 Current best estimates of the masses (M), radii (R), mean densities ($\bar{\rho}$), orbital periods (P), and eccentricities (e) of the TRAPPIST-1 planets compared to the Galilean satellites(Showman and Malhotra 1999)

Object	M/M_p $[10^{-5}]$[a]	R/R_p[b]	$\bar{\rho}$ (kg/m^3)	P (days)[a]	e $[10^{-3}]$[a]
TRAPPIST-1 b	4.596 ± 0.198	0.0859 ± 0.004	5425^{+265}_{-272}	1.510	3.05 ± 2.91
TRAPPIST-1 c	4.374 ± 0.151	0.0844 ± 0.0004	5447^{+222}_{-235}	2.421	0.55 ± 3.32
TRAPPIST-1 d	1.297 ± 0.025	0.0606 ± 0.0005	4354^{+156}_{-163}	4.049	5.63 ± 1.72
TRAPPIST-1 e	2.313 ± 0.043	0.0708 ± 0.0006	4885^{+168}_{-182}	6.101	6.32 ± 1.20
TRAPPIST-1 f	3.475 ± 0.052	0.0804 ± 0.0005	5009^{+138}_{-158}	9.207	8.42 ± 1.30
TRAPPIST-1 g	4.418 ± 0.057	0.0869 ± 0.0005	5042^{+136}_{-158}	12.35	4.01 ± 1.09
TRAPPIST-1 h	1.088 ± 0.062	0.0581 ± 0.0009	4147^{+322}_{-302}	18.77	3.65 ± 0.77
Io	4.497 ± 0.0001	0.026	3530	1.769	4.1
Europa	2.529 ± 0.0001	0.022	3010	3.551	10.1
Ganymede	7.988 ± 0.00023	0.037	1940	7.154	1.3
Callisto	4.504 ± 0.0026	0.034	1834	16.689	7.4

[a] From Table 2 of Agol et al. (2021)

[b] From Table 3 of Agol et al. (2021). Data for the Galilean satellites from Lieske (1998). Here, subscript ‘p’ refers to the primary. For the TRAPPIST-1 system, $M_p = 1.79019 \times 10^{29}$ kg (Grimm et al. 2018) and $R_p = 8.3003 \times 10^7$ m (Agol et al. 2021). For the Galilean satellites, $M_p = 1.898 \times 10^{27}$ kg and $R_p = 6.9911 \times 10^7$ m. Relative uncertainties in the orbital periods of the TRAPPIST-1 planets are less than 0.001% and are even lower for the Galilean satellites. The uncertainties in the radii of the Galilean satellites are about 0.4 km

or both (Elkins-Tanton and Seager 2008; Agol et al. 2021). Figure 10.1 illustrates a scaled comparison between the Galilean satellites and the TRAPPIST-1 system, using data from Table 10.1. Viewed in this way, the systems are quite similar. The sizes of the primary object (Jupiter and the TRAPPIST-1 star) are not dissimilar—the radius of TRAPPIST-1 is 1.16 Jupiter radii. The bodies also have similar orbital periods, between approximately 1.5 and 20 days. The similarities in orbital periods, secondary-to-primary mass ratios, and compositions suggest that solid tidal friction may be an important heat source. Even more striking is existence of an apparent mean motion resonance chain in both systems (see Sect. 10.3) which means that the orbital periods of the planets (or moons around Jupiter) can be expressed as ratios of small integers. Like the Galilean satellites, the orbital eccentricities of the TRAPPIST-1 planets will be replenished by gravitational interactions among them, allowing for ongoing tidal heating.

A notable difference between the systems is the difference in water content. The Galilean satellites, which accreted from a disk of material orbiting Jupiter (Canup and Ward 2002), formed slowly (Barr and Canup 2008) and far enough from the Sun for water to solidify, which resulted in their H_2O-rich compositions. If the TRAPPIST-1 planets formed in a location similar to their present location, where water is likely stable only for the outermost planets, this could explain their more rock-rich compositions. In drawing comparisons between the Galilean satellites and the TRAPPIST-1 planets, two questions naturally arise: could some of the

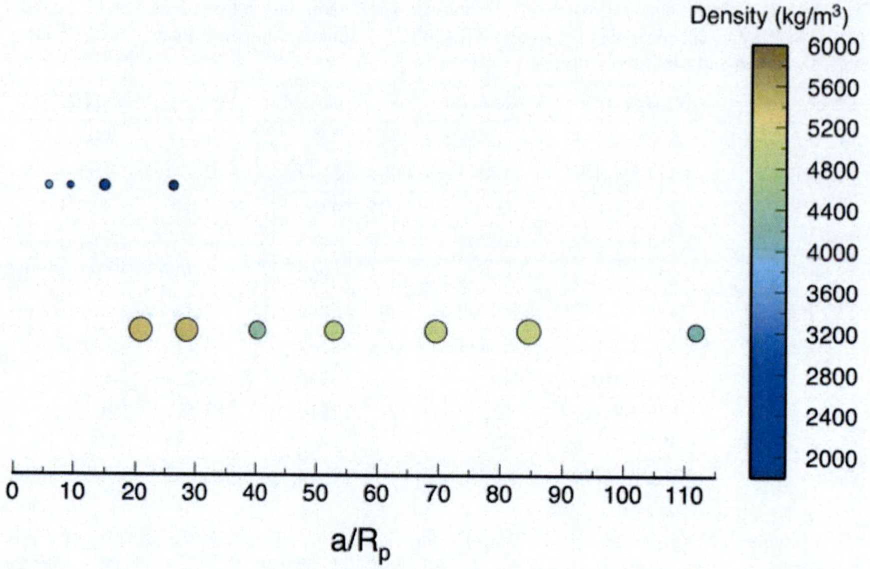

Fig. 10.1 Scaled comparison between the Galilean satellites (top row) and the TRAPPIST-1 planets (bottom row). Symbol location and size indicate the semi-major axis (a) scaled by the primary radius (R_p) and the body radius divided by the primary radius. Symbol color shows the mean density (see Table 10.1), with blue colors indicating low densities and more volatile-rich compositions, and beige colors indicating more rock- and metal-rich compositions. The radius of TRAPPIST-1 decreased by about an order of magnitude during the first Gyr of its evolution (Baraffe et al. 2015), so that TRAPPIST-1 b was originally near 4 R_* and the scale of the two systems was similar

TRAPPIST-1 bodies be as active as Io? Could Io and the other Galilean satellites be a close, observable analog for groups of tidally heated exoplanets?

Here, we explore the links between the geophysical and orbital settings of Io and the TRAPPIST-1 bodies to shed light on the similarities and differences between this remarkable moon and bodies outside our Solar System. Because of the number of bodies and their location in the habitable zone, the TRAPPIST-1 planets have been extensively studied with observations and theoretical models. There are not many other rocky exoplanets known to date that could be experiencing ongoing tidal heating or could have volcanic activity like Io. The possible geology and geophysics of tidally heated exomoons (moons of exoplanets) is even more speculative, as we are still waiting for the first confirmed exomoon discovery. For this reason, in this chapter, we will focus our attention on the TRAPPIST-1 system to discuss the use of Io as an analog for extrasolar bodies.

We discuss the simplest models for the possible interior structures for the TRAPPIST-1 bodies and compare them to Solar System objects. Then, we describe the orbital configuration of the TRAPPIST-1 system and the similarities and differences between that system and the Galilean satellites. We use a simple geophysical model, which accurately describes the heat flow of Io, to determine possible thermal states of the TRAPPIST-1 planets, then review arguments for and against Io-type volcanism on each body. We conclude by suggesting avenues for further study, both observational and theoretical.

10.2 Planetary Composition

Initial estimates of the masses and radii of the TRAPPIST-1 planets indicated rocky and perhaps mixed ice/rock compositions for each body (Gillon et al. 2017; Unterborn et al. 2017). However, the large error bars on the masses and radii meant that it was not possible to tightly constrain the composition of each planet. The red points in Fig. 10.2 show the current best estimates for the masses and radii of the planets, which are based on observations of the TRAPPIST-1 system over a time scale of several years (Agol et al. 2021). This more recent data indicates that most of the planets have more rock- and metal-rich compositions that previously thought (Gillon et al. 2017; Unterborn et al. 2017; Grimm et al. 2018). Additionally, there is a slew of processes that could affect the density-composition relationship for the

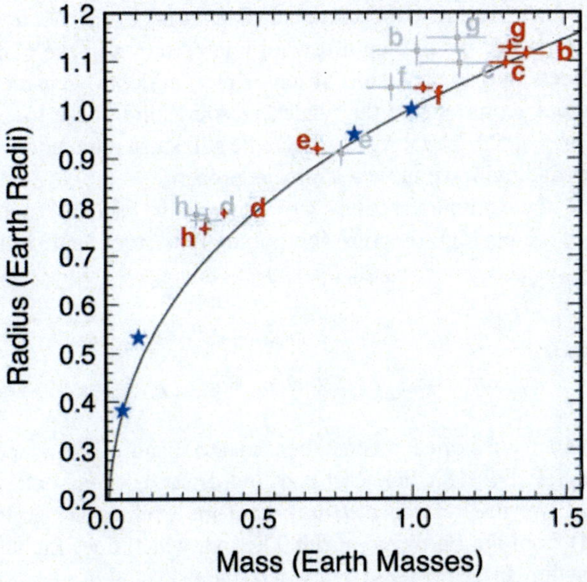

Fig. 10.2 Comparison between prior estimates of the masses and radii of the TRAPPIST-1 planets (gray) from Grimm et al. (2018) and the latest values from Agol et al. (2021) (red). Line indicates the mass/radius relationship for a pure rock composition (a uniform density sphere with $\rho = 5500\,\text{kg/m}^3$ and no compression), following our simple compositional model. Blue stars represent the terrestrial planets of our Solar System: Mercury, Venus, Earth, and Mars

planets, including the effect of a runaway greenhouse on the planet's radius, which could result in a thick atmosphere and no liquid or solid water (Turbet et al. 2020), the magma radius inflation effect (Bower et al. 2019), and the possibility that the planets may not have cores due to the oxidation state of source material (Elkins-Tanton and Seager 2008). Despite the additional observations, the masses and radii of the planets are still uncertain enough to permit a variety of different compositions, so when we model the interiors of the planets, we must consider a range of possible mean densities.

The simplest way of modeling the possible interior structures of the TRAPPIST-1 bodies is to assume each planet is composed of a mixture of ice I/liquid water (in the absence of a greenhouse), high pressure ice polymorphs (*hpp*), rock, and metal (Barr et al. 2018). This analysis does not take into account the possibility that some of the innermost TRAPPIST-1 planets may be too hot for water to be stable. In reality, for planets b, c, and perhaps d as well, H2O will be present in the form of water vapor ± supercritical water, not ice. We proceed with a model assuming that H2O is present as ice I/liquid water. However for planets b and c our estimated water contents should be understood as being upper limits (because water vapor has lower density than water ice.

We also assume each planet is fully differentiated, having undergone complete separation of ice/metal/rock into distinct layers. This seems likely given that the planets are similar in size and mean density to the Solar System's terrestrial planets, all of which are fully differentiated (Barr et al. 2018). More sophisticated modeling, focusing just on a straightforward interpretation of the planet's density, can include effects like compression of ice or rock at depth, density changes due to the temperature profile inside the planet, or solid/solid phase transitions in the rock (Unterborn et al. 2017). However, given the still substantial uncertainties in the planets' masses and radii, we favor a simple approach.

We compute the volume fraction (ϕ) occupied by each of four materials—ice I and liquid water, high-pressure ice polymorphs, rock, and iron—using the following:

$$\phi_{iw} + \phi_{hpp} + \phi_r + \phi_{iron} = 1, \tag{10.1}$$

$$\phi_{iw}\rho_{iw} + \phi_{hpp}\rho_{hpp} + \phi_r\rho_r + \phi_{iron}\rho_{iron} = \bar{\rho}, \tag{10.2}$$

where ρ are the densities of each constituent material and $\bar{\rho}$ is the mean density of the planet (Barr et al. 2018). For ice I and liquid water, we assume $\rho_{iw} = 1000$ kg/m^3. For rock we use reference values from the Preliminary Reference Earth Model (PREM) from the Earth, $\rho_r = 5500$ kg/m^3, which does include the effects of compression for an Earth-sized body (Dziewonski and Anderson 1981). For metal we use the PREM density for iron, $\rho_{iron} = 12{,}000$ kg/m^3 (Dziewonski and Anderson 1981). For the high-pressure ice polymorphs, we use a representative $\rho_{hpp} = 1300$ kg/m^3 intermediate between the densities of ice II, V, VI, and VII (Hobbs 1974).

In this simple model, any body with a mean density $\bar{\rho}$ less than that of rock can contain H_2O (Barr et al. 2018). For these bodies, we first determine ϕ_{iw}, which depends on the maximum depth at which ice I is possible, corresponding to a pressure of 209 MPa. The planetary radius at which this occurs, $R_{iw} = R_{pl} - z_{209}$ where $z_{209} \sim 209\,\text{MPa}/(\rho_{iw} g)$, where $g = GM_{pl}/R_{pl}$, and G is the gravitational constant. This gives $\phi_{iw} = 1 - (R_{iw}/R_{pl})^3$. With ϕ_{iw} constrained, we then solve for a suite of possible values for ϕ_{hpp}, ϕ_r, and ϕ_{iron} for each planet. We keep the iron content of the planet between 10% and 50% mass-fraction to keep the possible interior structures realistic.

The results are shown in Fig. 10.3 where colors indicate the mean of the H_2O mass fraction (including both the high pressure ice and the ice I/liquid water layer) for different M–R pairs. These M–R pairs show the most likely compositions for each planet because we have taken into account the probability distribution of mass and radius based on the error bars on both of these quantities. For each M–R pair we can calculate several possible interior models, so the color in the figure shows the mean of these H_2O values. Notice that for high-density cases (higher mass coupled with lower radius) there is less water and for low-density cases (lower mass coupled with higher radius) there is more water content in the same planet. The two smallest two planets (TRAPPIST-1 d and h) have the highest possible H_2O content, consistent with their distance from the parent star. TRAPPIST-1 b and c are closest to the star and accordingly are likely to have more rock-rich compositions. Taking into account only the planets' masses and radii, neglecting their warm surface temperatures, TRAPPIST-1 b and c could have H_2O mass fractions between 1% and 11%. Water will not stay in the form of ice on extremely hot planets and therefore the TRAPPIST-1 b and c water contents should be treated with special caution.

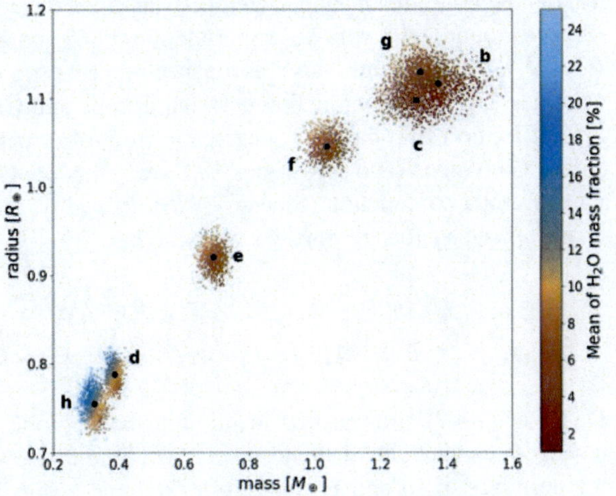

Fig. 10.3 Possible H_2O mass fractions for the TRAPPIST-1 planets. The color of each dot shows the mean H_2O mass fraction (by percent) for pairs of mass/radius values permitted by the latest observation (Agol et al. 2021). Black dots represent the mean radius and mass value for each of the planets

10.3 Orbital Configuration and Tidal Heating

10.3.1 The Galilean Satellites

Io is the innermost of the four Galilean satellites of Jupiter, four Moon- to Mercury-sized bodies whose combined mass ratio to that of Jupiter is $\sim 2 \times 10^{-4}$, similar to what is observed for the Saturnian and Uranian systems (Canup and Ward 2006). The mass of an individual Galilean satellite is on average 5×10^{-5} times the mass of Jupiter (Lieske 1998). The innermost three satellites are locked in multiple mean-motion resonances, and due to their particular configuration they are also locked in a so-called three-body resonance (e.g. Murray and Dermott 1999). Specifically the two satellites Io and Europa are locked in a 2:1 mean-motion resonance, and Europa and Ganymede are also locked in a 2:1 resonance. We designate the mean longitude $\lambda = \mathcal{M} + \varpi$, where $\mathcal{M}$ is the mean anomaly—the fraction of the orbital period that has elapsed since periastron and expressed as an angle—and $\varpi = \omega + \Omega$ is the longitude of perihelion; the angle Ω is the longitude of the ascending node with respect to a reference plane and reference direction (usually Jupiter's equator) and ω is the argument of perihelion. The mean motion of the satellites $n^2 = G(M_J + m_{\text{sat}})a^{-3}$. For Io and Europa, the observed rates of motion satisfy $n_I - 2n_E = 0.7395507361°$/day, and for Europa and Ganymede $n_E - 2n_G = 0.7395507301°$/day (e.g. Brown 1977) and the fact that $n_I - 3n_E + 2n_G = 0$ implies the existence of a three-body resonance. Observations of the motion of the satellites shows that

$$\lambda_I - 3\lambda_E + 2\lambda_G = 180°. \tag{10.3}$$

This three-body resonance is dubbed the Laplace resonance, whose argument librates around 180° with an amplitude of 0.066° and a period of about 2074 days (Lieske 1998). The three-body resonance prevents triple conjunctions amongst the satellites. During a conjunction between Europa and Ganymede, Io is on the other side of Jupiter; Europa is 60° degrees in front of or behind a conjunction between Io and Ganymede (and sometimes 180° away), while Ganymede is 90° away from an Io-Europa conjunction (Sinclair 1975). In reality, the motion of the satellites is characterised by four resonances, with angles

$$\begin{aligned} \phi = 2\lambda_E - \lambda_I - \varpi_I, \quad \theta = 2\lambda_E - \lambda_I - \varpi_E \\ \theta' = 2\lambda_G - \lambda_E - \varpi_E, \quad \psi = 2\lambda_G - \lambda_E - \varpi_G. \end{aligned} \tag{10.4}$$

De Sitter (1909) showed that in this four-body problem, the stable solutions are $\phi = 0$, $\theta = 180°$, $\theta' = 0$ and $\psi = 180°$. Due to the first and second resonances the perijoves of Io and Europa are locked and apsidally anti-aligned, so that we have $\dot{\varpi}_I = \dot{\varpi}_E \sim -0.7396°$/day (Peale et al. 1979) and $\varpi_E - \varpi_I = 180°$ i.e. the apses of Io and Europa are anti-aligned; conjunctions happen when Io is near perijove and Europa near apojove. The precession rate of both satellites has a secular contribution from mutual perturbations, a secular part caused by Jupiter's oblateness and a resonant contribution. Secular perturbations in general cause $\dot{\varpi} > 0$ while resonant action results in $\dot{\varpi} < 0$ (Murray and Dermott 1999), so that Io's and Europa's regression of the perijoves is probably dominated by the resonance. The secular precession rate due to Jupiter's shape is given by

$$\dot{\varpi}_{\text{sec}} = \frac{3nJ_2}{2}\left(\frac{R_J}{a}\right)^2, \tag{10.5}$$

where $J_2 \sim 0.015$ (e.g. Lainey et al. 2004) is one of the gravity coefficients of Jupiter, a is the semi-major axis and R_J is the equatorial radius of Jupiter. For Io $\dot{\varpi}_{\text{sec}} = 0.1290°$/day. The flattening of Jupiter causes ϖ to rotate anti-clockwise i.e. opposite to the direction of motion, so that the observed regression of ϖ_I must be caused by the resonance with Europa. The resonant contribution to the precession for Io is given by Murray and Dermott (1999)

$$\dot{\varpi}_{\text{res}} = \frac{n_I m_E}{e_I m_J}\alpha F(\alpha) \tag{10.6}$$

where $\alpha = a_I/a_E$. The general formula of $F(\alpha)$ for a resonant argument of the form $j\lambda' + (1-j)\lambda - \varpi$ is $F(\alpha) = -(j + \frac{1}{2}\alpha D)b_{1/2}^{(j)}(\alpha)$, where $D = \mathrm{d}/\mathrm{d}\alpha$ and a prime refers to the outer body. In the same way, when the resonant argument is $j\lambda' + (1-j)\lambda - \varpi'$ the coefficient becomes $H(\alpha) = \frac{1}{2}(-1 + 2j + \alpha D - 4\delta_{2j})b_{1/2}^{(j-1)}(\alpha)$. The functions and $b_{1/2}^{(j)}(\alpha)$ are Laplace coefficients (Murray and Dermott 1999). For Io we calculate $\dot{\varpi}_{\text{res}} = \varpi_I - \varpi_{\text{sec}} = -0.8686°$/day caused by the resonance, from which it follows that the forced equilibrium eccentricity is $e_I = 0.0043$, which is very close to the observed value of 0.0041 (Lieske 1998). The forced eccentricities for Europa and Ganymede from the multiple resonances are computed in a similar manner, with Europa receiving contributions from both Io and Ganymede (Sinclair 1975)

$$e_E = \frac{n_E}{\dot{\varpi}_{\text{res,E}}}[m_I H(\alpha) + n_E m_G \alpha' F(\alpha')] = 0.0101$$

$$e_G = \frac{n_G m_E H(\alpha')}{\dot{\varpi}_{\text{res,G}}} = 0.0007 \tag{10.7}$$

where $\alpha' = a_E/a_G$.

The forced eccentricity of Io is the source of its intense tidal heating and it has been suggested that the low amplitude of the libration of the resonant angle is due to the system being in equilibrium (Peale et al. 1979), wherein damping in Io is matched by damping in Jupiter. In the simplest tidal model the dissipation in Io is given by (Segatz et al. 1988)

$$\frac{\mathrm{d}E}{\mathrm{d}t} = -\frac{21}{2}\frac{k_2}{Q}\frac{n_I^5 R_I^5}{G}e_I^2 \tag{10.8}$$

where k_2 is the second degree Love number, Q is the tidal quality factor and R_I is the radius of Io; this formula applies to any secondary orbiting a primary. From long-term observational data Lainey et al. (2009) obtain $k_2/Q = 0.015$ for Io and $\dot{E} \sim 10^{14}$ W. If the energy were transported out of Io at the same rate, the associated surface heat flux would be 2.25 W m^{-2}, which is similar to the observed surface heat flux on Io (Spencer et al. 2000; Veeder et al. 2004).

Tidal dissipation in Io further causes an increase in the mean motion due to the damping of the eccentricity and the conservation of orbital angular momentum. We have (Peale et al. 1979)

$$\frac{\mathrm{d}n}{\mathrm{d}t} = \frac{3}{nma^2}\frac{\mathrm{d}E}{\mathrm{d}t}. \tag{10.9}$$

Tidal dissipation in a mean-motion resonance causes the inner and outer bodies to diverge from each other and their resonant angles and eccentricities collapse onto fixed points (Batygin and Morbidelli 2013; Delisle et al. 2012; Lithwick and Wu 2012). The Io-Europa pair could very well be in such a configuration because the resonant angles have a low libration amplitude and it appears that Io and Europa may actually be separating (Lainey et al. 2009). For a triple resonance the inner and outer body diverge while the evolution of the middle body depends on the relative masses (Batygin and Morbidelli 2013); observations appear to indicate a slow separation of Europa and Ganymede from Io so that the current configuration could not be primordial and the forced eccentricities may have been higher in the past.

10.3.2 Resonant Exoplanets

The resonant configuration of the Galilean satellites and their observed tidal evolution could be a proxy for tides in exoplanetary systems. One of the most studied resonant planetary systems is the TRAPPIST-1 system, which consists of seven roughly earth-sized planets in close orbit around a low-mass M-type star (Luger et al. 2017). Several of the planets lie in the habitable zone of the star, so much attention has been paid to the system as a possible abode for extraterrestrial life. However, from the perspective of Solar System studies, the planets are also interesting because they occupy several mean-motion resonances and have measurable non-zero eccentricities, similar to the satellite systems of the outer planets and the Galilean satellites in particular. The planets' mean densities, derived from estimates of their masses and radii, indicate a range of possible compositions, ranging from pure rock to mixtures of ice, rock, and iron (Grimm et al. 2018; Agol et al. 2021).

The TRAPPIST-1 planets are locked in many mean-motion resonances and three-body resonances (Luger et al. 2017). The inner two planets appear not to be part of any resonance (Grimm et al. 2018; Agol et al. 2021) although they are near the 8:5 b-c and 5:3 c-d resonances; confirmed resonances are the 3:2 d-e, the 3:2 e-f, the 4:3 f-g and the 3:2 g-h (Luger et al. 2017). Based on the dynamical solutions of Grimm et al. (2018), Brasser et al. (2019) show that most of the two-body angles of the first-order resonances librate, although numerical simulations with updated planetary parameters from TTV analyses show a decrease in the fraction of angles that could be librating (Agol et al. 2021). One of the resonant arguments, $3\lambda_h - 2\lambda_g - \varpi_h$, appears to circulate. Furthermore $\varpi_d - \varpi_e = 180°$ so that the apses of planets d and e are anti-aligned and the situation is very similar to the Io-Europa configuration. A frequency analysis on the motion by Brasser et al. (2019) indicates that $2n_d - 3n_e = 0.7235°$/day, and that the apses of planets d and e regress at the same rate i.e. $\dot{\varpi}_d = \dot{\varpi}_e = -0.7235°$/day; these rates of regression are comparable to that of Io and Europa, which is probably expected given the similar mass ratios between primary and secondary in both systems. The simulations from Brasser et al. (2019) show that all the periastra of planets d to h regress at this rate, confirming that all these planets are locked in a resonant chain (Luger et al. 2017). In a manner similar to Io's case we can compute the forced eccentricity of planet d from the resonance with planet e as

$$e_d = \frac{n_d m_e}{\dot{\varpi}_d M_*} \alpha F(\alpha) = 0.0043 \tag{10.10}$$

where $\alpha = a_d/a_e = 0.7608$. The coefficient $F(\alpha)$ is numerically different from that of the Galilean satellites because the TRAPPIST-1 planets are in a 3:2 resonance rather than in the 2:1. In a manner similar to Europa, the forced eccentricity of planet e is then calculated as

$$e_e = \frac{n_e}{\dot{\varpi}_e M_*}[m_d H(\alpha) + m_f \alpha' F(\alpha')] = 0.0068, \tag{10.11}$$

where now $\alpha' = a_e/a_f = 0.7600$. The forced eccentricities computed here agree well with the probability distributions obtained from long-term observations (Agol et al. 2021). Similar calculations can be performed for planets f, g and h, whose values once again agree well with the observations.

For planet d we cannot immediately calculate $\dot{E}$ because we do not know the tidal parameters k_2/Q and unlike (Lainey et al. 2009) we do not have decades of observations to fit the tidal parameters with. Naively substituting the k_2/Q value for Io we calculate $\dot{E}_d = 2.611 \times 10^{14}$ W, or a surface heat flux of 0.802 W m^{-2}, approximately a factor of four lower than that of Io. However, if planet d's tidal parameters are comparable to those of planets b and c, for which $k_2/Q \sim 10^{-4}$ (Brasser et al. 2019), then $\dot{E} \sim 10^{12}$ W and the surface heat flux could be

5 mW m^{-2}. Even so, the resonant forcing of the eccentricity should cause some tidal heating of planet d, which may have repercussions for its habitability. Similar arguments apply to planet e, but the tidal dissipation will be even weaker due to its greater distance to the star. We shall calculate updated surface heat fluxes using our interior models in Sect. 10.4 below.

10.4 Tidal Heating and Interior Modeling

In a tidally heated body, the interior temperature, the extent of melting, and the potential for volcanism, all depend upon the balance and feedback between internal heat generation by, for example, radiogenic and tidal sources, and heat transported by conduction, magma advection and possibly liquid- and solid-state convection (Schubert et al. 2001; Tackley 2001; Hussmann et al. 2002; Moore 2006).

The simplest method of estimating tidal heating in an orbiting secondary body is to assume that the interior of the secondary object has a Maxwell viscoelastic rheology (Love 1906; Peale and Cassen 1978), characterized by a shear modulus, μ, and viscosity, η. More complex rheologies can be employed, and give slightly different behaviors (e.g., Efroimsky 2012). However, given the large uncertainties in the compositions of the TRAPPIST-1 planets, we favor the Maxwell model because it has a simple behavior and relatively few parameters.

In a Maxwell viscoelastic material, heating is maximized in material whose Maxwell time, $\tau = \eta/\mu$, is comparable to the orbital period of the secondary, P (Love 1906). For ice I, the low-density phase of H_2O that floats on liquid water, the Maxwell time is comparable to the orbital period of Europa, 85 hours, if the ice is close to its melting point (Barr and Showman 2009). A body whose interior, or portion of its interior, has a viscosity and rigidity close to the values for which $P \sim \tau$, will undergo intense tidal heating. A body with a slightly higher viscosity, that is, one that is cooler, can be warmed by tidal heating until $P \sim \tau$. The body can experience inner melting. If the silicate portion of the body melts, this can lead to silicate volcanism (Peale 2003). In ice/rock bodies, this can lead to the formation and maintenance of liquid water oceans (Ojakangas and Stevenson 1989). Thus, the tidal heating rate in a solid body and its interior structure (namely: composition, temperature, and phase as a function of depth), depend on one another, and the body may reach an equilibrium structure where heat generation is equal to heat loss (Moore 2006).

We use a simple analytic model to calculate the response of the planet to the tidal forcing, expressed by the imaginary part of the k_2 Love number, Im(k_2). This value is used to calculate the surface heat flux from tidal dissipation F_{tidal}. In the case of Io, the value of F_{tidal} has been estimated by spacecraft observations to be around 1–3 W/m^2 (Spencer et al. 2000; Veeder et al. 2004). This provides us with an opportunity to test the realism of this simple model—if Io parameters are used and an Io-like heat flux is obtained, there is some confidence in the realism of this approach.

The heat generated from tidal dissipation is likely removed from the planets' deep interiors via solid-state convection, the dominant heat transfer mechanism in the terrestrial planets and ice/rock satellites in the Solar System. To determine the temperature in the planets' deep interiors, we balanced $F_{\rm tidal}$ and the convective heat flux ($F_{\rm conv}$), and found the value of temperature ($T_{\rm eq}$) for which the two are equal. If $T_{\rm eq}$ is above the solidus temperature for rock, the planet may possess a magma ocean either at its surface, or beneath a surface layer of solid rock.

10.4.1 Tidal Heating

We consider tidal heating as the dominant source of energy for heating the interiors of Io and all the TRAPPIST-1 planets. The magnitude of tidal heating expected in rock- or ice-dominated bodies at the orbital periods associated with the TRAPPIST-1 planets, and measured at Io, a few to ten watts per meter squared, is several orders of magnitude higher than the heat flows associated with radiogenic heating, which is typically of the order of tens of milliwatts per meter squared (Henning et al. 2009; Henning and Hurford 2014). In addition, the TRAPPIST-1 system is very old, 7.6 $\pm$ 2.2 Gyr (Burgasser and Mamajek 2017), so that the expectation is that its planets have a much lower abundance of long-lived radionuclides that heat up their interiors compared to the planets in the Solar System (Frank et al. 2014).

Details about our calculations of the tidal heat flux can be found in Barr et al. (2018) and Dobos et al. (2019). We use an analytic model that assumes the planet's viscosity and rigidity can be characterized by a single value that depends on planetary composition and is very strongly temperature dependent. We mimic the effect of multiple materials by determining the effective viscosity and rigidity for each layer and weighting them by their volume fractions. To approximate the material properties of a planet composed of several different materials, we calculate the volume fraction of each material contained in the planet: for a given material, i, the volume fraction $\phi = V_i/V_{tot}$, where $V_{tot} = 4/3\pi R_s^3$ is the total volume of the secondary body, the planet in the case of TRAPPIST-1 and the satellite in the case of the Galilean satellites. A single uniform viscosity and rigidity for the secondary is very roughly approximated by:

$$\eta \approx \phi_I \eta_I + \phi_{hpp} \eta_{hpp} + \phi_r \eta_r + \phi_{iron} \eta_{iron}, \tag{10.12}$$

where the ϕ values are the volume fractions of each material, based on our compositional modeling (see Sect. 10.2) and η is the temperature-dependent viscosity for each material (Barr et al. 2018). The rigidity is very roughly estimated in a similar way:

$$\mu \approx \phi_I \mu_I + \phi_{hpp} \mu_{hpp} + \phi_r \mu_r, \tag{10.13}$$

where μ is the rigidity of each material. A limitation of this method is that the incorporation of a small amount of highly viscous material (e.g., cold rock mixed with ice) can significantly affect the viscosity, which is unrealistic unless the volume fraction of the viscous phase is greater than about 25% (Friedson and Stevenson 1983).

Detailed information about the viscosities and rigidities for each material can be found in Barr et al. (2018). The viscosities for ice I, hpp ices, and rock are assumed to be Newtonian and temperature dependent. For rock, once the solidus temperature is achieved, the viscosity decreases with temperature until a "breakdown" temperature, at which the crystal/liquid mixture reaches the rheologically critical melt fraction (Renner et al. 2000) $\sim$1800 K, and the viscosity drops sharply to a value appropriate for liquid magma. Similarly, the rigidity for rock also decreases sharply as a function of temperature once the breakdown temperature is achieved. We do not include the effect of the iron core on the viscosity and rigidity of the planets because the deformation of the iron core is expected to be small over these tidal forcing timescales (Barr et al. 2018).

The amount of tidal heating depends on the k_2 Love number, which describes the change in the secondary's own gravitational potential in response to the tidal potential from the star (Love 1906). In a viscoelastic body, the Love number is a complex number, with the real part representing elastic (recoverable) deformation, and the imaginary part representing viscous (non-recoverable) deformation and dissipation of energy. The rate of tidal heating is related to k_2 (Segatz et al. 1988),

$$\dot{E}_{\text{tidal}} = -\frac{21}{2} Im(k_2) \frac{R_{\text{sec}}^5 \omega^5 e^2}{G}, \tag{10.14}$$

where R_{sec} is the radius of the secondary, ω is its orbital frequency, e is its orbital eccentricity, and G is the gravitational constant. This implies that uncertainties in the orbital eccentricity can affect the amount of tidal heating, which we have included in our model. This is similar to the expression for tidal heating (Eq. 10.8) but here we have explicitly evaluated k_2/Q in terms of the imaginary of the k_2 Love number. The value of Im(k_2) is given by (Henning et al. 2009),

$$-\,\text{Im}(k_2) = \frac{57\eta(T)\omega}{4\bar{\rho} g R_s}\left[1 + \left(1 + \frac{19\mu(T)}{2\bar{\rho} g R_s}\right)^2 \frac{\eta(T)^2\omega^2}{\mu(T)^2}\right], \tag{10.15}$$

where $\eta(T)$ and $\mu(T)$ are the temperature- and composition-dependent viscosity and rigidity for the secondary. The globally averaged tidal heat flux,

$$F_{\text{tidal}} = \frac{\dot{E}_{\text{tidal}}}{4\pi R_s^2}. \tag{10.16}$$

10.4.2 Convection and Conduction

Tidal heat is transported within the secondary by solid-state convection and then ultimately conducted across the body's crust to the surface where it is radiated away. We can estimate the possible mantle temperature by setting F_{tidal} equal to the heat flow due to stagnant lid convection (Barr et al. 2018),

$$F_{\text{conv}} = 0.53\left(\frac{Q^*}{R_G T^2}\right)^{-4/3}\left(\frac{\rho g \alpha k_{\text{therm}}^3}{\kappa_{\text{therm}}\eta(T)}\right)^{1/3}, \quad (10.17)$$

where Q^* is the activation energy in the rock flow law, k_{therm} is the thermal conductivity, and κ_{therm} is the thermal diffusivity, α is the coefficient of thermal expansion, and $R_G = 8.314$ J/mol-K is the gas constant (Solomatov and Moresi 2000). This assumes that most of the tidal heat will be produced in, and transported within, the rock mantle of the planet, so rock parameters are used to evaluate F_{tidal} (Barr et al. 2018).

To determine the equilibrium (or equilibria) between F_{conv} and F_{tidal}, we compute F_{tidal} and F_{conv} for a range of values of T, the temperature in the rock mantle of each planet. Where the heat flows match, there is a possible equilibrium.

Two equilibria are possible (Moore 2006). The left panel in Fig. 10.4 is a schematic illustration of how the heat fluxes from tides and convection vary as a function of mantle temperature. One equilibrium occurs at a temperature well below the solidus, corresponding to a completely solid planetary mantle ("solid mantle" in the left panel, and middle panel). In this case, convection can efficiently remove the tidal heat, even if the convective flow is relatively sluggish. However, this

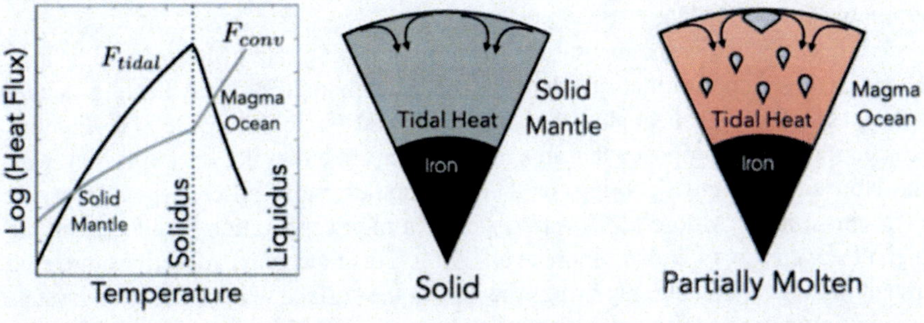

Fig. 10.4 (left) Schematic illustration of the two possible equilibria between tidal heating and convection illustrated by the two locations where the tidal heat flux (black) is equal to the convective heat flux (gray). In the sub-solidus equilibrium (middle), the planet is solid, but experiencing tidal dissipation. However, convection is capable of efficiently removing tidal heat, leading to a purely solid mantle. The right panel depicts a "magma ocean" equilibrium point, where the mantle experiences enough tidal heating to become partially molten. In this state, convection and melt transport remove heat from the interior. Our model can not distinguish between a magma ocean exposed to the surface and one that exists at depth, beneath a surface layer of solid rock. After Barr et al. (2018)

equilibrium is unstable—if the tidal heat generation increases, convection cannot remove the additional tidal heat (Moore 2003; Dobos and Turner 2015). A second equilibrium is possible, in which the planet is partially molten (left panel, and the right panel of Fig. 10.4). Tidal heat generation decreases as a function of temperature and melt fraction above the solidus because partially molten rock is less dissipative than warm, solid rock. However, the convective heat flux increases sharply as the presence of melt decreases the mantle viscosity. This high-temperature equilibrium is stable (Moore 2003)—as the tidal heating rate increases, the viscosity of the mantle decreases, permitting more efficient convective heat flow and resulting in cooling of the mantle.

10.4.3 Results

10.4.3.1 Io

The natural question about such a simplistic model is whether it can reproduce the heat flow for Io. Using the same approach we applied to the TRAPPIST-1 planets, we have calculated the tidal heat flux and interior mantle temperature for Io. We use the following parameters: mass $M = 8.93 \times 10^{22}$ kg, rock volume fraction $\phi_r = 0.95$, iron fraction $\phi_{iron} = 0.05$ (Schubert et al. 2004), semi-major axis $a = 421{,}700$ km, eccentricity $e = 0.0041$, and $R_s = 1821$ km, mass of the primary (Jupiter) $M_J = 1.8983 \times 10^{27}$ kg.

With these values, our model gives a tidal heat flux $F_{\text{tidal}} = 0.31\,\text{W/m}^2$, comparable in magnitude to that observed (Spencer et al. 2000; Veeder et al. 2004), and an equilibrium rock mantle temperature $T_{\text{eq}} = 1663$ K, above the rock solidus, and consistent with the inference of a partially molten mantle beneath Io's surface (Khurana et al. 2011).

Our result is below the measured global tidal heating rate of Io ($\sim 1 - 3\,\text{W/m}^2$). However, it is known that standard viscoelastic models of tidal heating cannot account for the high heat flux that was measured on Io (Moore 2003). For this reason, it has been proposed that Io's tidal heat flux is either not in equilibrium with the induced convective cooling, or there is another, more efficient heat transport mechanism in Io. Moore (2001) suggested that melt segregation could explain the high observed flux (see also Moore et al. 2007). This mechanism describes the rapid ascent of magma that quickly brings the heat to the surface while solid rock crystals are sinking down. Melt segregation provides a new equilibrium with tidal heating (in addition to convective cooling) which can explain high surface fluxes (see Figure 5.4 of Moore et al. 2007).

While melt segregation can explain the high tidal heating flux in Io, we have no reason to assume that the same mechanism is present also in the TRAPPIST-1 planets. On the contrary, their layered internal structure, if it includes H_2O, is likely to block such a process at the rock–ice boundary. Without a better understanding of their real internal structures and real tidal flux rates, it would be premature to refine our models to special cases which may or may not be applicable.

10.4.3.2 The TRAPPIST-1 Planets

Our tidal heating model finds an equilibrium between tidal heating and heat transport across the rocky mantle is possible for the inner five of the seven TRAPPIST-1 planets. The outer two TRAPPIST-1 planets are farther from the star and as a consequence, amount of tidal heat is always less than the amount that could be transported by convection. In these planets, the tidal heat could be concentrated in the icy layers of the planet, if present, or tidal heat is not an important heat source.

For the inner five planets, however, we find that enough tidal heat is generated to warm the planets' mantles considerably. The calculated tidal heat fluxes, together with the amount of tidal heating are listed in Table 10.2. The last column of Table 10.2 is the k_2/Q tidal parameter. For comparison, for the Earth $k_2 \sim 0.29$ (Kozai 1968) and the Q value for the solid Earth ~350 (Ray et al. 1996). This means that for solid Earth $k_2/Q \sim 8.3 \times 10^{-4}$. For Mars $k_2 \sim 0.17$ (Konopliv et al. 2016) and $Q \sim 80$ (Lainey et al. 2007), so that $k_2/Q \sim 0.002$. Note that the TRAPPIST-1 planets have similar values of the Earth and Mars to within an order of magnitude.

Table 10.2 also shows the equilibrium temperatures in the rock mantle. In all five planets the temperature is higher than the solidus temperature of rock (1600 K) which means that the rock starts to melt. However, the temperatures are not high enough to reach the liquidus temperature (2000 K), meaning that the rock does not become fully liquid. The rock mantle of these five planets are in a partly melted state, similar to what has been proposed for Io (Peale et al. 1979; Segatz et al. 1988; Khurana et al. 2011; Beuthe 2013). The current eccentricities of planets d to h are all forced eccentricities i.e. their eccentricities are not zero because of these planets are all in a resonance with each other, analogues to the eccentricities of Io and Europa not being zero either. The non-zero eccentricities of TRAPPIST-1d to TRAPPIST-1h is unexpected because tidal damping in the innermost planets should have damped them all to zero. One explanation is that the current eccentricities are likely to be a remnant of the planets' formation: during their formation the TRAPPIST-1 planets migrated starwards due to interaction with the protostellar disc and they were caught in mean-motion resonances at non-zero (i.e. forced) eccentricities (Ormel et al. 2017). As such, the system is still tidally damping these

Table 10.2 Equilibrium temperatures achieved due to a balance in tidal heating and convective heat flux in the inner TRAPPIST-1 planets (based on updated mass, radius and eccentricity estimates from Agol et al. (2021)), and for Io

Object	T_{eq} (K)	F_{tidal} (W/m^2)	k_2/Q [10^{-3}]
TRAPPIST-1 b	1676^{+15}_{-76}	$1.07^{+0.42}_{-1.07}$	$0.055^{+0.039}_{-0.055}$
TRAPPIST-1 c	1640^{+8}_{-40}	$0.39^{+0.06}_{-0.39}$	$0.66^{+0.12}_{-0.66}$
TRAPPIST-1 d	1635^{+15}_{-35}	$0.25^{+0.11}_{-0.25}$	$2.99^{+2.33}_{-2.99}$
TRAPPIST-1 e	1615^{+14}_{-15}	$0.17^{+0.06}_{-0.17}$	$7.31^{+1.87}_{-7.31}$
TRAPPIST-1 f	1604^{+8}_{-4}	$0.13^{+0.02}_{-0.13}$	$12.83^{+0.78}_{-12.83}$
TRAPPIST-1 g	–	–	–
TRAPPIST-1 h	–	–	–
Io	1663	0.31	2.11

forced eccentricities. This implies that the tidal heating of the planets has been greater in the past and that perhaps planets b to f have always been in a (partial) magma ocean state since their formation.

10.4.3.3 Planetary Habitability

To determine the habitability and atmospheric state of a planet, it is important to check whether it may be in the so-called runaway greenhouse state. Being in this state for a planet means that oceans entirely evaporate from the planet (Kopparapu et al. 2014). In the upper atmosphere, the H_2O molecules can separate by photodissociation to hydrogen and oxygen, which is unfavorable for life, since the hydrogen can easily escape to space from planets similar to the Earth. This can be a catastrophic process, because without the hydrogen, water may not reform later, not even if the temperature drops (Kasting et al. 1993). Consequently, a planet in a runaway greenhouse state is not considered habitable.

To determine whether the TRAPPIST-1 planets are in a runaway greenhouse state, we can calculate the limit of the runaway greenhouse flux, F_{RG}, above which the runaway process is triggered. We use the formulation of Pierrehumbert (2010) because it takes the size and mass of the planet into account through the surface gravity, g,

$$F_{RG} = o\frac{\sigma(l/R_{water})^4}{\ln\left(p^\star/\sqrt{2p_0 g/\kappa_0}\right)^4}, \tag{10.18}$$

where $o = 0.7344$ is an order unity constant, σ is the Stefan-Boltzmann constant, $l = 2.425 \cdot 10^6$ J/kg is the latent heat, $R_{water} = 461.5$ J/(kg K) is the gas constant for water vapor, $\kappa_0 = 0.055$ is the gray absorption coefficient, $p_0 = 10{,}000$ Pa is the reference pressure for absorption, $p^\star = p_{ref}\exp\left(\frac{l}{RT_{ref}}\right)$, $p_{ref} = 610.616$ Pa and $T_{ref} = 273.13$ K.

Figure 10.5 shows the tidal heating rates of all seven TRAPPIST-1 planets as a function of their absorbed stellar radiation. The large error bars in tidal heating are due to the uncertainties in planetary masses and orbital eccentricities. The error bars in the absorbed stellar radiation are due to the unknown albedos. These results suggest that due to the closeness to the host star (and to the additional tidal heating), planets b and c are very hot; they are in a runaway greenhouse state. This is represented in the figure as a red atmosphere.

Planets b and c could be remarkably similar to Io. Their tidal heating rates are relatively high, comparable to that of Io, which is accompanied by the strong stellar irradiation (that is a consequence of their closeness to the central star). Our calculations also revealed that their rock mantle layer is probably partially molten, resulting in a magma ocean. These findings imply that these two planets might be volcanically active, just like Io.

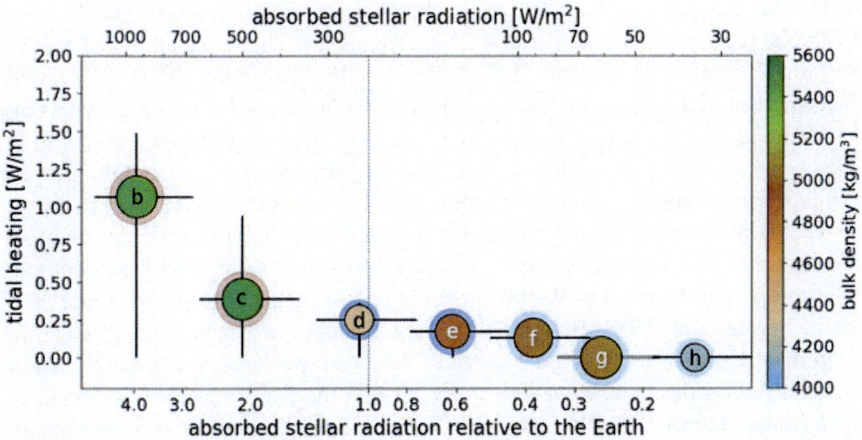

Fig. 10.5 Estimated tidal heating rates of the TRAPPIST-1 planets. The sizes of the planets are to scale. The representative bulk densities are shown as colours of the planets (see the colorbar); note that large deviations are possible due to their uncertainties (which are not shown). The colour of the atmospheres represent too hot (red), moderate (blue) and too cold (light blue) surface temperatures for life on the surface. For comparison the vertical dotted line shows the absorbed solar flux on Earth

We found that planets d and e might avoid the runaway greenhouse state and could have liquid water on their surfaces. In the absence of effects not included in this model, planet e could host a global surface ocean, because its density suggests large amount of H_2O on the surface which is most likely melted by the moderate stellar irradiation. This is supported by climate models which predict liquid water at least around the substellar point, if the planet is tidally locked (Vinson and Hansen 2017; Wolf 2017; Turbet et al. 2018; Sergeev et al. 2020). Because the albedo of these planets is unknown, we ran our calculations with three different values: 0.1, 0.3 (Earth-like) and 0.5. In the case of planet d, we found that it avoids the runaway greenhouse state only if its albedo is at least ~0.3 or higher (see also Barr et al. 2018; Dobos et al. 2019). The moderate flux on these planets is illustrated as a blue atmosphere in Fig. 10.5. While, in the context of our simple model, these two planets are not likely to be very similar to Io, they still have substantial tidal heating (~15 times the tidal flux of the Earth or higher) that could drive underwater volcanism through hydrothermal vents at the rock–ocean boundary.

10.5 Geological Consequences

Io's high heat flow and extensive volcanism are direct consequences of its strong tidal heating from Jupiter. Io's average energy output from tidal heating is on the order of 10^{14} W (Yoder and Peale 1981; McEwen et al. 2004; Veeder et al. 2004), and results in Jupiter's minimum Q value being decreased from 64,000 (Peale 1999) to 35,600 (Lainey et al. 2009) at Io's orbital frequency. The vast majority of heating within Io's interior is released via volcanism rather than lithospheric conduction. Indeed, most of the heat transferred to Io's surface is released at hotspots, with Loki Patera being singularly responsible for 10% of the moon's heat flow (Veeder et al. 1994). Because the partitioning of tidal heating between Io's solid and fluid layers is unknown, questions remain as to where tidal heating is dissipated in its interior.

The relatively high background temperature and the presence of vigorous volcanism at high latitudes could be explained by tidal heating being concentrated within Io's solid layers, i.e., the deep mantle and/or the asthenosphere (Peale et al. 1979; Segatz et al. 1988; Beuthe 2013) (see Fig. 4.11 in Chap. 4). In this case, a deep mantle source for Io's heat, or lateral heat transport within the mantle, could be achieved by the maintenance of a crystal-rich magma ocean (Keszthelyi et al. 1999) or by vigorous mantle convection (Tackley 2001). In the case of a deep mantle source, areas of intense volcanism at Io's surface could correspond to locations of high heat production (de Kleer et al. 2019c). The locations of paterae, active hot spots and volcanic centers across Io's surface suggest that tidal heating may be primarily concentrated in fluid layers in Io's upper mantle (Hamilton et al. 2013). In this case, such a melt layer would be global, intense volcanism could occur anywhere, not just in areas of enhanced heat production (de Kleer et al. 2019c), and tidal heating would be concentrated in both Io's fluid and solid layers (Tyler et al. 2015).

Exoplanets and exomoons that are on highly eccentric orbits around their primaries would experience enhanced tidal heating, internal magma production and associated volcanism that may be similar to, or exceed that of, Io (e.g. Dobos and Turner 2015; Quick et al. 2020). Even in the absence of tidal heating, exoplanets that orbit very close to their host stars (e.g., ultra-short period exoplanets) would experience extreme melting leading to the formation of surface magma oceans. Corot-7b, Kepler 78-b, Kepler 10-b and 55 Cancri e are examples of such worlds (Léger et al. 2011; Rouan et al. 2011; Batalha et al. 2011; Pepe et al. 2013; Demory et al. 2016). These planetary bodies may also have global magma oceans (Chao et al. 2021). As could be the case for Io (Khurana et al. 2011), global magma oceans in the interiors of exoplanets or exomoons could produce significant induced magnetic fields. Moreover, volcanic species that would be vented from the surfaces of tidally heated extrasolar worlds (e.g., volatiles such as SO_2 and refractory elements), could be detected in transit spectra (Henning et al. 2009; Kaltenegger and Traub 2009; Oza et al. 2019), as trailing exospheres (Mura et al. 2011), or dust tails produced by the vaporization of surface materials. As such, they would provide clues to the interior composition of rocky exoplanets or exomoons (Budaj et al. 2015; Van Lieshout et al. 2016; Gaidos et al. 2019). Furthermore, variations in tidal stress fields as they orbit their primaries could cause periodicity in volcanic activity on strongly tidally heated extrasolar worlds (de Kleer et al. 2019a).

10.6 Prospects for Future Observations

Variations in exoplanet eccentricity may affect the timing and strength of volcanism at the surfaces of tidally heated exoplanets. Recent studies have attempted to predict the timing and strength of volcanism at Io's surface based on variations in its eccentricity as it orbits Jupiter. Observations of the Ionian volcano Loki Patera, which is responsible for 10% of Io's heat flow (Veeder et al. 1994), suggest that Io's volcanic activity may vary with its orbital phase. Io experiences the greatest amount of tidal heating and volcanic activity at apojove, when its average eccentricity is the highest (de Kleer et al. 2019a,b). In addition, periodicities in tidal forcing and eruption strength are likely to exist on Io, with recent eruptions occurring on 440–475 day periods (de Kleer et al. 2019b; Rathbun and Spencer 2019). If volcanism on tidally heated extrasolar worlds operates similarly to volcanism on Io, then this would suggest that extrasolar volcanism may also be periodic, and that tidally locked planets may experience the greatest amount of internal heating and elevated volcanism at apoapsis. Observing how Io's volcanic activity varies with orbital phase and periodicity would therefore shed light on how similar variations might affect volcanic activity and tidal heating on extrasolar planets and moons; the results of the aforementioned studies could then be used as a baseline to predict the time during their orbits in which volcanic eruptions are likely to be the strongest, and hence are most detectable, on extrasolar words.

Additionally, observations that shed light on where tidal heating is concentrated in Io (i.e., in solid or liquid layers, or both), that aid in constraining the relationship between the distribution of this heat and volcanic activity at the surface, and measurements of Io's induced magnetic field in an effort to determine if it has a global magma ocean would further aid us in understanding the extent to which Io can be viewed as an exoplanet analog. Many of these investigations would be carried out by the proposed Io Volcano Observer (IVO) mission (McEwen et al. 2020). Finally, observations of Io's plasma torus (e.g., see Bagenal 1994), its link to volcanism at the surface, and to Jupiter's magnetic field, could aid in our interpretation of UV observations of plasma tori around volcanically active extrasolar worlds (Kislyakova et al. 2019). Transit spectroscopy measurements support the observability of exo-tori, since the observed sodium lines in WASP 76b and WASP 121b can be explained by the presence of an exo-torus, that might be fueled by the volcanic activity of an exomoon (Gebek and Oza 2020; Hoeijmakers et al. 2020).

Owing to their bright infrared flux and short orbital periods, exoplanetary bodies with global magma oceans may be among the most detectable and characterizable low-mass exoplanets in the coming decades (Henning et al. 2018). Sulfur dioxide and other sulfur compounds that may be indicative of widespread volcanism on tidally heated extrasolar worlds can be detected via transit spectroscopy (Kaltenegger and Traub 2009; Kaltenegger et al. 2010; Oza et al. 2019; Gebek and Oza 2020). In addition, owing to their high surface temperatures, exoplanets with surface magma oceans may also be detected in infrared emission and by the transit and

Doppler RV methods, both of which are biased towards detecting close-in planets (Chao et al. 2021). Peters and Turner (2013) proposed that even exomoons could be detected with direct imaging at infrared wavelengths. These exo-Ios may be detected by infrared telescopes, with which intense surface heating and variability of surface temperature between day and night could be revealed via infrared phase curves. Similarly, Forgan (2017) suggest observing the combined phase curve of an exoplanet–exomoon pair with the James Webb Space Telescope (JWST) in the infrared wavelength range. These detections mostly favor close-in and/or tidally heated planets or their moons.

Finally, we note that, similar to the gas giant satellite systems in our Solar System, gas giant planets in other solar systems also probably host families of moons (e.g., Barr 2016). These moons may also be subject to tidal dissipation, just like Io, and if they orbit a giant planet which itself orbits close to the parent star, the moons may feel additional gravitational perturbations and/or be subject to loss by evaporation (Cassidy et al. 2009). Thus, in addition to providing a natural laboratory for the study of tides in exoplanets themselves, Io may provide a laboratory for the study of the moons of giant exoplanets, as well.

Acknowledgments We are grateful to the editors of this volume for the invitation to contribute to this book. We also thank Edwin Kite and Apurva Oza, who provided thoughtful comments that significantly improved the clarity of the chapter. Author Barr acknowledges support from NASA Habitable Worlds Grant 80NSSC20K1555. Lynnae C. Quick acknowledges support from NASA's Habitable Worlds Program [WBS 811073.02.36.01.78]. Author Dobos has been supported by the Hungarian National Research, Development, and Innovation Office (NKFIH) grant K-131508. The COFUND project oLife has received funding from the European Union's Horizon 2020 research and innovation programme under grant agreement No 847675.

References

Agol, E., et al.: Refining the transit-timing and photometric analysis of trappist-1: Masses, radii, densities, dynamics, and ephemerides. Planet. Sci. J. **2**(1), 1 (2021)

Bagenal, F.: Empirical model of the Io plasma torus: Voyager measurements. J. Geophys. Res. Space Phys. **99**(A6), 11,043–11,062 (1994)

Baraffe, I., Homeier, D., Allard, F., Chabrier, G.: New evolutionary models for pre-main sequence and main sequence low-mass stars down to the hydrogen-burning limit. A&A **577**, A42. https://doi.org/10.1051/0004-6361/201425481 (2015)

Barr, A.C.: Formation of exomoons: a solar system perspective. Astron. Rev.,**12**(1–4), 24–52 (2016)

Barr, A.C., Canup, R.M.: Constraints on gas giant satellite formation from the interior states of partially differentiated satellites. Icarus **198**(1), 163–177 (2008)

Barr, A.C., Showman, A.P.: Heat transfer in Europa's icy shell. Pappalardo, Robert T., William B. McKinnon, and Krishan Khurana, eds. Europa. University of Arizona Press, p. 405–430 (2009)

Barr, A.C., Dobos, V., Kiss, L.: Interior structures and tidal heating in the trappist-1 planets. A&A (2018). https://doi.org/10.1051/0004-6361/201731992

Batalha, N.M., et al.: Kepler's first rocky planet: Kepler-10b. Astrophys. J. **729**(1), 27 (2011)

Batygin, K., Morbidelli, A.: Dissipative divergence of resonant orbits. AJ **145**(1), 1 (2013). https://doi.org/10.1088/0004-6256/145/1/1

Beuthe, M.: Spatial patterns of tidal heating. Icarus **223**, 308–329 (2013). https://doi.org/10.1016/j.icarus.2012.11.020

Bower, D.J., Kitzmann, D., Wolf, A.S., Sanan, P., Dorn, C., Oza, A.V.: Linking the evolution of terrestrial interiors and an early outgassed atmosphere to astrophysical observations. Astron. Astrophys. **631**, A103 (2019)

Brasser, R., Barr, A.C., Dobos, V.: The tidal parameters of TRAPPIST-1b and c. MNRAS **487**(1), 34–47 (2019). https://doi.org/10.1093/mnras/stz1231

Brown, B.: The long period behavior of the orbits of the galilean satellites of Jupiter. Celestial Mech. **16**(2), 229–259 (1977). https://doi.org/10.1007/BF01228604

Budaj, J., Kocifaj, M., Salmeron, R., Hubeny, I.: Tables of phase functions, opacities, albedos, equilibrium temperatures, and radiative accelerations of dust grains in exoplanets. Mon. Not. R. Astron. Soc. **454**(1), 2–27 (2015)

Burgasser, A.J., Mamajek, E.E.: On the age of the TRAPPIST-1 system. ApJ **845**(2), 110 (2017). https://doi.org/10.3847/1538-4357/aa7fea

Canup, R.M., Ward, W.R.: Formation of the Galilean satellites: conditions of accretion. Astron. J. **124**, 3404–3423 (2002). https://doi.org/10.1086/344684

Canup, R.M., Ward, W.R.: A common mass scaling for satellite systems of gaseous planets. Nature **441**, 834–839 (2006). https://doi.org/10.1038/nature04860

Cassidy, T.A., Mendez, R., Arras, P., Johnson, R.E., Skrutskie, M.F.: Massive satellites of close-in gas giant exoplanets. Astrophys. J. **704**(2), 1341 (2009)

Chao, K.-H., deGraffenried, R., Lach, M., Nelson, W., Truax, K., Gaidos, E.: Lava worlds: From early earth to exoplanets. Geochemistry **81**, 125735 (2021)

de Kleer, K., Nimmo, F., Kite, E.: Variability in Io's volcanism on timescales of periodic orbital changes. Geophys. Res. Lett. **46**(12), 6327–6332 (2019a)

de Kleer, K., et al.: Io's volcanic activity from time domain adaptive optics observations: 2013–2018. Astron. J. **158**(1), 29 (2019b)

de Kleer, K., et al.: Tidal heating: Lessons from Io and the jovian system-final report. Final report for the Keck Institute for Space Studies, (2019c)

De Sitter, W.: On the periodic solutions of a particular case of the problem of four bodies. KNAW proceedings, **11**, 682–698 (1909)

Delisle, J.B., Laskar, J., Correia, A.C.M., Boué, G.: Dissipation in planar resonant planetary systems. A&A **546**, A71 (2012). https://doi.org/10.1051/0004-6361/201220001

Demory, B.-O., et al.: A map of the large day–night temperature gradient of a super-earth exoplanet. Nature **532**(7598), 207–209 (2016)

Dobos, V., Turner, E.L.: Viscoelastic models of tidally heated exomoons. ApJ **804**, 41 (2015). https://doi.org/10.1088/0004-637X/804/1/41

Dobos, V., Barr, A.C., Kiss, L.L.: Tidal heating and the habitability of the TRAPPIST-1 exoplanets. A&A, **624**, A2 (2019). https://doi.org/10.1051/0004-6361/201834254

Dziewonski, A.M., Anderson, D.L.: Preliminary reference Earth model. Phys. Earth Planet. Interiors **25**, 297–356 (1981). https://doi.org/10.1016/0031-9201(81)90046-7

Efroimsky, M.: Tidal dissipation compared to seismic dissipation: In small bodies, Earths, and super-Earths. ApJ **746**, 150 (2012). https://doi.org/10.1088/0004-637X/746/2/150

Elkins-Tanton, L.T., Seager, S.: Coreless terrestrial exoplanets. Astrophys. J. **688**(1), 628 (2008)

Forgan, D.H.: On the feasibility of exomoon detection via exoplanet phase curve spectral contrast. MNRAS **470**(1), 416–426 (2017). https://doi.org/10.1093/mnras/stx1217

Frank, E.A., Meyer, B.S., Mojzsis, S.J.: A radiogenic heating evolution model for cosmochemically Earth-like exoplanets. Icarus **243**, 274–286 (2014). https://doi.org/10.1016/j.icarus.2014.08.031

Friedson, A., Stevenson, D.: Viscosity of rock-ice mixtures and applications to the evolution of icy satellites. Icarus **56**(1), 1–14 (1983)

Gaidos, E., Hirano, T., Ansdell, M.: Monitoring of the d doublet of neutral sodium during transits of two 'evaporating' planets. Mon. Not. R. Astron. Soc. **485**(3), 3876–3886 (2019)

Gebek, A., Oza, A.V.: Alkaline exospheres of exoplanet systems: evaporative transmission spectra. MNRAS **497**(4), 5271–5291 (2020). https://doi.org/10.1093/mnras/staa2193

Gillon, M., et al.: Seven temperate terrestrial planets around the nearby ultracool dwarf star TRAPPIST-1. Nature **542**, 456–460 (2017). https://doi.org/10.1038/nature21360

Grimm, S.L., et al.: Astron. & Astrophys., 613, A68 (2018)

Hamilton, C.W., Beggan, C.D., Still, S., Beuthe, M., Lopes, R.M.C., Williams, D.A., Radebaugh, J., Wright, W.: Spatial distribution of volcanoes on Io: Implications for tidal heating and magma ascent. Earth Planet. Sci. Lett. **361**, 272–286 (2013). https://doi.org/10.1016/j.epsl.2012.10.032

Henning, W.G., Hurford, T.: Tidal heating in multilayered terrestrial exoplanets. ApJ **789**, 30 (2014). https://doi.org/10.1088/0004-637X/789/1/30

Henning, W.G., O'Connell, R.J., Sasselov, D.D.: Tidally heated terrestrial exoplanets: viscoelastic response models. ApJ **707**, 1000–1015 (2009). https://doi.org/10.1088/0004-637X/707/2/1000

Henning, W.G., et al.: Highly volcanic exoplanets, lava worlds, and magma ocean worlds: An emerging class of dynamic exoplanets of significant scientific priority. Preprint. arXiv:1804.05110 (2018)

Hobbs, P. V.: Ice Physics (Oxford: Oxford University Press), (1974)

Hoeijmakers, H.J., et al.: Hot exoplanet atmospheres resolved with transit spectroscopy (HEARTS). IV. A spectral inventory of atoms and molecules in the high-resolution transmission spectrum of WASP-121 b. A&A **641**, A123 (2020). https://doi.org/10.1051/0004-6361/202038365

Hussmann, H., Spohn, T., Wieczerkowski, K.: Thermal equilibrium states of Europa's ice shell: implications for internal ocean thickness and surface heat flow. Icarus **156**, 143–151 (2002). https://doi.org/10.1006/icar.2001.6776

Kaltenegger, L., Traub, W.A.: Transits of Earth-like planets. ApJ **698**, 519–527 (2009). https://doi.org/10.1088/0004-637X/698/1/519

Kaltenegger, L., Henning, W., Sasselov, D.: Detecting volcanism on extrasolar planets. Astron. J. **140**(5), 1370 (2010)

Kasting, J.F., Whitmire, D.P., Reynolds, R.T.: Habitable zones around main sequence stars. Icarus **101**, 108–128 (1993). https://doi.org/10.1006/icar.1993.1010

Keszthelyi, L., McEwen, A., Taylor, G.: Revisiting the hypothesis of a mushy global magma ocean in Io. Icarus **141**(2), 415–419 (1999)

Khurana, K.K., Jia, X., Kivelson, M.G., Nimmo, F., Schubert, G., Russell, C.T.: Evidence of a global magma ocean in Io's interior. Science **332**(6034), 1186–1189 (2011)

Kislyakova, K.G., et al.: Detecting volcanically produced tori along orbits of exoplanets using UV spectroscopy. Preprint. arXiv:1907.05088 (2019)

Konopliv, A.S., Park, R.S., Folkner, W.M.: An improved JPL Mars gravity field and orientation from Mars orbiter and lander tracking data. Icarus **274**, 253–260 (2016). https://doi.org/10.1016/j.icarus.2016.02.052

Kopparapu, R.K., Ramirez, R.M., SchottelKotte, J., Kasting, J.F., Domagal-Goldman, S., Eymet, V.: Habitable zones around main-sequence stars: dependence on planetary mass. ApJl **787**, L29 (2014). https://doi.org/10.1088/2041-8205/787/2/L29

Kozai, Y.: Love's number of the earth derived from satellite observations. PASJ **20**, 24 (1968)

Lainey, V., Duriez, L., Vienne, A.: New accurate ephemerides for the Galilean satellites of Jupiter. I. Numerical integration of elaborated equations of motion. A&A **420**, 1171–1183 (2004). https://doi.org/10.1051/0004-6361:20034565

Lainey, V., Dehant, V., Pätzold, M.: First numerical ephemerides of the Martian moons. A&A **465**(3), 1075–1084 (2007). https://doi.org/10.1051/0004-6361:20065466

Lainey, V., Arlot, J.-E., Karatekin, Ö., van Hoolst, T.: Strong tidal dissipation in Io and Jupiter from astrometric observations. Nature **459**(7249), 957–959 (2009). https://doi.org/10.1038/nature08108

Léger, A., et al.: The extreme physical properties of the corot-7b super-earth. Icarus **213**(1), 1–11 (2011)

Lieske, J.H.: Galilean satellite ephemerides E5. A&AS **129**, 205–217 (1998). https://doi.org/10.1051/aas:1998182

Lithwick, Y., Wu, Y.: Resonant repulsion of Kepler planet pairs. ApJL **756**(1), L11 (2012). https://doi.org/10.1088/2041-8205/756/1/L11

Love, A.E.H.: A Treatise on the Mathematical Theory of Elasticity. Cambridge University Press (1906)

Luger, R., et al.: A seven-planet resonant chain in TRAPPIST-1. Nat. Astron. **1**, 0129 (2017). https://doi.org/10.1038/s41550-017-0129

McEwen, A.S., Keszthelyi, L.P., Lopes, R., Schenk, P.M., Spencer, J.R.: The lithosphere and surface of Io. In: Jupiter: The Planet, Satellites and Magnetosphere, pp. 307–328 (2004)

McEwen, A., et al.: Io volcano observer (IVO): Does Io have a magma ocean? In: Lunar and Planetary Science Conference, vol. 2326, p. 1648 (2020)

Moore, W.B.: NOTE: The thermal state of Io. Icarus **154**(2), 548–550 (2001). https://doi.org/10.1006/icar.2001.6739

Moore, W.B.: Tidal heating and convection in Io. J. Geophys. Res. (Planets) **108**, 5096 (2003). https://doi.org/10.1029/2002JE001943

Moore, W.B.: Thermal equilibrium in Europa's ice shell. Icarus **180**, 141–146 (2006). https://doi.org/10.1016/j.icarus.2005.09.005

Moore, W.B., Schubert, G., Anderson, J.D., Spencer, J.R.: The interior of Io, p. 89 (2007). https://doi.org/10.1007/978-3-540-48841-5_5

Mura, A., et al.: Comet-like tail-formation of exospheres of hot rocky exoplanets: possible implications for corot-7b. Icarus **211**(1), 1–9 (2011)

Murray, C.D., Dermott, S.F.: Solar System Dynamics. Cambridge University Press (1999)

Ojakangas, G.W., Stevenson, D.J.: Thermal state of an ice shell on Europa. Icarus **81**, 220–241 (1989). https://doi.org/10.1016/0019-1035(89)90052-3

Ormel, C.W., Liu, B., Schoonenberg, D.: Formation of TRAPPIST-1 and other compact systems. A&A **604**, A1 (2017). https://doi.org/10.1051/0004-6361/201730826

Oza, A.V., et al.: Sodium and potassium signatures of volcanic satellites orbiting close-in gas giant exoplanets. Astrophys. J. **885**(2), 168 (2019)

Peale, S.: Origin and evolution of the natural satellites. Annu. Rev. Astron. Astrophys. **37**(1), 533–602 (1999)

Peale, S.J.: Tidally induced volcanism. Celestial Mech. Dynam. Astron. **87**, 129–155 (2003)

Peale, S.J., Cassen, P.: Contribution of tidal dissipation to lunar thermal history. Icarus **36**, 245–269 (1978). https://doi.org/10.1016/0019-1035(78)90109-4

Peale, S.J., Cassen, P., Reynolds, R.T.: Melting of Io by tidal dissipation. Science **203**(4383), 892–894 (1979). https://doi.org/10.1126/science.203.4383.892

Pepe, F., et al.: An earth-sized planet with an earth-like density. Nature **503**(7476), 377–380 (2013)

Peters, M.A., Turner, E.L.: On the direct imaging of tidally heated exomoons. ApJ **769**, 98 (2013). https://doi.org/10.1088/0004-637X/769/2/98

Pierrehumbert, R.T.: 2010, Principles of Planetary Climate (Cambridge, UK: Cambridge University Press)

Quick, L.C., Roberge, A., Mlinar, A.B., Hedman, M.M.: Forecasting rates of volcanic activity on terrestrial exoplanets and implications for cryovolcanic activity on extrasolar ocean worlds. Publ. Astron. Soc. Pacific **132**(1014), 084,402 (2020)

Rathbun, J., Spencer, J.: Io's loki volcano: An explanation of its tricky behavior and prediction for the next eruption. In: Lunar and Planetary Science Conference, vol. 2132, p. 2402 (2019)

Ray, R.D., Eanes, R.J., Chao, B.F.: Detection of tidal dissipation in the solid Earth by satellite tracking and altimetry. Nature **381**(6583), 595–597 (1996). https://doi.org/10.1038/381595a0

Renner, J., Evans, B., Hirth, G.: On the rheologically critical melt fraction. Earth Planet. Sci. Lett. **181**(4), 585–594 (2000)

Rouan, D., Deeg, H.J., Demangeon, O., Samuel, B., Cavarroc, C., Fegley, B., Léger, A.: The orbital phases and secondary transits of kepler-10b. A physical interpretation based on the lava-ocean planet model. Astrophys. J. Lett. **741**(2), L30 (2011)

Schubert, G., Turcotte, D.L., Olson, P.: *Mantle Convection in the Earth and Planets*, Cambridge University Press, (2001)

Schubert, G., Anderson, J.D., Spohn, T., McKinnon, W.B.: Interior Composition, Structure and Dynamics of the Galilean Satellites, in Jupiter. The Planet, Satellites and Magnetosphere pp. 281–306 (2004).

Segatz, M., Spohn, T., Ross, M.N., Schubert, G.: Tidal dissipation, surface heat flow, and figure of viscoelastic models of Io. Icarus **75**, 187–206 (1988). https://doi.org/10.1016/0019-1035(88)90001-2

Sergeev, D.E., Lambert, F.H., Mayne, N.J., Boutle, I.A., Manners, J., Kohary, K.: Atmospheric convection plays a key role in the climate of tidally locked terrestrial exoplanets: insights from high-resolution simulations. ApJ **894**(2), 84 (2020). https://doi.org/10.3847/1538-4357/ab8882

Showman, A.P., Malhotra, R.: The Galilean satellites. Science **296**(5437), 77–84 (1999)

Sinclair, A.T.: The orbital resonance amongst the galilean satellites of Jupiter. Celestial Mech. **12**(1), 89–96 (1975). https://doi.org/10.1007/BF01228627

Solomatov, V., Moresi, L.-N.: Scaling of time-dependent stagnant lid convection: Application to small-scale convection on earth and other terrestrial planets. J. Geophys. Res. Solid Earth **105**(B9), 21,795–21,817 (2000)

Spencer, J.R., Rathbun, J.A., Travis, L.D., Tamppari, L.K., Barnard, L., Martin, T.Z., McEwen, A.S.: Io's thermal emission from the galileo photopolarimeter- radiometer. Science **288**, 1198–1201 (2000). https://doi.org/10.1126/science.288.5469.1198

Tackley, P.J.: Convection in Io's asthenosphere: Redistribution of nonuniform tidal heating by mean flows. J. Geophys. Res. **106**, 32,971–32,982 (2001). https://doi.org/10.1029/2000JE001411

Turbet, M., Bolmont, E., Leconte, J., Forget, F., Selsis, F., Tobie, G., Caldas, A., Naar, J., Gillon, M.: Modeling climate diversity, tidal dynamics and the fate of volatiles on TRAPPIST-1 planets. A&A **612**, A86 (2018). https://doi.org/10.1051/0004-6361/201731620

Turbet, M., Bolmont, E., Ehrenreich, D., Gratier, P., Leconte, J., Selsis, F., Hara, N., Lovis, C.: Revised mass-radius relationships for water-rich rocky planets more irradiated than the runaway greenhouse limit. Astron. Astrophys. **638**, A41 (2020)

Tyler, R.H., Henning, W.G., Hamilton, C.W.: Tidal heating in a magma ocean within Jupiter's moon Io. Astrophys. J. Suppl. Ser. **218**(2), 22 (2015)

Unterborn, C.T., Desch, S.J., Hinkel, N., Lorenzo, A.: Constraining the compositions of the TRAPPIST-1 planets to trace snow lines and migration in M dwarf disks, ArXiv e-prints (2017)

Van Lieshout, R., et al.: Dusty tails of evaporating exoplanets-II. Physical modelling of the KIC 12557548b light curve. Astron. Astrophys. **596**, A32 (2016)

Veeder, G.J., Matson, D.L., Johnson, T.V., Blaney, D.L., Goguen, J.D.: Io's heat flow from infrared radiometry: 1983–1993. J. Geophys. Res. Planets **99**(E8), 17,095–17,162 (1994)

Veeder, G.J., Matson, D.L., Johnson, T.V., Davies, A.G., Blaney, D.L.: The polar contribution to the heat flow of Io. Icarus **169**, 264–270 (2004). https://doi.org/10.1016/j.icarus.2003.11.016

Vinson, A.M., Hansen, B.M.S.: On the spin states of habitable zone exoplanets around M dwarfs: the effect of a near-resonant companion. MNRAS **472**(3), 3217–3229 (2017). https://doi.org/10.1093/mnras/stx2100

Wolf, E.T.: Assessing the habitability of the TRAPPIST-1 system using a 3D climate model. ApJL **839**, L1 (2017). https://doi.org/10.3847/2041-8213/aa693a

Yoder, C.F., Peale, S.J.: The tides of Io. Icarus **47**(1), 1–35 (1981)

Chapter 11
Outstanding Questions and Future Observations of Io

Alfred S. McEwen, Amanda F. Haapala, Laszlo P. Keszthelyi, and Kathleen E. Mandt

Abstract Io, the world with the greatest tidal flexing, volcanic and tectonic activity, and mass-loss in our solar system, is a prime target for study. There are numerous outstanding science questions described in prior chapters of this book and summarized in this chapter. Spacecraft missions such as *Juno* and *JUICE*, along with Earth-based telescopes such as *JWST* and *ALMA*, will acquire important Io observations over the next 15 years. However, a mission designed for Io science is needed for key advances. Such a mission would contribute to understanding the early evolution of terrestrial planets, tidally-heated exoplanets and ocean worlds, and magnetospheric physics across the galaxy. There may be an opportunity in 2025 to propose an Io mission in NASA's New Frontiers program. An Io orbiter could provide the best geophysical measurements but would be very challenging deep inside Jupiter's gravity well and high radiation zone. Alternatively, the Discovery-class *Io Volcano Observer (IVO)* concept, a Jupiter orbiter, could be augmented to support 15–20 rather than 10 close encounters with Io and accommodate additional science instruments and perhaps small satellites for special high-risk measurements.

11.1 Introduction

Io is a natural laboratory for understanding planetary processes that drive the evolution of many worlds (Fig. 11.1), including the terrestrial planets (especially their early histories), ocean worlds with differentiated silicate interiors such as

A. S. McEwen (✉)
Lunar and Planetary Laboratory, University of Arizona, Tucson, AZ, USA
e-mail: mcewen@pirl.lpl.arizona.edu

A. F. Haapala · K. E. Mandt
Johns Hopkins University Applied Physics Laboratory, Laurel, MD, USA
e-mail: Amanda.haapala@jhuapl.edu; Kathleen.Mandt@jhuapl.edu

L. P. Keszthelyi
U.S. Geological Survey, Astrogeology Science Center, Flagstaff, AZ, USA
e-mail: laz@usgs.gov

R. M. C. Lopes et al. (eds.), *Io: A New View of Jupiter's Moon*, Astrophysics and Space Science Library 468, https://doi.org/10.1007/978-3-031-25670-7_11

Fig. 11.1 A better understanding of Io would advance our understanding of many worlds

Europa and Enceladus, and tidally heated exoplanets such as the TRAPPIST-1 system (Barr et al., Chap. 10). Future exploration of Io is motivated by Io itself as a spectacular world, as a place where active processes and tidal heating can be studied to better understand many worlds, and because exploration of Io will motivate future generations of scientists (Keane et al. 2021a, b).

Preceding chapters have summarized Io's uniqueness in terms of geologic activity and magnetospheric interactions, activity that is relatively easy to observe. But there are many open questions, especially about Io's interior, because past exploration near Io has been from spacecraft missions designed for non-Io objectives. These outstanding questions are summarized below, followed by future exploration concepts that would address the most important outstanding questions.

11.2 Outstanding Questions

11.2.1 Formation and Evolution of Io

Several sources for Io's bulk materials have been hypothesized: from the circumplanetary accretion disk during the very end stages of gas accretion onto Jupiter (Canup and Ward 2002), more remote material associated with formation of Saturn

(Ronnet et al. 2018), or material in a dense circumplanetary disk (Peale and Canup 2015). In all cases the source material is assumed to be of solar composition, with fractionation of volatiles. The Galilean satellites are in regular orbits and have a compositional gradient of decreasing density (greater ice contents) from Io (3530 kg m^{-3}) to Europa (3010 kg m^{-3}) to Ganymede (1940 kg m^{-3}) to Callisto (1830 kg m^{-3}) (McKinnon et al., Chap. 3). There are three ways this compositional gradient could have formed (Bierson and Nimmo 2020): (1) a temperature gradient in the circum-Jovian disk, preventing ice formation or preservation closer to Jupiter, (2) ice loss from higher impact velocities closer to Jupiter, and/or (3) tidal heating that later removed volatiles from Io and Europa. Bierson and Nimmo (2020) favor ice loss from impacts, while Bierson and Steinbrügge (2021) found that tidal heating alone does not explain the removal of volatiles from Io and Europa. These models might be tested by measuring the D/H ratio of Europan water plumes by Europa Clipper. Another recent study found that atmospheric processes over a hot (1250 K) surface could dry out Io (Mousis et al. 2021). Clearly, more work exploring volatile loss at Io is needed.

Models predict a strong radial migration of planetesimals and satellites by aerodynamic and tidal forces in the circum-Jovian disk. Canup and Ward (2006) argued that today's satellites are the final survivors of a history in which an earlier generation of satellites formed but were lost because they migrated into Jupiter. An alternative model (Shibaike et al. 2019) stops the inward migration by a cavity in the gas disk around Jupiter opened by the strong magnetic field. One idea to explain the compositional gradient is to invoke inward movement of the "snow line" (inner boundary of the region where H_2O ice condenses) in the final growth phases of the satellites (Canup and Ward 2006). A formation scenario for the TRAPPIST-1 system that has various ice mass fractions of planets may also apply to the Galilean satellites (Ormel et al. 2017). Thus, study of the Galilean satellites and exoplanet systems are synergistic.

Once formed and initially differentiated into core and mantle, the evolution of Io over geologic time is largely unknown. Thermal-orbital coupling could have resulted in Io and Europa oscillating between hot and cold mantle states (Hussmann and Spohn 2004). There are no ancient terrains preserved at the surface of Io, but two methods to better understand Io's evolution (de Kleer et al. 2019) are to (1) measure the current rate of orbital migration of all Galilean Moons to understand whether or not the Laplace resonance is in equilibrium; and (2) measure isotope ratios of gases escaping from Io and also understand the mass loss processes as a constraint on Io's chemical differentiation. A better understanding of the state of Io today would also improve boundary conditions for models of Io's evolution.

11.2.2 Tidal Heating and the Interior of Io

There is no doubt that Io is the most tidally heated world in our solar system (Keane et al., Chap. 4). Therefore, Io provides an unparalleled opportunity to study tidal heating, especially in a body that contains a significant amount of melt. How tidal dissipation is partitioned between the solid and fluid portions of Io's interior is a major outstanding question. While the theory of dissipation in solids and liquids is reasonably well developed, fundamental questions remain when it comes to interconnected partial melts, or melts with solids that can interlock. Such materials have complex rheologies that are highly dependent on strain rate and possibly strain direction and strain history as well. Much can be learned by laboratory experiments but, as with rock mechanics, scale may be important. Io is the best location to put laboratory and theoretical models to the test. Those models can then be applied to other tidally heated worlds, including ocean worlds like Europa, with increased confidence.

While tidal heating mechanism is the pre-eminent unknown about Io's interior, major questions remain about Io's core. What are its dynamics given that the heat is being injected into Io's mantle and thus little if any heat is expected to be flowing from the core into the mantle? The extreme level of geologic activity on Io may allow much of Io's mantle to be cycled through the core-mantle boundary, potentially allowing reactions between the metal and silicate parts of Io to have run more completely than elsewhere in our solar system. How sulfur is partitioned between the core and the silicate part of Io is an open question with implications that run through the entire body and its geochemical history.

There are also deep questions about how melt is transported through the mantle. Theoretical models for heat transport within Io have steadily increased in sophistication but they are largely unconstrained by observations. Continued work is needed in this arena, with models focusing on producing observable predictions and observations focusing on sensing the interior of Io. For the latter, new and improved geophysical measurements are obviously needed, but important tests can be related to the petrology of the melts that make their way to the surface.

11.2.3 Geology and Composition of Io

While various geomorphological features have been categorized and their spatial distribution analyzed (Williams et al., Chap. 5), some of the most fundamental aspects of Io's geology remain unknown. Perhaps the most vexing is the fact that Io's extremely young surface does not provide a view into Io's geologic past. Time is the dimension on which classical geology is most focused, but on Io it is difficult to infer much about what happened before the *Voyager* flybys in 1979. A close second for most challenging open question is the composition of Io—there is actually no definitive proof that Io is generally chondritic like the rest of the Solar System and some suggestion that it might not be (Keszthelyi et al., Chap. 7). The proportions of sulfurous and silicate materials in the crust are also effectively unconstrained. Determining the composition of Io's erupting lavas is one of the most valuable potential future measurements.

The lack of some of the most fundamental information about Io's geology has not precluded the development of a wide range of hypotheses about the Ionian rock cycle. These ideas revolve around the interplay between volcanism and tectonism. The extreme volcanism must result in extreme tectonic activity and the towering mountains confirm this. However, most models rely on temporal variations in volcanism and/or tectonism to explain the observed spatial variations in features at the surface. This is perhaps the most promising avenue to gaining some insight into the geologic history of Io. Future observations will need to focus on determining the ways in which tectonism and volcanism affect each other in the present to allow inferences to be drawn about the past.

11.2.4 "Hot Spots" and Thermal Emission from Io

Io is a favorite target for observation at visible to radio wavelengths because surface temperatures range from near 80 K at the poles (best seen at >30 microns) to >1200 K where lava is erupting (bright even in visible wavelengths), and all temperatures in-between as the surface changes and lava spreads and cools (de Kleer and Rathbun, Chap. 6). The global distribution of hot spots has intriguing implications for the internal tidal dissipation, although better geophysical measurements (gravity, magnetics, orbital motions, topography) are needed to support those interpretations.

A major outstanding question is how much of Io's heat is transported to the surface via conduction rather than high-temperature (and easy to observe) hot spots. Well-designed remote sensing data with global coverage at visible to IR wavelengths at many times of day to accurately map both the Bond albedo and thermal inertia could, in theory, enable interpretation of low-temperature anomalies as conducted heat (Rathbun et al. 2004). But the question would remain as to whether the conducted heat is from cooling lava flows (perhaps buried under SO_2 or other sulfurous materials) or from subsurface intrusions. This distinction is important because the thermal contribution from surface eruptions can be measured and modeled (Davies 2007) to understand the global heat flow and its regional patterns, whereas subsurface intrusions may go undetected. The presence of hundreds of caldera-like depressions, if analogous to calderas on Earth, suggests underlying subsurface magma chambers. Likewise, long-lived eruptions such as at Prometheus are difficult to explain without magma chambers (Leone et al. 2009). How to address this question is a challenge. One innovative idea is to use passive radar (with Jupiter as the source) to probe Io's crust (Steinbruegge et al. 2021). Passive radar can only probe the sub-Jupiter hemisphere, but that includes Loki Patera and many other volcanic centers. Seismic data would be extremely valuable for Io (de Kleer et al. 2019), but is a major challenge to implement in the high radiation environment.

Key questions about Io's thermal emission are summarized by de Kleer and Rathbun (Chap. 6). A better understanding of different types of eruptive centers and their global distributions may be one of the most important next steps. For example, eruptions more directly connected to the mantle versus those associated with long-lived magma chambers may have different distributions, perhaps related to tidal heating. Other important questions are "How does Io's heat flow vary with time?", "What is the nature of Io's polar volcanoes?", "What is Io's dominant magma composition and how does it vary?", and "What eruption properties are directly controlled by tides?"

11.2.5 The Plumes and Atmosphere of Io

Io's atmosphere is one of the most unique in the solar system. The atmosphere has three sources—volcanic plumes, sublimation, and sputtering—with sublimation being the primary source of observed material. Although a great deal has been learned about Io's atmosphere since *Galileo*, many questions remain (de Pater et al., Chap. 8).

The structure and dynamics of the atmosphere are driven by a cycle of condensation onto the surface and sublimation from the surface that is not well understood. The temperature profiles and wind patterns have been difficult to constrain with remote observations, particularly temperatures close to the surface and how wind patterns can be species-specific. In particular, understanding is limited on what happens to the atmosphere during eclipse (Tsang et al. 2016; de Pater et al. 2020). Because the atmosphere contains condensable and non-condensable species and, in some places, active plumes, there has been much discussion about how quickly the atmosphere collapses during eclipse, or what reactions occur in the atmosphere and with the surface.

Io's volcanic plumes also present many puzzles. Their impact on the chemistry and dynamics of the global atmosphere, beyond being an initial source of volatiles, is not well understood. The composition of material outgassed in plumes is also of interest, and adds to the number of questions. Species of particular interest are Na and K. They are derived from NaCl and KCl, which may originate in volcanic plumes. However, ALMA observations of these species find that they are not co-located with SO_2 or SO (Moullet et al. 2015; Redwing et al. 2020). These patterns, along with the observed Na/K ratios have implications for volcanic composition and temperatures that have not fully been explored. Furthermore, NaCl and KCl could condense to the surface and later be sourced by sputtering. As Io's plumes erupt into the sublimated atmosphere, the interactions between the plumes and the atmosphere and the surface are of great interest.

The role of sputtering in contributing to the atmosphere is not well constrained. Because sputtering requires energetic particles to reach the surface, it is only possible away from volcanic plumes at high latitudes, at night, and during eclipse because sublimated and plume-driven atmospheres would block enough energetic particles to prevent production of sufficient sputtered material to influence the atmosphere. This is connected to how the atmosphere interacts with the space environment of Jupiter's magnetosphere, which impacts atmospheric loss processes.

Finally, many questions remain about constituents that may be present in the atmosphere, especially in the plumes, that have not yet been detected. This includes species that may be expected (Wurz et al. 2021) like CO, H_2S, OCS, S_2O, ClO, CS, NaOH, SiO, and CS_2 as well as species like the noble gases that cannot be detected remotely but are important for understanding the formation and evolution of Io. Additionally, isotope ratios of the species known to be present and any unexpected species are critical for advancing understanding of Io now and in the past.

11.2.6 The Magnetosphere and Plasma Environment Around Io

Io is the main source of plasma for Jupiter's magnetosphere (Bagenal et al., Chap. 9), but many questions remain about how Io influences the magnetosphere, as well as the impact of the magnetosphere on Io. Although we know that volcanic activity, atmospheric loss, torus density, and auroral activity are all interconnected, we know very little about how they are connected because of the lack of simultaneous observations. Many questions remain because Earth-based observations or spacecraft-based monitoring are too limited.

We know that Io's atmosphere introduces neutral material to the magnetosphere, forming neutral clouds, but understanding of the composition of the neutral clouds is limited to species that are easy to observe remotely. Additionally, the neutral cloud structure, as well as the interaction between the neutral clouds and the plasma in the magnetosphere, is not well constrained. Furthermore, the input of neutrals from Io's atmosphere to these clouds has been observed to vary on 2- to 3-month timescales (Delamere et al. 2004), but a clear connection between volcanic activity and variability of the torus has not yet been made (e.g. Yoshikawa et al. 2017; Roth et al. 2020). Neutrals introduced to the magnetosphere from Io are ionized to form a plasma torus. Although the structure of the plasma torus is reasonably understood, the underlying physical processes that control the structure and activity of the torus are not clear.

Finally, Jupiter's magnetosphere and the plasma interacts with Io. Energetic charged particles in the magnetosphere impact the atmosphere and surface. This leads to sputtering loss of the atmosphere, one of many types of escape relevant to Io, and sputtering production of atmospheric species from the surface. However, the total and variability of atmospheric loss rates, as well as the role of plasma in heating the atmosphere, are not well understood.

11.2.7 Io as an Analog for Exoplanets and Exomoons

Io is the best solar system analog for tidally-heated exoplanets and exomoons (Barr et al., Chap. 10). In particular, the TRAPPIST-1 exoplanetary system has seven planets orbiting close to the parent star, with tidal heating modeled as sufficient to sustain magma oceans in several of the bodies. The orbital periods and eccentricities are also similar to those of the Galilean satellites, and both systems have mean-motion resonances. Barr et al. (Chap. 10) describe a simple tidal heating model appropriate for the TRAPPIST-1 system, and apply that same model to Io to show that the predicted heat flow is within an order of magnitude of the measured heat flow. In particular, TRAPPIST-1 planets b and c could be "remarkably similar to Io."

Exomoons like Io (Fig. 11.2) must exist in a variety of exo-Jupiter systems based on theories of satellite formation, and may be discoverable from their sodium and potassium signatures (Oza et al. 2019). A better understanding of Io will aid understanding such Io-like exomoons.

Barr et al. (Chap. 10) noted several areas where future observations are needed. Observing how Io's volcanic activity varies with orbital phase and periodicity could be used to predict the time during exoplanet orbits in which volcanic eruptions are likely to be the strongest and most detectable. Observations that better inform Io's internal structure and melt distribution are needed. Observations of Io's plasma torus and neutral clouds and links to volcanism could aid interpretation of potential exo-

Fig. 11.2 Artist's rendition of a volcanic exo-Io undergoing extreme mass loss. The hidden exomoon is enshrouded in an irradiated gas cloud shining in bright orange-yellow, as would be seen in a sodium (Na D2) filter. Patches of sodium clouds are seen to trail the moon's orbit, possibly driven by the gas giant's magnetosphere. ©University of Bern, Illustration: Thibaut Roger

torus detections. Another important point is that bright infrared fluxes and short orbital periods make volcanic exoplanets and exomoons with global magma oceans among the most detectable and characterizable low-mass exoplanets (Henning et al. 2018).

We note one other important measurement to better understand tidally-heated exoplanets and exomoons: the rate of orbital migration of the Galilean satellites. The results of Lainey et al. (2009) are often cited and used to model orbital evolution and tidal heating, but the models are based on assumptions that may be incorrect (de Kleer et al. 2019).

11.3 Future Observations of Io

11.3.1 Telescopic Observations

Earth-based observations of Io have been invaluable and will continue to be so. Telescopic observations are not limited to telescopes on Earth, but also those in orbit around Earth or reside nearby (like the Hubble Space Telescope and the James Webb Space Telescope). The most obvious benefit comes from monitoring temporal variations in the volcanoes, atmosphere, and magnetosphere. There is a long history of obtaining these types of data and it is important to extend the time series with continued observations. But telescopic observations can do much more than provide "more of the same." Fundamental new insights could be obtained by new observations with higher temporal and spectral resolution or extending to wavelengths that have not been well covered. For example, high temporal resolution multi-band imaging of outburst eruptions could provide critical constraints on Ionian eruption temperatures, providing major insights into the composition of the lavas and processes that are taking place deep within Io. Star occultations by Io constrain Io's orbit (Morgado et al. 2019), and thus its orbital energy loss by tidal friction.

There is an important synergy between telescopic observations and spacecraft that travel to Io. The visits by spacecraft are few and far between, and telescopic observations are essential to put those data into context. However, the ability to interpret the telescopic data relies on the higher spatial resolution and *in-situ* measurements that can be obtained by missions to the Jovian system. It is also important to consider that observation time with telescopes is obtained through intense competition, and supporting an active mission provides powerful justification for more intense monitoring campaigns.

11.3.2 Future Spacecraft Observations

The ***Juno*** mission is designed to study the interior and magnetosphere of Jupiter (Bolton et al. 2017), but has capabilities useful for the study of Io, such as monitoring hot spots (Mura et al. 2020), mapping surface compositions (Tosi et al. 2020), microwave observations (perhaps getting heat flow constraints), and studying magnetospheric interactions (Bagenal et al., Chap. 9). *Juno*'s extended mission includes two relatively close (1500 km) passes by Io, and nine other orbits that pass within 100,000 km, in years 2023–2024 (Fig. 11.3). The main Io objectives of the *Juno* extended mission are gravity science to help constrain models of Io's interior, and improved understanding of magnetospheric interactions. Imaging at visible,

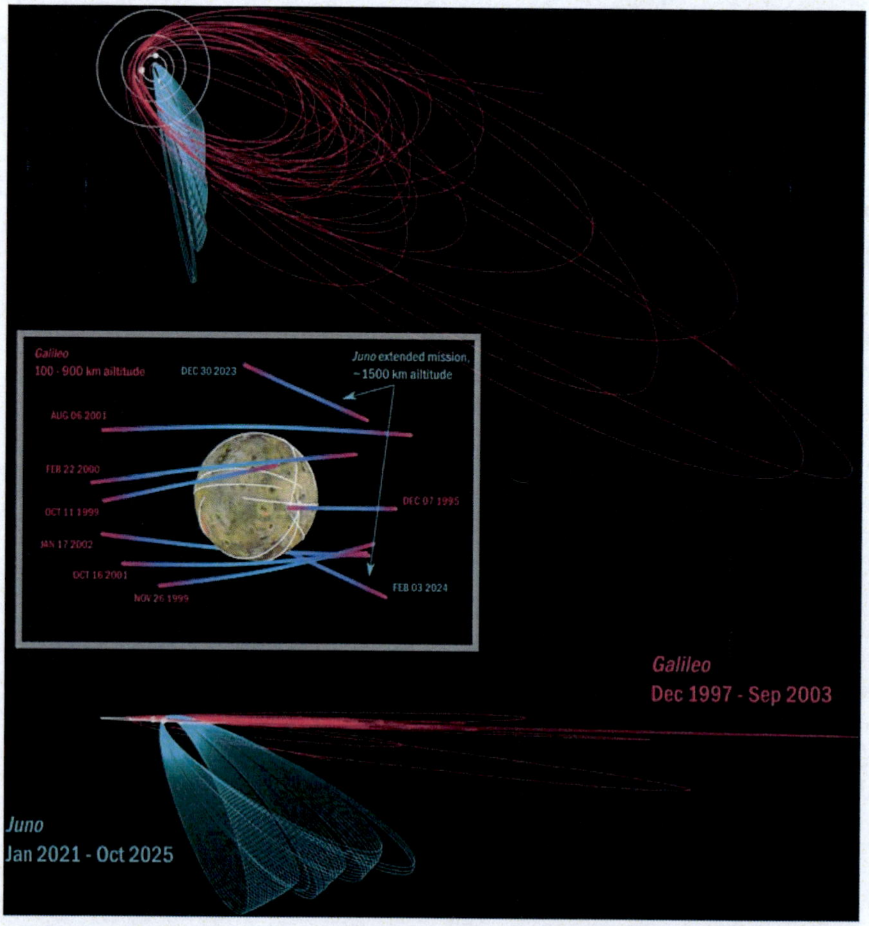

Fig. 11.3 Past (*Galileo*) and currently planned future (*Juno*) close flybys of Io provide important Io science, however, a dedicated Io mission is required to answer the many open questions

near-IR, and microwave wavelengths will also be acquired at times when Io passes through the instrument fields of view while the spacecraft maintains Earth/Sun pointing. The best possible resolutions are ~0.7 km/pixel in the IR, ~1.9 km/pixel in the visible, and hundreds of km in the microwave, based on a preliminary trajectory.

The **Jupiter Icy Moons Explorer (*JUICE*)** mission of ESA (Grasset et al. 2013) will acquire distant monitoring of Io with a suite of remote-sensing instruments at visible and IR wavelengths. Of special interest to *JUICE* is understanding the stability of the Laplace resonance (Dirkx et al. 2017). *JUICE* can acquire precision astrometry from radio science during close flybys of Callisto, Ganymede, and Europa. If *Juno* acquires useful astrometry as well, that will increase the time baseline for measuring rates of orbital migration, key to testing a model for whether or not the Laplace resonance is changing. However, a mission making close flybys of Io well after the *Juno* flybys is required to complete this experiment, otherwise the errors on Io's location over time negate a significant result (Dirkx et al. 2017; de Kleer et al. 2019). Understanding the orbital evolution and resulting tidal heating of the system may be one of the most fundamentally important results for understanding Io, Europa, and Ganymede.

The ***Europa Clipper*** mission focuses on understanding the habitability of Europa (Pappalardo et al. 2021). Observations of Io are currently planned only for calibrations, but if science observations of Io are allowed in the future, then the spacecraft and instruments could acquire higher-resolution visible and near-IR observations of Io than can *JUICE*. In addition, the mission includes a thermal-IR instrument that is not present on *JUICE* or *Juno*. *Europa Clipper* can also potentially contribute radio science astrometry of Callisto, Ganymede, and Europa, extending such data from *Juno* and *JUICE*.

NASA's **SIMPLEx** program exists to send small spacecraft to deep space destinations. Small spacecraft cannot easily accommodate substantial radiation shielding, so any mission near Io would be short-lived. High-risk science from cubesats is conceivable at Io, such as probing the dense portion of erupting plumes (Keane et al. 2021b).

Io mission concepts as part of **ESA's Voyage 2050** were presented by Thomas (2021), with an emphasis on following the energy and mass. The mission concepts are similar to those discussed below for NASA's Discovery and New Frontiers programs, but with additional instruments for measuring energetic particles and atmospheric processes. A strong possibility could be an ESA contribution to a NASA New Frontiers mission.

The **Chinese National Space Agency** is considering two Jupiter mission concepts, one focusing on Callisto, and another that would explore the Jupiter system including several close flybys of Io (Blanc et al. 2020; Li et al. 2021).

11.4 New NASA Mission Concepts

NASA currently plans the next **Discovery** Announcement of Opportunity (AO) to be released no earlier than 2025, and the **New Frontiers** 5 AO in 2023. Io is a candidate destination for both PI-led mission opportunities.

In Discovery, the ***Io Volcano Observer (IVO)*** has now been proposed three times (McEwen et al. 2021). *IVO* advanced to a Phase A study in 2020 and was found "selectable" but was not given a new start. The *IVO* concept focused on understanding tidal heating as a fundamental planetary process, including determining if Io has a magma ocean. An inclined Jupiter orbit enabled ten close encounters (200–1000 km) with Io, carefully designed to observe Io at the right combinations of true anomaly to test for a magma ocean via both gravity science and measurements of Io's libration amplitude. Only a dedicated Io mission can provide the measurements needed for unambiguous results on Io's interior state of melt. Waiting for the eccentric orbit to precess around Jupiter twice led to a 3.5-year mission duration, and migration of the subsolar longitude at closest approach enabled near-global mapping at <300 m/pixel (visible narrow-angle camera), plus near-global (>70% day and night) thermal IR mapping at <2.5 km/pixel. Measuring the induced magnetic signature with a magnetometer and plasma instrument was another key objective to understand Io's interior. A neutral mass spectrometer was included to provide the first such measurements near Io, and a wide-angle camera was included for topographic mapping.

"Io Observer" is a candidate mission for New Frontiers (NF) 5. The science objectives of Io Observer, as identified by the second decadal survey, are fully met by the Baseline *IVO* mission, but not by the Threshold mission. Only the Threshold mission is formally guaranteed to happen. The New Frontiers mission class would also enable much more robust accomplishment of the identified objectives, e.g.: Ka-band telemetry for better radio science and >2 times faster data return; enhanced radiation design to survive a higher total ionizing dose; increased number of orbits passing close to Jupiter, e.g., 15–20 rather than 10 Io encounters within the required 3.5-year period. There is also a long list of additional science instruments that would return valuable data from Io while potentially remaining within the NF cost cap (Table 11.1).

More ambitious NF Io missions are conceivable (Keane et al. 2021b). For example, penetrators could be carried and released to provide seismometer and heat flow data, but how to return the data is a major challenge. Alternatively, a small satellite could be released to make a close pass inside an erupting plume, much closer than considered safe for the main spacecraft, to measure gas compositions and surface properties that may go undetected at higher altitudes (Wurz et al. 2018). An Io orbiter (described below) is probably not feasible within NF.

Table 11.1 Spacecraft science instruments of value for future Io missions

Instrument	Science	Platform	Instrument	Science	Platform
Narrow-Angle Camera	Monitor volcanism, map landforms, measure libration amplitude, measure lava temperatures	Jupiter orbiter with close Io flybys (*IVO*)	Near-IR Spectroscopy	Composition of surface materials, silicate lava eruption processes	Jupiter orbiter (*JUICE, Europa Clipper, Juno*)
Thermal Mapper or Spectrometer	Measure heat flow, determine silicate lava composition, monitor volcanic eruption style	Jupiter orbiter with close Io flybys (*IVO*)	Dust Mass Spectrometer	Composition of solid particles	Jupiter orbiter with close Io flybys
Wide-Angle Camera	Topographic mapping, context for high-res images	Jupiter orbiter with close Io flybys (*IVO*)	Laser Altimeter	Tidal amplitude of potential magma ocean, shape of Io	Io Orbiter
Magnetometer and Plasma Instrument	Induction signature of mantle melt, magnetospheric interactions	Jupiter Orbiter with close Io flybys (*IVO*)	X-ray and Gamma-Ray Spectrometers	Elemental compositional mapping	Io Orbiter
Neutral Mass Spectrometer	First comprehensive detection of neural species near Io	Jupiter orbiter with close Io flybys (*IVO*)	Seismometer	Understand Io's crustal and deeper interior structures	Lander, penetrator, orbiter (in theory)
Gravity Science	Tidal amplitude of potential magma ocean, rate of orbital migration	Jupiter orbiter with close Io flybys (*Juno, IVO*)	Interferometric Synthetic Aperture Radar (InSAR)	Measure surface deformation at cm scales	Io orbiter with high power
Radar Sounding	Crustal structure, magma chambers	Jupiter orbiter with close Io flybys or Io orbiter	In-situ Heat Flow	Key unknown in Io's volcanic and tectonic processes	Penetrators
UV Spectroscopy	High-energy processes at Io and Jupiter system	Jupiter orbiter (*JUICE, Europa Clipper, Juno*)	Energetic Particle Detector	Properties of energetic particles impacting Io's surface, plumes and atmosphere	Jupiter orbiter with close Io flybys (*IVO*)

11.5 Io Orbiter Mission Challenges

While an Io Orbiter could deliver rapid and more complete geophysical science return compared with a flyby mission, such a concept would face major challenges. Assuming launch on a Falcon Heavy Expendable vehicle into a 2-year ΔV-EGA (Earth gravity assist), a lift mass of just over 8000 kg is expected. A capture sequence at Jupiter could include either an Io- or Ganymede-aided Jupiter Orbit Insertion (JOI) maneuver, followed by a Perijove Raise Maneuver (PRM) to target a subsequent flyby of Ganymede. While the propellant required is comparable for either scenario, an early Io flyby could provide useful information and reduce JOI ΔV by ~300 m/s, while incurring a radiation penalty of ~60 krad. A pump-down sequence is required to further reduce orbital energy before insertion into orbit about Io. While traditional trades of ΔV versus time-of-flight offer flexibility to balance mission constraints, the added challenge of the extreme radiation environment significantly impacts a long duration pump-down.

Long-Duration Pump-Down, Lower Orbit Insertion ΔV
The minimum energy transfer to Io from a Galilean moon is the Hohmann transfer from Europa to Io, which corresponds to a roughly 1.87 km/s minimum Io-relative velocity (V_∞). To achieve this transfer, a relatively long duration pump-down is required, initially exploiting transfers between Ganymede and Callisto to reduce Ganymede V_∞ to ~1.9–2.0 km/s. Perijove remains relatively high during this phase, oscillating between ~11 and 16 R_J. Additional flybys of Europa further reduce energy, and the spacecraft ultimately enters onto a Hohmann transfer to Io. Because a number of transfers between different moons are required, the potential for multi-revolution transfers could increase dwell time, exacerbating the radiation incurred prior to Io orbit insertion. An orbit insertion maneuver of ~260 m/s inserts the spacecraft into a polar orbit about Io. The Total Ionizing Dose (TID) of radiation is expected to be ~2.5–3.5 Mrad, depending on the moon-transfer geometries and assuming a Radiation Design Margin (RDM) of RDM×2.

Short-Duration Pump-Down, Higher Orbit Insertion ΔV
Alternatively, a shorter duration pump-down could be designed, using flybys of Ganymede to reduce the orbit to a 3:1 resonance with Ganymede. Including flybys of Europa provides a marginal reduction in Ganymede V_∞ as the spacecraft transitions to a 1:1 Ganymede resonance. Transferring to Io without further energy reduction reduces TID, but yields a higher Io-relative V_∞ of ~4.4 km/s. Additional radiation penalties from a prolonged tour can be avoided by ending the tour here and performing the Io orbit insertion maneuver, which requires >2600 m/s. TID is expected to be reduced by a factor of 3–4× by avoiding the longer tour, however, planning for a maneuver of this size comes with a number of challenges. Although unlikely to yield a feasible concept, this option may be useful to inform the lower bound on tour TID.

While these two tour designs are expected to roughly bound the design space, a spectrum of options exist in between and may provide the optimal balance of ΔV

Table 11.2 Comparison of example long- and short-duration tours

	Long-duration tour	Short-duration tour
No. flybys	19	10
No. leveraging maneuvers	2	0
Time-of-flight (days)	721	454
TID (krad, RDM×2)	2670–3580	630–1250
Leveraging ΔV (m/s)	230	0
Io Orbit Insertion ΔV (m/s)	260	2615

versus TID for future concepts. For comparison, example long- and short-duration moon tours beginning from the same post-PRM Ganymede-flyby state are shown in Table 11.2. Here, a range of possible TID values is given to encompass uncertainty in the need for multi-revolution transfers to accommodate phasing between distinct moons. In Fig. 11.4a, b, Jupiter-centered orbit plots appear in an inertial frame for each tour type. The path charted by each tour in orbital period and perijove (r_p) space, i.e., Tisserand graphs (Heaton et al. 2002), appears in Fig. 11.4c.

Science Orbit

In the orbiter phase, a roughly 2-h, 100–200 km altitude orbit is flown for the science campaign (Fig. 11.5). Jupiter acts as a significant perturbing force during orbital operations, driving a chaotic dynamical environment in the vicinity of Io. Without active control, the orbiter could be expected to survive ~7–21 d before impacting Io's surface, with longer durations likely achievable assuming sufficiently accurate modeling of Io's gravity (Ferreira et al. 2022). Ensuring the success of the orbit insertion under numerous sources of error and uncertainty would be critical to minimizing the risk of the orbiter phase. While challenging, an autonomous navigation and station-keeping demonstration at Io could serve as a useful testbed for extending such capabilities to bodies under planetary protection, such as Saturn's moon Enceladus.

A short campaign of roughly 7 days is required to complete the highest-priority science measurements with sufficient margin and resiliency to potential anomalies degrading the science data return. A daily TID of ~120 kRad (RDM×2) is expected while in Io-orbit (not accounting for shielding by Io). Further study is needed to quantify the shielding required to safely deliver the spacecraft through the moon tour and Io orbit phases. The recent NASA Heliophysics Mission Concept Study COMPASS may provide useful insight into design options to accommodate the challenges of operating in an extreme radiation environment. Because the science campaign is short, and the flight system faces certain death by radiation, a relay spacecraft is likely required to downlink the science data. Utilizing a nearby operational spacecraft during this campaign would be difficult to ensure, suggesting the need for a dedicated relay as part of the Io Orbiter mission.

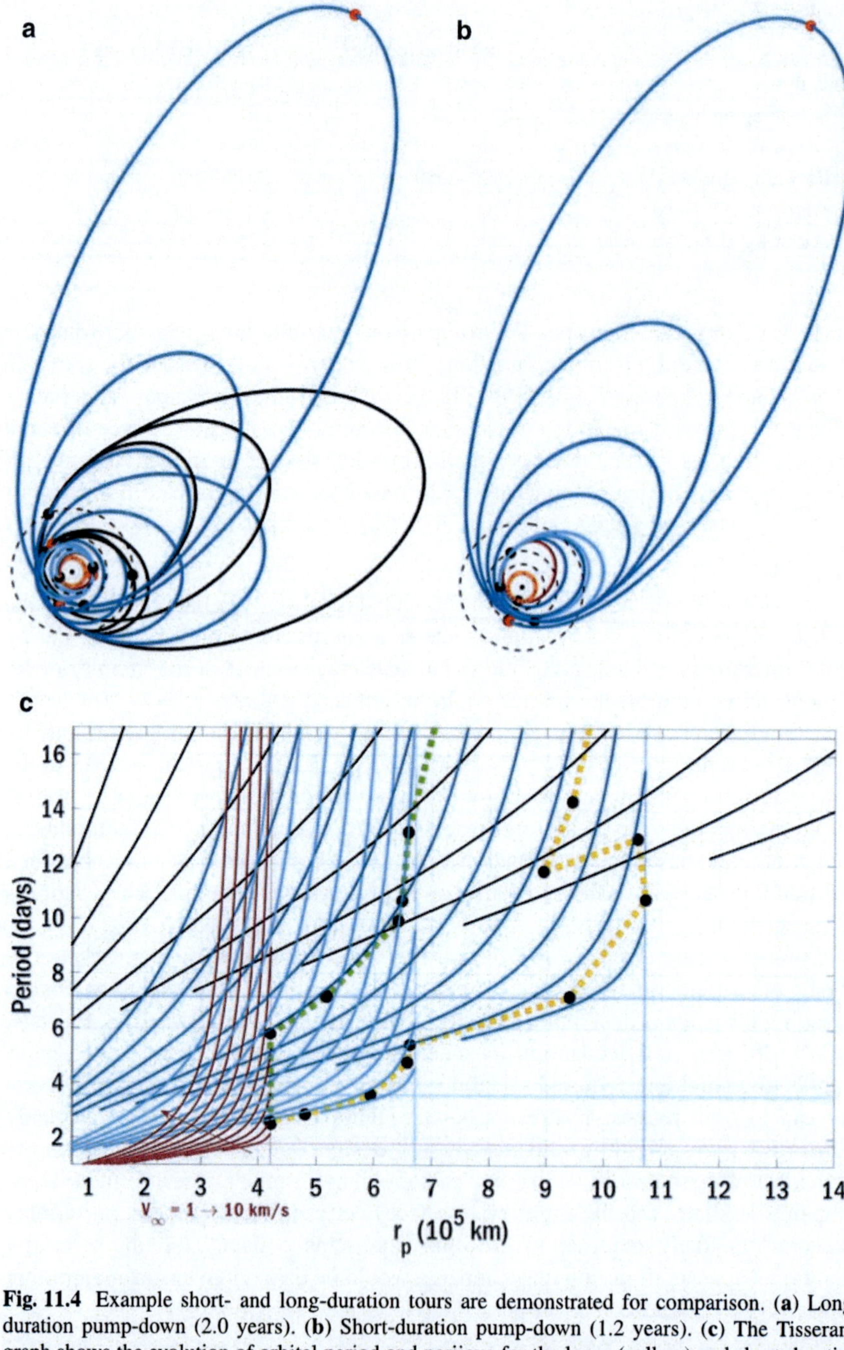

Fig. 11.4 Example short- and long-duration tours are demonstrated for comparison. (**a**) Long-duration pump-down (2.0 years). (**b**) Short-duration pump-down (1.2 years). (**c**) The Tisserand graph shows the evolution of orbital period and perijove for the long- (yellow) and short-duration (green) tours near the end of each pump-down. Outgoing flyby conditions are plotted as black dots for each tour. Curves of constant V_∞ over the range 1–10 km/s are plotted for Callisto, Ganymede, Europa, and Io in black, dark blue, light blue, and red, respectively. Horizontal and vertical dotted lines of the same colors indicate the period and average orbital radius for each moon

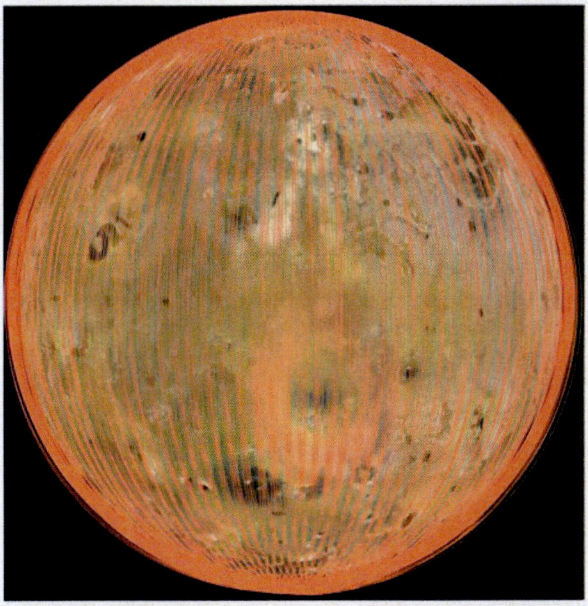

Fig. 11.5 Example ~90 km altitude orbit propagated for 25 d with no orbit maintenance, assuming Jupiter zonal gravity harmonics to degree 6, and Io gravity harmonics to degree and order 2 (IAU-Io frame)

11.6 Summary: Top Ten Questions (Fig. 11.6)

How did Io form? Which of the many scenarios describe by McKinnon (this volume, Chap. 3) is most correct? Did Io initially have substantially more volatiles, especially H_2O? If so, how was it lost? What was Io's initial composition?

Is the Laplace resonance in equilibrium? If it varies periodically, is the present-day rate of tidal heating relatively high or low, and what are the implications for the evolution of Europa and Ganymede?

Does Io have a magma ocean? If not, what is the distribution of magma inside Io? What are the implications for tidal heating? What can we learn from Io about magma oceans on the terrestrial planets and exoplanets?

What processes lead to lava eruptions at >10^5 m^3 s^{-1}? Do such eruptions directly tap a magma ocean below a >20 km thick lithosphere? Does the Ionian process help us to understand high effusion rate eruptions in Earth's past, such as those closely associated with mass extinctions? There appear to be two very different types of volcanic eruptions on Io: persistent and outbursts, and two fundamentally different ways in which magma traverses the lithosphere. This

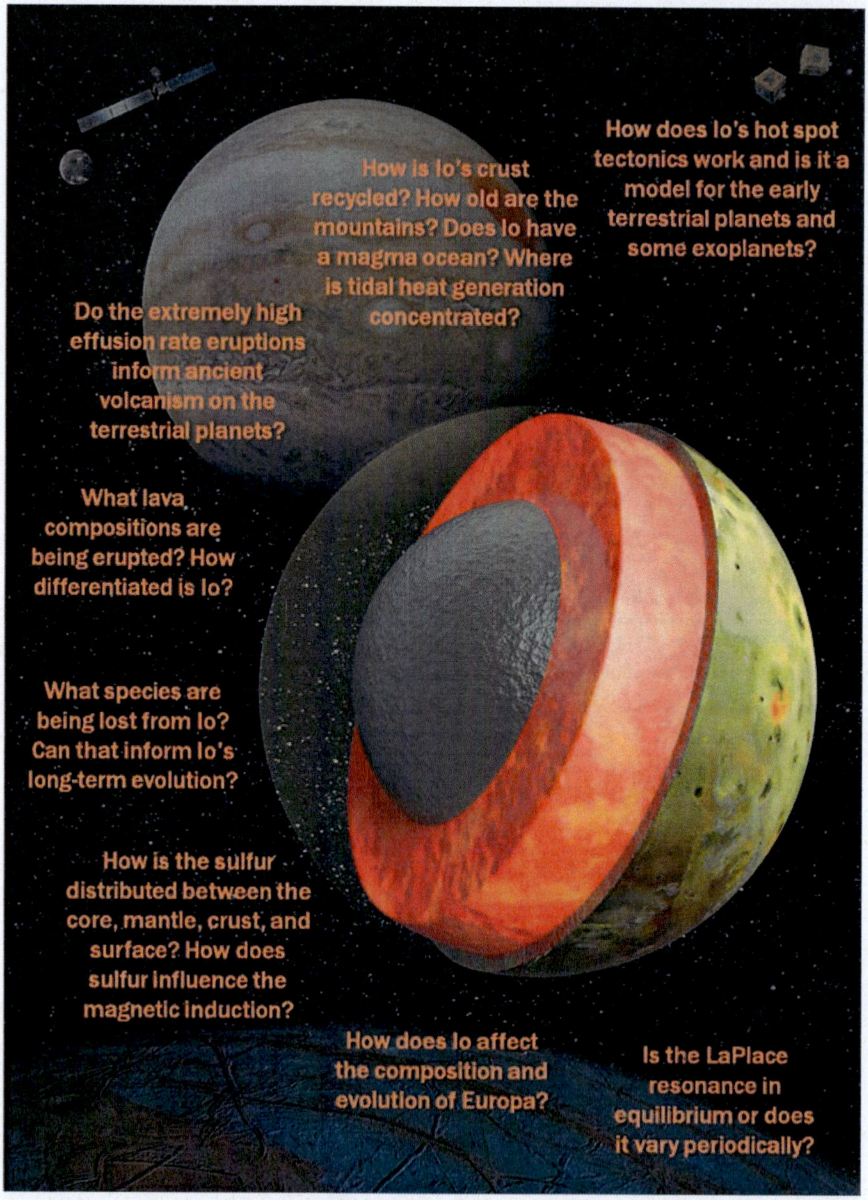

Fig. 11.6 Key questions for future Io exploration. Base figure from the Keck Institute of Space Studies (de Kleer et al. 2019)

has strong implications for volcanism on any planetary body with a thick cold lithosphere, including Mars and the Moon, and maybe Mercury and some parts of the Earth and Venus. The simplest idea, that persistent eruptions are fed by shallow magma chambers with long-lived conduits while outbursts are fed by dikes reaching down to the base of the lithosphere, has a host of unresolved issues. Answering this question is tantamount to determining if/how the heat pipe model works.

What is the crustal composition of Io? The ubiquitous cover by sulfurous compounds hides the silicate composition from spectrometers, although thermal IR spectroscopy of fresh and still-warm surfaces (preventing SO_2 condensation) is promising. Better determination of peak eruption temperatures provides a complementary constraint. Based on current theories, the composition of the deeper crust should be similar to that of erupting lavas, and is expected to be either mafic or ultramafic, and this distinction is of key importance to understanding the degree of mantle melting. A spacecraft must get close to Io to fully resolve many fresh lava surfaces at thermal-IR wavelengths.

How long-lived are the mountains? Are they dynamically uplifted as they collapse from mass wasting, or are they stable features for more than 10^3 years? Are they typical sections of crust (stacks of lava flows) or something else? High-resolution imaging and topography are needed to address this question. How does the tectonic branch of Io's rock cycle compare to the volcanic branch in terms of km^3/year? The volume of Io's mountains is of order 1–10 million km^3. The lava output is around 1000 km^3/year so the amount of rock being moved tectonically could be similar to the volcanism if the mountains are only 1000–10,000 years old.

What gases are escaping from Io's plumes? We know there is abundant SO_2 in the plumes, but measuring the full composition with a mass spectrometer may lead to key surprises that can solve mysteries about Io's complex atmosphere and mass loss as well as about Io's crustal composition and magma chambers.

How sulfur-rich is Io's surface and crust? Although bright flows can be interpreted as sulfur flows (Williams et al., Chap. 5), they might be sulfurous coatings over silicate lava. Do paterae form in a sulfur-rich layer kilometers thick? SO_2-rich fluids could both lubricate faults and move heat far more efficiently than thermal conduction. This could significantly alter models for the stresses and temperatures within the lithosphere. Significant amounts of sulfur in the crust are also difficult to reconcile with an iron-rich core which means it places interesting constraints on the bulk composition of Io, the degree of mixing within the mantle, and how Io formed. Unfortunately, this is a particularly difficult question to address with remote observations (spectral and geophysical) but important progress can be made by examining exposures on steep cliffs and active sulfurous volcanism.

How does volcanic activity drive Io's mass loss? Io loses ~1 t/s of mass to Jupiter's magnetosphere through a combination of processes (Bagenal et al., Chap. 9), all related directly or indirectly to volcanic activity. However, controversy persists about direct cause and effect between the variability of Io's volcanism and variability in the plasma torus or neutral clouds. A better understanding of these processes is needed to interpret observations of other planetary systems, such as potential sodium clouds that may reveal the presence of exoplanet satellites (Oza et al. 2019).

How does Io affect Europa? The thermal-orbital histories of Io and Europa are linked (Hussman and Spohn 2004), so if Io is unusually active or inactive today, Europa may likewise be relatively active or inactive. Sulfurous materials escaping from Io (Bagenal et al., Chap. 9) and impact ejecta (silicate rocks) escaping Io (Alvarellos et al. 2008) are deposited on Europa, perhaps improving Europa's habitability.

Acknowledgements We thank K. de Kleer, V. Lainey, and R. Lopes for constructive reviews.

References

Alvarellos, J.L., Zahnle, K.J., Dobrovolskis, A.R., Hamill, P.: Transfer of mass from Io to Europa and beyond due to cometary impacts. Icarus. **194**(2), 636–646 (2008). https://doi.org/10.1016/j.icarus.2007.09.025

Bierson, C.J., Nimmo, F.: Explaining the Galilean satellites' density gradient by hydrodynamic escape. Astrophys. J. Lett. **897**, L43. 15 pp. (2020). https://doi.org/10.3847/2041-8213/aba11a

Bierson, C.J., Steinbrügge, G.: Tidal heating did not dry out Io and Europa. Planet. Sci. J. **2**(3), id.89 (2021) 7 pp

Blanc, M., et al.: Gan De: science objectives and mission scenarios for China's mission to the Jupiter system. In: 22nd EGU General Assembly, held online 4–8 May 2020, id.20179 (2020)

Bolton, S.J., et al.: The Juno mission. Space Sci. Rev. **213**(1–4), 5–37 (2017)

Canup, R.M., Ward, W.R.: Formation of the Galilean Satellites: Conditions of Accretion. Astron. J. **124**, 3404–3423 (2002)

Canup, R.M., Ward, W.R.: A common mass scaling for satellite systems of gaseous planets. Nature. **441**, 834–839 (2006). https://doi.org/10.1038/nature04860

Davies, A.: Volcanism on Io: A Comparison with Earth. Cambridge University Press, Cambridge (2007)

de Kleer, K., Park, R., McEwen, A., et al.: Tidal Heating: Lessons from Io and the Jovian System. Final Report for the Keck Institute of Space Studies. https://www.kiss.caltech.edu/final_reports/Tidal_Heating_final_report.pdf (2019)

de Pater, I., Luszcz-Cook, S., Rojo, P., Redwing, E., de Kleer, K., Moullet, A.: ALMA observations of Io going into and coming out of Eclipse. Planet. Sci. J. **1**, 60 (2020) 25 pp

Delamere, P.A., Steffl, A., Bagenal, F.: Modeling temporal variability of plasma conditions in the Io torus during the Cassini era. J. Geophys. Res. Space Phys. **109**(A10) (2004) CiteID A10216

Dirkx, D., Gurvits, L.I., Lainey, V., et al.: On the contributions of PRIDE-JUICE to Jovian system ephemerides. Planet. Space Sci. **147**, 14–27 (2017). https://doi.org/10.1016/j.pss.2017.09.004

Ferreira, T.C.F.C., Prado, A.F.B.A., Giuliatti Winter, S.M., Ferreira, L.S.: Mapping natural orbits around Io. Symmetry **14**(7), 1478–1496 (2022). https://doi.org/10.3390/sym14071478

Grasset, O., et al.: JUpiter ICy moons Explorer (JUICE): an ESA mission to orbit Ganymede and to characterise the Jupiter system. Planet. Space Sci. **78**, 1–21 (2013)

Heaton, A.F., Strange, N.J., Longuski, J.M., Bonfiglio, E.P.: Automated design of the Europa orbiter tour. J. Spacecr. Rocket. **39**, 1 (2002)

Henning, W.G., et al.: Highly volcanic exoplanets, lava worlds, and magma ocean worlds: an emerging class of dynamic exoplanets of significant scientific priority. arXiv preprint arXiv:1804.05110 (2018)

Hussmann, H., Spohn, T.: Thermal-orbital evolution of Io and Europa. Icarus. **171**(2), 391–410 (2004)

Keane, J., et al.: The science case for Io exploration. Planetary Science and Astrobiology Decadal Survey 2023-2032 white paper e-id. 178. Bull. Am. Astron. Soc. **53**(4), e-id. 178 (2021a)

Keane, J., et al.: Recommendations for addressing priority Io science in the next decade. Planetary Science and Astrobiology Decadal Survey 2023-2032 white paper e-id. 179. Bull. Am. Astron. Soc. **53**(4), e-id. 179 (2021b)

Lainey, V., Arlot, J.-E., Karatekin, O., van Hoolst, T.: Strong tidal dissipation in Io and Jupiter from astrometric observations. Nature. **459**(7249), 957–959 (2009). https://doi.org/10.1038/nature08108

Leone, G., Davies, A.G., Wilson, L., Williams, D.A., Keszthelyi, L.P., Jaeger, W.L., Turtle, E.P.: Volcanic history, geologic analysis and map of the Prometheus Patera region on Io. J. Volcanol. Geotherm. Res. **187**(1), 93–105 (2009)

Li, L., et al.: Gan De: a mission to search for the origins and workings of the Jupiter system. In: 3rd COSPAR Scientific Assembly. Held 28 January – 4 February 2021. Abstract B0.7-0016-21 (oral), id.253 (2021)

McEwen, A.S., Keszthelyi, L.P., Mandt, K.E., the IVO Team: The Io Volcano Observer (IVO). In: 52nd Lunar and Planetary Science Conference, held virtually, 15–19 March 2021. LPI Contribution No. 2548, id.1352 (2021)

Morgado, B., et al.: First stellar occultation by the Galilean moon Europa and upcoming events between 2019 and 2021. Astron. Astrophys. **626**, id.L4., 10 pp (2019). https://doi.org/10.1051/0004-6361/201935500

Moullet, A., Lellouch, E., Gurwell, M., Moreno, R., Black, J., Butler, B.: Distribution of alkali gases in Io's atmosphere. In: AAS DPS Meeting #47, Abstract 311.31 (2015)

Mousis, O., Marcq, E., Aguichine, A., Leblanc, F., Mandt, K., Schneeberger, A., Turbet, M.: Io's dry body shaped by atmospheric instability. In European Planetary Science Congress. EPSC 2021 (2021)

Mura, A., et al.: Infrared observations of Io from Juno. Icarus. **341**, 113607 (2020)

Ormel, C.W., Liu, B., Schoonenberg, D.: Formation of TRAPPIST-1 and other compact systems. A&A. **604**, A1 (2017)

Oza, A.V., et al.: Sodium and potassium signatures of volcanic satellites orbiting close-in gas giant exoplanets. Astrophys. J. **885**, 168 (2019)

Pappalardo, R., et al.: The Europa Clipper Mission: understanding icy world habitability and blazing a path for future exploration. Planetary Science and Astrobiology Decadal Survey 2023-2032 white paper e-id. 255. Bull. Am. Astron. Soc. **53**(4) (2021) e-id. 255

Peale, S.J., Canup, R.M.: The origin of the natural satellites. In: Schubert, G. (ed.) Treatise on Geophysics Planets and Moon, vol. 10, 2nd edn, pp. 559–604. Elsevier, Boston (2015)

Rathbun, J.A., Spencer, J.R., Tamppari, L.K., Martin, T.Z., Barnard, L., Travis, L.D.: Mapping of Io's thermal radiation by the Galileo photopolarimeter-radiometer (PPR) instrument. Icarus. **169**(1), 127–139 (2004)

Redwing, E., de Pater, I., Luszcz-Cook, S., Moullet, A., Rojo, P., de Kleer, K.: Observations and analysis of NaCl and KCl in Io's atmosphere. American Geophysical Union, Fall Meeting 2020 P031-06 (2020)

Ronnet, T., Mousis, O., Vernazza, P., Lunine, J.I., Crida, A.: Saturn's formation and early evolution at the origin of Jupiter's massive moons. Astron. J. **155**, 22413 pp (2018). https://doi.org/10.3847/1538-3881/aabcc7

Roth, L., et al.: An attempt to detect transient changes in Io's SO_2 and NaCl atmosphere. Icarus. **350**, 113925 (2020)

Shibaike, Y., Ormel, C.W., Ida, S., Okuzumi, S., Sasaki, T.: The Galilean Satellites formed slowly from pebbles. Astrophys. J. **885**(1), 79 (2019) 19 pp

Steinbruegge, G., et al.: PRIME — A Passive Radar Sounding Concept for Io. Planetary Science and Astrobiology Decadal Survey 2023-2032 white paper e-id. 271. Bull. Am. Astron. Soc. **53**(4) (2021) e-id. 271

Thomas, N.: A comprehensive investigation of the Galilean moon, Io, by tracing mass and energy flows. Exp Astron.. Online First. (2021). https://doi.org/10.1007/s10686-021-09768-y

Tosi, F., et al.: Mapping Io's surface composition with Juno/JIRAM. J. Geophys. Res. Planets. **125**(11), e06522 (2020)

Tsang, C.C.C., Spencer, J.R., Lellouch, E., Lopes-Valverde, M.A., Richter, J.J.: The collapse of Io's primary atmosphere in Jupiter eclipse. JGR. **121**, 1400–1410 (2016)

Wurz, P., et al.: An impacting descent probe for Europa and the other Galilean moons of Jupiter. Earth Moon Planet. **120**(2), 113–146 (2018). https://doi.org/10.1007/s11038-017-9508-7

Wurz, P., Vorburger, A., McEwen, A., Mandt, K., Davies, A., Hörst, S., Thomas, N.: Modelling of Io's Atmosphere for the IVO Mission. In: vEGU21, the 23rd EGU General Assembly, held online 19–30 April 2021, id. EGU21-778 (2021)

Yoshikawa, I., et al.: Volcanic activity on Io and its influence on the dynamics of the Jovian magnetosphere observed by EXCEED/Hisaki in 2015. Earth Planets Space. **69**(1), 110 (2017) 11 pp

Index

R. M. C. Lopes et al. (eds.), *Io: A New View of Jupiter's Moon*, Astrophysics and Space Science Library 468, https://doi.org/10.1007/978-3-031-25670-7

D

E

F

G

H

I

J